The Imagined Empire

The Imagined Empire
Balloon Enlightenments in Revolutionary Europe

Mi Gyung Kim

UNIVERSITY OF PITTSBURGH PRESS

Published by the University of Pittsburgh Press, Pittsburgh, Pa., 15260
Copyright © 2016, University of Pittsburgh Press
All rights reserved
Manufactured in the United States of America
Printed on acid-free paper
10 9 8 7 6 5 4 3 2 1

ISBN 13: 978-0-8229-4465-2
ISBN 10: 0-8229-4465-0

Cataloging-in-Publication data is available from the Library of Congress

Cover art: Louis Joseph Watteau, *La Quartozième expérience aérostatique* (1785).
Cover design by Joel W. Coggins

This book is dedicated to the Korean students who sacrificed their lives to fight against military dictatorship.

Bird

If it were possible not to be human,
and to choose my rank in the animal kingdom,
do you know, my friends, what I would like to be?
I would like to be a bird, not a bird of prey,
but a swallow, or a bird of paradise.
I would like to be a bird, I repeat,
for the force and the expanse of its vision.
How I would love to glide over towns and steeples!
To see the forests like carpets of greenery!
To catch the vast roundness of the firmament!
Not to lose the image of brilliant golden clouds!
To be able to lift my body with astonishing speed,
To sustain and balance it in the air for a flexible and complex flight!
To emit in all directions gentle and sonorous vibrations from my throat!
Ah! What pleasure it would be to leap toward the pompous day-star,
to plunge myself in it, to play in its rays!
My eye, because of its structure, would not be tired but rejoice in it:
I would contemplate all the radiant colors
that make an enchanting picture on the earth's surface.
I would make long voyages in free space,
around all republics and kingdoms,
traversing the seas and visiting the islands.
For the evening, I would choose a refuge high up, sheltered;
In the morning, I would leave at dawn,
always in the ecstasy of the delightful view.
I would have before me only the rich perspectives
of a nature vividly colored.

Louis-Sébastien Mercier
Mon Bonnet de Nuit **(1784)**

CONTENTS

ACKNOWLEDGMENTS *xi*

PROLOGUE: Flying through Time *xvii*

INTRODUCTION: A People-Machine *1*

COLOR PLATES follow page 14

PART I. INVENTION IN THEATRICAL POLITY 23

   1. A Rupture of the Equilibrium *29*

   2. Balloon Transcripts *53*

   3. True Columbus *77*

PART II. PHILOSOPHICAL NATION *103*

   4. Balloon Spectators *109*

   5. Fermentation and Discipline *131*

   6. Provincial Citizens and Their Nations *151*

   7. The Fall of a National Artifact *173*

PART III. MATERIAL EMPIRE *195*

   COLOR PLATES follow page 198

   8. Modern Atlantis *209*

   9. Crossing the Channel *235*

   10. A Liminal Geography *261*

EPILOGUE: Revolutionary Metamorphoses *283*

NOTES *295*

GLOSSARY *343*

BIBLIOGRAPHY *345*

INDEX *415*

ACKNOWLEDGMENTS

This project had a distinct beginning and a few unexpected turns, which involved many people. I owe special thanks to Thomas Broman who invited me to a panel on news in the spring of 2003. Unable to find suitable material from my research in the history of chemistry, my attention turned to ballooning. I had noticed the enthusiasm jumping off the pages while sifting through the fifteen-volume *Registre* of the Dijon Academy—something I had not seen in the procès-verbaux of the Paris Academy of Sciences. Preliminary research turned up Charles C. Gillispie's exquisite book, *Montgolfier Brothers* and a short bibliography by Gaston Tissandier. In the summer I took a copy to the Bibliothèque nationale de France and ordered a microfilm of all available pamphlets. The enthusiastic response from a large audience at the History of Science Society meeting in the fall, which included an editor of *Endeavour* who published the presentation, lured me into a full-scale research.

Even more surprising was a series of funding that came to support this project. A fellowship from the National Humanities Center for the 2006–2007 academic year allowed me to survey a broad range of secondary literature fetched by unfailing librarians Eliza Robertson and Jean Houston. The entire staff and director Geoffrey Harpham provided an unusually supportive environment, while fellow historians read a couple of early chapters in our weekly discussion group and offered valuable comments. A Mellon sabbatical fellowship from the American Philosophical Society and a residential fellowship at the (Columbia University's) Institute for Scholars, which were granted for a separate project on Dijon Academy, allowed me to spend a semester in Paris and to explore the archives more systematically for both projects than I could manage during the annual summer trips. The Institute's director Dr. Danielle Haase-Dubosc wrote elegant letters of introduction and Dr. Mihaela Bacou made everything else easier. Madame Pouret at the Archive of the Academy of Sciences introduced me to Mr. Dégardin at the Musée de l'Air et de l'Espace. Their superb image collection, in addition to the Fonds Montgolfier, made me appreciate the project's potential. The staff also alerted me to their balloon lady, Dr. Marie Thébaud-Sorger, who had recently finished an excellent dissertation on French ballooning. She graciously sent me a copy and pointed out an earlier dissertation by Dr. James Martin Hunn, which proved quite valuable in understanding the Bordeaux case. Marie's exhaustive research on French ballooning allowed me to conceive a more comparative project with the help of a faculty fellowship at the Humanities Research Center of Rice University. A grant from the National Science Foundation (#0924054) made it possible for me to explore various European archives and spend a term at the Department of History and Philosophy of Science at the University of Cambridge. A project of this scale would not have been possible without these fellowships and the teaching release from North Carolina State University. I am deeply grateful to the mentors and colleagues who wrote numerous letters of support, especially Simon Schaffer and my Doktorvater M. Norton Wise, as well as the fellowship institutions and their referees. Their encouragements kept up my spirit wandering through the infinity of fragments that refused to coalesce into a meaningful story.

I have incurred substantial debt to the colleagues who read raw chapters over the past decade—Katherine Hayles, Thomas E. Kaiser, Sarah Maza, Brent Sirota, Philip Stewart, and especially Jeffrey S. Ravel and Simon Schaffer who read the entire manuscript at an early stage and suggested further venues of exploration. Kenneth Loiselle readily shared his expertise on Freemasonry. Presentations at various institutions and conferences helped me to focus and refine this project. Triangle Groups in French History and Culture and Intellectual History Seminar provided a kind of intellectual home for the past two decades. I thank Malachi Hacohen, Lloyd Kramer, Anthony LaVopa, Martin A. Miller, Donald M. Reid, Steven Vincent, and James Winders for sustaining these groups. I also wish to thank Bernadette Bensaude-Vincent, Hasok Chang, Karine Chelma, Bruce Hunt, Jenny Rampling, and especially Mary Terrall for the invitations to their institutional seminars, as well as those who answered my call to the conferences and panels. Three dealt with the public—at 2009 and 2013 meetings of the Society for French Historical Studies and at the History of Science Society's 2008 annual meeting (organized with Michael R. Lynn). Three more dealt with geography—Enlightenment Geographies at Rice University, Material Geographies at North Carolina

State University and Machine Geographies at the History of Science Society's 2012 annual meeting. I wish to thank David Ambaras, Robert Batchelor, Ann Blair, Jill Casid, Joyce E. Chaplin, Alex Csizar, Peter Dear, Lauren Dubois, Sarah Ellenzweig, Marie-Claude Felton, Jan Golinski, Lisa Jane Graham, Florence Hsia, Matthew L. Jones, Betty Joseph, Thomas E. Kaiser, William Kimler, Keith Luria, Sarah Maza, David Mazella, Catherine Molineux, Dorinda Ourtram, Alexander Regier, Andrea Rusnock, J. B. Shank, Mary Sheriff, Richard Slatta, Philip Stern, Rajani Sudan, Anoush Terjanaian, Mary Terrall, John Tresch, Elvira Vilches, Steven Vincent, Timothy Walker, Charles Walton, Simon Werrett, and John Zammito for their participation.

My debt to the archivists and librarians is immeasurable. They made me feel welcome and lent their expertise without reserve. I owe more than I can say (or remember) to the staff at the Bibliothèque nationale de France, various municipal and departmental archives in France, Huntington Library, Musée de l'Air et d'Espace, National Library of Scotland, National Library of Ireland, Princeton University Library, the City Library of Birmingham, the Library of Congress, and the National Air and Space Museum. Mr. Alexander Bakker and Madame Danielle Ducout at the Dijon municipal library offered me indispensable support while I gradually formed a different image of the town and its academy, which fundamentally altered my understanding of science's role in the Enlightenment and the relationship between Paris and the provinces. Special thanks are due to Mrs. Sjoukje Atema at the Haags Gemeentearchief, Ms. Gilles Bernasconi at the Lyon municipal archive, Dr. Leonard Bruno at the Library of Congress, Mr. Joachim W. Frank at the Staatsarchiv Hamburg, Dr. Martina Maříková at the Prague city archives, Dr. Brian Riddle at the National Aerospace Library, Prof. Denis Reynaud at the Lyon Academy, and Dr. Marie-Hélène Reynaud at the Musée des papeteries Canson et Montgolfier in Annonay for their extraordinary effort to secure relevant material. Dr. Mary Ruwell at the Gimbel Collection and Dr. Tom D. Crouch at the National Air and Space Museum supplied many illustrations, which allowed me to construct a visual narrative. My research trips would have been dreary without the kindness and hospitality of Dr. Jean Bart in Dijon who readily shared his local knowledge, Dr. Michel Dürr who opened the Lyon Academy archive whenever I needed it, and Dr. Antoine Villesuzanne who helped me navigate through Bordeaux.

The first draft of the manuscript was completed by the spring of 2011 in time for the promotion process, of which a substantial portion of the first chapter was published as "Invention as a Social Drama" in *Technology and Culture* 54 (2013): 853–87. I wish to thank Professor Barbara H. Smith for helping me to put together a formal book proposal. Abby Collier read the entire manuscript and took it on. She has worked diligently to enlist excellent reviewers and mediate the conversation for two years. Without her dedication, this book would have struggled even longer to appear in print. The production team led by Alexander Wolfe at the University of Pittsburgh Press has done a superb job. I can only marvel at their patience and professionalism in dealing with a large number of illustrations and an under-prepared manuscript. Mary Terrall and Emma C. Spary offered immensely helpful comments from divergent angles. Mary's practical and sympathetic advice on how to lead the readers and Emma's knowledge of secondary literature proved invaluable in gauging potential audience responses. Colin Jones also read the entire manuscript of a stranger who chanced upon him and offered valuable comments. I am deeply impressed by their openness to and serious investment in this project.

My family always understood why this project was important to me. My husband Jung-Goo Lee sacrificed his career to stay with me in North Carolina for the duration of this project. Our son, Anthony Chang-Bock Lee, made me realize how much he matured during my balloon-crazed years when he hid a broken ankle from our communication so that I wouldn't shorten my research trip. He came home just in time to read over the entire manuscript. This book has been a family endeavor for a greater cause than our material or social needs.

Figure P.1. Albrecht Dürer, *Icarus*, in Friedrich Riederer, *Spiegel der wahren Rhetorik* (Freiburg, 1493). Gimbel Collection XL-1-4650.

PROLOGUE

Flying through Time

> Here Icarus fell; these waves beheld his fate, . . .
> Here the flight ended; here the event took place,
> Which those unborn will yearn to emulate. . . .
> His name now echoes loud in every wave,
> across the sea, throughout an element;
> Who ever in the world gained such a grave?[1]

In Ovid's poems of Greek mythology, Icarus was the son of the great Athenian inventor Daedalus who made artificial wings with feathers and reeds to escape from king Minos of Crete. Beewax and strings held the wings together in a gentle curve like the birds'. The father warned the son to fly at a medium altitude to avoid the sun melting the wax or the sea dampening the feathers. Exhilarated by the flight, however, Icarus soared higher and higher. The wax melting, he fell into the sea. A father no longer, Daedalus cried over the feathers scattered on the waves and flew away.

Ovid's tale of flight was one most expressive of his subversive desire. A poet of Augustan Rome, he aspired to become a genius that "forever trespasses upon" human limits for eternal fame.[2] The Golden Age—free peoples without arms and Nature in an eternal spring—expressed the poet's inalienable desire for a homeland without tyranny and savagery, perhaps a realm of free love and liberty. The two inventor-exiles, Daedalus the artificer and Ovid the poet who universalized the theme of metamorphosis, became entwined in European memory as the harbingers of a deeply transgressive desire.

In an imperial province a century later, the itinerant lecturer and satirist Lucian would expose the hypocrisy of kings, nobles, and philosophers in his "Icaromenippus." Menippus with an eagle's eye could observe from the moon a myriad of atrocious crimes that filled human lives. Men and cities appear to the imagined philosopher-voyager as so many anthills at war against one another, each putting up a discordant performance to offer plenty of food for laughter in its vagaries. He assumes a superior position vis-à-vis the earthly philosophers or the "brawling censors," who dwell in the separate word-mazes of farcical names such as Stoics, Epicureans, and Peripatetics. While sitting in judgment on others, they are "idle, quarrelsome, vain, irritable, lickerish, silly, puffed up, arrogant, and, in Homeric phrase, vain cumberers of the earth."[3] They publicly praise fortitude and temperance and condemn wealth and pleasure, while privately indulging in all the earthly delights. The Lucianic gaze configured the *cosmic spectator*, a textual figure who could unleash a biting satire of the earthly homeland from a safe distance. The spectator would offer an enduring mode of resistance in European history.

The story of Daedalus, a human inventor trapped in a game of gods, became a recurrent trope of humanity's boundary existence. In order to appreciate what the flying machine meant for the Enlightenment public, we must take stock of the successive cultural translations of Ovid that helped shape modern states, their heroes, and their citizens. The Enlightenment was a transformative period that nurtured an intense desire for human emancipation and diverse forms of modernity based on classical ideals.[4] Like many other classics, Ovid's poetry was printed throughout the Enlightenment to constitute a stable diet and a shared cultural memory for the rapidly growing reading public.[5]

The self-conscious celebration of the balloon as an invention in "modern natural philosophy"—a philosophical flying machine entirely different from the previously imagined mechanical flying devices—took place on the eve of the French Revolution.[6] Such claims of modernity served to erase the balloon's connection to the mythological past and the chimerical flying dreams, which transformed it into a scientific artifact that would build a modern aerial empire. In order to understand the balloon's extraordinary agency in mobilizing the Enlightenment public and the populace, however, we must bridge the long-standing bifurcation in the historiography between aeronautics and mechanical flying. The mythical heritage and temporal metamorphoses of the flying machine helped ignite the enlightened passion of balloon enthusiasm.

In taming Ovid's fantastic and immoral tales for the Catholic spiritual empire, late medieval theologians developed an allegorical reading that chimed with the newly dominant Aristotelian natural philosophy.[7] Roger Bacon (ca. 1214–1292), a Franciscan friar who listed flying engines among the known inventions, was a devout scholar who sought to devise a universal science that would reform Christian theology along the Aristotelian outline.[8] The spiritual ascent trumped the physical in *The Divine Comedy* by Dante Alighieri (1265–1321). The Florentine poet appropriated the theme of metamorphosis for his Christian epic in an effort to universalize the spiritual condition of humanity. The pilgrim's journey was a carefully constructed script for a spiritual ascent and resurrection that would reverse the Ovidian descent from the Golden Age.[9] The *Divine Comedy* fostered a body of commentaries and illuminated manuscripts, which constituted the "Christian Ovid." During the medieval "Age of Ovid" that standardized Greek myths, the fate of Icarus became a moral tale against overblown human ambitions. In *Ovid moralisé*, Daedalus became more the envious murderer of his talented nephew Perdrix than an ingenious inventor and grieving father.[10] His flying dream would become an outstanding chimera in European cultural memory.

Renaissance Humanists and artists had to reinvent Ovid's characters in order to express their desire for an earthly metamorphosis or self-fashioning.[11] The Florentine artist-engineer Leonardo da Vinci (1452–1519) may be characterized as a Renaissance Daedalus in his restless curiosity and deep-seated desire for a physical ascent. His training in the architect Andrea del Verrocchio's workshop taught him painting, sculpture, theatrical machinery, festival decorations, and architectural designs. The spiritual and the physical came together in the theatrical productions of *Ascension* and *Annunciation* with intricate contrivances to lift, suspend, and lower the actors playing Christ or the angels.[12] Leonardo's digressive talents displaced him from the Medici court (where he was a beautifully dressed singer) to Milan where he worked on the military projects of Ludovico Sforza, Il Moro. He designed an ornithopter, perhaps to conquer space in the incessant warfare that formed the Italian city-states.[13]

Leonardo's dream of flight grew in conversation with Machiavelli during a migratory period in service of Cesare Borgia, the model for Machiavelli's *Prince*.[14] Working on the canal project that would connect Florence to the sea, his soaring imagination came back to the great bird that would fill "the universe with amazement" and bring eternal glory to its inventor. His obsession with flight reveals the existential restlessness of a captive genius.[15] Leonardo was an inventor who sought to materialize Ovid's poetic invention that mapped a hidden desire for metamorphosis to attain power, glory, and liberty in an imagined empire.[16]

The superhuman ascent—physical or spiritual—became an open quest during the militant centuries that followed Columbus's symbolic possession of America.[17] As the "modern Argonauts" penetrated every corner of the Earth and the modern "Daedalian heads" such as Copernicus, Kepler, and Galileo made the Earth "stand and go at their pleasures," the Oxford scholar Robert Burton (1577–1640) struggled to come to terms with the expanding, bewilderingly diversified cosmography that confined him to an ever-shrinking

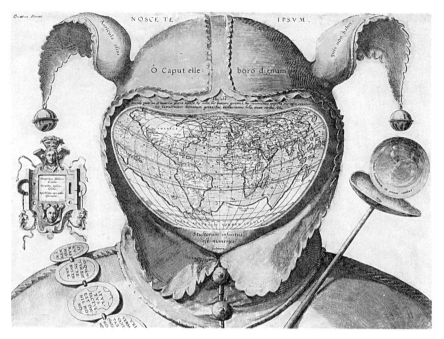

Figure P.2. *Fool's Cap* (ca. 1590).

part of the growing universe. He wished to fly like a long-winged hawk to visit celestial spheres, roam about the continents and oceans, and penetrate the center of the Earth in the company of mythological figures (including Lucian's Icaromenippus) and modern explorers to discern the truths of ancient, Christian, and modern tall tales.

The imaginary flight above, across, and beneath the globe posed perturbing questions: What is the center of the Earth? Whence proceeds that variety of manners and characters of nations? Whence comes this variety of complexions, colors, plants, birds, beasts, and metals peculiar almost to every place? Were they created in six days? Were they ever in Noah's ark? Why are the heavens so irregular?[18] In *Anatomy of Melancholy* (1621), Burton captures a modern malady suffered by a learned individual placed at a confusing juncture of history when no ancient, theological, or modern authority could provide a complete set of answers to the vexing questions on God's wisdom and humanity's place in the universe. The quest for an intelligent human empire was as much a psychological need for the early modern self (perhaps no wiser than the foolish jester shown in figure P.2) as it was an economic need for the emergent nation-empires.[19]

The infinite universe also served as a foil for utopian dreams in the cosmic tales that began to proliferate, ironically, after the Catholic condemna-

Figure P.3. Frontispieces: (left) Francis Godwin, *Man in the Moone* (1638); (right) Savinien Cyrano de Bergerac, *Comical History* (1687).

tion of Galileo in 1633.[20] Following the first English translation of Lucian's lunar voyage in 1634, imaginary voyages such as Francis Godwin's *The Man in the Moone* (1638) (see figure P.3) and John Wilkins's *Discovery of a New World* (1638) familiarized the figure of the cosmic voyager and otherworldly polities. In France where the civil war (the *Fronde*) did not topple the Catholic order, Cyrano de Bergerac (Savinier de, 1619–1655) challenged the absolutist state in his *Histoire comique* (1657).[21] He depicts a naturalistic lunar world, constructed out of eternal matter according to the cold logic of reason, where the inhabitants are free to imagine, make love indiscriminately, procreate, and invent new worlds (see figure P.3). Perfect rationality requires a complete inversion of the rules for establishing philosophical and religious truths. Cosmic tales mapped the emergent geography of early modern Europe—still a marginal, patchwork continent aspiring to the greatness and refinement of the Eastern empires—to shape diverse utopian dreams that would build Christian empires.[22]

Cosmic imagination produced dystopian satires as well as utopian fictions.[23] Imaginary flying through mythical time and among distant planets helped resurrect the cosmic spectator who could reflect on earthly ills from a safe distance and illuminate the political and ecclesiastical oppression.[24]

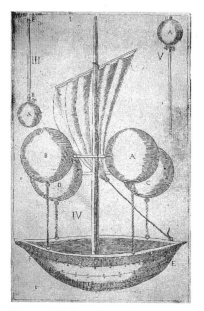 

Figure P.4. Imaginary flying vessels: (left) Francesco Lana, *Prodromo dell'arte maestro* (1670); (right) Eberhard Werner Happel, *Grösste Denkwürdigkeiten der Welt* (1689).

The cosmic view from nowhere resurrected the Lucianic gaze as religious wars and constitutional conflicts forged the early modern European states. The cosmic spectator could establish timeless and universal standards for truth and morality. Though intended as a critical figure, such claims of universal morality also helped valorize European claims to the Roman/Christian imperial legacy, often expressed as a quest for the "universal monarchy."[25]

The wings of Icarus cast a long shadow across European history (see figure P.1 and plate 1).[26] Although Christian moralization, Renaissance reveries, and early modern cosmic tales made the flying machine a dangerous chimera, it grew in the utilitarian sensibility and military reality of the Enlightenment. European empires and their centralizing administration required a mastery of space.[27] Francesco Lana de Terzi's design of a flying boat (see figure P.4) often surfaced in imagined and real endeavors along with a variety of imaginary wings. In 1709 Bartolomeu Lourenço de Gusmão demonstrated a balloon for the king of Portugal, perhaps the *Passarola* that circulated in drawings (see plate 3), although it has been obscured in history by the Montgolfiers' invention.[28] In *Gulliver's Travels* (1726) by Jonathan Swift, the flying island of Laputa functions as the ultimate control mechanism of an absolute ruler. As the counselor Kiper (to Frederick the Great of

Prussia) reminded the inventor Melchior Bauer, a successful flying device was "worth more than a kingdom; for by this means the King could make the whole world subject to him."[29]

Imperial dreams often found a noble façade in the name of science and progress. Young Rousseau entertained the possibility of flying in his unpublished manuscript *Le Nouveau Dédale* (written ca. 1742), perhaps inspired by the hapless marquis de Bacqueville's attempt to fly across the Seine (see plate 2). A new route in the air would provide a superior means of travel that could penetrate the most remote continents for the good of humanity. Such "noble intentions" would justify even the most "chimerical projects."[30] The marquis d'Argenson (René-Louis de Voyer, 1694–1757) regarded the art of flying as one of the most important discoveries to be made in his century. Large flying vessels would bring about a speedy and convenient transport of merchandise. Aerial armies would render fortresses useless and expose treasures and women, the defense of which would require a new "secretary of the State for aerial forces." Natural philosophy had to lead humanity to this discovery.[31]

To make flying possible, European dreamers had to transform the baroque field of cultural production and its mediators into a new set of heroic characters.[32] Olympian gods and their subsidiary kings yielded their places of honor to rational philosophers, intrepid voyagers, and ingenious artificers during the Enlightenment. Human ambassadors could visit planets in an airship fashioned after Lana's design, as illustrated in *Die Geschwinde Reise auf dem Luft-Schiff nach der obern Welt* (1744) by Eberhard Christian Kindermann (see figure 1.1). Flying gladiators (British and German) engage in an epic fight in Richard Owen Cambridge's mock-heroic poem *The Scribleriad* (1751). Winged humanoids roam about their subterranean world in *The Life and Adventures of Peter Wilkins* (1751) by Robert Paltock (see figure P.5). By the mid-eighteenth century, a diverse repertoire of flying devices and exotic creatures populated European imagination to push the boundaries of the human empire.

The flying machine and the cosmic spectator conjured up a mirage of the scientific empire that would discipline the earthlings into rational and universal citizens. In Voltaire's *Micromegas* (1752), the cosmic voyager embodies a search for the truth and ideal polity. The flying machine in *The History of Rassela* (1759) by Samuel Johnson offers a link to the outside world and knowledge. In *Le philosophe sans prétention* (1775) by Louis-Guillaume de La Folie (1739–1780), the supposed cosmic visitor from Mercury reveals the imperfection of human knowledge (see figure II.1). Nicolas-Edme Rétif de la Bretonne (1734–1806) translated Ovid's *Metamorphoses* for himself, after four years of Parisian debauchery, to recover his "native purity" and

Figure P.5. Imagined wings: (left) Richard Owen Cambridge, *Scribleriad* (1751); (right) Robert Paltock, *Life and Adventures of Peter Wilkins* (1751).

"upright soul." He conjured up a multiracial republican empire that guarantees equality and justice for all in *La Découverte australe par un Homme-volant ou le Dédale français* (1781) (see figure 1.4).[33] In imagining a physical ascent through flying, the enlightened inheritors of Daedalus dreamed of a moral ascent to a just world.

In order to appreciate the imagined aerial empire born of the Enlightenment culture, therefore, we must broaden the range of historical actors and include material artifacts and literary characters. In contrast to the human actors whose actions are largely shaped by the short-term cultural forces such as social status, family upbringing, education, and profession, things (real or imagined) and characters (mythological, literary, and theatrical) could conjure up deep cultural memories and long-range cultural forces that bind them together as a historical collective—a spatiotemporal complex of humans and things that act in unforeseen ways. Things and characters could function as enduring, yet transformative tropes to interweave transcultural and transtemporal experiences of human collectives. For this reason, they could implement a process of internal colonization and integration almost effortlessly without explicit ideological articulation. Domination without hegemony, as Ranajit Guha has outlined, requires that the dominated internalize the structural oppression or, in Pierre Bourdieu's term, naturalize their cultural disposition.[34]

Spectacles sustain cultural hegemony, as Guy Debord has argued, by materializing all facets of ideological systems—the impoverishment, enslavement, and negation of real life—to implement an "effective dictatorship of illusion." For those living under this "social hallucination" which erases the dividing line between self and world, only an alternative form of spectacle could communicate idealism, or a reflexive critique of their "abnormal need for representation."[35] A scientific spectacle, seen as commanding Nature's forces to focus mass veneration, could induce voluntary submission without resistance from all walks of society. It could also stage an authoritative vision of a utopian world in which everyone could dream and act freely. The liminality of the scientific spectacle—designed with imperial intentions, yet conducive to emancipatory dreams—helped forge modern mass public and nation-empires.

The chimera of flying broke into performance in June 1783, just a few years before the French Revolution, to stir up the dream of republican monarchy for the mass public. A successful balloon ascent provided a universal (shared) occasion to process divergent civic sentiments: it enabled the sovereign to project his glory, the nobility to exhibit their valor, the literary public to transmit Enlightenment ideals, and the populace to rejoice in a miraculous spectacle. It enacted the cultural memory of metamorphosis to disrupt the intricately balanced "state-body" of the ancien régime and to visualize a potentially republican nation.[36] Machines can congregate diverse collectives and function as a measure of humans and their civilizations in their travels, contacts, and translations over space and time.[37] This book is an effort to place the ephemeral balloon affair in the Enlightenment experience of deep time and universal geography with the hope to unpack its cultural and political agency in forging European nations and empires.

Figure Int.1. The ascent of Jacques-Alexandre-César Charles and Marie-Noël Robert from the Tuileries Garden on December 1, 1783. Tissandier Collection. The spectators are shown to observe good order by carefully avoiding the lawn. The palace in the background, where the balloon was made, would be destroyed by the Paris commune in 1871.

INTRODUCTION

A People-Machine

On August 27, 1783, carriages choked the streets leading to the Champ de Mars, then a military training ground at the edge of Paris, where a balloon ascent was in preparation. The police had secured the entire park and surrounding roads to direct traffic and to prevent accidents.[1] The eighty-seven-year-old duc de Richelieu walked to the site with his guards clearing the road before him. The princes of blood—the duc de Chartres (future Philippe Égalité), the comte de Provence (future Louis XVIII), and probably the comte d'Artois (future Charles X)—paraded through the crowd in their fashionable attire.[2] Women clad in muslin robes and covered by large hats (*chapeaux à la malborough*), or in "revolting" modern costumes, presented a "truly curious and amusing spectacle."[3]

Even a royal procession had never "attracted a greater gathering of society from all estates and conditions."[4] Benjamin Franklin estimated the crowd at fifty thousand and others, at three hundred thousand.[5] Throngs of people lined the streets and the roofs along the Seine. As in popular entertainment scenes, the crowd comprised "all orders of citizens," which included grand seigneurs, ministers, princes, savants, artists, and the populace. The governor of the École militaire brought his students with "every apparatus of a great ceremony."[6] The enclosure held about twenty thousand souls. People packed the surrounding field to make a colorful "canvas . . . decorated with the immense multitude of the curious."[7]

The spectacle as "a locus of illusion" can enforce a system of cultural hegemony that would sustain a political structure without violence.[8] More persuasive because they seem less despotic, as Jean-Jacques Rousseau intuited, science, literature, and art can camouflage the "iron chains with which they are laden" and make civilized men "love their slavery."[9] Universally acclaimed as a "majestic" spectacle, the balloon ascent could have expanded the absolutist state-machine. The magnificent artifact decorated with royal emblems might have substituted for the king's body to multiply the theatrical relations of the court (dominance and subjugation) around the nation and thereby to integrate the cultural nation as an extended version of the court society. By congregating a mass public whose veneration focused on its scientific performance, however, the balloon floated the specter of an alternative, potentially republican, nation. It became a national artifact that could destabilize the theater-nation centered on the king's body.

In Louis XIV's court, the king decided the texts, décor, costume, and heroes of the court theater to constitute a symbolic body that represented the state or the "king-machine" in Jean-Marie Apostolidès's term.[10] In the baroque court, princely power was materialized in the clothes and jewels heaped and dangled on the royal body, as Stephen Greenblatt notes, to create "a realm of matter so rich, detailed, and intense" like a brilliant sun over a seascape.[11] Symbolic capital circulated through the material fashion—wigs, clothes, jewels, furniture, and so on—that marked the bodies of power and their spatial relations. An alluring geography of bodies and things disciplined Versailles courtiers to internalize their scripted roles and thereby to transmit the court hierarchy outward to the nation in fashionable displays. This *material geography* of distinction—status and power expressed and recognized through the spatial arrangement of differentially costumed bodies—inscribed the absolutist polity.[12] Emulation was a game that entrapped everybody who wished to find a place in the king-machine.

Pomp served power at Versailles to constitute a *theatrical polity*, not power pomp as in the "theater-state" whose sovereignty consisted in its exemplary function as a "microcosm of the supernatural order." In Clifford Geertz's story, Negara as the seat of Bali rule had to provide "a paragon, a faultless image of civilized existence" to shape the world around it into similes of its excellence.[13] As a living theater, in contrast, Versailles staged curiosity and pleasure as the primary means of political persuasion and cultural integration by domesticating the courtiers through a never-ending play. It seduced them by festivals, spectacles, luxury, pomp, pleasures, vanity, and effeminacy to occupy their minds with worthless things and to relish trifling frivolities.[14] A fine-tuned symbolic economy of pleasure defined the court society. The "triumph of pleasure" propagated through the Parisian royal theaters—the Opéra (Académie royale de Musique), the Comédie

Française, and the Comédie Italienne (Opéra comique)—which monopolized public entertainment.[15]

The king's body defined the absolutist theatrical polity to place his subjects in a complex arrangement of subordinate bodies. A "perfect courtier" mastered the art of refinement or "falsehood," which allowed him to become a willing slave to the prince and a lord to the others. He played capriciously, constantly adjusting his plans and goals, to participate in this serious, yet sad game guided by vanity and self-interest. Nobody was a greater slave than an assiduous courtier, Jean de La Bruyère (1645–1696) observed, who served not one but many patrons to advance his position.[16] The courtier who performed his role with precision could be easily replaced, much like a machine part. The "fall of the favorite" was a routine mechanism that demonstrated the king's putatively absolute power.[17] A well-functioning absolutist state-machine would be a perfect automaton, which accorded precisely defined places and functions to all subjects, as illustrated in the Salzburg mechanical theater, while excluding the populace.[18]

How and why such an intricately balanced state-machine fell apart is a question that has long haunted French historians in their effort to identify the economic, social, religious, cultural, or political causes of the French Revolution.[19] In modifying the Marxist notion of the bourgeois revolution, revisionist historians have broadened the explanatory repertoire with keen attention on the transformation of the public.[20] Despite the rich historiography stemming from Jürgen Habermas's notion of the public sphere, however, the path of Enlightenment from the literate public to the illiterate populace remains obscure. To fill this lacuna, we must trace how the fashionable "public"—who read books and journals, attended concerts and lectures, and frequented cafés and conversational soirées—expanded to include the illiterate "people" whose public expression can be found only through rites, festivals, ceremonies, carnivals, and riots.[21] As Harold Mah points out, the unspecified spatial expansion of the Enlightenment public (or public sphere) into the mass subject is a historiographical fantasy that undermines historians' capacity to understand the revolutionary crowd.[22] A social body is formed not by "the universality of wills," notwithstanding Rousseau's wishful formulation, but by "the materiality of power."[23]

The role of science and technology in setting an ideological agenda requires a careful assessment in this regard. While historians have traced the revolutionary ideology to Enlightenment thought and the cult of reason to science,[24] Parisian scientific institutions played an important instrumental and symbolic function in sustaining the absolutist polity and its imperial machinery.[25] Charles C. Gillispie thus characterized the relationship between academic scientists and the state as an instrumental transaction of technical expertise and patronage. In objecting that science provided a model of rational authority that could counter any despotic regime, Keith Michael Baker sought to reinforce the alliance between mathematical reason and revolutionary thought.[26] The rhetoric of liberal Enlightenment is difficult to reconcile with the practice of absolutist science, however, especially when one focuses on the use of mathematics and measurements in the royal institutions. Ken Alder's exquisite study of interchangeable guns has shown how mathematical education cultivated a strong quest for technocracy, which persisted through the revolutionary political changes.[27]

In order to evaluate the complex relationship between science and polity in revolutionary France, we must pay attention to the other kinds of scientific knowledge (other than mathematics) that appropriated material powers for "popular" consumption and probe how the boundaries between the scientific public and the illiterate populace became porous.[28] Natural philosophy offered spectacles of active powers to the enlightened audience, as Simon Schaffer has argued persuasively, to shape their moral, aesthetic, and political sentiments. Emma C. Spary has shown how the production and consumption of coffee and liquors shaped Parisian science and culture.[29] How these fashionable urban sciences related to political culture is nevertheless a difficult historiographical issue, especially if we wish to include the populace and their role in shaping mass politics. In his pioneering study of mesmerism, Robert Darnton argued persuasively that popular sciences occupied the center of public attention in the 1780s when the intensifying censorship of political news and libels created "a curious calm before the storm."[30] He relied on the layers of elite discourse to unearth radical thought, however, which strengthened the revisionist historiography of discursive contestations. How to characterize the crowd as legitimate political actors remains a vexing problem.

Unlike other scientific spectacles that targeted the fashionable society (*le monde*), the balloon ascent also attracted the populace, which in turn invited state control and public propaganda. The concerted effort by the state and the elite public to discipline the crowd engendered a mass public—a transitional collective between the literate Enlightenment "public" and the modern mass subject that supposedly encompasses the plebian. In other words, the balloon public was conceived as a means of expanding state control over the illiterate populace. Balloon spectators in their variegated composition and unprecedented number should offer us an exceptional opportunity to understand the prerevolutionary crowd but for the silence in public reports and the absence in printed images of common "people."

Balloon historians have not yet considered its theatrical relation with the mass audience that forged its historical agency. Charles C. Gillispie's exquisite account of the Montgolfiers' invention made a qualitative leap from the nineteenth-century triumphalist accounts, but his focus remained on the balloon and its technical progress. More recently, Marie Thébaud-Sorger's sophisticated sociological probe and Michael R. Lynn's geographical coverage have considerably enriched the balloon historiography and its relevance to the eighteenth-century consumer revolution, but they do not consider mass audience as a serious political agent.[31] Neither have literary scholars utilized the insights from the reader-response and reception theories to characterize the enormous balloon public and their situational agency.[32]

People set out, filled with hope, for the majestic balloon ascent.[33] As a venerated scientific spectacle, it blurred the boundary between the educated public and the populace to engender a "contact zone," which refers in Mary Louise Pratt's definition to the social space "where disparate cultures meet, clash, and grapple with each other, often in highly asymmetrical relations of domination and subordination." By configuring this space of encounter and its constitutive power relations, historians can discern the strategies of "anti-conquest" that naturalize such asymmetric relations of power as well as the strategies of resistance that challenge the status quo.[34] The Enlightenment "public" did not automatically develop into the modern mass public as a consequence of political, industrial, or commercial revolutions. The transformation required cultural resources, state intervention, and the technologies of mass control.

A place of memory could translate a historical imaginary—a story of the past that legitimizes political regimes and practices by utilizing history and historiography—into a political one.[35] The spontaneous gathering of the citizens of all estates at the Champ de Mars, the mythical place of origin for the French nation, would have presented a vivid image of the "nation above kings."[36] Seen as the site of the original parlement or "general assembly of the nation"—one that could legislate laws, deliberate on matters of the state, declare war, and elect kings—this place of collective memory had shaped aristocratic and judicial resistance to the absolutist regime.[37] Interpreted as a democratic assembly, albeit tempered by aristocratic power, the originary gathering lent itself to utopian imaginaries.[38]

The emergent "state-body" swirling around the patchy balloon visualized an imagined, potentially republican, nation. Forging a republican nation was a topic that had dominated café politics during the American War, gaining urgency with the peace talks.[39] The new republic of America offered an "imaginary recourse against" the ancien régime, as François Furet saw it, "to invent a new historical memory, free from persecution and injustice."[40] Balloon festivals coordinated material resources, human actors, administrative control, and publicity mechanisms to instantiate a Janus-faced mass action precariously poised between the carnival and the riot. If the stratified barriers on the ground marked the ancien régime social hierarchy defined by social rank and capital, the aerial vista opened a powerful egalitarian vision: everybody was "equal" in the air.[41] The "chimera of equality" was the most dangerous of all beliefs in a civilized society, according to Denis Diderot.[42]

Egalitarian fraternity reigned at the Champ de Mars, according to an imaginative provincial satirist, where men and women of all professions and social status embraced their opposites: women their husband's friends, men their neighbors, the learned the ignorant, physicians theologians, mathematicians poets, musée members academicians, and so on. Three priests, two philosophers, four financiers, one housewife, five "bourgeois," and two fishwives supposedly fell into his arms.[43] Such temporary liberation from the prevailing sociopolitical order and such suspension of all privileges, norms, and prohibitions used to merge "the utopian ideal and the realistic" in the traditional carnival, according to Bakhtin, to instantiate a "true feast of time . . . of becoming, change, and renewal." For a brief moment, the people would enter "the utopian realm of community, freedom, equality, and abundance." While the carnival lasts, people live in it, free and hopeful for their world's revival and renewal.[44]

The balloon ascent may be seen as a "politically significant *mise-en-scène*," a modernizing carnival that brought an immense crowd of diverse composition to the same site for a briefly intensified celebration of the nation's scientific accomplishment and technological future.[45] It focused disparate energies and activities of the science-minded public on a single machine and activated an experimental, ephemeral form of nation-making that infused enthusiasm into the emergent citizenry. Subversive words did not simply trickle down from the published literature, as Arlette Farge reminds us, but attached themselves to the discussion of the things, spectacles, and events seen by actual people. These acts of appropriation shaped popular culture and opinion.[46] As a metagenre of cultural performance, which demonstrated the authority of science and induced universal veneration, the balloon ascent wove historically, spatially, and socially differentiated forms of symbolic action into a new whole.[47]

A mass collective at the Champ de Mars might have intensified the subterranean longing for an emancipation from the absolutist polity, or the king-machine, when tales from the Bastille—friendly rats and all—drove home the oppression and fear that sustained despotism.[48] In staging a venerated mass spectacle, the balloon constituted a *people-machine*—a composite body of the nation whirling around a fragile, patchy machine, which included the populace and thereby blurred the intricate social hierarchy that sustained the absolutist polity. As such, the balloon floated the vision of an egalitarian polity that could free the citizens from their servitude, or an alternative to the king-machine. It seemed to answer Louis-Sébastien Mercier's (1740–1814) call for a public without social boundaries gathered at a "superb public place that was capable of containing the whole body of the citizens."[49] Rousseau had wished to counter the artifice of royal theaters and festivals that bored the rich and disheartened the poor by an open communal festival where gentle equality simulated natural order.[50]

As a flying machine that brought an indocile natural element under human mastery, the balloon floated the vision of a new golden age when a philosopher-king would govern France and nations would compete on their merits rather than on their strength.[51] One cannot but wonder what it meant for the "people," often despised by the elite philosophes and fashionable socialites, to witness a scientific wonder that seemed to bring human existence closer to the realm of the gods.[52] What kind of theatricality did this profound moment of mass absorption engender?[53] Interpretations of this extraordinary moment would differ among historians, especially because of its chronological proximity to the French Revolution.[54]

The people of Paris have attracted historians' attention mostly for their poverty, marginality, and instability to become "a legendary and mythological historical subject" as the crowd of the French Revolution.[55] As such, their historical agency has been limited to making the revolutionary violence real. Historians have debated whether the revolutionary violence reflected the people's hostility to modernization or their impatience at its slow progress, stoked by the emancipatory dreams of Enlightenment.[56] As Micah Alpaugh has recently shown, however, revolutionary marches in the beginning were mostly peaceful demonstrations.[57] In trying to explain the Terror, historians have lost sight of the hegemonic system of power built on science's promise of rational progress, the exuberant mood at the end of the American War, and the administrative technologies that shaped the nation's material culture and imperial aspirations for the subsequent generations. Notwithstanding the ongoing scrutiny of the prerevolutionary public, we do not as yet possess an adequate category for the balloon public/crowd that included most of the adult urban population.

In their search for the structural and ideological causes of the French Revolution, historians have neglected the glorious moment of military and scientific victory in 1783, which promised a peaceful empire as well as internal cohesion—a vision of "true union" in the body politic for the public good, as Montesquieu had envisioned it.[58] The balloon spectacle in its capacity to provide "total justification for the conditions and aims of the existing system" opened a liminal realm—a transitional realm between normal social structures that could engender new possibilities.[59] The mass spectacle of hope can offer an exceptional window to the tenuous, hidden connection between the Enlightenment public and the revolutionary crowd. Invisible natural fluids attracted insignificant bodies to visualize a nation of equal citizens by utilizing, ironically, the resources of the ancien régime. Once we move away from the ideological caging of the Enlightenment and the French Revolution,[60] balloon fever at the conclusion of the American War offers a promising subject in exploring the relationship between science, Enlightenment ideals, the French Revolution, modernity, and the European nation-empires.[61]

By situating balloon mania in prerevolutionary political culture, we can consider the importance of material agency in mediating between elite thought and mass action, which in turn facilitated human agency and ideological articulation.[62] An ensemble of public, administrative, and commercial technologies stabilized the balloon's scientific status and philosophical virtue so as to constitute a national artifact. An archeology of this monumental, yet ephemeral, artifact would alert us to an emergent system of scientific hegemony that coordinated state power, elite knowledge, and material artifacts to enlist the uneducated populace as rational citizens. The *scientific imperium* would also blur the boundary between the nation and the empire. The nation-state in French elite desire was an imagined empire with plastic boundaries, rather than an imagined community of citizens as the philosophes wished for.[63]

The Imagined Empire: Balloon Enlightenments in Revolutionary Europe aims at an archeology of mass silence, a genealogy of the mass public, and a material geography of European Enlightenment to uncover how the flying machine—both imagined and real—stirred utopian visions and patriotic sentiments in revolutionary Europe.[64] The balloon staged the vision of a moral empire built on scientific prowess—a vision that had previously been nurtured through Aristotelian philosophy (for the Catholic empire) or Newtonian mathematics (for the British Empire).[65] By rehabilitating a machine's agency vis-à-vis that of philosophy and the theoretical sciences in forging imperial cultures and polities, the book configures a "history of the present" which, in Michel Foucault's vision, would intensify the "insurrection of subjugated knowledges" and expose the vulnerability of "global, totalitarian theories."[66] If we wish to characterize science as a communicative action without boundaries and abandon the term "popular science," as James A. Secord proposes, we must understand how machines and material artifacts communicate and translate science for a mass audience.[67] Unlike philosophy or mathematics, spectacular machines could reach the populace without the layers of mediation. By configuring a machine's agency and geographical reach, we can probe its relevance to mass politics and global history.

Our ability to conduct an archeology of subjugated knowledges and silent actors is severely limited by the "archive"—the collection of documents carefully selected and preserved for posterity. History also depends on what historians choose to write and how they interpret extant documents, the bulk of which were produced and preserved by dominant groups. Foucault's archeology of silence thus aimed at uncovering the "broken dialogue" between the extant and the extinct modes of representation to uncover the rules of producing the successful knowledge-truth-power complex that came to dominate the world. Access to these formative rules depends on excavating alternative historical actors and their statements.[68] The archeology of mass silence poses an insurmountable challenge, therefore, in identifying the subterranean layers of discourse. Popular culture remains an "umbrella term for practices rooted primarily in oral exchange, local settings, and the vernacular," which has long limited the attempts to write the "history from below."[69]

In the case of ballooning, the populace is absent in most images, reports, and histories: the journalistic "public transcript" screened out dissenting voices and unsightly people to consolidate a broad consensus on its meaning for the public good. As "the *self*-portrait of dominant elite as they would have themselves seen," the public transcript in James C. Scott's definition is designed to naturalize their power by creating "the appearance of unanimity among the ruling groups and the appearance of consent among subordinates."[70] In order to write a critical (rather than monumental or antiquarian) history of ballooning in Friedrich Nietzsche's conception, therefore, we must scrutinize "the archive" that has perpetuated a glorious memory of the French aerial conquest and unearth a diverse "repertoire" of literary and material performances that shaped the mood on the streets.[71]

The archeology of mass silence in this book employs a dual strategy in interpreting the sources. On the one hand, it brings to light the few extant pamphlets that have been written out of history. In contrast to the number-ridden reports and the myth-enacting poetry published in censored newspapers, the pamphlet literature on ballooning often conjured up diverse characters. Identified by their titles or professions, these pamphlet characters voiced their social, political, and cultural standpoints.[72] Such a representational field configured an inward gaze that, by criticizing French society and polity, undermined the outward vision of a vast technological and cultural empire.[73] As literary performances, these pamphlets fractured the balloon's scientific identity and representative political function. They offer a glimpse of the imagined machine polity engendered by this ephemeral, yet monumental artifact. In other words, these pamphlets harbor the memories, political claims, and identities of the "anti-balloonists," which produced the hidden transcripts on ballooning—the "offstage speeches, gestures, and practices that confirm, contradict, or inflect what appears in the public transcript."[74]

On the other hand, the historian must scrutinize the triumphalist sources themselves in an attempt to discern the literary and visual technologies that silenced the populace. Historians have long tried to circumvent the lack of subaltern sources by tapping official archives such as the Inquisition and the police records.[75] A small sampling of transgressive historical actors would not capture the balloon's extraordinary visibility and representativeness, however, which seemed to unite a crowd of diverse composition and unprecedented size in a shared dream.[76] In order to make the balloon relevant to mass Enlightenment and revolutionary festivals, we must understand how the royal administrators deployed censorship and policing mechanisms to subsume dissident political imaginaries. Without acknowledging such hegemonic operation of the ancien régime, which sustained its fine-tuned equilibrium and a moderately progressive vision of a benevolent monarchy, political significance of the balloon's theatricality will remain hidden.

Spontaneous balloon festivals also lend themselves to a genealogy of the French Revolution.[77] A genealogical approach in Foucault's terminology seeks to identify the stabilized techniques of a power-knowledge regime that turn human bodies into objects of knowledge to constitute the "body politic." For example, Foucault focused on the prison, its organization, and its machinery of disciplining the body for an embodied "history of the modern soul."[78] Similarly, an analysis of the machine polity can focus on the ideal machine that coordinates a repertoire of stable technologies to consolidate a distinct knowledge-power complex. Machines as cultural artifacts articulate sociopolitical power in more diverse forms than the human body to provide distinct models for alternative hegemonic systems. Their mobility also allows the historian to consider politics on a larger scale. As Michael Adas has argued, machines function as the measure of men and their civilization across asymmetric cultures.[79]

A genealogy of scientific spectacles can also help us discern a repertoire of stable technologies that work across a representational divide.[80] A sudden discursive change may take place not necessarily because of individual actors' intentions and interests but through the configuration of a new object on which the technologies of control operate. By delineating the power relations that transformed a patchy machine into a spectacular scientific artifact of mass veneration, we can identify the balloon public as a lost historical entity that nevertheless remained in dialogue with the revolutionary crowd and their scientific spectacle—the guillotine. A scientific killing machine designed to be humane and democratic for the revolutionary cause, the guillotine became a substitute spectacle (and national artifact) for the balloon with intimate emotive impact.[81] Despite their divergent emotional and political outcomes, these two monumental machines shared an ensemble of scientific, administrative, social, and literary technologies that stabilized their hegemonic status.

New machines can articulate alternative polities. The baroque polity utilized automata as a hidden source of power to materialize princely authority as in the court masque and the water gardens.[82] The automaton as an intricately designed machine also specified the role each person had to play in the state-machine and symbolized a mode of existence within a complex, interrelated whole. For this reason, a machine's capacity to coordinate an alternative collective became a critical resource under volatile political circumstances, as Steven Shapin and Simon Schaffer have shown in *Leviathan and the Air-Pump* (1985). In Restoration context, the Royal Society of London advocated a parliamentary polity by envisioning a consensual form of life around the air-pump. The glass machine's physical fragility, coupled with its social need for credible spokespersons, made a paradoxically convincing argument for the parliamentary constitution of experimental philosophy and legitimate polity.

A century later in France, the balloon weakened the king-machine by opening a liminal zone in which the existing technologies of scientific demonstration, social control, and political machination could no longer contain the audience or the meaning of a mass spectacle. By radically expanding the public for scientific spectacles, the balloon projected an alternative nation that would mobilize royal power for the citizens' liberty and happiness. The effort to utilize the balloon for a regime of scientific hegemony and to discipline its crowd as a mass public ultimately failed despite the refined administrative technologies in communication, transport, and policing. Exactly who might assume the sovereignty of the nation became a matter of contention. After a brief reign of the guillotine and the Napoleonic Empire, a variety of "romantic" machines would take up the transformative task.[83]

European modernity was inextricably interwoven with the machines that could design alternate worlds. In an archeology of the Peacock Island near Berlin, for example, M. Norton Wise and Elaine M. Wise have unearthed the changing visions of the Prussian monarchy, which transformed it into an English garden powered by a steam engine in the post-Napoleonic period. In its spectacular visibility, the engine house symbolized human activity that would form an ideal nature and a powerful nation. According to John Tresch, nature became an interactive entity that involved human consciousness and action through the romantic machines and their translations of philosophy and aesthetics for the mass public, which charted the contentious path of Parisian modernity. The transformative power of machines shaped progressive social visions and utopian dreams for a harmonious state-machine that would realize the human potential. Modernism in the fin-de-siècle Europe would also depend on machine-mediated "physiological" aesthetics, as Robert M. Brain's elegant study indicates. Specific configurations of measuring instruments, techniques, and living substances produced reliable elements of physiological knowledge to cultivate avant-garde arts.[84]

The story begins in part I, "Invention in Theatrical Polity," with a genealogy of the material public sphere—the domain of public spectacles that mixed the enlightened public and the uneducated populace—which developed only limited genres of discursive articulation such as standardized newspaper reports and entertaining pamphlet literature. In order to appreciate the political significance of this material domain, we must bring to light the technologies of social control that monitored its boundaries and how a fragile scientific machine destabilized them. Drawing on the Montgolfiers' extraordinary archives, chapter 1 characterizes ancien régime inventors as liminal figures prone to the dreams of a spectacular ascent through their genius in the way literary, artistic, and scientific talents had been rewarded.[85] The Montgolfiers' spectacular ascent from the automata-like existence makes poignant sense only against the sensational failure of Jean-Pierre Blanchard's flying carriage a year earlier. The first balloon ascents near Paris, chronicled in chapter 2, engendered the public transcript, which characterized the "aerostatic machine" as a useful scientific artifact that would serve the nation and humanity. By identifying the hegemonic apparatus that sustained the theatrical polity, this chapter unveils a complex public domain supported by the royal institutions, variegated patronage networks, the public press, the postal system, subscription mechanisms, and individual aspirations. This mixed public domain transformed a useless provincial invention into a potentially useful national artifact. Subsequent Parisian human ascents, discussed in chapter 3, shifted public attention to the "intrepid" human actors, or the aeronauts who became authentic folk heroes. The "Assumption" of the philosopher-voyager Jacques Alexandre César Charles mesmerized the crowd whose "universal" admiration inaugurated a new body of the nation. The apotheosis of the philosophical aeronaut—devoted to the patrie and humanity—signified a triumph of the common people who wielded the power to select their own heroes. The mass theater commanded by this venerated philosophical Columbus, or "true Columbus," pushed the absolutist theatrical polity precariously toward the unknown.

The balloon floated the hope for a republican nation, which would be ruled by a philosophical majesty and populated by happy citizens. In part II, "Philosophical Nation," the analysis moves away from the marvelous artifact and its representatives to the silent spectators and their imagined nations. By paying attention to diverse "hidden transcripts," chapter 4 stratifies the balloon audience and their efforts at transculturation, which were integrated as a consensual vision of benevolent monarchy. Rétif de la Bretonne's subversive fantasy of republican empire is shown to share similar sentiments, for example, with the Orleanist push for a philosopher-king. The administrative efforts to control mesmerism and ballooning in the spring of 1784, delineated in chapter 5, attest to a subversive thread of the material politics that utilized natural fluids to mobilize the populace. Both the open balloon theater and the private mesmeric séance lay outside the well-policed public culture of royal theaters. Provincial fermentations caused serious concerns about their disruptive potential. Even the successful ascents in Lyon and Dijon, analyzed in chapter 6, coordinated elite patrons, material technologies, administrative resources, and local publics to strengthen regional patriotism rather than produce docile royal subjects. Whether provincial citizens would commit to a unified nation depended in part on the balloon's capacity to project universal aspirations. A failure, as in the Bordeaux attempt discussed in chapter 7, invariably caused a riot that threatened the status quo. The balloon's fall from the royal and public grace may be attributed in part to its potential to incite popular enthusiasm and unrest. It transgressed the carefully maintained boundary between the literate "public" and the illiterate "people" to float the vision of another nation.

The geography of balloon spectacles structures part III, "Material Empire," to map the mediated cultural translations and patriotic revolutions that produced a patchwork of European modernity. Although the travels across cultural and linguistic boundaries shaped the balloon's identity as a French artifact, contestations over its scientific promise and political associations exposed a heterogeneous European modernity that was as patchy and fragile as the balloon. Simply put, the balloon carried French civilization abroad, excised of its philosophical aspirations. The slow effervescence of London ballooning, chronicled in chapter 8, contrasted sharply with its instant stabilization as a scientific artifact in Paris. London ballooning became neither royal nor public but commercial. The contrasting geographies of science and spectacle in the two imperial capitals reflected their divergent polities. Interdependency among the geographically proximate powers often generates competitive differentiation and individualization through complex cultural interaction and resistance.[86] The competition to cross the English Channel became fierce, as recounted in chapter 9, but the London public did not invest in the French aeronaut Blanchard's historic crossing. Nor could he turn a profit with his Aeronautic Academy. As the British press noted, Blanchard's "French" strategy of appealing to affluent patrons did not work in London.

When British native ballooning began to mushroom, Blanchard launched an itinerant business on the continent, as discussed in chapter 10. He staged ascents mostly along the northeastern French border zone while occasionally venturing out to the German-speaking cities. The infrastructure of the Grand Tour—mapped by a host of London hotels and the routes of transport such as rivers, canals, and highways—played a significant role in charting the geography of ballooning.[87] Nevertheless, the cost and the mechanism of preparation limited potential sites.[88] Blanchard's continental itinerary, determined by the prosperity and aristocratic pretensions of target audiences (often Freemasons), suggests a liminal geography of the balloon Enlightenment—the French cultural empire expanding through aristocratic pleasure and mass veneration, which paradoxically strengthened German patriotic resistance to French civilization. The Masonic ideal of universal fraternity, which facilitated balloon travels and translations across cultural and political borders, helped build nationalistic empires.[89] A process of deliberate cultural and political appropriation transformed the balloon's meaning and function in the emergent European nations.

The epilogue sketches out the revolutionary metamorphoses of French balloonists and their patrons to muse upon the intersecting historical agency of machines and humans. The imagined flying machine had carried an emancipatory desire for millennia to open a liminal moment of intense mass veneration—when people are detached from the dominant system and free to imagine alternative social arrangements.[90] The machine's ability to fly across time—or to enact deep historical memory—manifested as its spatial capacity to consolidate a mass public. This spatiotemporal nexus of a mass spectacle would resurface through the Federation movement, which began in the provinces and culminated in a grand "festival of the nation." A "superb balloon in the colors of the nation," stripped of royal emblems, rose from the Champ de Mars at the end of the weeklong Fête de la Fédération on July 18, 1790.[91] As the revolution turned violent, the emotive power of mass spectacle would be carried on by the guillotine—a scientific killing machine that accentuated the terror of the sublime.[92] A genealogy of scientific machines can illuminate the continuity as well as the discontinuity across the revolutionary divide.

Plate 1. Jacob Peter Gowry, *The Flight of Icarus*, after a sketch by Peter Paul Rubens (1636).

Plate 2. An imagined rendering of the marquis de Bacqueville's attempt to fly across the Seine in 1742. NASM A19750726000cp02. The flying figure is fashioned after that of Besnier's flight, published in *Journal des Sçavans* (1678).

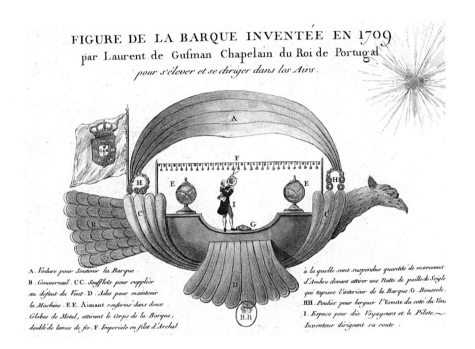

Plate 3. Bartolomeu de Gusmão's airship *Passarola*, Lisbon, 1709. Gallica, BNF.

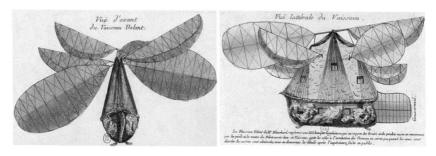
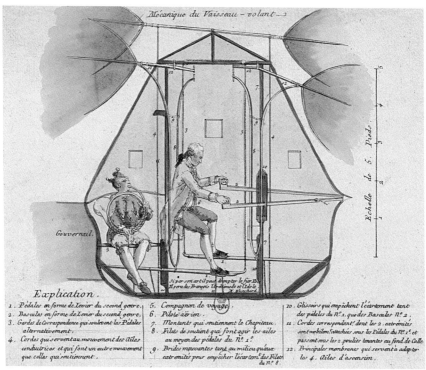

Plate 4. Jean-Pierre Blanchard's flying carriage, front, lateral, and internal view. Gallica, BNF. Made in light solid wood, it looks like a bird, a nose imitating the head and a rudder forming the rear. The exterior is covered by varnished carton used in making carriages. It has six wings of the same size, each 10 feet long and 6 feet wide. *Mémoires secrets,* May 6, 1782.

Plate 5. Tethered ascension from the Réveillon factory on October 17, 1783. Tissandier Collection.

Plate 6. *Le Flesselles* for the Lyon ascent on January 19, 1784. Gallica, BNF.

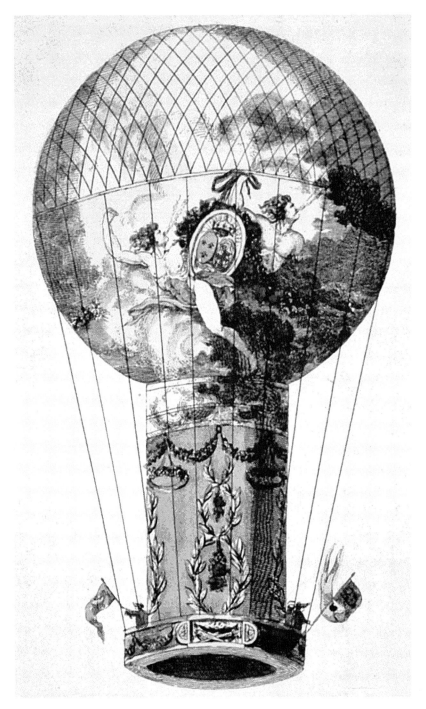

Plate 7. Pilâtre de Rozier's Calais balloon. Gimbel Collection XL-5-4623.

Plate 8. Consumer items. Both fans feature Blanchard's balloon, while the two objets d'art depict the royal Montgolfière. NASM T20140072052 (fan), A19750721000cp05 (objet), T20140072070 (watch), T20140072054 (fan), A20020121000cp02 (plate), and A19750701000cp05 (plate).

PART I

INVENTION IN THEATRICAL POLITY

The king-machine dominated public culture of the ancien régime and molded its theatrical polity and sociability. We must ask, therefore, how its hegemonic theatricality—one that prescribed specific functions to all subjects, except for the populace, as if they were a part of a well-designed automaton—opened a liminal space for the balloon to float the specter of an alternative machine polity.[1] Victor Turner's notion of "social drama" is particularly helpful in this regard. In order to characterize a community's movement through time, he privileges the moments of liminality, or public reflexivity, when the members are liberated from the dominant "cultural script" to formulate "a potentially unlimited series of alternative social arrangements." Since this period of ambiguity constitutes "in principle a free and experimental region of culture, a region where not only new elements but also new combinatory rules may be introduced," Turner characterizes liminality as a storehouse of possibilities that engender a new structure in milieus detached from mundane life. In his scheme, "yesterday's liminal becomes today's stabilized, today's peripheral becomes tomorrow's centered."[2] A dominant culture generates within itself an unstable zone that harbors visions of alternative systems.

In order to understand the rise of ballooning in prerevolutionary France, then, we must understand the ancien régime culture of spectacles that maintained its polity, sociability, and culture of invention. Rousseau observed in *Lettre à d'Alembert sur les spectacles* (1758) that theatrical consciousness permeated all social relations and brewed a "universal desire for reputation and

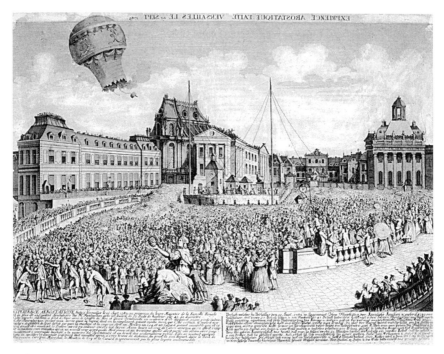

Figure I.1. "Aerostatic Experiment made at Versailles by M.M. Montgolfier on September 19, 1783, in the presence of their Majestics of the royal family and more than 130,000 Spectators with a Balloon 57 feet in height and 41 feet in diameter. This superb machine of azure color decorated with the royal symbols and diverse ornaments in gold color had the volume of 37,500 cubic feet and would weigh 3,192 livres when filled with the atmospheric air. Filled with the vapor that weighs less than a half of common air, it would obtain a rupture of the equilibrium of 1,596 livres. The machine and the cage containing a sheep, a rooster, and a pigeon weighed 900 livres all together, which left 696 pounds of lifting capacity. At one o'clock, a cannon shot announced the beginning; a second, after eleven minutes, notified that it was full and a third, that it was ready for departure. It rose majestically to a great height to surprise the spectators and cause tumultuous public acclamations. It remained there in equilibrium for eight minutes and came down slowly to the Carefour Marechal in the woods of Vancresson, 1,700 toises removed from its point of departure. The sheep, the rooster, and the pigeon did not experience even the slightest inconvenience." Gimbel Collection XL-2-1062.

This iconic image printed with French and German subtitles detailing factual information illustrates the desired public image of the royal balloon witnessed by orderly spectators. They form a seamless body of the nation without any visible distinction between orders. The layers of soldiers that surrounded the stage are hardly visible.

distinction."³ To be seen and to look at others—the habit of theatricality—was the most important genre of social activity in French absolutist polity. The centrality of the theater in molding public culture forced even the reform-minded philosophes to appropriate it, rather than condemn it. Denis Diderot (1713–1784) pioneered in *Le Fils naturel* (1757), for example, the bourgeois drama that staged ordinary characters and their plight to induce tears, rather than laughter or terror as in classical tragedy. By rethinking theatrical relations, the new drama would enlist the audience in creating a new social space that privileges genuine sentiment and public virtue.⁴ The theater had to provide a medium of human transformation, rather than a mechanism of human subordination and slavery.

Parisian royal theaters that monopolized public performances became a contested site of national consciousness during the Maupeou Revolution (1771–1774)—a top-down reform effort by the chancellor Maupeou (René Nicolas Charles Augustin de, 1714–1792) that incited feverish judicial and patriotic opposition to ministerial despotism and underlined the need for "a reconstitution of the body politic."⁵ Jeffrey S. Ravel argues that the privileged royal theaters as "a laboratory of the nation" should be more central than marginal fair theaters and boulevard playhouses to "a fully integrated study of eighteenth-century French political culture because they combine, in the same space, the rituals and concerns of the court, the ideas of philosophes and others, and the everyday actions of Parisians."⁶ The audience demanded that the royal theaters stage a "national spectacle."⁷

The theater-nation excluded the lowly people (*menu peuple*), which agitated the Rousseauean writers to radicalize the theater. Louis-Sébastien Mercier believed that the theater could transform the populace into enlightened citizens and foster "true morality & healthy politics" for the nation. The true *poëte* was a "man of the universe," not a slave to the "frivolous & petty taste" of aristocrats. A "truly patriotic" writer would lead the uneducated people out of their vices, errors, and prejudices. In order to speak for the oppressed, he had to possess a simple soul and a moral language, reduce the inhumane distances among the citizens, and lead public opinion.⁸ A reformed theater could cultivate "veritable patriotism" among the ignorant people and shape them into citizens.

> Nevertheless, the most active & prompt means of arming the forces of human reason invincibly & of enlightening a great mass of people all at once, would certainly be the theater: it is there that . . . simple & enlightened eloquence could instantly awaken a dormant nation: it is there that the *majestic* thought of a single man would enflame all souls by an electri-

cal commotion; it is there, finally, that legislation would encounter fewer obstacles and effect the greatest tasks without effort & *without violence*.[9]

By the 1770s Parisian theaters had lost their primary function to represent the king-machine, then, to provide room for the alternative theaters that sprang up on the streets of Paris.[10] Material spectacles—or the marvelous performances of bodies and things designed to amuse common people—opened a liminal public sphere that mixed all orders of the society, which raised the specter of another world.[11] Material spectacles diffused the hegemonic intention of royal theaters through unlettered entertainments and blurred the social hierarchy by giving a location to the shared curiosity for unmediated pleasures.[12] Acrobats, animal shows, and automata drew a broad spectrum of spectators who participated in the performative moment that pushed the boundaries of the human. The "savant" animals (monkeys) simulated human intelligence.[13] The automata exhibited illimitable human ingenuity. Wolfgang von Kempelen's chess player, the "most amazing automaton" that embodied "the most daring idea a méchanicien could ever conceive," spent the summer of 1783 at the Café de la Régence competing with the best Parisian players.[14]

The material sphere largely escaped literary discussion, yet it served to blur the boundary between the literary "public" and the illiterate "people." The transgressive agency of material spectacles refashioned the Parisian geography of pleasure for a malleable popular culture—seemingly indifferent to concrete political concerns yet poised to generate new forms of political action. Material spectacles harbored the potential, in other words, to transform the existing culture of emulative consumption that normally helped the populace internalize the existing social hierarchy. It was precisely because the ancien régime maintained its hegemony through theatrical performances that a marvelous spectacle could project an alternative polity.

True to the Ovidian theme of metamorphosis, inventors working in this milieu of mixed performances sought to please their patrons while harboring an inalienable desire for another world. Such duality of servile existence and chimerical desire was systemic. Savants served the absolutist regime yet worked for the symbolic capital in the Republic of Letters.[15] Cooks toiled in the hidden corners of aristocratic households yet strove for public status by inventing a "modern cuisine" that would build an empire of taste from "the Arctic to the Antartica."[16] Only in this context of material performance and transgressive audience can we understand why respectable provincial manufacturers invented a machine that would disrupt the intricately balanced automaton that was the ancien régime. It also helps explain why the French public and royal administrators accepted the balloon as a scientific artifact, while staging it as a majestic spectacle.

Machines do not initiate a revolution or determine its course, but the frivolous French balloon and the hard-working English steam engine, often characterized as their respective national inventions, offer an illuminating contrast between the political and economic cultures of these world empires. Britain developed an industrial economy, culture, Enlightenment, and revolution.[17] In France, the balloon ascent became a majestic spectacle by utilizing the material, social, and literary technologies of the ancien régime—a large quantity of expensive silk, high quality paper, exotic coating material, and industrial acid; royal police and parks; a well-established mechanism of public subscription and witnessing; and the newspapers and the postal system, which spread the news all over France and beyond.[18] The balloon as a people-machine offered an experimental form of theater that helped project a utopian republic-empire.

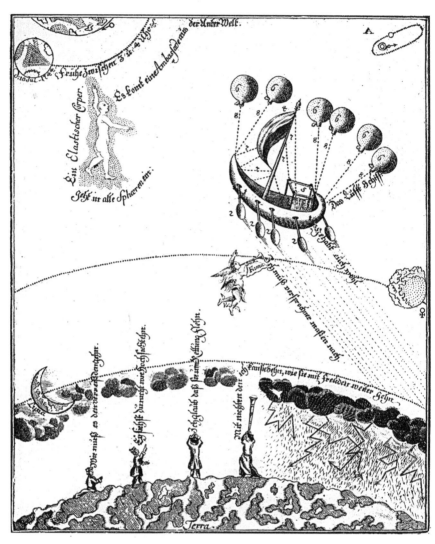

Figure 1.1. Illustration from Eberhard Christian Kindermann, *Die Geschwinde Reise auf dem Lufft-Schiff nach der obern Welt* (Berlin, 1744). It shows a flying boat after Lana's design, with a human being in elastic body rising up to other planets as an earthly ambassador.

1

A RUPTURE OF THE EQUILIBRIUM

The English, mighty proud nation, claim the empire of the sea;
The French, light nation, seize the empire of the air.

"Impromptu de Monsieur"

On a late November evening in 1782, a middle-aged man of large muscular frame sat alone in front of a fireplace in his boarding room at Avignon.[1] A dreamer with wandering habits, Joseph-Michel Montgolfier had led an itinerant life along the Meridian from Paris to Montpellier, circling his hometown situated midway near Lyon. His father, Pierre, was a successful papermaker in Vidalon and had in vain provided various educational and business opportunities for his restless son. Looking into the crackling fire and up the wall, perhaps reflecting on his dreams and frustrations, Joseph caught sight of a print depicting the siege of Gibraltar, the strategic Mediterranean fortress that had induced Spain to join forces with France in the American War (1776–1783). Impenetrable by land or sea, the British garrison had withstood more than three years of bombardment. Just two months earlier, five thousand British soldiers had successfully blocked a joint attack on land and sea by fifty Spanish and French ships in the formation of floating batteries, six hundred guns, and seventy thousand men (see figure 1.2).[2] This sad example of "French levity," as Jean le Rond d'Alembert characterized it to Frederick the Great in Berlin, was on everybody's mind.[3] Joseph supposedly pondered: "But could we not get in through the air? Smoke rises in the chimney. Why can't we gather it to make a usable force?" Thus began, as this legend has it, the history of aeronautics.[4]

The military rhetoric fashioned an "arch-patriotic" invention.[5] France and Britain had been engaged in global warfare for nearly a century.[6] The three-year siege of Gibraltar may have recalled from Joseph's memory *Le*

Figure 1.2. The grand assault at Gibraltar, September 13–17, 1782. Gallica, BNF.

Siège de Calais, a play that sparked a surge of patriotic passion after the Seven Years' War (1757–1763).[7] The French government engaged in extensive espionage to catch up with British industrial might, as her citizens called for a patriotic invention that would lift sagging French spirits and strengthen the nation's place in global politics.[8] The costly American War only accentuated British naval superiority, inducing the dream of an aerial empire: Romans had to build roads to conquer the world and Britons ruled the sea. Why couldn't the French become "the sovereign power of the air?"[9] The public bought into this timely rhetoric—that the balloon could become a more efficient and economical "murdering device." Would it not have cost much less to launch twenty hundred-foot balloons over Gibraltar rather than exposing fifty ships to the sharp-shooting cannon?[10]

The figure of the lone inventor in a deep reverie spoke to a modern myth of genius. In Diderot's vision, a genius would require solitude and silence to discern the contemplative truth of Nature.[11] As the future revolutionary André Chénier (1762–1794) extolled in his poem *L'Invention*, a modern inventor in his quest for beauty, truth, and immortal glory had to be a solitary navigator in the vast Newtonian universe to expand the "empire" of human knowledge. He would emulate classical authors without becoming their "slave imitator" in order to demonstrate Nature's secrets and hidden riches, establish unifying laws, and render abstract truths accessible to common people.[12] The Montgolfiers would be characterized as "genius-observer[s] . . . born with the taste for useful knowledge" who devoted their leisure time to natural philosophy, observed cloud formations and movements, and sought to "imitate Nature in one of her greatest and most majestic operations" with a large envelope and light vapor. This idea required genius, the balloon promoter Barthélemy Faujas de Saint Fond (1741–1819) opined to flatter, and its execution demanded courage and an organized mind. The legend would only grow in time to enshrine a methodical inventor, "one of the most remarkable in his century" endowed with the "great force of imagination and the spirit of invention."[13]

The public persona of a romantic genius who made a useful invention by deciphering the truth of Nature captures the liminality, or the in-between status, of ancien régime inventors. Attuned to the existing culture of invention, they sought privilege and profit by advertising their virtue and utility. They could rise above their station, however, only through the emergent myth of genius. Arts and sciences offered a path toward true nobility in the Enlightenment, as illustrated in the careers of Voltaire, Maupertuis, and Buffon.[14] Scientific nobility surpassed in public recognition the paper titles conferred on manipulative courtiers and frivolous courtesans, or the genealogically reconstructed high nobility that Mercier identified as "the greatest enemies of our national mores." In his view, they "outraged a proud nation" that did not recognize or need them.[15]

Although Diderot sought to exclude the inventors bent on exclusive rights and profit making from the rank of genius-artists, others advocated that an inventive artisan (*artiste*) required imagination and genius in addition to skilled hands.[16] The aerostatic machine as a useful invention that could contribute to the public good and patriotic virtue offered the Montgolfiers a unique opportunity to attain the glory and nobility normally denied to middling families. The promise of utility bolstered an invention's moral value, won the hearts of a discerning public, and bestowed immortal glory on the inventor.[17] The Montgolfiers' public identity as natural philosophers (*physiciens*) set an important precedent for the new category of inventors deemed scientific (*savant-inventeur*), a category that was not yet in common use unlike that of the philosophical chemist (*savant-chimiste*). The Montgolfier brothers' metamorphosis from provincial papermakers into nationally acclaimed scientific inventors can illuminate the possibilities and limitations of the absolutist theatrical polity and its culture of invention.

PROVINCIAL PAPERMAKERS

The Montgolfier family name was a prominent one in the French paper industry. Joseph's father Pierre (1700–1793) inherited a nearly industrial-scale facility at Vidalon-le-Haut in 1743, while his uncle Antoine (1701–1779) received one at Vidalon-le-Bas. Situated on the Deûme River, which provided clear water most of the year to ensure long production cycles, Vidalon and the neighboring Johannots' factories supplied the high quality "Lyon paper" to a broad geographical region that included most urban centers in France and some in neighboring countries.[18]

A pious patriarch in vigorous health, Pierre managed his factory community with unbending authority and austere discipline, tempered by paternalistic care.[19] Making paper involved washing and sorting old rags (*chiffons*), fermenting them, cutting them into inch-long filaments, and hammering them into pulp.[20] On workdays, noise from the factory filled his tiny kingdom—deafening thunder from wheels and cylinders, hammers crushing the pulp, splashing falls forcing beams and joists, foremen's whistles, workers' pitched dialects, and women and children singing while sorting and cutting rags—blending into the sounds of nightingales and the winds hissing through the woods.[21] Pierre Montgolfier ran a tight ship, supervising every aspect of production while tending to his subjects' moral disposition. He governed them with wisdom and goodness as a "just and respected king."[22]

The consumption of luxury paper depended on quality and cost, which called for steady innovation. Although Pierre worked hard to build up his reputation and market share,[23] Dutch papermakers had surpassed the French with the Hollander beater that shredded fresh linen and dispensed with the fermentation stage.[24] In 1751 Pierre with his competitor Mathieu Johannot toured Swiss factories employing the device. Their effort to adopt it resulted only in a significant financial loss.[25] The threat to an industry in which France had enjoyed a considerable lead over Britain also attracted administrative attention.[26] Daniel-Charles Trudaine, director of public works, encouraged Nicolas Desmarest (1725–1815) to visit French and Dutch factories in 1768.[27] Desmarest had been appointed as an inspector of manufactures by the enlightened intendant of Limoges, Anne-Robert-Jacques Turgot (1727–1781) whose laissez-faire principles did not preclude a measure of paternalistic intervention for the public good (*le bien public*).[28]

When the government sent out a detailed survey to paper manufacturers, Pierre prepared an inventory of his factory and detailed memoirs on paper production.[29] André-Timothée Isaac de Bacalan at the Bureau de Commerce complimented this "man enlightened by hereditary experience and chemical knowledge"[30] and introduced him to Desmarest who was preparing reports on the Dutch method.[31] As a leading papermaker, Pierre Mongolfier charted a complicated path between state control and market demand. A kind of mediated competition guided by government incentives and subsidies was the norm rather than the exception in the ancien régime economy.[32]

Paper products shaped a layered social topography for the Montgolfiers who had to negotiate with government officials at Versailles, urban distributors, elite clientele such as Pierre-Augustin Caron de Beaumarchais (1732–1799), local dignitaries, and a homebound workforce.[33] For the expanding family, social mobility would have meant a generational progression into respectable professions in the church, medicine, or law.[34] Pierre Montgolfier saw to his sons' education, favoring religious careers as conventional wisdom and family tradition dictated.[35] Pierre's modest prosperity and the imperial economy nevertheless steered many of his children toward less conventional paths, often sheltered by his childless brother Jacques (1722–1805) who had initially settled in Paris as a paper merchant.

A family geography in a global empire can be spatially large. As Emma Rothschild has shown, a provincial British family with lesser means could spread out to the empire's far-flung corners and, upon their financial success, be integrated into the gentlemanly class.[36] The Montgolfiers proved more cautious in their career choices, but no less daring in their dreams. Their exposure to the Parisian Enlightenment, as well as their hopes and frustrations in pursuing diverse itineraries of learning, explains in part their inventive energy for a chimerical machine.

Pierre's eldest sons, Raymond (1730–1772) and Jean-Pierre (1732–1795), attended the local Collège d'Annonay. While Raymond was groomed to take over the factory, Jean-Pierre went on a *tour de France*, a journeyman's rite of passage that allowed broad training in his métier and ingrained a taste for liberty and fraternity. It also shaped a crazy-quilt world that juxtaposed the real and the imagined, as seen in an extant autobiography of the contemporary glazier Jacques-Louis Ménétra (1738–1812).[37] When Jean-Pierre ended up in Paris, his uncle handed over the paper business upon becoming receveur général of the archdiocese. Even after a respectable marriage, Jean-Pierre could not settle into a life of domestic peace and economic prosperity.[38] He wandered through the Parisian scene, frequenting the cafés. When the business suffered as a consequence, he returned home where Pierre set up an atelier for making playing cards. Jean-Pierre's younger brother Alexandre-Charles (1737–1794), the abbé, was raised by his maternal uncle Jean-Jacques Duret, canon of the Annonay church. Educated at the Collège d'Autun in Paris, the abbé became the headmaster, which would have given him hope for a stable career.

Pierre's next two sons proved more restless and migratory, which reflects a growing generational rift between their aspirations and Pierre's modest means. Our inventor Joseph-Michel (1740–1810), sent to the Jesuit Collège de Tournon for a religious vocation, taught himself arithmetic, chemistry, and mechanics. He ran away from the school and made dyestuff to sell in the nearby market. He loved to travel, mostly on foot, learning a wide range of (mostly technical) subjects along the way.³⁹ Maurice-Augustin (1741–1788) was sent at a tender age to a curé in a neighboring village for religious instruction, but he ran away to Paris. He lived in penury and perhaps as a "non-entity" like the Burgundian writer Rétif de la Bretonne who passed a "shameful" period (1755–1764) in Paris as a libertin journeyman printer.⁴⁰ Maurice soon moved on to the East Indies, the Cape of Good Hope, and St. Domingue, becoming involved in the commerce of animals and briefly managing a sizable coffee plantation.⁴¹ Joseph and Maurice's migratory impulse repeatedly dashed Pierre's reasonable expectations for his sons' financial independence.

Losing his wife, Anne Catherine Manon Duret, in 1760, Pierre made several arrangements to streamline the household. He took over the Vidalon-le-Bas paper mill for fourteen thousand livres to save it for his nephew Antoine-François (1733–1785) and installed Joseph and Maurice, along with their elder sister, Marianne. With the marriage of Raymond to Claudine, the daughter of the Lyon merchant (of silk dye) Antoine Devant, Pierre legally settled one of the Vidalon factories on him, but without specifying which.⁴² While this complicated arrangement lasted, Joseph enjoyed a degree of independence to work on perfecting the art of papermaking. The Vidalon household also grew unexpectedly. When the Collège d'Autun was absorbed into the Collège de Louis-le-Grand in 1763, the abbé came home with a small annual pension of four hundred livres.⁴³

When the two Vidalons became untenable with Raymond's illness and the fear of competition, Pierre joined them as a single society in 1764. Joseph resumed his itinerant life, frequenting the Parisian cafés that brewed coffee, specialty liquors, conversations, poetry, and sciences to shape the enlightened urban figure.⁴⁴ Maurice left for the Isle-de-France (now Mauritius) where a booming plantation economy attracted young fortune-seekers. He would return home—only after a long journey seriously compromised his health—with the draconian attitude of a plantation manager who saw skilled journeymen as "a race . . . of true hacks."⁴⁵ He would have brought home tales of exotic landscape, fast fortune, and ruthless exploitation. As the disenchanted writer Jacques-Henri Bernardin de Saint-Pierre (1737–1814) observed in *Voyage à l'Ile de France* (1773), Europeans in their pursuit of happiness depopulated America for the land and Africa for the slaves.

The youngest son Étienne-Jacques (1745–1799) was sent shortly before his mother's death to the Collège de Sainte-Barbe in Paris, an illustrious institution for the preparation of young men for the religious vocation.⁴⁶ He changed course to attend the studio of Pierre-Louis Moreau-Desproux (1727–1793), a prominent neoclassical architect who became maître des bâtiments to the city of Paris in 1763.⁴⁷ Working in this capacity until 1783,

Moreau also designed festival constructions for the city and worked on the new Opéra at the Palais Royal (1764–1770).[48] Étienne would have watched the theater construction at a close range during his student years, which might help explain his later quest for the spectacular.

The Montgolfière (as the hot-air balloon was often called) would combine the spectacular and the mathematical. Parisian curricula in architecture also taught mathematics, drawing, design, and mechanics in order to fuse "clear and exact knowledge of theory" and "an infinity of knowledge that . . . can be passed on only through practice."[49] It was a consequence of the educational reform instituted in the 1740s by Jacques-François Blondel (1705–1774) at the École des Arts and in public lectures,[50] coupled with more radical theories espoused by Étienne-Louis Boullée (1728–1799) and Claude-Nicolas Ledoux (1736–1806).[51] Étienne would have studied architecture at least for two years before he qualified as an inspecteur des bâtiments in January 1767. Pierre sent three hundred livres to settle him "honorably" in the new profession.[52]

THE ALLURE OF PARIS

The Montgolfier brothers were in Paris when the imperial capital developed an active public culture of science. The long-established lectures in botany, anatomy, and chemistry at the Jardin du roi were free and open to the public, along with Jean-Antoine Nollet's new course on experimental physics (1753–1770) at the Collège de Navarre. By the mid-century, the city brimmed with science entrepreneurs—public lecturers, brokers, apothecaries, inventors, industrialists, and so on. At the apogee of this chaotic activity stood the Academy of Sciences, an institution that conferred ultimate prestige on the men of science.[53]

Paris offered layers of scientific education. Antoine-Laurent Lavoisier (1743–1794), an ambitious bourgeois youth wishing to leave his name for posterity, took public courses in botany, chemistry, and physics in addition to his mathematical education at the Collège de Mazarin before he was admitted to the Academy. Not many would have been able to afford the education Lavoisier received from an illustrious list of Academicians, but this did not preclude the generation's exposure to and enthusiasm for all things scientific. Louis-Sébastien Mercier graduated from the same Collège as Lavoisier in 1763, although their careers diverged markedly.[54] Jean-Paul Marat (1743–1795), the future revolutionary from Boudry, took a variety of public courses in Paris (1762–1765) before he became a physician in London. Charles C. Gillispie has found among the Montgolifer papers folders on mechanics, astronomy, an essay against the Leibnizian concept of monads, and an abstract of Buffon's Époques de la nature.[55]

If science shaped Parisian minds, Jean-Jacques Rousseau (1712–1778) cultivated their sentiments. His epistolary novel *La nouvelle Héloise* (1761) cast the entire generation of reading teenagers in the role of St. Preux searching for his Julie, while deploring the mask of dissimulation required in navigating the theatrical polity.[56] The novel prompted young Lavoisier to write a play and the comte d'Antraigues (1753–1812) to read Rousseau's political writings.[57] The career of Simon-Nicolas-Henri Linguet (1736–1794) as a paradoxical writer, or a "thinker-in-action," also began with his engagement with Rousseau.[58]

Letters from Étienne Montgolfier's friends offer us a rare glimpse of Rousseau's formative presence among the literate Parisian youths of modest means. Destined for diverse métiers and adventures in the French empire but living in a city where human malice hurt them more deeply than they could bear, they formed intimate friendships through regular rendez-vous and correspondence that cultivated sincere sentiments à la Rousseau. They exchanged news, books, and journals of their lives, offered reflections on them "article by article," missed their "corporeal company" when apart, and expected each other to keep up the correspondence "with exactitude" to maintain their "spiritual company."[59]

Étienne's hometown friend Pierre-Antoine Monneron (1747–1801), whose family figured prominently in the Annonay Masonic lodge called La Vraie Vertu, abandoned Paris and his mistress for Mauritius with a promise to secure a passage for Étienne.[60] He found the solitary pedestrian journey as monotonous as the "passage of air," which betrays his feelings of insecurity and worthlessness as a young globetrotter in search of an elusive fortune. Traveling in a small carriage to Orléans and Nantes, he found humanity and solace in some—a middle-aged engineer-machinist traveling to Mauritius, a poor abbé who claimed close friendship with Jean-Jacques (Rousseau) and recited healthy extracts from his works, and a man who could recount extraordinary tales from his five voyages to India.[61] Monneron traveled for two years (from May 4, 1767) around Bengal on board the *St. Jean-Baptiste*, working for the Compagnie des Indes, which lost its monopoly rights on July 24, 1769.[62] Étienne must have wondered about the imperial adventures that lured away both his hometown friend and his brother.

Mahuet and Louis le fils, Étienne's classmates at Moreau's studio, offer rare insights on the Parisian life of middling youths.[63] Mahuet's letters contain casual references to metamorphoses, current plays, festivities, and Rousseau's *Émile* (1762). They offer a privileged window to his fluid identity and uncertain aspirations. Mahuet spent hours at a public garden observing the passers-by, the "automata that find pleasure in vegetating" and "pursue curiosities out of boredom and ignorance."[64] He vowed not to become an automaton, however limited his prospects in life. While sampling diverse Parisian amusements, these educated youths slated for particular métiers often found themselves suffering from invincible ennui in an emotional void. When insipidity took over, their hearts grew insensible. They yearned for a passion that would awaken their souls and rekindle a sense of hope, yet they feared the disruption of their carefully planned ordinary lives. Gentle devoted friendship sustained their mundane existence and fragile hope.

The new métier took Étienne away from his Parisian friends. In March 1768, the canon Sauvé of the Faremoutiers abbey (about fifty kilometers east of Paris) contacted him to commission a chapel. Moreau paid a visit in June, and construction began soon afterwards.[65] Étienne arrived in August at an annual salary of 1,200 livres, out of which he paid the abbess 450 livres for lodging and meals. Mahuet and Louis le fils followed him in letters, constantly asking about his well-being, news, and love affairs. They waited impatiently for his return while filling his orders for books (*Émile* among them), Parisian clothing, and other necessities. Their close friendship bordering on passionate love withered as construction dragged on. Étienne also began to work on Réveillon's new paper mill nearby in Courtalin-en-Brie, which forged an important connection for his balloon endeavors later on.

BECOMING A PAPERMAKER

Pierre's plan for succession faltered when Raymond died after a long illness on July 31, 1772. Whoever took over the factory had to shoulder the responsibility for an extended family. Pierre settled on Étienne, passing over Jean-Pierre and the abbé residing in Vidalon, as well as Joseph and Maurice who ran paper mills in the nearby Dauphiné. It is not difficult to guess why the innovation-minded patriarch chose his youngest son. Étienne had received a more systematic training in mathematics and sciences. He also kept up Parisian attire and appearances and could mingle comfortably with his social superiors.[66] Giving up his imaginary role as a St. Preux looking for his Julie, Étienne settled into the daily routine of a papermaker and married Adélaïde Bron, a convent friend of his sister Thérèse, on May 17, 1774. The abbé had rescued her by pleading in the papal court of Clément XIV.[67] Involuntary consignment to the convent was a common fate for under-endowed daughters and dishonored wives, which inspired Denis Diderot's (yet unpublished) novel *The Nun*.[68]

The Montgolfiers maintained a palpable presence in the local society of enlightened men, which included aristocrats and ecclesiastics. Landowners such as Pierre-Marie-Christophe Bolliuod de Brogieux (1735–1826) often hosted intimate gatherings to discuss various issues ranging from taxation to natural philosophy.[69] Étienne and his wife entertained guests at home with music, playing the piano and the violin, which put Adélaïde's convent education to good use. Étienne would also amass a library of more than a thousand volumes. Masonic brotherhood helped.[70] Having joined a lodge in Paris, the abbé became a founding member of the Annonay lodge La Vraie Vertu in 1766 and rose steadily in the ranks. With a seat at the seneschal's court, he counted among the local notables. Raymond joined a Lyon lodge called the Choix des Hommes Libres as well as the Société de Bonne Vente de Lyon, both in 1766, perhaps to build a business network.[71] It would have been logical for Étienne to follow suit when he took over the paper factory, although his Masonic activities can be traced only after the Parisian balloon ascent.[72]

A shared family domain can become an exceptionally productive site for experiments, debates, and innovation.[73] Living in the "obscurity of the province," Étienne formed a close partnership with Joseph in scientific and technical projects. An "affected man who conducted his business by the rules of mathematics," Étienne had the "aura of an academic . . . and was ruffled by haphazard experiment."[74] An "opinionated and curious researcher and assiduous reader of all scientific works that fell into his hands," Joseph traveled extensively for meditation, observation, calculation, conversation, and open learning.[75] Joseph's curious mind and Étienne's scientific training complemented each other productively when it came to tackling various technical projects. They kept abreast of the new machines like the steam engine and the fashionable sciences like pneumatic chemistry.

The discovery of the "different kinds of air" in the 1770s played a crucial role in bridging the traditional divide between chemistry as the material culture of lowly apothecaries and natural philosophy as the scholarly quest for contemplative truth. Lavoisier's *Opuscules physiques et chimiques* (1774), which appropriated British pneumatic studies for French academic use, was widely distributed to the provincial academies to multiply chemistry courses in the provinces.[76] In 1777, the newly launched daily *Journal de Paris* advertised forthcoming translations of Priestley's *Experiments and Observations on the Different Kinds of Air* in a new section "Physique," which made pneumatic studies a part of natural philosophy in public recognition, rather than the empirical art of chemistry. A careful differentiation between the physicien (natural philosopher) and the mécanicien (machinist) would also glorify the Montgolfiers' accomplishments as scientific inventors.

Parisian scientific fashion attracted provincial curiosity through personal networks and the print media. Although the Montgolfiers missed out on firsthand experience, Joseph could interrogate his cousin Mathieu-Louis-Pierre Duret (1758–1841) on the Parisian lectures. Alessandro Volta's *Lettres sur l'air inflammable* appeared in 1778, while Priestley's *Experiments and Observations on Different Kinds of Air* (1774) were fully translated by 1780. Inflammable air (our hydrogen) with its noticeable levity became a new scientific fashion, especially after Tiberius Cavallo at the Royal Society of London made soap bubbles with it to offer a new repertoire in demonstration lectures. He apparently concluded that one should be able to sustain heavy bodies in the air with an impermeable envelope, but his trials (with thin paper, pig and fish bladders, and so on) came to nothing.[77]

Exactly when the Montgolfiers formed a project to make an "ascending machine" is not known, notwithstanding their cousin's retrospective claim that Joseph's "two grand subjects of meditation" in the late 1770s became navigating in the atmosphere and raising water.[78] A rational choice would have led him to making an efficient pump.[79] Hydraulic machines powered early modern industrial development and supplied water to palaces and cities.[80] The most magnificent technological marvel of the ancien régime was the Machine de Marly, which supplied water to royal palaces. The discussion about how to improve its performance intensified from the mid-century.[81] The woefully inadequate water supply for Paris also prompted in 1762 Antoine Deparcieux's ambitious proposal to build an aqueduct from the Yvette River, an expensive solution that saw completion only in Mercier's utopian Paris in the year 2440.[82] New treatises such as the marquis de Ducrest's (Charles-Louis, 1747–1824) *Essai sur les machines hydrauliques* (1777) appeared in the market. The Bordeaux Academy offered a prize competition on the means of supplying clean water abundantly in the least expensive way.[83]

James Watt's new steam engine was first applied in France to procuring water for urban customers. The Périer brothers formed the Compagnie des Eaux in 1778 as a major financial speculation with the duc d'Orléans's patronage to pump water from the Seine for household uses, an expensive solution that worsened the "corruption of the air," according to a critic who proposed instead a canal as "a simple and natural means" to the same end. The *Journal de Paris* nevertheless endorsed the Périers' experiment in 1779, ostensibly to fulfill the "duty" to transmit to posterity all that is in the interest of humanity. The Chaillot steam engine was inaugurated in the summer of 1781, which prompted even the caustic Mercier to praise it as "an innovation that carries a character of greatness and national utility."[84] Such spectacular advancement could not have escaped the Montgolfiers who routinely kept track of new machines.

With the entire family's livelihood hanging on the paper business, Étienne worked hard to transform Vidalon into a technologically competitive facility. The administrative effort to outpace Dutch competition intensified with Turgot's appointment as controller general in 1774. His protégé

Nicolas Desmarest began an active campaign to modernize the paper industry, a timely concern judging from the acute paper shortage in 1777 following a quarto edition of the *Encyclopédie*.[85] To build a pilot mill, he went back to Holland with the Dutch-speaking artisan-draftsman Jean-Guillaume Écrevisse (1734–1787) and returned with a full sketch.[86] Desmarest visited Vidalon during his tour of French facilities and lobbied the Languedoc estates with a memoir that forecast a "revolution in this genre of manufacture" that would boost "national commerce."[87] He was in Montpellier by January 1780 for a delicate negotiation, paying personal visits to influential men.[88]

Over the Johannots' counter-lobbying and a complaint even from Antoine-François Montgolfier that lowering the price through mechanization could "wipe out the commerce of his fellow tradesmen [*confrères*],"[89] the Vidalon factory was granted a subsidy of eighteen thousand livres to develop a pilot mill with Dutch machinery and to teach all other paper manufactories in their province.[90] As the richest province in the kingdom whose elite maintained a measure of autonomy from the Crown by generously financing royal projects, Languedoc had an established track record in promoting public projects such as the Canal du Midi, which connected the Mediterranean to the Atlantic.[91]

After obtaining the grant, Desmarest kept the Montgolfiers on their toes. Écrevisse arrived in May 1781 under strict orders to keep an exact journal of the construction process.[92] Production slowed for months while the new mill (22.5 meters long by 12 meters wide) emerged to turn out high-quality writing paper and reams for drawing, sheathing, and wrapping. The violet wrapping paper for sugar cones ended the Dutch supply to the royal sugar refinery in Montpellier. The family still had to train the workers and develop new paper products.

The effort to institute a machine-centered order of production caused intense conflicts with the journeymen, which prompted a lockout in November 1781 that lasted two months.[93] Étienne had to maintain a tight grip on the workers during the transition, while Joseph was exploring other professional paths toward economic independence.[94] The Montgolfiers' triumph in this battle did not even guarantee their long-term prosperity, given the large number of siblings the factory had to support in an increasingly competitive business environment. Nor could they expect a path of spectacular social advancement notwithstanding their commanding position in the paper industry. Their success placed them instead in a liminal region between the artisanal mode of production and industrial capitalism. An impractical flying machine could not have been the Montgolfier family's top priority in 1782.

PARISIAN ICARUS

What focused the Montgolfiers' attention on the flying machine, then? The most likely catalyst was the sensational failure of Blanchard's flying carriage. Jean-Pierre Blanchard (1753–1809) was the second son of an all-purpose village artisan at Petit Andelys, a small town ninety kilometers northwest of Paris. He may have attended the Collège de St. Nicaise at Rouen, but he developed an interest in mechanical inventions such as the clock, the wool-spinning device, the loom, and the horseless carriage, which apparently took him from the Andelys to Rouen (about thirty kilometers) in three and a half hours.[95]

Blanchard emerged from this obscure background when he demonstrated on February 2, 1779, a hydraulic machine that pumped water in just two minutes from the Seine to the Château Gaillard, a medieval castle situated 180 meters above. Blanchard traveled to promote his invention—to Chancellor Maupeou exiled at Thuit, the duc de Penthièvre at the Château de Vernon, and the marquis de Marcien at Grenoble. Although he could not secure their patronage, apparently due to the high cost (seven hundred louis d'or, according to his recollection), he must have attracted enough attention to find his way to Paris. One of his supporters obtained permission from the director of royal buildings, the comte d'Angiviller (Charles Claude de la Billarderie, 1730–1809), to demonstrate a horseless carriage.[96] It was advertised for regular performances at the Champs Elysées near the large entertainment complex, the Colisée.[97]

Blanchard announced a public demonstration of his flying carriage (see plate 4) only two years later, under the protection of the abbé de Viennay (Jean-Baptiste-Charles Pineau, 1712–?), a conseiller at the Paris parlement and financial administrator of the Benedictine abbey of Turpenay at grande rue Taranne. Catholic abbeys were rich institutions and the Benedictines were enlightened monks who took an avid interest in the arts and sciences.[98] On display at Viennay's residence, Blanchard's flying carriage attracted the fashionable society and received favorable press coverage for nearly a year.[99] Étienne Montgolfier, who was in Paris during January 1782 to present a memoir to the intendants of commerce, could not have missed it.[100] It was rumored that the comte d'Artois and the duc de Chartres each promised four thousand louis upon its success, probably competing to outshine each other, although some considered it a chimera.

The demonstration scheduled at the Viennay residence for May 5 drew a large crowd but the machine could not take off due to heavy rain.[101] Blanchard sought to control the damage with a lecture, citing his "wise reflections" on the marquis de Bacqueville's chef d'oeuvre—the artificial

Figure 1.3. Diverse utilities of the flying vessel mentioned in *Journal de Paris*, May 23, 1782. Library of Congress. On the left page is shown its personal utility for a debtor pursued by his creditors. On the right page are shown other types of disreputable persons who might wish to take advantage of the new invention such as a frivolous abbé, an astrologer, and gallant men visiting forbidden women.

wings designed to fly over the Seine, which had dropped him into the river in 1742 (see plate 2). Regardless, the skepticism over Blanchard's "chimera" only intensified to cripple his career as a fashionable inventor (see figure 1.3).[102]

That a provincial inventor—recently arrived in the capital—abandoned a hydraulic pump in favor of a flying machine speaks volumes about the fast-moving, novelty-driven Parisian entertainment culture. One cannot but recall the scene in Rétif de la Bretonne's *La Découverte australe* (1781) when the protagonist Victorin ventures into the château where his beloved Christine lives. He is introduced as the inventor of a self-propelled carriage, which allows him to mingle with society women. The status of inventor does not demolish the social barrier that prohibits his marriage to Christine. Victorin thus secretly perfects artificial wings and carries her away to an inaccessible mountaintop where the "air of liberty" reigns (see figure 1.4), and they establish an "empire" that expands to the Southern Islands. Artificial wings

Figure 1.4. Victorin taking flight (left) and taking up Christine (right) in Nicolas-Edme Rétif de la Bretonne, *La Découverte australe par un Homme-volant* (1781). Gallica, BNF.

became a fictional transport to an imagined republic. Rétif's republican empire is a mixed regime with the royal couple, which compromises his clear desire for an egalitarian polity.[103] Blanchard the inventor and Rétif the author illustrate the restless Rousseauean generation that roamed the streets of Paris searching for an extraordinary purpose in their ordinary lives. We know much less about the hopeful inventors than the malcontent writers.[104]

PROVINCIAL DAEDALUS

The Montgolfiers could not have missed the debate over the possibility of flight that continued well into the summer of 1782. The Academician Joseph Jérôme Lefrançois de Lalande (1732–1807) declared it impossible, based on the authoritative calculation of another Academician Charles-Augustin de Coulomb (1736–1806), while the young engineer François-Nicolas Martinet (1760–1800) proposed to postpone the final judgment.[105] Even with his failure to persuade the august Academicians, Blanchard blazed a trail for the aspiring provincial inventors—one paved with princely patronage and public approbation.

When Joseph emerged from debtors' prison in June 1782 (his time there caused by an entanglement in the Johannots' family dispute), he began a serious research on the fluids specifically lighter than air.[106] The specific levity of inflammable air had become a research frontier in natural philosophy thanks to Henry Cavendish, Joseph Priestley, and Alessandro Volta.[107] Soap bubbles filled with inflammable air was a standard repertoire in public lectures. The balloon itself was not a novel idea, at least in imagination, judging from François Lana's design in the 1670s.[108] The Avignon professor Pierre-Joseph Galien had also speculated in *L'art de naviger dans les airs* (1757) that the layers of air with different densities could support aerial navigation. His imagined vessel of "immense capacity, made with an envelope of cloth or leather, & full of air . . . much lighter than the atmospheric air" had much "affinity" with the Montgolfière.[109]

In November, Joseph took a business trip to Avignon, a papal enclave that enjoyed relative freedom from censorship and nurtured a thriving business in banned books and counterfeits.[110] He made a small parallelepiped model out of old taffeta and thin wood in his room. When he burned a few pieces of paper in it, it rocketed to the ceiling. Joseph hurried home to continue the experiment with Étienne on a larger scale and in relative privacy. In order to make a three-foot cubic chamber, they commandeered pieces of old silk clothing from their wives and even the taffeta (*soie de Florence*) that they had reserved for new dresses.[111]

The next balloon, measuring nine feet in diameter, was made of paper. It rose high enough to be lost from view and descended gently without sustaining any damage, which allowed several more trials.[112] The Montgolfiers' effort to use inflammable air apparently led nowhere because of the quantity of the gas needed, its prohibitive cost, and the difficulty of handling it. Although they experimented with a variety of gases to fill the envelope, according to Joseph's later claim, they decided on burning straw instead, for the sake of utility and economy.[113] The lack of funding would have deterred other inventors from pursuing the chimera but the Montgolfiers could draw on their expertise in handling paper and labor.

As the experimental balloon became an object of local gossip, the family debated on how to secure their priority. They desired recognition of merit, glory, privilege, and profit from this expensive venture, which required a delicate balance of secrecy and publicity.[114] A small-scale demonstration for a select audience could jeopardize their priority, as the Lyon architect Jean-Antoine Morand (1727–1794) found out too late. After exhibiting his model hydraulic machine to an audience chosen by the comte d'Angiviller, he had to plead with the Academy's commission for a testimonial that it differed from another inventor's.[115]

Given the relative simplicity, uncertain utility, and unavoidable visibility of their artifact, the Montgolfiers could not follow a similar route to obtain a marketing privilege or initiate other procedures that would take more time and expense. [116] As soon as their machine showed promise, therefore, Étienne asked Desmarest to announce it to the Paris Academy or in some journal to fix the date of invention and their priority. It was an established procedure for an inventor to apply for the Academy's approval before contacting the Bureau de Commerce, which in turn would advise the Royal Council on granting an exclusive marketing privilege (*privilège*) for a limited term.[117] Desmarest, a manufacturing inspector in charge of the paper industry, was an ideal intermediary for a provincial papermaker. Étienne promised him a machine that could send signals over great distances, deliver news to a besieged town, and allow experiments on atmospheric electricity—a "much more brilliant" machine that "will cause a greater sensation than all our experiments on papermaking," judging from Blanchard's case.[118]

The lofty Academician was not convinced by this urgent plea. Since he did not understand their "ascending machine," Desmarest responded nearly three months later, "a good drawing and a detailed description" were necessary to illustrate its uses, as in the case of Dutch cylinders. He was still waiting for their report on the *papier vélin*, an expensive English commodity that he wanted to substitute by French products. Réveillon would respond to this call and secure the title of royal manufactory for his factory at the faubourg St. Antoine.[119] From Desmarest's point of view, a flying machine could only be a distraction from the Montgolfiers' proper métier in papermaking.

Undeterred by academic skepticism, Joseph and Étienne hurried their experiment to stage a demonstration in Paris, an irrational resolve that would weaken their competitive edge in the paper industry. They nevertheless desired honor and glory, which also promised greater social mobility. Numerous calculations and more trials followed, while the family took care of the paper business. Joseph and Étienne estimated the balloon's weight, lifting force, flight time, and speed of ascent and descent to design a large linen globe, backed inside and out by thin paper, cut in *fuseaux* and sewn together by eighteen hundred buttons.[120] The finished machine measured thirty-five feet in diameter, five hundred pounds in weight, and twenty-three thousand cubic feet in volume with a central opening of sixty-four square feet. It was first flown at the Château de Brogieux to the delight of its owner Bollioud de Brogieux, a confrère at the Masonic lodge La Vraie Vertu. Another trial was made from the Château du Colombier (owned by the industrialist Pierre Lioud) at night with a lantern attached to the globe, which terrorized the passers-by who thought it to be an "apparition of the devil."[121]

In Paris, new airs and steam became all the rage by the spring of 1783.

Figure 1.5. Pilâtre de Rozier exhaling inflammable air. Tissandier Collection.

Inflammable air caused a stir, thanks to Jean-François Pilâtre de Rozier (1754–1785), a science broker who illustrated a path of social mobility open to bright minds willing and adaptive to the theatrical polity.[122] Originally from Metz, his fortune improved quickly when he became a client of the comte de Provence (commonly called Monsieur, as the eldest brother of the king), who patronized a new musée—the Cabinets de Physique, de Chymie, & d'Histoire Naturelle. The Musée de Monsieur drew resounding endorsements from the Académie des sciences, the Académie française, the Observatoire, the Société royale de médecine, and the École royale vétérinaire. With their instrument collections, which allowed for rare and daring experiments, Pilâtre's lectures became a continual source of instruction and amusement that attracted the fashionable society.[123]

Always seeking the spotlight, Pilâtre was soon embroiled in a public controversy over the deadly gas emanating from infected places such as cesspools, beer vats, and sewers. Having designed a respirator, a mask made of rubber-treated silk (*taffeta gommé*) connected to a tank of atmospheric air by a tube running from the nose, Pilâtre sought to prove that the deadly gas was mephitic rather than inflammable gas, as had been claimed by the abbé de Fontana. He also executed an entertaining experiment in which he inhaled inflammable gas from a bladder and exhaled it through a glass tube on candlelight to produce a jet of green flame (see figure 1.5).[124] When this caused a controversy, he had to repeat the demonstration for expert witnesses including the duc de Chaulnes (Marie-Joseph-Louis d'Albert d'Ailly, 1741–1792) and Faujas de Saint-Fond.[125]

The Périer brothers also set a brilliant example. They read memoirs on the Chaillot steam engine to the Paris Academy on February 22 and March 12, 1783, after which they asked for a commission to examine the model constructed for the duc de Chartres. Following the Academician Coulomb's report on March 19, the elder Périer was elected to the class of mechanics. The Montgolfiers could not have missed the public euphoria over their "distinguished talents" and "patriotic zeal" as "good citizens."[126] When the Périers installed a new steam mill at Nîmes, Joseph made a detailed observation during his business trip in May.[127] The Périers' prominent career, patronized by the duc d'Orléans family, was advertised as a service to the public good. In comparison, efforts by Jouffroy d'Abbans (1751–1832) to develop a steamboat in Lyon, where he set up an atelier in January 1782, went largely unnoticed. For inventors, the path to glory and prosperity seemed to require a savvy combination of princely patronage and public approbation.

A PUBLIC EXPERIMENT: ANNONAY, JUNE 4, 1783

Lacking access to Parisian resources, the Montgolfiers settled on a "public experiment" for the Vivarais estates to confer "authenticity and éclat to their discovery."[128] In other words, they resorted to the provincial elite and their pride as a source of legitimacy. As Shapin and Schaffer have demonstrated through the Royal Society's effort in Restoration England, stabilizing a fragile machine to produce consensual facts required maintaining its physical integrity, establishing its philosophical credibility, and engineering its publicity—material, social, and literary technologies in their terms. Contextual variations of these requirements in authenticating scientific facts can tell us much about the particular experimental site and relevant values. Scientific truth has a social history.[129]

In order to secure their priority in a timely fashion for an all-too-visible machine, the Montgolfiers transported the fragile balloon to the market town of Annonay. On June 4, they risked the rain to float the thirty-five-foot paper balloon, made of linen backed inside by several layers of paper and fortified by an outside net, at the Cour des Cordeliers. It rose gradually to a considerable height until it appeared the size of an average lantern and descended slowly after nine and a half minutes. Landing on a wall, it caught a fire that consumed its entire body in less than five minutes. The following day, Joseph and Étienne petitioned to the Vivarais Estates during their last day of meeting for a formal approbation to fix the date of their first public experiment.

The Montgolfiers made a grand artifact to impress the august audience, claimed their authorship and expertise through quantitative measurements and theoretical calculations, and sought to demonstrate their commitment to the public good. The report of the experiment, included in the procès-verbal of the Vivarais Estates to protect their "honor" in having imagined a potentially useful machine, carried exact measures of its size and performance. The globe, twenty-eight thousand cubic feet in volume and thirty-five feet in diameter, rose to five hundred toises, stayed in the air for about ten minutes, and descended slowly at a distance of seventeen hundred toises from the point of departure. The brothers probably supplied the numbers to authenticate their ability to predict its performance according to their "theory." The invocation of theory often served as a code that asserted the authority to represent an engineering project's merit and utility to an elite audience.

In other words, the Montgolfier brothers sought to address not only the utilitarian requirement but also a potential scholarly disdain toward the unschooled hands ignorant of scientific principles.[130] More accurate information on their "diostatic machine" followed, with the intent to prove their ability to perform exact calculations and to predict their machine's technical performance. They claimed that considerations of economy made it necessarily imperfect. The abundant rain must have impeded the machine, they reasoned, by dispersing the gas and adding to the weight. If it was made "quite exactly" and sufficiently large to carry materials and men (who could replenish the dispersed gas), it should maintain a prescribed height.[131] Numbers served as a universal measure of a machine's technical capacity as well as the inventors' authorship and authority.[132]

BACK TO PARIS: HONOR, GLORY, AND PROFIT

The administrative report finally carried balloon news and Étienne to Versailles and Paris. The route mattered in authenticating the Montgolfiers' claim as original inventors and forging their machine's public identity as a scientific artifact. At Versailles, the controller general, the marquis d'Ormesson (Henri-François de Paule Lefèvre, 1751–1808), received the procès-verbal of the Annonay experiment properly witnessed by the Vivarais Estates. He relayed it to the Paris Academy's secretary, the marquis de Condorcet (Jean-Antoine-Nicolas de Caritat, 1743–1794), for an official report that would evaluate its utility.[133] In this administrative communication, the Montgolfière metamorphosed from an "ascending" machine, designed to perform work, to an "aerostatic" or scientific machine. In contrast to the "vulgar" word "balloon," the "aerostatic globe" referred to the "science of weights" that defined the equilibrium of a body in the atmosphere.[134]

The Academy duly appointed a commission. Étienne had also sent a letter to Desmarest to solicit financial support. He wished to render "the government *or the nation* public depositories of this new invention" so that it could be made useful to his fellow citizens.[135] This casual distinction between the royal government and the nation (of fellow citizens) reflects a growing public discourse on the nation as an entity somewhat distinct from the royal court and its bureaucratic appendage.[136]

The balloon commission initially consisted of Charles Bossut, Lavoisier, Desmarest, Jean-Baptiste Le Roy, and Condorcet. They recognized the Montgolfiers' "truly ingenious theory" for which "the evidence was extremely good" and desired a demonstration in Paris.[137] This routine demand by the Academy often silenced the inventor, who could not afford the expense or trouble, as it happened with Jouffroy d'Abbans. He demonstrated in vain his steamboat on the Saône River near Lyon on July 15, 1783, when Étienne passed through the city on his way to Paris.[138] The balloon commission, now expanded to include Mathieu Tillet, Mathurin-Jacques Brisson, and Louis-Claude Cadet de Gassicourt, issued a cautious report on July 19.[139]

The Academy soon went into summer recess but the commission followed the progress of ballooning. As a royal institution, the Academy could not ignore the sovereign's wish to honor the inventor who could project royal glory to the public.[140] As the highest scientific tribunal in the Republic of Letters, the Academy sought to observe proper procedures, to curtail vulgar enthusiasm and curiosity, and to present an authentic account that would merit the utmost public confidence. Lavoisier, albeit worried that it might derail the Academy's proper functions, was keenly interested in utilizing the balloon mania to advance his overhaul of theoretical chemistry.[141]

News of the Annonay ascent began to circulate in Parisian gossip from late June. Some regarded it as "nearly impossible . . . that any object of whatever size . . . can be made to rise by itself into the air."[142] Early reports in public newspapers were not encouraging. The *Affiches, Annonces, et Avis divers, ou Journal général de France* printed a skeptical letter dated June 21, 1783, ostensibly from a landowner near Annonay, of an expensive apparition that cost nine hundred livres. Terrified peasants regarded it as a falling moon that presaged Judgment Day. The potentially neck-breaking venture was foolhardy, the letter writer opined, when Mathieu Johannot built "the most spacious and possibly the most handsome paper mill in France" to produce beautiful paper. The *Mercure de France*, a conservative newspaper with over ten thousand subscriptions, reprinted this ill-humored letter on July 26.[143] The *Journal de Paris* whose founding editor was Antoine-Alexis Cadet de Vaux (1743–1828), brother of the Academician Louis-Claude Cadet de

Gassicourt (1731–1799), soon printed a detailed report that was more accurate on the aspect of physical dimensions. The report identified their gas as inflammable air, however, which caused a considerable confusion over the "gas of Montgolfier."[144]

Étienne arrived alone at the capital. Parisian socialites would have found Joseph "a bit strange in his ideas and social bearing."[145] He took up lodging with his uncle Jacques whose position in the archdiocese provided access to a few notables. Jean-Baptiste Réveillon offered his spacious paper factory just outside of the city wall for the construction. Family expectation of financial advantages, rather than "merit and celebrity," weighed heavily on Étienne's shoulders as he nervously went about establishing contacts with Parisian luminaries.[146] He worried incessantly and the family responded to his letters courier by courier, offering advice as best they could and keeping him apprised of the day-to-day work at Vidalon.

Alerted to Lana's precedent, Étienne asked one of his friends to hunt down the source. Lana's *Prodromo dell'arte maestro* (1670) seems to have been rare, but his ideas circulated in other contemporary sources. Étienne's friend Fontenelle dutifully made multiple trips to his parents' library until he found a description in Jérôme Richard's *Histoire naturelle de l'air et des météores* (1770). He copied the passage "word by word"—that the Italian Jesuit had "proposed to construct a quite thin copper globe filled with extremely rarified air." It should have floated in the air to produce quite an astonishing phenomenon to the unsuspecting observers, but the art was never perfected to execute this idea.[147]

Étienne also feared that the Périers would forestall his Parisian experiment and complained about the delay caused by the Annonay ascension. His brothers persuaded him that the "public experiment" made the family name famous and brought the experiment to the minister's attention, forcing the Academy to examine the matter seriously. It would serve as "a bulwark raised against all counterfeiters and other hornets that would like to devour your honey."[148] Inventors of Périers' stature, who made the most perfect machine in France for two million livres, had no reason to poach another's chimera, especially one that was sent to the Academy.[149] Étienne was soon assured of official interest by the controller general and the Academy.[150] The family sought to calm Étienne's nerves by reasoning against rumors and mobilizing their Masonic allies to mend their public image.[151] For the most part, they took care of the factory so that Étienne could concentrate on securing a victory in Paris.

The Montgolfier family's ability to invent an ascending machine, to claim their exclusive authorship of a scientific artifact, and to cultivate public persona as patriotic inventors required a delicate balance of courting royal favors and garnering public recognition. They kept track of scientific fashions and technological needs, socialized with local notables and Masonic brothers, lobbied provincial estates, communicated with the royal administration, and competed for symbolic distinction. Their "mixed regime" that produced both paper and scientific knowledge had to negotiate a variety of social and spatial barriers.[152] As hard as the Montgolfiers tried to improve their competitive edge in the European market through innovation and cost reduction, they strove most strenuously for the symbolic distinction that would enhance the family name. Such a move jeopardized their manufacturing interest in the efficient production of high-quality paper. While they appreciated the civic utility of hydraulic or heat pumps, they chose to make a spectacular flying machine and to authenticate it through a public demonstration to establish their priority. Their ability to see it as a sensational spectacle, before the Parisian entertainers could, attests to their illimitable desire for an escape from their automata-like existence. This desire would resonate among the middling public. The rupture of gravitational equilibrium that put humans in the realm of gods would make them hope for a social emancipation of similar magnitude.[153]

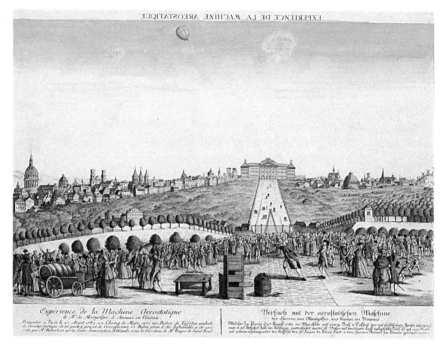

Figure 2.1. "Experiment of the aerostatic machine of M.M. de Montgolfier of Annonay in Vivarais, repeated on August 27, 1783, at the Champ de Mars, Paris, with a taffeta balloon of 36 feet and 6 inches in circumference. This balloon soaked in elastic gum and filled with inflammable air was executed by M.M. Robert by a national subscription under the direction of Mr. Faujas de Saint Fond." Gimbel Collection XL-4-1160.

In this singular illustration of the first Parisian ascent, no visual trace exists of the enormous crowd or the soldiers that enclosed the site. The École militaire is figured prominently at the center in the background with the road open to the experimental site, as if to form a military-scientific axis. The hydrogen-producing arrangements (including the lead drawer that had failed) are placed in the front to characterize the event as a scientific experiment. Perhaps intended for the foreign courts, an excerpt from the *Journal de Paris* is printed below in French and German, as translated above, which erased the role of Jacques-Alexandres-César Charles.

2

BALLOON TRANSCRIPTS

Public culture of Parisian science depended on the novelties, or fashionable commodities, coveted by elite patrons and staged by their brokers to fashion worldly sciences. The wealthy and curious were well served in this imperial capital full of artificial pleasures.[1] Since the Regency (1715–1723) of Philippe d'Orléans, which defined the enlightened "century of frivolity," the Parisian quest for pleasure had focused on living in the moment.[2] Witty conversations, alluring theatrical previews (*parades*), and dazzling spectacles induced a "profoundly subversive attention to the present."[3] Culture brokers forged a trickle-down economy of pleasure between the highborn idlers and the vulgar multitude. The commodification of pleasure formed diverse publics that circulated through the sites of the Enlightenment—soirées, cafés, salons, theaters, Masonic lodges, and royal public parks where ideas, texts, images, news, and rumors circulated across the social spectrum. Public space of the ancien régime had sprouted upon royal spaces such as palaces, gardens, theaters, and lecture halls. The Louvre housed various royal academies and the Palais des Tuileries hosted concerts and the annual salon, while the Palais Royal opened to cafés, restaurants, and boutiques, which fostered mixed publics and perhaps revolutionary ideas.[4] Royal gardens served all Parisians except the most destitute, as we have seen with Étienne's teenage friends.

Urban science required complex networks of patronage, physical sites, personal liaisons, and media outlets that formed a socially progressive public, notwithstanding the political stasis.[5] While the reform-minded Academicians such as Condorcet at the Mint and Lavoisier at the

Gunpowder Administration laid the foundation of technocracy, scientific demonstrations and journals nurtured a fashionable public for science.[6] Lectures on natural history, chemistry, mathematics, and experimental physics competed for the elite audience. Science with its promise of utility helped the royal administration's effort to police public opinion and engineer civic consciousness in the deteriorating public "culture of calumny."[7]

The musées, often supported by Freemasons (such as the Loge des Neuf Soeurs), promoted civic interests in the arts and sciences by offering a mélange of lectures and activities to a mixed audience ranging from princes to artisans.[8] Most prominent were Claude Mammès Pahin de la Blancherie's Salon de la Correspondance (and the accompanying journal *Nouvelles de la République des Lettres et des Arts*), Antoine Court de Gébelin's Musée de Paris, and Pilâtre de Rozier's Musée de Monsieur. Less formal than royal academies but more organized than soirées or cafés, the musées engendered a public culture of science that mixed social ranks, métiers, and cultural genres.[9]

Polite science for the fashionable "public" excluded the uneducated "people."[10] In contrast, the balloon attracted the populace to project an alternative polity—a nation of citizens that would support universal science, infinite progress, and a glorious empire. Its capacity to mobilize crowds made for a problematic political experiment. How the spontaneous balloon crowd perceived such a potent artifact would have been a serious concern for the power elite who had to contend with its ambiguous meaning and mass affect. Newspapers as a public- and nation-forming technology shaped the balloon's hegemonic agency.[11] Journalistic facts orchestrated public opinion to produce a public transcript on ballooning that expressed a broad consensus on its meaning for the public good. In other words, journalistic balloon transcripts performed serious political work to suppress or conceal dissident subcultures and malcontents and to consolidate a domain of material appropriation, public mastery, subordination, and ideological justification in the name of science. Communication is a means of creating public space.[12]

An archeology of mass silence thus requires a careful analysis of what was made public by whom as well as what remained hidden and why. After a brief uprising of the radical press following Louis XVI's accession in 1774, repression had consolidated the publishing market under Charles-Joseph Panckoucke (1736–1798). The *Journal de Paris* as the only daily newspaper in the kingdom played a central role in crafting and propagating a hegemonic balloon transcript. While complying with censorship control (through Jean-Baptiste-Antoine Suard, Panckoucke's brother-in-law), it also maintained strong ties to the scientific establishment (through Cadet de Vaux).[13]

Its reports show a striking uniformity in the kind of information the editors sought to convey, the authorial voice they assumed, and the audience they envisaged. They presented the first Parisian ascents—a public hydrogen balloon set loose at the Champ de Mars on August 27 and a royal Montgolfière staged at Versailles on September 21, 1783—as scientific experiments that would serve the nation and humanity. The volatile mixture of an immense crowd, a "majestic" artifact, and the unpredictable weather appeared as concise descriptions of the balloon's technical accomplishment. Other public newspapers often copied this information that was also distributed via artfully constructed images of the scene, which conveyed a sense of the harmonious nation.

Censored newspapers thus filtered balloon news to shape a hegemonic transcript of the balloon's scientificity and public utility. The rhetoric of useful science would tame the new machine for the absolutist polity to configure a progressive nation of royalist citizens. This journalistic transcript silenced the populace and erased their presence to maintain a representative public sphere that glorified royal authority and power. The balloon's potential as a (re)public-forming spectacle must be read, ironically, from the silence of the people—then at the scene, afterwards in the published accounts, and ever since hidden in the archives.[14]

PUBLIC CHARLIÈRE: Expert Science and Public Patronage

News of the Annonay ascent put the wheels of the Parisian entertainment machinery in motion. Faujas de Saint-Fond set up a subscription at the Café du Caveau to "replicate" the Montgolfiers' experiment. Public subscription was a well-established mechanism for funding expensive ventures such as Diderot's *Encyclopédie*, Buffon's *Histoire naturelle*, and Mesmer's *Société universelle d'harmonie*.[15] Faujas could count on well-heeled patrons who took an avid interest in scientific novelties such as Blanchard's flying carriage.[16] Without even so much as a newspaper notice, as the *Journal de Paris* pronounced after the fact, princes, ministers, academies, men of letters, and artists rushed in their subscriptions to prove their ardor and zeal for useful and brilliant experiments.[17]

Barthélemy Faujas de Saint-Fond, a middle-aged naturalist recently arrived in the capital after a career in the Grenoble parlement, meant to utilize the first Parisian ascent as a spectacular means of social advancement.[18] As a seasoned lawyer, he knew which strings to pull. He had approached the comte de Buffon (Georges-Louis Leclerc, 1707–1788), the intendant of the Jardin du roi, with a gift of volcanic substances for the Cabinet du roi. Buffon gave him a position at the Jardin, which launched Faujas's career as a naturalist. Ambitious and cunning, Faujas promoted his endeavor as "the first *national* subscription" that attracted the most illustrious names. In other words, Faujas's "nation" consisted of court personages and the fashionable society, which excluded most ordinary citizens, not to mention the populace.[19] He would use the balloon affair to win the favor of the queen's faction at the court.

The Café du Caveau, paneled with large mirrors and decorated with the busts of musical revolutionaries—Christoph Willibald Gluck, Antonio Sacchini, Niccolò Piccinni, André Grétri, and so on—was a hot spot for newsmongers.[20] Situated in the garden of the Palais Royal (residence of the duc d'Orléans), it attracted "the politically-inclined, the literati, news-carriers, projectors, movers, speculators and the pretentiously idle."[21] Café sociability shaped an egalitarian, yet undemocratic public space for the "citizens without sovereignty." Those living under the absolutist monarchy did not seek a radical subversion in political order, according to Daniel Gordon's insightful analysis, while trying to secure a social space where they could interact freely.[22]

The garden of the Palais Royal catered to a middling audience for all kinds of knowledge. One could learn physics, poetry, chemistry, anatomy, languages, natural history, and other sciences pressed out "like grapes, by the teeming interests" all around. Idlers debated pointless literary questions a thousand times, Mercier noted, molding a generation of timid writers.[23] Women in their best attires strolled in this most famous promenade in Paris, lined with beautiful chestnut trees imported from India, and staged during the night by soft lighting from the shops (see figure 2.2).[24] The restaurants and cafés served as a hub for current news since the prince's legal exemption from censorship offered a safe haven for the underground press. Dubbed "the capital of Paris," "a temple of vice," or "a pretty Pandora's box," the garden became the birthplace of the French Revolution.[25]

The hydrogen balloon brought expert science into the public domain, which broke the pattern of high science practiced in royal institutions. Faujas enlisted Jacques-Alexandre-César Charles (1746–1823), a public lecturer of physics, and the royal engineers Anne-Jean (1758–1820) and Marie-Noël (1760–1820) Robert.[26] Having received the Annonay news with some regret, Charles seized the opportunity without realizing that he was making a pact with a skilled broker who, once crossed, would bring him interminable angst. The team chose cutting-edge chemistry and materials, which differentiated their "aerostatic globe" from the provincial papermakers' "ascending machine."

Raising a substantial amount of capital normally required the invention's business potential, as in the Périers' Chaillot engine and Jouffroy d'Abbans's steamboat.[27] Despite the unusually high expense for an ephem-

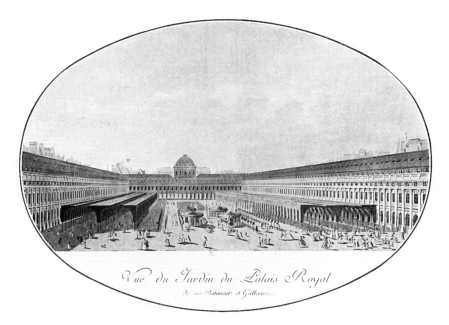

Figure 2.2. Garden of the Palais Royal with galleries, Louis Nicolas de Lespinasse (1785). Gallica, BNF.

eral spectacle, Charles decided on inflammable air, the lightest gas that defined frontier chemistry. Lavoisier had just replicated Henry Cavendish's experiment on June 24, 1783, reaching the conclusion that water was composed of inflammable air and vital air (oxygen), which would soon become a critical piece of evidence for the Chemical Revolution.[28] Procuring a large quantity of concentrated sulfuric acid to produce the gas would have been difficult for the Montgolfiers. Large manufacturers of the acid were located at Javel, near Paris, and in Rouen where John Holker's factory was.[29] The availability of concentrated sulfuric acid would become a serious material constraint in making hydrogen balloons.

The hydrogen balloon also required an impermeable envelope, which depended on chemical knowledge. The inability to contain inflammable air with untreated fabric was a reason the Montgolfiers cited for burning straw instead, but the cost of silk fabric would have been prohibitive as well. Charles's team used patterned silk (*taffeta chiné*) soaked in rubber, an imported substance drawn from the global economy of green gold, as Londa Schiebinger characterizes the Enlightenment commerce of botanics.[30] French academic research on this tropical substance (*Caoutchouc*), originally found in the Peruvian province of Emeraudes and later along the Amazon, had continued since Charles Marie de La Condamine's report in 1751.³¹ The Academician Pierre-Joseph Macquer dissolved it in ether in 1768, which allowed a variety of commercial applications. The Robert brothers apparently developed a method of dissolving rubber in turpentine to coat leather, which made it impermeable to boiling water and acid. According to François-Joseph Bélanger, architect to the comte d'Artois, this was a proprietary knowledge that should merit the esteem of all European savants.³² Other methods must have been available, judging from Pilâtre's rubber mask and Faujas's contestations.³³

The heroic effort that produced the first hydrogen balloon in a month was well documented by Faujas in *Description des Expériences de la machine aérostatique* (1783), which was meant to secure his reputation in public opinion and for posterity.³⁴ His carefully selected documents complemented the reports in the *Journal de Paris* and transformed an ensemble of marginal figures, institutions, and resources into an epoch-making scientific endeavor. Dedicated to the comte de Vaudreuil (Joseph-Hyacinthe-François-de-Paule de Rigaud, 1740–1817), a close associate of the duchess de Polignac and the comte d'Artois (Marie Antoinette's faction), it was a calculated career move for Faujas. Historians' uncritical reliance on his story ever since has substantially distorted the historiography in his (and the Montgolfiers') favor. In order to correct this imbalance, we must listen to hidden transcripts. In a contrasting report sent to the *Journal encyclopédique*, for example, Charles claimed credit for the balloon's design while attributing the construction to the Roberts and the subscription to Faujas.³⁵

Charles designed a relatively small balloon (twelve feet in diameter), but he still had to produce nine hundred cubic feet of hydrogen, an unprecedented task. His team had to sew the parts of the envelope together while the fabric was wet with rubber solution (*gomme élastique*) and afterward pass more solution over the seams to seal it airtight.³⁶ After checking the solidity, impermeability, and flexibility of the envelope, the team began to fill it, assisted by a number of amateurs including Charles's students and Aimé Argand (1750–1803), a friend of the Montgolfiers. By advertising the presence of amateurs, the experimenters could maintain the allure of a public artifact with open access. Pouring five hundred pounds of sulfuric acid over one thousand pounds of iron filings to make enough hydrogen gas produced intense heat, caused dangerous complications, and invited serious skepticism.

On August 23, a test flight in Charles's small garden caused numerous Parisians, estimated at thirty thousand, to crush against the guards. They had to leave the doors to the streets open. "All of Paris" soon converged on Charles's residence at the Place des Victoires (near the Palais Royal) to see the balloon suspended at about thirty feet above the ground.³⁷ The team decided to change the venue to the Champ de Mars.[38] Purportedly to safeguard it from a peril that could be caused by an importunate public, they transported the half-inflated balloon overnight with police escort. In other words, the Charlière (as the hydrogen balloon was sometimes called and not always with a positive meaning) became a public artifact with the help of the royal police. The police chief Jean-Charles-Pierre Lenoir (1732–1807) was a highly visible member of the fashionable society and enlightened administrator who would have appreciated the value of a scientific spectacle. He may not have foreseen the size of the crowd a balloon spectacle could attract.[39]

CHAMP DE MARS, AUGUST 27, 1783

On August 27, the operation began at dawn and was almost finished by noon, although the final stage of filling the balloon was delayed until the public could witness the process. Public approbation mattered to the entertainers tied to their purse strings. The *Journal de Paris* announced the event in the morning, which mobilized the crowd. At 3 PM, they admitted the four thousand ticket holders into the enclosure.[40] The Academy stationed two astronomers on the rooftops of the Observatoire (south of the Jardin du Luxembourg) and the Garde-Meuble at the Place de Louis XV (nowadays Place de la Concorde) to triangulate the height of the ascent.[41] The paucity of extant illustrations (other than the propagandistic one shown in figure 2.1) means that the phenomenal success of the first Parisian balloon ascent was a surprise even to the merchants of novelties who had to predict consumer demands and prepare the prints in advance.

The experiment began at 5 PM with two cannon shots to alert peripheral spectators and off-site astronomers. Shining because of the rain that coated it, the taffeta globe in yellow and orange stripes rose gently to about fifty toises and accelerated rapidly and majestically into the air. In the intensifying rain, all spectators eagerly directed their eyes, opera glasses, and telescopes to follow it. Vigorous and repeated applause spread through the crowd, as if to hasten its course. At about one thousand toises, as Benjamin Franklin reported to Joseph Banks, president of the Royal Society of London, the globe "enter'd the clouds, when it seem'd to me scarce bigger than an orange, and soon after became invisible."[42] Clouds separating, it was seen once more a few moments later, inducing another round of applause from the crowd.[43]

How the spectators behaved at the scene did not make news in the *Journal de Paris* or Faujas's self-serving transcript. In order to discern the collective mood, we must draw on expatriate newspapers and personal correspondences. While some dispersed "all well satisfied and delighted with the Success of the Experiment," most lingered on despite the rain that

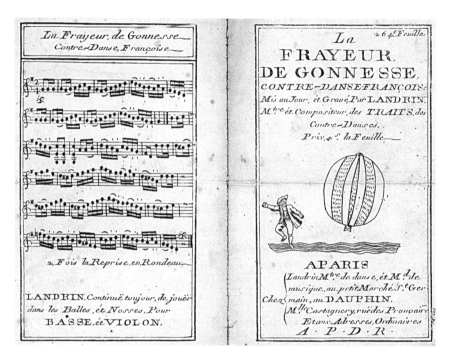

Figure 2.3. *La Frayeur de Gonesse, contre-danse françois.* Tissandier Collection.

seemed to intensify with the recurrent applause, ready for a moment of relief from the intense emotion and confusing spectacle on the ground—"a thousand beauties elegantly dressed fighting to tear themselves away from the whirlpool that was pulling them in, their stylish hairdos being ruined by the weather."[44]

The École militaire became an island of people unsure of their destination. Groups emerged here and there. Savants entertained questions but "the insatiable desire to know multiplied at each instant—What is the weight of the machine? What is the true proportion of the weight of the atmospheric air compared to the weight of inflammable air? At which height did it disappear? To which height was it elevated? What can be the duration of its course? Which effects would lead to its fall? In what state would it come down? What are the advantages this experiment can offer? Can one steer this machine, if one perfected it? &c. &c."[45]

The Charlière, filled to capacity for an impressive show, exploded at a higher altitude. Its fall on a field near Gonesse, a village about four leagues northeast of Paris, created a problematic contact zone with the rural populace. French savants had long campaigned to naturalize their authority over the "ignorant" artisans and populace and to extend their theoretical control over the practitioners' skill. Turgot's scientific solution to the cattle plague in 1775, which brought a royal troop into the affected villages for a quarantine and slaughter, was only one example of such skewed encounters.[46]

After a ground battle with frightened peasants, according to the *Journal encyclopédique*, the balloon remnant was eventually brought to the abbé Barrière, a local administrator of military pensions. He found two pieces of oiled paper inside. One from the comte de Vergennes (Charles Gravier, 1717–1787), minister and secretary of the state at the department of foreign affairs, instructed the finder to notify him immediately. The other from Charles begged the imagined reader, "some sincere amateur of science," to accept a friendly greeting sent by air and to secure the machine. The anticipated fall of an expensive machine in the midst of the rural population had put the minister and the scientist in a defensive predicament. The abbé duly dispatched a brief note with detailed information on the circumstances of the descent.[47] The battered balloon followed on a cart to make a procession through Paris.

The imagined encounter between the terrifying technological wonder and the peasant Don Quixote took on a narrative and visual life of its own to induce reverberating laughter in Parisian soirées and street theaters (see figure 2.3).[48] In the most contrived illustration (see figure 2.4), the gun configures an imagined contact zone that exaggerates the peasants' ignorance. The story printed in the *Mercure de France*, though more dramatic, only mentions stones and a knife.

> At a quarter to six in the evening of the same day, it was noticed by two carters working near Gonesse. It began to descend & grew larger in their view. Thunderstruck by this unaccustomed phenomenon, the carters stopped work, released their horses, and started to run. Nevertheless, as they were naturally courageous, though confused by fear, they dared to look behind them. The machine had landed, but it continued to move, bounce and turn in all directions. The carters then remembered the threats of a neighboring shepherd, and thinking that it was one of his tricks, they resumed their running even faster than before. But in vain: a gust of wind sent the monster rolling in their direction, and it rapidly overhauled them. It became necessary to fight. Thus chased, and with no time to lose, they armed themselves with some stones and hurled them with furor. The animal, still shaking and rolling, dodged the first blows. Finally, however, it received a mortal wound, and collapsed with a long sigh. Then a shout of victory rose; new courage reanimated the fighters. The bravest of the two, like another Don Quixote, approached the dying monster, and with a trembling hand plunged his knife into its breast.[49]

Figure 2.4. "General Alarm of the inhabitants of Gonesse occasioned by the fall of the aerostatic balloon of Mr. de Montgolfier. This balloon, 38 feet in circumference and made in taffeta soaked in elastic gum and filled with inflammable air made from iron and vitriolic acid, rose from the Champ de Mars, Paris, on August 27, 1783, at 5 o'clock in the evening in the presence of more than 300,000 persons. Heavy rain at the moment of its departure did not deter its accelerated ascent into the clouds. One presumes that it reached more than 20,000 feet in altitude where it burst due to the atmospheric air acting on inflammable gas. It fell after 45 minutes near Gonesse, 10 miles from the Champ de Mars. The inhabitants thinking that it was the skin of a monstrous animal attacked it with stones, pitchforks and flails. A local curé was summoned to reassure his fearful parishioners. This most beautiful experiment in natural philosophy was finally attached to a horse and dragged more than a thousand toises through the fields." Tissandier Collection.

Again following the description in the *Journal de Paris*, the Charlière is identified as belonging to the Montgolfiers.

The balloon forced Parisian institutions to conceive the nation as an administrative unit.[50] In order to "prevent the terrors it could occasion among the people" and to safeguard future balloons, the police distributed a "Notice to the people" with information on the principle of ballooning and first flights. It also alerted them to anticipate other floating objects that might look like an obscured moon. It was only a harmless machine that could some day have useful applications.[51] In order to allay "the astonishment & even the terror the apparition in the air" could induce, the *Journal de Paris* also took an extraordinary measure of sending several thousand copies to remote villages. A letter of the intendant general of the postal system, the baron d'Ogny (Claude Jean Rigoley, 1725–1798), accompanied them to order broad local distribution. These measures would have occasioned public readings for the illiterate populace, thereby fueling balloon mania across the cultural nation.[52] Local newspapers soon picked up the scent, which made the *Journal* the most reliable source of balloon news.

HIDDEN TRANSCRIPTS: Montgolfists versus Charlists

Public newspapers underplayed the intense competition between the Mongolfists and the Charlists, which frequently erupted as heated debates at the Palais Royal and circulated as vicious gossip in underground newspapers (*nouvelles à la main*). What became public and what remained hidden of this partisan bickering that has also been erased from the triumphalist historiography on ballooning can tell us how public newspapers maintained the "placid surface" of scientific hegemony by accommodating the existing social hierarchy.[53] Since the aerostatic globe demonstrated science's capacity to master Nature and the people, a proper distribution of credit would ensure a harmonious public sphere for the philosophical nation.

The *Journal de Paris* served as the central clearinghouse of balloon news and the public arbiter of credit. Its neutrality and transparency were easily compromised, however, by the sources of information and their respective interests. Those juggling for status and power craved public credit for this marvelous artifact. The *Journal* initially attributed all glory of this "sublime experiment" to the Montgolfiers—the scientific inventors who applied mathematics, physics, and chemistry to the industry for the public good. The *Journal* also publicized Faujas's leadership in organizing the "first national subscription" and directing the operations by suggesting the use of inflammable air and rubber.[54] This claim raised suspicions that Faujas had authored the article "for the simple purpose of puffing himself." This was "exactly in his character," according to Charles Blagden who had intelligence (probably from the duc de Chaulnes) that Faujas had no claim to the scientific aspect arranged by Charles and the Roberts.[55]

In contrast, the *Journal de Paris* characterized Charles merely as a willing contributor to diverse experiments, while blaming the mid-air explosion on his "chimerical pretensions."[56] The Roberts rose to Charles's defense, accepting the responsibility for causing the explosion by filling the balloon to make it look agreeable. They vigorously denied Faujas's claim to have directed operations, however, while acknowledging Charles's expertise and leadership. Faujas responded with libelous attacks on Charles's character—

that Charles sought to appropriate the Montgolfiers' glory, contrary to the subscribers' strong wish to repeat the Annonay experiment, and that he denied Étienne's access to the Champ de Mars among "the most illustrious in the Nation."⁵⁷ The Roberts rebutted in stronger terms: the theory and execution of their machine had nothing in common with the Montgolfiers'.⁵⁸

These pointed exchanges in the representative public newspaper were mild in comparison to the story that surfaced in underground newspapers. While his experiment was advertised as a repetition of the Montgolfiers' experiment, the accusation in the *Correspondance secrète* ran, Charles took on "the most absurd presumption" to "appropriate the entire glory" as its inventor. He thus insinuated to his students that this discovery was nothing new, which caused a stir. When the subscribers demanded a retraction, he vouched on his honor and conscience that he had "discovered and demonstrated the Montgolfiers' procedure." This assertion apparently ignited a "general fermentation," unleashing a torrent of epithets that characterized Charles as a "manipulator," a "merchant of natural philosophy," and so on. The subscribers, while defending Faujas's knowledge (*lumières*) and integrity (*honnêteté*), also accused the Roberts of financial irregularity. The subscription had already signed up five hundred persons, far beyond the initial goal. It dragged on, they claimed, to enlist more than twelve hundred subscribers at five louis each. Such "cupidity" was undignified, their judgment went, for "a man of talent" whose true compensation should be glory. The Roberts protested in vain that the financial arrangement was Faujas's responsibility.⁵⁹ By inflating the report in the *Journal de Paris*, the *Correspondance secrète* stirred a cloud of suspicion on Charles's character and expertise.

Vague financial suspicion became a general "fermentation" by another rumor that Charles had denied access to Étienne Montgolfier and Faujas. The *Mémoires secrets* charged that Charles's "imagination" had been inflamed by the large number of spectators to usher out Faujas, "the most zealous and useful" figure in this affair and accompanied by "the author of the discovery," as if they had a "diabolic intention" to ruin his experiment or if their mere presence had the "magical faculty" of weighing down the balloon. Faujas's reasoned pleas were in vain until the spectators intervened in this "revolting" scene and expressed their "just indignation" and "outrages" toward the "ungrateful & presumptuous manipulator" who deserved to be lost with the balloon.⁶⁰

The unrelenting assault on Charles's merit and integrity would plague his camp for months in a crescendo, which hints at an organized campaign much like a court intrigue—most likely between the queen's faction and that of the duc de Chartres. Faujas's subscription became "the source of all persecution," as Charles realized belatedly.⁶¹ Rampant rumors of the rivalry raised serious passions between the partisans, which worked against Charles, or Charlatan ("Charles-attend").[62] True or false, the colorful language of moral indignation toward Charles's "air of pretension" indicates that Faujas fanned such partisan animosity.

While it would be easy to sympathize with the persecuted, our interest in this subterranean bickering is not in determining the winner either then or now. Charles was not completely innocent of scientific pretension or jealousy. More problematic, though, was his ineptitude in dealing with intricate machinations in a manner that would support the public transcript. Without appreciating that the Montgolfiers had been certified as original inventors by the Academy and the royal administration, he boasted that his balloon was an entirely different and much better machine than theirs. In a report to the *Journal encyclopédique*, Charles asserted that the knowledge of artificial airs and their differential levities were known to everybody who had "the least tincture of chemistry" and that this knowledge should "naturally present the idea of an aerostatic science." Since the soap bubble experiment was well known to savants, as seen in François Chaussier's experiment at the Dijon Academy, the Montgolfiers' demonstration only instructed "the ignorant multitude." During the construction of Charles's balloon, Faujas was no more than a bothersome onlooker. Étienne Montgolfier in his visit only discouraged the collaborators with his indifferent glance, and so on.[63] By airing his petty jealousy against the virtuous inventors and their spokesman, Charles violated the code of honor that underlined the culture of calumny.[64] He also trivialized the fragile hope that these modest provincial inventors stirred in public opinion.

A hidden transcript often emerges at a site where the subordinate feels safe from persecution. In his lectures to more sympathetic audiences, Charles openly lamented that his balloon had become a "vain spectacle" only to "eternalize" the Montgolfiers' glory. He also offered an alternative story of invention that would have appealed to the scientifically informed public —that the balloon was a simple application of the pneumatic chemistry developed by Stephen Hales, Joseph Priestley, Jan Ingenhousz, Alessandro Volta, Felice Fontana, Antoine-Laurent Lavoisier, Henry Cavendish, Carl Wilhelm Scheele, and Tiberius Cavallo. Their research had isolated individual gases, analyzed the atmospheric composition, and revealed the "singular levity" of inflammable gas. By containing it in the soap bubble, physiciens could detonate it in midair.

Charles's story gave prominence to pneumatic chemistry, an expert domain of the Paris Academy of Sciences, which made his claim as a theoretical inventor plausible. In other words, his story of invention reflects more closely what the Parisian scientific community must have thought of the Montgolfiers' claim. Lavoisier and other discerning figures hid such sentiment under the public transcript in order not to swim against the tide of public opinion. To air such criticism of the inventors under royal protection would have pegged them as calumniators unworthy of public respect.

Charles did not heed this tactful decorum, which only served to denigrate his reputation. He claimed that the savants should have drawn a "natural" conclusion that humans themselves could rise in the air by enclosing inflammable air in a "light and impermeable prison." This was all one needed to invent "a new bird" or, rather, "an aerial fish," which had somehow eluded other physiciens. While this "incontestable" principle offered the "theory" of aerostatics, he did not pursue an "authentic demonstration" in public because others labeled it as a "chimerical theory" until Blanchard's flying carriage prompted him to demonstrate the insufficiency of mechanical means. Although he performed preliminary calculations, he could not finance a public demonstration.[65]

Perhaps wishing to protect the Montgolfiers' reputation, the *Journal de Paris* did not print Charles's story. We should regard it with caution as well. While he certainly possessed the theoretical, mathematical, and technical expertise to make a superior balloon than the Montgolfiers', the invention of this sensational machine required much more than a theory. Even if the balloon did not require as much investment as Watt's steam engine or Jouffroy's steamboat, making a hydrogen balloon required expensive fabric (silk), exotic coating material (rubber), effective glues, and intensive labor, which all came together only because of the Annonay ascension. Becoming an original inventor of a useless machine required material and social resources beyond the means and wits of an ordinary public lecturer.

The intensity of partisan spirits suggests a powerful patron, most likely the duc de Chartres whose populist rhetoric often crossed the queen. His sponsorship of the Charlière would have been interpreted as a deliberate stance against the royal Montgolfière, a situation Faujas could have exploited to his own benefit.[66] Antoine de Rivarol (1753–1801) was soon commissioned to publish an anonymous polemic *Lettre à Monsieur le Président de **** (1783), which supported Charles's priority and indicted the Montgolfiers' and Faujas's vainglory.[67] Rivarol mocked the Montgolfiers' "chance" discovery unguided by scientific theory, their sensational approach to win over the public, and their limited knowledge and success. He characterized them as men "without geometry, without mechanics, & without chemistry" who brutalized natural philosophy. Their exceedingly simple discovery in the century of light could not compare with Columbus's hard-won discovery that had opened a new world, educated barbaric peoples, and established colonies. Rivarol ridiculed Faujas as the "quite distinguished physicien" of the café, a "ridiculous and incompetent tribunal" where idlers passed time disparaging the works of genius. Faujas might have steered the globe with his eyes, but not the experiment, a claim at odds with those of five hundred thousand witnesses.[68] In addition, Rivarol lamented the balloon's uselessness. His pessimistic objections to the possibility of aerial navigation induced lengthy rebuttals.[69]

The ruinous cycle of news-making, which turned one of the most glorious moments in French history into a mud-wrestling match between the nation's most talented inventors, speaks volumes about the culture of calumny that would later drive revolutionary politics into the Terror.[70] Notwithstanding his earnest intent to destroy Charles's reputation, Faujas did not send his libels to public newspapers. Such voluntary suppression offers a telltale sign of how sacred the symbolic capital of the balloon and its representatives had become, as if the logic of *lèse-majesté* applied to the majestic machine and its guardians as well. The destabilization of divine kingship, especially through the libels that lacked reform agenda, had dissociated the notion of majesty from the royal body without an alternative locale.[71] In this period of uncertain sovereignty, the balloon's miraculous performance and the inventors' virtue became symbolic resources too precious to be wasted for those who dreamed of a beautiful nation. The "hidden transcripts" on ballooning comprised not just the voices of the severely oppressed and silent but also all variations to the public transcript. Such voluntary harmonization of public opinion ensured a sacred public trust in the authority of science. An archeology of calumny and hidden transcripts can reveal the fundamental values such as honor and merit that sustained the public space of the ancien régime.[72]

Bitter contestations, largely hidden from the provincial and European public, reveal a Parisian public sphere operating on the premise of *absolute* public opinion—a mirror image of the absolutist polity. In other words, the rise of public opinion as the ultimate tribunal for political legitimacy during the Enlightenment did not necessarily cultivate proro-democratic dispositions of free opinions and debates but absolutist politics by other means. In their focus on cultivating egalitarian sociability, the "citizens without sovereignty" did not fundamentally challenge the status quo in the political system. The imagined tyranny of public opinion invited endless intrigues just as intense as the courtly ones. The hidden transcripts in the underground newspapers and pamphlets allow us a glimpse of the "culture of calumny and honor" engineered by contending politiques wishing to control royal power and public opinion.[73]

A LIMINAL COURT: Middling Ascendants

Public enthusiasm and partisan disputes put more pressure on Étienne Montgolfier, now perceived as a royal (or the queen's) inventor. According to the Parisian chronicler Siméon-Prosper Hardy, the public press emphasized the balloon's potential contribution to "the general good of the society."[74] Physiciens came in high demand.[75] The excitement was more pronounced in underground newspapers. The *Mémoires secrets* noted public impatience in waiting for the grand machine under construction by the original inventor for a royal fête at Versailles.[76] *Grimm's Correspondance* reported that "among all our circles, at all our suppers, in the antechambers of our lovely women, as in our academic schools, all one hears is talk of experiments, atmospheric air, inflammable gas, flying carriages, aerial voyages."[77] Science entrepreneurs conjured up new projects in rapid succession.

Most popular was the miniature balloon made in goldbeater's skin and filled with inflammable air, first flown on September 11, 1783, by the baron de Beaumanoir, an amateur spectator at Charles's balloon construction site.[78] Even with the high cost and inconvenience, the miniature balloon quickly became a fashionable commodity.[79] Benjamin Franklin purchased one for his grandson and secretary William Temple Franklin on September 15. Released in his bedchamber, "it went up to the ceiling and remained rolling around there for some time." He sent it to Richard Price so that his British friends might repeat the experiment.[80] The merchants of novelties began to sell various balloons ranging in price from three livres for a nine-ounce model to eight hundred livres for a twelve-foot model.

The Montgolfier family pressured Étienne relentlessly to obtain practical advantages from his momentary fame. Perhaps, he could ask Franklin for a business deal, as Réveillon did soon after the peace talks began?[81] The departure of Écrevisse created an "absolute necessity" to secure a competent replacement. Perhaps, he could obtain privileges to the Compagnie des Indes through the Navy secretary marquis de Castries (Charles Eugène Gabriel de La Croix, 1727–1801)?[82] Étienne could not have been indifferent to his family's financial woes since they had not been paid fully by the Languedoc Estates, much less for the cost of the royal balloon. He was surrounded, however, by luminaries accustomed to exacting subservience from their entertainers with empty compliments and courtiers more than willing to comply. A hint of self-interest would have ruined the public transcript on the patriotic invention and its virtuous inventors at a time when even Marie Antoinette was treading carefully to curb the public perception of frivolity. Étienne hid his family's financial struggle from public view, lest it should tarnish the aura of the savant devoted to the pursuit of truth and the public good. Burdened with Étienne's long absence, the family had to suffer their economic loss in silence. Theatrical polity rewarded those who could display their virtue, honor, or wealth but ridiculed those in straitened circumstances.[83]

Étienne was hard at work, surrounded by a close circle of supporters including the apothecary Antoine Quinquet (assistant to Antoine Baumé) and the Genevan inventor Aimé Argand. Réveillon's factory at the faubourg St. Antoine (east of the Bastille) provided a spacious yet access-controlled site that placed the middling ascendants among the ancien régime notables.[84] Originally a paper merchant, Jean-Baptiste Réveillon (1725–1811) had purchased Titon's outlandish country house complete with wainscots, paintings, sculptures, and a gallery (see plate 5).[85] Just outside the city wall, the Folie Titon became an ideal location for the production and distribution of wallpaper and circumvented Parisian guild restrictions.[86] Aside from their past association, Réveillon had much to gain by graciously offering his factory and workers to Étienne.[87] He had applied for the title of royal manufactures, citing his invention of papier vélin, which was granted to the paper factory in July and to the paper mill in December.[88]

By controlling access to the inventor and his fashionable machine, Réveillon exercised tangible power that could help his business deals. To be seen in choice company and to converse upon fashionable topics were essential modes of participation in the theatrical polity. The Montgolfière was a sensational novelty that attracted socialites of all ranks of society: Princes, bishops, and royal administrators such as the police chief Lenoir had automatic access. The retired statesman Lamoignon de Malesherbes spoke through the comte d'Antraigues, one of Étienne's compatriots who kept watch along with the comte de Boissy d'Anglas (François-Antoine, 1756–1828).[89] The journalist Pascal Boyer asked for a spot so that he could instruct the public on this "brilliant discovery" through the *Mercure de France*.[90] Étienne's old friends and their families claimed his obligation in vain.[91]

The Réveillon factory became a liminal court that blurred the boundary between the court society and the nation, which invited new political imaginaries. In lieu of the royal body, the royal balloon emerged at the center of an illustrious congregation of the fashionable society. Its size and design changed as Étienne came to reflect on his monumental task—to construct a glorious object that would project the sovereign's power and authority.[92] Joseph the dreamer urged him to make a more impressive (rather than useful) balloon and to send up a larger animal (a cow rather than a sheep) to stage "the most glamorous show."[93] In contrast to Charles's simple stripe design, the royal Montgolfière was a spectacular artifact, beautifully painted in azure blue and gold fleur-de-lis, symbols of French monarchy since the Middle Ages. It was 40 feet in diameter and 70 feet in height, and had three tiers of geometrical shapes—a 24-foot prism placed on an 18.5-foot truncated cone and topped by a 27.5-foot pyramid—thought "more convenient for penetrating the air."[94]

An inspection by the Academicians on September 12, a necessary step to certify the balloon's scientific status, turned into a disaster. Ever conscious of economy, Étienne had made it in coarse linen backed by oiled paper without coating the fabric.[95] The size necessitated an assembly in open air by numerous workers. Folding and moving the envelope indoors every evening required at least twenty men and caused much anxiety.[96] The night before the demonstration, they had to leave the finished balloon (weighing one thousand pounds) outside, hanging between two high poles. The rain that began during the night steadily increased as the morning progressed. Étienne rose at 5 AM to dry the balloon and inflated it with fire.[97] Around 9 AM, the Academy's commission and Benjamin Franklin arrived.[98] Increasing the fire, they inflated the balloon fully to get it five feet off the ground, but a storm broke and tossed it about. Mindful of his upcoming obligation at Versailles, Étienne refused to cut the cords as Argand advised. Soaked with rain, the balloon came down in colorful shreds, dissolving two months of work.[99]

Étienne missed the birth and baptism of his daughter Hélène-Rose.[100] In order to keep his "word given to Versailles," he immediately set out to make another balloon that would not be vulnerable to fire, wind, or water in good *toile de Rouen* coated with Succin.[101] This close-knit blend of linen and cotton, mostly used for decorative purposes, would become the fabric of choice in making the Montgolfière.[102] Worried about the approaching winter if Étienne could not bring funds home, the abbé again urged him to extract some "earthly good" from his fame—be it a commission for the paper business, direct financial compensation, or a lucrative market share. Pilâtre's flattery as if Étienne was an Olympian god was as useless as the incense.[103] Étienne had no choice but to press on, considering the cost already incurred and not yet reimbursed.

The full workforce of the Réveillon factory, thirty-nine male workers and a host of seamstresses, began at 9 PM Saturday and finished on Thursday morning. They had become practiced hands. The new balloon was smaller, measuring 41 feet in diameter, 57 feet in height, and 37,500 cubic feet in volume. Étienne spared no expense and took every precaution to prevent another mishap. The new balloon cost 2,700 livres, a considerable increase over 1,600 livres for the first one.[104] It was painted not only outside but also inside, mixing in the earth of alum for heat resistance. With the weather

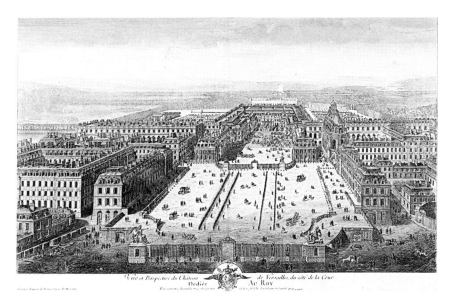

Figure 2.5. Château de Versailles seen from the side of the main gate, which shows the Cour des ministres where the balloon ascent took place on September 19, 1783. Gallica, BNF.

finally cooperating, they tested the new machine successfully for a small crowd including the Academicians.[105]

A ROYAL MONTGOLFIÈRE: Versailles, September 19, 1783

The Versailles fête was organized by the Menus-Plaisirs du Roi, a branch of the royal household in charge of ceremonies and entertainments, overseen by the four Gentlemen of the King's Chamber (*Premier Gentilhommes de la Chambre*).[106] They gave orders to construct "a kind of theater" in the courtyard, just inside the main gate of the palace and open to public view (see figure 2.5). An octagonal wooden platform, at least forty feet in width and from six to seven feet in height, was built with a fifteen-foot opening to receive the fumes from the iron stove underneath. It was draped all the way around with tight-knit fabric to contain the smelly smoke. On the stage stood two tall masts, each tied down by four ropes and a cord running between them, to raise the balloon as it filled.[107] The Menus sent a wagon to the Réveillon factory in the evening of September 18.

The next morning, carriages packed the road to Versailles. Étienne left Paris at 5 AM. When he arrived at Versailles, the duc de Duras (Emmanuel-Félicité, 1715–1789) ushered him to the king's rising ceremony (*levée*) to present an overview of the experiment. The honor of this ceremonial reception, normally reserved for select courtiers, could not have escaped Étienne as he walked afterward to the launch site through a passage lined by royal guards. They also surrounded the vast theater in double circles to place the savants, amateurs, and all courtiers in the outer circle. The familiar representation of this ascent (part 1 frontispiece), presumably prepared for the German audience, hides the body of soldiers that layered the audience along with the populace standing outside the courtyard. Most images did not represent the crowd at all.

The court was packed by 7 AM with the members of Faujas's nation: "the greatest, the most illustrious, and the most knowledgeable in the nation gathered to pay a solemn homage to the sciences under the eyes of the august court that protected and encouraged them."[108] Outside, spectators numbering from 120,000 to 130,000 lined windows, roofs, and avenues. In the enclosure under the stage, Étienne received a stream of visitors—the comte d'Artois, Monsieur and Madame, and the royal couple along with their entourage—who listened attentively to his explanation.[109] After the mass, the king appeared on the grand balcony with his entourage. The queen and her ladies sat under a large tent. The entire court was seated by 11:30 AM at a safe distance from the stage.

Dressed in black, Étienne commanded the scene with calm composure to appear as a modest, yet skilled physicien.[110] When the king gave the signal at 1 PM, they drew a round of cannon shots and lit the fire on the straw, throwing in the clippings of animal horns, old leather, and hair in an effort to increase fumes, which only created a mystery over "the gas of Montgolfier."[111] Consuming twenty-four pounds of straw and five pounds of chopped wool, the balloon filled in seven minutes and began to tug at the ropes held down by at least fifteen men. They attached a wooden cage containing a sheep, a rooster, and a pigeon, which was designed to test if the air at a high altitude could support life without rupturing blood vessels.[112] With the second round of cannon shots, they increased the fire. At the third round, everyone let go of the ropes at once. The machine rose "majestically," pulling up the cage. The massive volume of the azure envelope decorated with royal monograms and crowned by yellow ornaments resembled "a new planet."[113] Somewhat tilted by a sudden gust of wind, it continued to rise as majestically as ever.

After the takeoff, Étienne proceeded to the royal apartments and found Louis XVI tracing the balloon with his field glasses. Many dignitaries invited him to dinner, but he attended one offered by the controller general for the Academicians. Afterwards, he had to make several rounds and visit the ladies in Madame d'Ossun's apartment, ending the evening with an audience with the queen. Privately to his wife, Étienne was "willing to admit . . . the pleasures of this day," which made him forget "the work, the trouble, and the worry" it occasioned him.[114] The balloon traveled about 1,700 toises in 10 minutes, rose to 240 toises, and descended gently in the woods of Vaucresson, no doubt to Étienne's great relief. Faujas caught up with the machine as it descended, along with the abbé d'Espagnac, the chevalier de Lorimer, the architect Alexandre Théodore Brongniart, and Pilâtre de Rozier. The animals were found intact. Having "braved death and danger" for the general good of society, they would soon acquire voices in a fable well calculated to ridicule man's cowardice.[115]

The family conveyed their sense of relief and elation, especially since Pierre had just unleashed a temper tantrum for being kept in the dark. In an unusually light mood, the abbé imagined making and selling mini-balloons filled with inflammable air. He was fast learning chemical vocabulary in the local circle around Brogieux, the master "phlegmatic physicien." The family's "small committee" also scripted Étienne's future maneuvers. Royal honors such as the Cordon noir or the title of royal manufactory were worth no more than the smoke in the balloon and would "only serve to excite the envy and jealousy" of their business associates. If Étienne took a place among them "without any distinction," his modesty would honor him more than the "vain title." Since one could not always control what one received, Étienne should include Joseph in any forthcoming distinction. The title of royal manufactory should be bestowed on their eighty-four-year-old father who had acquired the most brilliant distinction in the commerce for seventy years, raised a large family, and provided all possible education for his sons. While relaying the family's decision on public posturing, Jean-Pierre as the eldest son also advised practical measures—a lump sum and a stable occupation for Joseph, a lucrative abbey for the abbé, and a government commission for the family business. These were but routine means of social mobility in the absolutist polity. When Étienne remained silent, the restless family committee schemed further to procure government posts.[116] As respected provincials, they could not have known the complex machination at Versailles.

Balloon fever shot up a notch after the Versailles ascension, which induced stringent media control. All Parisian heads seemed to be "in fermentation . . . filled with the wind."[117] Popular enthusiasm brewed trouble, which was not reported in the *Journal de Paris*. On September 25, a failed commercial venture at the Redoute Chinois in the Saint-Laurent Fair broke into a riot, sounding an alarm for the authorities. The proprietor Roger Timothé Regnaud de Pleinchesne had teamed up with the famous firework master Petroni Sauveur Balthasar Ruggieri, an *artificier du roi*, to stage a small balloon for profit. The Redoute, dubbed a "new temple" for "absolute idleness," was an entertainment complex containing a café, a dance hall, oriental swings, a kiosk, a restaurant set up in several tents, and a garden lighted with multicolored lanterns, all decorated in the popular Chinese vogue to attract novelty-seeking customers.[118] Down on their luck since the 1770 accident during the dauphin's marriage celebrations, Ruggieri was gathering information from Faujas and Lavoisier to stage "a sound and attractive aerostatic balloon" with fireworks that could be seen as far away as Versailles.[119] No doubt to raise the profile of his commercial endeavor, Regnaud sent an invitation to Étienne, wishing to pay "a tribute and a homage" to the author of the "sublime discovery" that had inspired a "national sentiment."[120] The event drew a large crowd, rumored to have paid from eight to nine thousand livres at thirty-six sous each. When the balloon caught fire during Ruggieri's attempt to fill it, a tragic-comic scene ensued among the spectators set up in the nearby cemetery by the church valets. Vigorous demands for refund invited police intervention.[121] The incident was kept out of the public press, however, probably to maintain the balloon's public identity as a scientific and virtuous artifact in the interest of social order.

Aristocratic frivolity was also excised from the hegemonic public transcript on ballooning. On October 1, 1783, the duc de Crillon celebrated the birth of two Spanish princes with an elaborate fête at the Ranelagh garden near the Château de la Muette. All "ambassadors, ministers, foreign princes, and the persons of highest quality of the court and of the city" were invited. The program included theatrical performances accompanied by an orchestra, fireworks, and a hydrogen balloon ascent. Six feet and four inches in diameter, made of many pieces of goldbeater's skin glued together, the globe carried a lantern with clearly marked inscriptions: "Vive Charles, vive Louis. . . . Their names fly in the air, they embellish the universe." Deliberately held at the height of two or three toises for several minutes, it allowed the guests to read the inscription before it was set free at about 1 AM accompanied by "superb music." It went straight up into the calm night sky, its light diminishing gradually "till it appear'd no bigger than one of the Stars," as Benjamin Franklin reported to Joseph Banks.[122] The ascent produced "an effect impossible to describe" for the guests who enjoyed this "image of happiness."[123] This fashionable event, which mobilized the entire court society, was not reported in the *Journal de Paris*, despite the broad coverage in other European newspapers.

The first balloon ascents around Paris illustrate the fusion of royal and public resources that materialized a composite body of the nation in prerevolutionary France. The balloon drew indiscriminately on royal largesse and public resources to become a venerated scientific artifact that could float an alternative machine polity. If staging the patchy machine required police control and protection, packaging it as a national artifact depended on the public transcript that highlighted its scientific performativity and the inventors' virtue. The *Journal de Paris* offered a convenient medium for the scientific, administrative, and political elite of the ancien régime to forge a consensual public opinion and a hegemonic public conduct shorn of unruly enthusiasm. It suppressed the bitter competition between the dueling balloon factions, riots, and aristocratic parties to present the aerostatic globe as a glorious French invention. A rationally reconstructed genealogy from Robert Boyle to the Montgolfiers permeated public discourse and idealized them as scientific inventors rather than the mere papermakers who had chanced upon a discovery.[124] Such careful rehabilitation of their public identity would prove crucial in the French balloon mania that captured the hearts and souls of underemployed middling professionals—countless abbés, officers, science lecturers, reporters, malcontents, and so on—who desired a dramatic escape from their hopeless lives.

It would be tempting to contrast the first two Parisian ascents as visualizing competing polities—republican versus absolutist. Both utilized administrative and public resources, however, to shape the balloon's public identity as a useful scientific artifact within the absolutist theatrical polity. In contrast to the royal theaters whose complicated machinery remained hidden to enhance their political efficacy, a variegated cluster of moneyed, titled, or educated elite mobilized and displayed all of their resources and gadgets—paying for the subscription, commandeering proper sites of ascension and the police, publishing technical knowledge and production process, and so on—to stabilize the balloon's physical integrity and moral value. According to a pamphleteer, such illimitable "enthusiasm" from powerful sectors silenced the editors of public newspapers and condemned the anti-balloonists as "ignorant" enemies of useful sciences, reason, and good sense. Not wishing to offend powerful patrons of these marvelous machines, they followed "the majority of suffrage" in public opinion and thereby succumbed to a general error.[125] The enthusiasm for science, a kind of furor born from reason, seduced the enlightened audience with an intense emotion much like religious enthusiasm to suppress dissenting voices in the name of the public good—an absolute moral mandate in public opinion that consolidated the figure of philosophical majesty.[126]

Figure 3.1. Jacques-Alexandre-César Charles receiving a wreath from Apollo. Engraving by E. A. Tilly (1788). Tissandier Collection.

3

TRUE COLUMBUS

> When Charles and Robert, full of noble audacity,
> Leap into the sky on the wings of the wind,
> What honor can reward their glorious efforts?
> They marked their place between men and gods.
> "Vers de M. le Vicomte de Ségur à MM. Charles et Robert"

The new ocean called for new Argonauts. The brave Grecian youths who set out in the *Argo*, the fifty-oared ship that was strong enough to withstand buffeting winds and waves yet sufficiently light to be carried on their shoulders, enacted the classical past for the triumphant nation. The aeronaut's daunting task would require a similar vessel, perhaps one that could traverse "from planet to planet."[1] An amateur of fine arts offered a graceful design in the form of a galloping Pegasus that combined mobility, safety, and constant posture for the aerial cavalier: the body for gas, the head for bows, the wings to regulate elevation and speed, the tail as the helm, and loaded feet as the ballast.[2]

The aerial Argonaut would serve the nation not simply by his physical strength and moral courage but also by his ability to harness Nature, which lent her wings to genius, the courtier-poet Paul-Philippe Gudin de la Brénellerie (1738–1812) declared.[3] He had to possess scientific knowledge as well as a brave heart to build an aerial empire. Translating the myth involved a temporal metamorphosis that turned familiar characters and deep memories into abiding moral commitments and political allegiances to shape the topography of public opinion.[4]

More problematic was the allusion to the contaminated figure of Columbus. The abbé de Raynal (Guillaume-Thomas, 1713–1796) lamented in the *Histoire des deux Indes* (1770) that Columbus the modern Argonaut had set off a destructive revolution in commerce, politics, and mores. Raynal questioned whether past and future revolutions in commerce would produce virtuous citizens, highlighting the immorality of slave commerce in particular.[5] Louis Antoine de Bougainville's circumnavigation of the globe under the protection of the foreign minister, the duc de Choiseul (Étienne-François, 1719–1785), and the subsequent publication of his *Voyage autour du monde* (1771) were designed to revamp French imperial efforts after the defeat in the Seven Years' War. They received a mixed reaction among the educated public, as seen in Diderot's *Supplément au voyage du Bougainville* (1772).[6] Forecasting the enslavement of Tahitians that would destroy their happiness, Diderot impugned French imperial excess in the 1780 edition of Raynal's *Histoire des deux Indes*, which invited an official condemnation by the Paris parlement and forced Raynal into exile.[7]

A modern myth has to make sense of the problematic past and shape the uncertain future.[8] Notwithstanding the imperialist "look of power" deployed in global explorations, the Columbus topos had sustained a genre of transgressive hope in European cultural imagination.[9] The American dream grew during the war, with the charming Franklin in the Parisian salons and the gallant Lafayette on American battlefields.[10] The war became "a sort of Pandora's Box out of which poured a cloud of books and articles advocating equality, republicanism, liberty, and constitutionalism."[11] Whether the discovery of America had been beneficial or damaging to humanity remained a troubling question of the time.[12]

The aerial empire called for a "true Columbus," one who would seek the truth of Nature and tame it for the benefit of humanity. The honorable title aroused noble dreams and zealous aspirants "who to enjoy the honour of being the first Travellers through the air, are willing to . . . run ten risks to one of breaking their necks."[13] There was "hardly a man in Paris" who did not vie for this "honor of being the first flyer," as London newspapers exaggerated so as to encourage English competition.[14] Interpreted as a vehicle of scientific exploration, the balloon transformed the compromising figure of Columbus into a virtuous philosopher-voyager who would serve the nation and humanity—a timely metamorphosis that might have softened the rising criticism of global empires.

TETHERED DREAMS: Floating above the Bastille

After the Versailles ascent, Étienne resumed work at the Réveillon factory located east of the Bastille, the formidable prison that symbolized despotic power—one that the journalist Linguet wished to strike down by lightning to preserve the monarchy (see figure 3.2).[15] Mercier had also wished to demolish the "hideous castle," or the place of "royal vengeance," and to erect a temple of clemency on the spot.[16] The balloon attracted not just the fashionable society but also the curious populace and thieves to create a

Figure 3.2. "Be Free, Live." Frontispiece to Simon-Nicolas-Henri Linguet, *Mémoires sur la Bastille* (1783). On the statue is written, "To Louis XVI, on the place of the Bastille." To the right, *Ces souffrances inconnues & ces peines obscures, du moment qu'elles ne contribuent point au maintien de l'ordre par la publicité, & par l'exemple, deviennent inutiles à notre justice. Déclaration du 30 août 1780* (These unknown sufferings and obscure pains, since they do not contribute to the maintenance of order by publicity or by example, become useless to our justice. Declaration of August 30, 1780).

potential trouble spot, although public newspapers did not report their presence.[17] Unnerved by the riot at the Redoute Chinoise, the police wished to eliminate any pretext for large gatherings. The faubourg Saint-Antoine was an experimental region that provided a working model of the laissez-faire economy and harbored a diverse population outside city governance, as Réveillon would realize only through the revolutionary violence.[18] His palatial factory stood next to the Bastille as a shining example of incomprehensible wealth.

Depending on the outcome, the balloon could either offer a timely distraction or trigger a riot. The recent bankruptcy of the Caisse d'escompte had produced many malcontents predisposed to violence.[19] Established in 1776 to boost commerce with low-interest loans, the Caisse issued paper notes that circulated only in Paris but were guaranteed for real coins. Entangled in wartime finance schemes, it stopped payments on September 26, 1783. The shareholders had to freeze the notes still in circulation (amounting to 33 million livres by one estimate), while waiting for a comprehensive report.[20] The royal treasury barely escaped the same fate by publishing a decree that guaranteed the notes. Louis XVI had to withdraw the plan to reform the notorious Tax Farm, which would lead to d'Ormesson's resignation and Calonne's appointment as controller general in early November, 1783.[21]

Public clamor spun a complicated web of obligations for the inventor and his promoters. It also fueled Étienne's growing anxiety over the hydrogen competition. While the Montgolfiers' reputation as original inventors remained intact in the public press, more scientifically informed persons such as Benjamin Franklin suspected that Charles and the Roberts were better "men of science and mechanic dexterity" who could introduce material improvements not yet seen with the Montgolfière.[22] The Academy's research focused mostly on the hydrogen balloon, although Lavoisier and Jean-Baptiste Meusnier (1754–1793) followed Étienne's work closely, engaging him in scientific conversations.[23] Étienne sought validation from the Academicians, paying a visit to Buffon, to stabilize his public persona as a scientific inventor.

Réveillon found it difficult to control access and maintain the balloon's scientific identity. Persons of all ranks, except for the royal couple, flocked to his factory to catch a glimpse of the historic venture as soon as test flights began in early October. Réveillon had to announce publicly that the experimental site was reserved exclusively for the physiciens who might help establish and confirm the *"theory* of the Montgolfiers' discovery."[24] By invoking the inventor's need for reflection in learned company to perfect a potentially useful art, Réveillon sought in vain to keep the undesirables away.[25] He also helped Étienne to make an impressive balloon and advertise it among the fashionable society, or his target audience.

The satire of lost uncle, exceptionally published in the *Journal de Paris* and circulated in color illustrations (see figure 3.3), can be interpreted as a part of the Montgolfists' continuing propaganda against the hydrogen faction. Although the anonymous writer indicted all novelties and their makers as dangerous and useless in a letter dated September 23, 1783, he attributed the loss of his uncle to the hydrogen balloon. In order to fill it conveniently, his uncle—a small, scrawny physicien with a large head and face, thick lips, flat nose, and round shoulders—used to employ two syringes. One day, he had a cup of café au lait with a visitor and became violently ill. Having found him unconscious, the letter writer applied the syringes to administer medicaments, which filled him with inflammable air instead. His uncle flew out the window in his robe and was lost, although his bonnet and wig were discovered on the way to Normandy.[26]

Semi-public trials at the Réveillon factory provided a splendid occasion for the fashionable society to bear witness to a historic event. The second royal balloon measured seventy feet high and forty-six feet wide, weighed

Figure 3.3. "The Aerostatic Man or My Poor Uncle." Gimbel Collection XL-15-4628.

sixteen hundred pounds, and encompassed sixty thousand cubic feet. It was painted in azure blue and gold emblems—with the signs of zodiac and the fleur-de-lis around the dome, ceremonial royal initials and blazing suns around the equator, and open-winged eagles lining the bottom.[27] After a few minor accidents, it rose to three hundred feet (the length of the four cords) to demonstrate the "point of perfection" it had attained. The "rupture of the equilibrium" was sufficient to carry two people.[28]

On October 15, 1783, with the Academicians Dominique Cassini and Jean-Baptiste Le Roy ready to take measurements, Pilâtre ascended alone at 4:30 PM. With the ropes attached, he floated for four minutes and twenty-five seconds at the height of eighty feet.[29] He had earned this privilege by working tirelessly for Étienne's favor, opening subscriptions for a print to "fix the epoch of the discovery of the aerostatic machine," a gold medal, and a new machine.[30] Falling short, he chose to make the medal and commissioned the famous sculptor Jean-Antoine Houdon, a member of the Loge des Neuf Soeurs.[31] The second trial on October 17 attracted from two to three hundred carriages and a large gathering that included the police chief Lenoir and the archbishop of Paris, but it proved less successful. Étienne halved the gallery to reduce the weight.[32] Ever the publicist, Pilâtre sent out

TRUE COLUMBUS | 81

invitations to all subscribers of the Musée de Monsieur by special couriers to "elude the prohibitions of the police" for the last trial on October 19.³³

Tethered ascents might have symbolized the usual mode of dual existence for the ancien régime citizen—chained to his duty in the standing order, yet struggling to rise above his station. The prospective aeronaut Jean-François Pilâtre de Rozier offered a perfect example of such ambition. Born to Mathurin Pilâtre, a retired military officer, he was educated (through the generosity of his father's friend Viollet) at the Benedictine Collège royal de St. Louis in Metz, trained briefly to become a lowly surgeon, and apprenticed to the local apothecary and academician Jean-Baptiste Thyrion.³⁴ An avid reader of scientific literature, young Pilâtre was soon admitted to the local academy's meetings. He could have seen the royal intendant Charles-Alexandre de Calonne (1734-1802) who would become controller general in 1783. The marquis de Lafayette (Gilbert Du Motier, 1757–1834) was also stationed in Metz. Mathurin Pilâtre ran Le Croissant Rouge, an entertainment complex for officers with a billiard room, a dozen rooms, a restaurant, and a large bar open late into the night.³⁵

This combination of family background and scientific training prepared young Pilâtre for an extraordinary career. He was introduced to the duc de La Rochefoucauld-Liancourt (François-Alexandre-Frédéric, 1747–1827) who explored the local flora with him for two months. In October 1775, Pilâtre left for Paris where he found a steadfast patron in the royal physician François Weiss and worked for the royal apothecary Pierre-François Mitouart. He attended public courses, began to publish memoires, worked briefly as an apothecary to the prince de Limbourg, taught chemistry for the Société d'émulation de Reims through the Academician Baltazar-Georges Sage's recommendation, and presented memoires to the Academy of Sciences on dyes, electrical experiments, and noxious gases.³⁶ After Weiss died in 1777, Pilâtre opened a chemistry course in the Marais with his instruments and married his daughter whose inheritance allowed his entry into the fashionable society. As we have seen, the Musée de Monsieur made him a fashionable scientist. To maintain his status, Pilâtre circled Étienne with a full repertoire of flattery on his lips.

The balloon, rather than the royal body, occupied the center of Faujas's nation gathered at the Réveillon factory. Science, rather than religion, constituted its symbolic order. More than two thousand persons jammed the courtyard on Sunday, October 19, while a crowd packed the surrounding streets. The entire spectrum of the fashionable society bore witness to a successful human ascent, albeit tethered, which included the Polignac family, the duc de Chartres, and the comte de Dillon, the last riding up to about forty feet.³⁷ Pilâtre then rose "pompously into the air amongst public acclamations." In the third ascent, he was accompanied by Giroud de Villette, one of Réveillon's business associates, who subsequently vouched for the balloon's "real utility" in military operations and Pilâtre's courage, agility, and talents at maneuvering the machine. In the fourth ascent, the marquis d'Arlandes (François Laurent, 1742–1809) accompanied Pilâtre to twelve hundred feet.[38] Public newspapers reported that the Montgolfière could "be elevated and lowered at will" without any risk to the aeronauts, albeit lacking the means of horizontal steering.[39]

The Réveillon trials meant significantly more than a mere prelude to the public demonstration because of their illustrious audience that effectively constituted the nation in elite imagination. The most beautiful illustration of Parisian balloon ascents was set on this occasion with the royal balloon standing in place of the king (see plate 5). The public transcript and image of these trials were meant to validate the standing order and its representative public gathered at this displaced court. Only after these trials witnessed by the entire body of Parisian notables could Étienne ask the controller general for reimbursement of his expenses. Pilâtre acquired instant fame, sharing the spotlight with Étienne, and easy access to the court society through Faujas's connections.[40]

The vertical rise of middling ascendants in full view of the nation's elite must have induced a sense of inversion in social hierarchy. A satirist substituted a balloon for the ascending Christ in Joseph Benoît Suvée's *Résurrection*, a painting exhibited in the 1783 salon at the Palais des Tuileries.[41] Whether anybody noticed the symbolism of a colorful paper balloon rising high above the grim walls of the Bastille is unknown.

INTREPIDS: Château de la Muette, November 21, 1783

Étienne might have relaxed a little, except for his family's endless array of "foolish or sensible, silly or serious" prompts to take advantage of the "sensation in the capital" and the "commotion in France."[42] Joseph dreamed of obtaining prodigious power through the use of steam and electricity to pump water. If nothing came out of the balloon venture, the abbé counseled, perhaps Étienne should approach the queen for a "simple and noble" favor to fund Joseph's engine project? He could not afford to be an "impractical" philosopher: modesty was a virtue that would make friends and invite him to dinners, but it could ruin one's estates. Étienne's silence induced the abbé to cajole, threaten, and preach in vain.

Étienne treaded cautiously. He had to maintain a respectable public image as a royal inventor, a nonexisting vocational category that defied the ancien régime social hierarchy. After the royal intendant Jacques de Flesselles (1730–1789) issued a public subscription on October 22, Étienne kept track of Joseph's project in Lyon, lest a blunder should crush the family's prospects.[43] Perhaps to secure his family's financial future, he delayed the public flight until after he read Joseph's memoir to the Academy's *rentrée* on November 15. Étienne already had permission to use the grounds of the Château de la Muette, the residence of the royal children. Their governess the duchess de Polignac had persuaded Louis XVI through Marie Antoinette.[44] Louis XVI suggested sending up condemned prisoners, but those vying to become the first aeronauts in history thought it a travesty to confer such momentous honor on beings so base.[45] The announcement on November 19 of Charles's plan for ascension hurried Étienne's pace.

On November 21, 1783, the crowd was smaller, although the flight was not kept a real secret. Franklin saw a "vast concourse of gentry," while Faujas noticed his nation—"the most superb" assembly of the fashionable elite. Étienne had bypassed public newspapers, Faujas would later explain, wishing to avoid "pompous announcements, which so humiliate the sciences, when the ignorant or the mediocre promise with certainty" what they could not guarantee.[46] It would not have been a royal wish, either, to draw a potentially riotous crowd to his children's residence.

At noon, a cannon shot announced the moment when they began to fill the balloon. Eight minutes later, the marquis d'Arlandes and Pilâtre de Rozier boarded the circular gallery. They intended to test the machine tethered, but the wind pushed it sideways toward an alley, tugging the ropes. Franklin feared for the men "in danger of being thrown out, or burnt," but the balloon was soon brought back to the stage using the ropes still attached. The envelope, torn in various places (one as long as six feet), had to be repaired on the spot. A few clusters of uneducated and ungrateful men began to murmur, according to Faujas, while all those "distinguished by rank or knowledge," including several ladies, got to work.

After an hour and a half, the Montgolfière finally took off "in the most majestic manner." The aeronauts lowered their hats to salute the spectators who became silent, overwhelmed with "a mixed sentiment of fear and admiration." Such intense emotion "at the moment when one saw this globe . . . rise little by little majestically into the air" was difficult to describe, as the *Grimm's Correspondance* relayed the scene to the European courts:

> The emotion, the surprise and the kind of anxiety caused by a spectacle so rare and so new was carried to the point that several ladies were taken ill when they saw our modern Titans pass by the hillside, first glide through [*planer*] the depth of the valley, subsequently rise to nearly 500 toises above the château, stop, and rise still more, to travel toward Paris, and disappear gradually. How can one paint this globe gliding over the city,

always at a height of about 4,000 feet: The people, who knew nothing of this experiment and did not know that this globe carried two men, filled the streets, running with cries of admiration which might have been converted into cries of fear if they could have guessed the audacious intrepidity of the two voyagers.[47]

The successful ascent sparked exuberant celebrations. A courier was dispatched to apprise the royal couple at Fontainebleau of this "memorable event." The multitudes in Paris saw the balloon pass over them.[48] Public newspapers began to report on this royal fête. It was important, the *Affiches, Annonces, et Avis divers, ou Journal général de France* extolled, that such an epoch-making experiment in the history of sciences was executed under the auspices of the dauphin and witnessed by those entrusted with the primary education of this prince to cultivate his taste for science. The "desire to please him, this vow of all French," gave the hearts of d'Arlandes and Pilâtre "force and intrepidity." The wings of Daedalus that had existed only in Horace and Ovid became "a truth that fills us with astonishment, *without making us fear the lot of Icarus.*"[49] The globe opened "the most surprising epoch since the car of the prophet Elias."[50] Embellished by Greek heroes and biblical prophets, the public balloon transcript was meant to naturalize Louis XVI's rule with scientific truth.

VIRTUAL FLIGHT

Human flight fundamentally changed the public experience of ballooning. No longer a frivolous spectacle, the balloon offered a veritable theater with the heroic navigators who could engage the audience in their sublime experience. Like the cosmic giant in Voltaire's *Micromegas* (1752), the aeronauts could offer a reciprocal view of the spectators who had become an object of pity from a safe distance.[51] In a letter written to Faujas ostensibly to "fix public opinion," the marquis d'Arlandes claimed his "natural" right as the first aeronaut in history to publish a first-person narrative. His account was carefully constructed to exhibit his knowledge of the landscape as a military captain trained in reading maps and handling crises. As if charting his course on a three-dimensional map, he narrated an accurate path of the flight that offered an aerial view of the landscape divided by the serpentine course of the Seine and dotted with familiar landmarks and neighborhoods. Flying between and above these landmarks, he and his "brave companion" had to tend the fire, extinguish sparks with wet sponges, and handle a tear in the envelope that threatened an impending peril. A heroic tale of adversity, purpose, and aestheticized leisure invited the reader to follow the aerial voyage with heightened sensibility. Such affective engagement of the spectator could no longer be reduced to numerical data. Nevertheless, the first-person narrative published in the *Journal de Paris* was excised of religious sentiments or a sense of the sublime.[52]

The marquis's seemingly transparent narrative that admitted his own shortcomings was culturally coded. It drew a sharp contrast between a noble spectator of Nature's grandeur, intent on relieving the sufferings of lesser human beings, and a subordinate commoner bent on making it through the task at hand. The marquis generously acknowledged the mere professor's capacity, courage, and intelligence that made him a suitable, if surprising, choice for the voyage. When a tear in the envelope threatened an impending peril, however, it was the marquis who sprang into action. Pilâtre merely kept the fire alive and maneuvered around windmills. The scene of landing was no less humiliating for him. While the marquis leaped out of the machine, Pilâtre supposedly crawled out from under the collapsed mass in his shirtsleeves, having cast off his redingote, which was torn to pieces by the peasants before the police arrived. Ostensibly worried about a chill, d'Arlandes sent him to the nearest house. When the "persons of distinction" arrived with carriages, the improperly garmented commoner stayed behind, allowing the marquis to receive a hero's welcome alone at the court. According to the *Mémoires secrets*, both aeronauts had emerged "as black as coal-sellers" from their flight.[53]

The distribution of symbolic capital became an issue once more, which caused Étienne publicly to defend Pilâtre's title as "the first aerial Argonaut." Pilâtre had the zeal, drive, courage, and intelligence to persuade the Academy for a manned flight, to defend it against ridicule, to try to build a machine on his own, and to master the art of maneuvering it. D'Arlandes had merely offered his assistance with constructing the launch site. With Étienne's steadfast support, the Academician Le Roy proposed a subscription to forge a gold medal for the two aerial voyagers, while others urged Faujas to open another subscription to erect a "historical monument" at their landing site.[54] In his turn, Pilâtre presented the Montgolfiers' medal (see figure 3.4) to the court on November 30, 1783. The ceremony at Duras's quarters became a perfect occasion to show off his knowledge, to solicit patronage, and to float his project of crossing the English Channel.[55] This camaraderie between the modest inventor and the heroic aeronaut would not last in the theatrical polity. The "fall of the favorite" was a routine mechanism in sustaining the king-machine.[56]

Spectacular human ascents entranced all French men and women, according to Linguet who espoused a radical reform while preserving the monarchy.[57] In his expatriate *Annales politiques, civiles et littéraires*, this self-appointed representative of public opinion rehashed the pros and cons swirl-

Figure 3.4. "Étienne and Joseph Montgolfier, born in Annonay, Vivarais: Collaborating inventors of an aerostatic globe." Tissandier Collection.

ing around the balloon in an attempt to validate its utility. The promise of great power at moderate cost made it an ideal masterpiece of mechanics. Rapid improvement in design promised many uses, particularly in opening an uncongested aerial passage that would eliminate forced labor (*corvée*) in road construction and repair. Now that the Montgolfiers had succeeded in maneuvering it vertically, every "Parisian Prometheus" should concentrate on steering it horizontally for a "true conquest of nature." Linguet could envisage a beautiful era when new chariots would instantly display their masters' status—eagles for the king and army generals, docile and courageous falcons for military men, diligent swallows for those working in the commerce and the arts, badly dressed crows for magistrates, and so on. On festival days, one would witness an aerial spectacle of a thousand cars in all forms and colors. Their seductive variety would please the spectators, unlike the ruthless carriages that endangered pedestrians and exacerbated plebian misery.[58] Linguet's enthusiasm captures how the promise of utility forged a hegemonic spectacle by enlisting a broad range of opinion makers across the political spectrum. The balloon nurtured their eclectic political imaginaries, a patchwork of republican and monarchial dreams that would somehow bring about happiness for all citizens.[59]

INTRIGUE, CALUMNY, AND HONOR

Charles was not far behind in his effort to humiliate the competition, but he had to fight his way through the intensifying libels and a royal prohibition. Seeing that "everybody was an accuser," Charles vowed privately to present to the nation a grand spectacle that would fix his reputation, an "experiment . . . of much greater importance than" all others.[60] He designed an intricate machine that would serve as the prototype of all subsequent hydrogen balloons. To prevent an explosion, he added a valve on top of the balloon to release hydrogen gradually without admitting the atmospheric air. A cord from the valve ran through the interior of the balloon to the gondola. An opening near the bottom would release the gas automatically when the internal pressure mounted. By jettisoning the ballast, the aeronaut could ascend at will until the balloon reached a desirable altitude and sail along without further effort.[61] The twenty-six-foot silk balloon was not soaked in but varnished with rubber to reduce the weight.[62]

The team also took measures to protect Charles's reputation. The endeavor was advertised as a sequel to the Champ-de-Mars experiment "under the direction & according to the *theories* of M. Charles."[63] The subscription, headed by the police chief Lenoir, was issued in the Roberts' name to raise ten thousand livres.[64] Notwithstanding these precautions, Charles boasted to anybody who would listen that the public clamored for his "taffeta bladder" a hundred times more than their previous support of the parlement during the Maupeou Revolution—a vainglorious claim that reveals his sense of insecurity.

Charles was not a master of word games to win over public opinion. He called Faujas a "double dealer" and the Versailles Montgolfière "the shit [*la belle cacade*]," an insult for which Blagden at the Royal Society of London could not find any decent English translation. Charles also mobilized friends and patrons, which produced Rivarol's diatribe against Faujas. As Blagden understood the situation from the duc de Chaulnes, a frequent visitor to the Royal Society, balloon enthusiasm or "madness" took "a turn more *characteristic of the nation*" to induce "a most violent party-spirit" that deployed ridicule and invectives "without mercy."[65]

Charles's impertinent protests raise a tantalizing possibility that he had a powerful patron working behind the scene. He obtained permission to construct his balloon at the grand salon of the Tuileries where the annual salon usually took place (see the building in figure I.1). The authorities did not welcome the request even from the "author of the first [Parisian] aerostatic balloon." If they accorded "one folly," the reasoning went, the demand would be endless. They nevertheless relented, citing that the governor of the Tuileries Garden, the marquis de Champcenetz (Louis-Edmond Quentin, 1760-1794), had already given permission for a launch there.[66] The sudden reversal points to an intervention, quite likely by the duc de Chartres. This populist prince, heir to the duc d'Orléans fortune, might have seen the hydrogen balloon as a way of promoting his public image. A scientific balloon would have served his family legacy of patronizing the arts and sciences. He was also the grand master of the Grand Orient since 1771. Freemasons regarded science as a kind of rational religion that would bring about a more egalitarian society. This princely agitation to appropriate the hydrogen balloon would not have pleased the queen's faction that Faujas catered to.[67] Neither the king nor the queen could compete with the prince in mobility or financial resources without appearing frivolous.

Unlike the royal Montgolfière that was made and flown outside of Paris, the new Charlière emerged as a public artifact at the city center. On display at the Tuileries from November 19, it enchanted even "the most ignorant." In addition to the twenty-six-foot balloon made of extremely smooth taffeta in orange and yellow, the spectacle included a wicker gondola and a small balloon made of green taffeta. Elegantly shaped, the gondola was covered with gold and blue wallpaper, wrapped by rolls of scalloped taffeta and tied by silk ribbons and gold garlands.

The distribution of tickets began on November 26 and attracted such a crowd that they had to call in the guards. Accusations of profiteering imme-

Figure 3.5. "The Golden Age." This is probably a satire of Charles's ascent from the Tuileries, judging from the background building. NASM A19680057000cp2.

diately surfaced in various newspapers since the "most brilliant" experiment by the original inventor, conducted just five days earlier, forecast that all Parisians would turn out for Charles's ascension whose success was now little in doubt. As usual, the underground *Mémoires secrets* turned most abusive: the "excessive" price notwithstanding, there was a "furor" to obtain the tickets. Given the size of the Tuileries Garden, the "singular favor" of staging an ascent there should yield an "enormous benefice" (see figure 3.5) for "Robert the devil."[68]

Public scrutiny and malicious gossip gave Charles's camp much grief, but they could only persevere. We are fortunate in that one of the collaborators has left a precious diary that captures the team's mood.[69] On November 27, the balloon was tied to the first tree in the main promenade of the Tuileries Garden, around which was formed an enclosure to serve as the workshop. Immense jars of sulfuric acid arrived from Javel the next day, as did barrels of iron filings. With the garden closed to the public, they began to produce hydrogen gas under icy conditions for an ascent scheduled for November 29. The Academician Balthazar Georges Sage (1740–1824) brought calculations for the best proportion of iron filings, sulfuric acid, and water that would reduce heat. Although they used an efficient design—a dozen barrels in a ring format (see figure 3.6), which allowed time for cooling and recharging each barrel after its turn—the operation still took a long time.

With the balloon less than a quarter full by the evening, Charles decided to postpone the ascension until Monday, December 1, and admitted the public to see the operation. Numerous guards surrounded the enclosure to push back overzealous spectators.[70] Sage wrote to the police chief Lenoir and the baron de Breteuil to report the necessary delay. Vicious rumors started immediately, doubting the team's technical ability or suspecting their profit motive. The following day, an accident nearly aborted the whole endeavor. One of the workers, M. Coustou, noticed that a lighted lamp had been left near one of the tubes carrying hydrogen. He jumped to remove it, which agitated the air and mixed the evolving hydrogen with the atmospheric air to cause an explosion. Although he was not seriously hurt, the accident brought back all apprehensions about safety and put the team on alert.

The most serious blow to Charles's struggle for public approbation came from the royal administration. A rumor began to circulate on November 30 that the king had issued an order prohibiting the aeronauts from ascending. Once confirmed, the news hit the collaborators like a "flash of lightning." The Academician Le Roy naïvely suggested suspending the work and asking for the Academy's approval, which was "universally rejected." When the collaborators went to see Charles at his residence, they received a "vivid impression" of his crushed spirit not knowing where to turn after so much

Figure 3.6. A ring of barrels for producing hydrogen. Gallica, BNF.

public scrutiny and injustice. He was also barred from publicizing the royal prohibition, which could only mean that the administration wished to cripple his reputation and honor. If Charles failed to keep his promise without a proper cause, it would have put him in permanent public disgrace. In other words, royal authority was invoked to commit an act of cruel calumny, "a kind of murder to attack the honor and reputation of someone, which are held to be dearer than one's life" in the ancien régime society.[71]

Regaining his composure with extreme difficulty, Charles urged the operators to fill the globe so that the public could at least see that it was ready for departure. Perhaps, they could find a way to save at least a part of their glory. During the night, however, Breteuil summoned him to issue an explicit prohibition. Crestfallen, Charles could only grumble: "who would believe them?" The sudden injunction, along with the secret manner in which it was delivered, hints at an intrigue by the queen's faction. Staging a publicly subscribed balloon at the heart of Paris could only invite trouble, either by upstaging the royal balloon in success or by causing a riot in failure.

HILARITY: Tuileries, December 1, 1783

The next morning, the globe was nearly full by 7:30 AM and the weather seemed perfect. When the aeronauts arrived, Sage fetched the police chief Lenoir with a plan to visit Breteuil and ask him to rescind the "frightful order, if it were possible." Breteuil evaded the question, alluding to a mysterious letter sent to Champcenetz, not seen by anybody but certainly existing, which could lift the king's order. While Charles struggled with intense anguish, several persons came to talk over the situation. Some suggested going to Versailles to "pierce these impenetrable veils," but there was no time to make the journey. The duc de Chartres eagerly offered to purchase the machine and reimburse the public, but Charles refused, knowing that he would still lose his reputation. Frustration and anxiety reigned at the launch site with the Montgolfists' bad-mouthing, the fear about safety mounting, and the rumor of royal interdiction casting a serious doubt on the success of the flight. By 10 AM, people began to arrive in numbers. The rumor that nobody was going to ride in the machine caused significant turmoil. Murmurs rose among the spectators that they had been duped: the Roberts and Charles were much more skilled in stealing from their wallets, playing on the word *voler* that means both to fly and to steal, than the Montgolfiers who only knew how to fly in the air.[72]

The crowd was truly immense. More than half the Parisian population had converged on the Tuileries. Soldiers greeted them at the gates, admitting those with valid tickets and turning the others away. They also patrolled the garden walls and surrounding avenues to keep the undesirables away (see figure 4.1). At the center of the basin in the garden stood a wooden stage surrounded by water and a semicircular amphitheater with the seats reserved for the dignitaries—princes, ministers, distinguished foreigners, academicians, and the one hundred subscribers who had paid four louis each for the privilege. Soldiers surrounded this inner enclosure, erecting a barrier between the dignitaries and the crowd.

As the morning progressed, the park overflowed with over ten thousand persons who had paid three livres each for admission. Spectators ballooned outside the garden walls, jamming the winding streets, packing the windows and balconies of nearby buildings, perched on rooftops, and lining both sides of the Seine from the Pont-Neuf to the Hôtel des Invalides. Englishwoman Ann Francesca Cradock and her party had to leave their carriage and force their way across the Pont-Royal. The Duke of Cumberland, younger brother of George III and the newly minted Grand Master of the

Premier Grand Lodge of England, nearly suffocated in a similar march.[73] As Benjamin Franklin reported, "never before was a philosophical experiment so magnificently attended."[74] Nor had any royal procession, public ceremony, or festival seen such a massive gathering of people.

As the ever-increasing crowd waited restlessly for the marvelous sight that had been the focus of Parisian gossip since August, the fog lifted to unveil a sunny day.[75] After a round of cannon shots, the operators led the balloon to the basin in the midst of universal applause. With another round of cannon shots, Étienne Montgolfier released the small balloon as a public testimonial of the two camps' reconciliation. It rose vertically and vanished as a small dot in about fifteen minutes. A "simple allegory," as Charles explained to the public, to cool the inflamed rivalry.[76] It was also designed to test the wind direction.[77]

The crowd applauded and then waited, ever more impatiently, to see the large balloon that would make a "majestic and imposing" sight. While they prepared the main balloon, rumors magnified the royal prohibition to harden Charles's resolve. Surely, even the king could not countermand the will of such an immense crowd? If not, at least the police chief Lenoir would understand the potentially explosive situation, should he fail to fulfill his promise? The aeronauts resolved in the end to disobey "the order that would cover them with shame in the eyes of four hundred thousand witnesses who already accused them of having solicited" the royal prohibition.[78]

The audacity of Charles's decision—to defy what was conveyed to him as a royal command—speaks for the power of the crowd at the scene. He did not have a choice if he wished to salvage his honor and livelihood. He thus "interpreted" the prohibition as "nothing but a paternal counsel."[79] The possibility of a riot gave him a reasonable excuse to keep his word to the public. At 1:30 PM, the aeronauts stepped into the gondola that had been generously provisioned for a long flight.[80] Friends and family embraced younger Robert tenderly and bid him farewell, uncertain of his safe return. Lest the public should interpret this as their anticipation of an impending danger, Charles ordered the embrace to be cut short. In a show of gallantry, the two aeronauts opened a bottle of champagne and saluted the crowd.

When the collaborators released the restraining cords, the melon-shaped globe went straight up, shining in the sun. A deep silence befell the crowd. Many women became ill. Only when the aeronauts waved pennants, one red and the other white, did the crowd come back to "vivid sensation and applause." Even the royal Swiss guards joined in this universal cheer by throwing their swords into the air. The almost religious rapture and subsequent outburst of emotions in the crowd—joy, fear, terror, and enthusiasm —have come down to us in vivid narratives.[81] Louis-Sebastien Mercier, a critical spirit who chronicled Parisian life with a sense of irony, inscribed this intense moment of veneration with a rare hint of enthusiasm:

> This swarm of people was in itself an incomparable sight, so varied was it, so vast and so changing. Two hundred thousand men, lifting their hands in wonder, admiring, glad, astonished; some in tears for fear the intrepid physicists should come to harm, some on their knees overcome with emotion, but all following the aeronauts in spirit, while these latter, unmoved, saluted, dipping their flags above our heads; what with the novelty, the dignity of the experiment; the unclouded sun, welcoming as it were the travelers to his own element; the attitude of the two men themselves sailing into the blue, while below their fellow-citizens prayed and feared for their safety; and lastly the balloon itself, superb in the sunlight, whirling aloft like a planet, or the chariot of some weather-god—it was a moment which can never be repeated, the most astounding achievement the science of physics has yet given to the world.[82]

As the globe rose in "a silence concentrated by emotion and surprise," Charles felt not pleasure but happiness and a "moral sentiment" of "hilarity," free from the "frightful torments of persecution and calumny." Happiness in Enlightenment sensibility was a moral sentiment that tuned personal pursuits and desires to the public good. In order to present the greatest spectacle to the nation, he had remained silent for three months while enduring unreasonable hatred and prejudice. Even the spectators who witnessed the utmost geometrical precision and expert execution in making the globe still raised a "thousand puerile objections" to its safety, which inspired terror in others; all was passion, blind to reason; an accident nearly aborted the project; then came the ministerial order to stop the experiment at the potential cost of his reputation and honor to cause a bottomless despair; his detractors even accused him of having solicited the order at the halls of Versailles, presumably for personal profit without risking his life, and so on.[83]

Charles could do nothing else but take off, convinced of its success and cognizant of public interest. All troubles and emotional toll were now below him. Much like the cosmic spectator in Lucian, Cyrano de Bergerac, and Voltaire, Charles could look down on the spectators with a sense of pity for their earthbound existence. This "moral sentiment" of hilarity soon gave way to an even more vivid sensation—an admiration for the "majestic spectacle" presented by a mass of heads below and a clear sky before him. When the globe reached a desired height measured by the barometer, Charles kept it stationary until it floated out of the spectators' view in about fifty-six minutes, which was announced by a cannon shot.

The hydrogen balloon offered a "mobile theater" that allowed the aeronauts to interact with spectators.[84] They would shout "Long Live the King!" and the entire countryside would roar with cries of joy, solicitude, alarm, and admiration: "My good friends, do you not have fear! Are you not sick! God, that is beautiful. We pray to God to preserve you!" Touched by such "tender & true interest," the aeronauts waved flags and asked where they were, what time it was, and about local dignitaries such as the prince de Conti. Weaving a path across the Seine, they passed several landmarks and villages to arrive at Nesle, about nine leagues from Paris, after three and a half hours of flight. The duc de Chartres greeted them, having followed the balloon on horseback with a hundred chevaliers and his Masonic friends, the duc de Fitz-James (Charles William, 1712–1787) and the Duke of Cumberland (Henry Frederick, 1745–1790). A spontaneous festival broke out among the peasants who ran after the balloon like "the children chasing butterflies."[85]

Charles promised to return in half an hour and soared alone into the sky (see figure 3.7). The globe had lost too much inflammable air to carry two passengers. In order to read both the thermometer and the barometer placed at opposite ends without tipping the gallery, he had to kneel down at the center and stretch his body and arms to the front, holding in his left hand a watch and paper, and in his right hand, a pen and the valve string. Rising to a height of at least fifteen hundred toises in ten minutes, he suddenly went from spring to winter temperature, which gave him a sensation of very dry, unbearable cold.[86] The sun reappeared to bathe the balloon once more in a dark aerial plane. Soon, the moon rose to illuminate the clouds rising from below. These "vast phantoms crept up slowly from all sides" as if to contemplate the "new inhabitant of the space." Charles thought about steering the globe by stirring the air, as if "to disturb this imposing silence that reigned all around." A sudden pain in his ears awoke him from this "inexpressible rapture and contemplative ecstasy." He descended after traversing a league and a half in thirty-five minutes. The younger Robert met him with M. Farrar who offered to lodge these "sky-men."[87]

THE APOTHEOSIS OF A PHILOSOPHER-VOYAGER

The aeronauts left the next morning and walked on foot to Beauvais where they caught a stagecoach to Paris. Notwithstanding the inglorious ground transport, Charles and Robert made a triumphant Parisian entrée to command "public veneration." When they arrived at the Place des Victoires in the evening, an immense crowd lifted them into the air. Fishwives showered them with "compliments, bouquets, & the laurels of this first body of people." Musicians played and everyone shouted "Long live Charles!"[88] All adverse sentiments changed suddenly into admiration. After some rest, Charles

Figure 3.7. Charles rising from Nesle in the presence of the duc de Chartres and his Masonic friends, the duc de Fitz-James and the English gentleman Mr. Farer. Tissandier Collection.

went over to the Palais Royal with the marquis de Lafayette, the folk hero of the American War, to thank the duc de Chartres. The pair must have made an indelible impression on the crowd, although no public account relayed their joint appearance. When the emptied globe arrived around 8 PM with the "car of triumph," a procession of peasants and musicians with cockades accompanying the "chariot" traversed Paris by torchlight.[89]

Human ascension from the heart of Paris brought this long-fermenting city to a boiling point, but the exuberant celebration on the street did not make public news. The *Journal de Paris* only declared that "the nation should be proud of" the discovery that offered such a "majestic and imposing" spectacle. It printed a brief procès-verbal without details, ostensibly not wishing to yield to "public curiosity," and decided to wait for the mathematician Meusnier's report commissioned by the Academy.[90] The disparity between street celebrations and the journalistic public transcript points to the social engineering that supported a façade of scientific hegemony.

Public demand soon broke the media reserve. The *Journal de Paris* printed a part of Charles's account delivered to his fee-paying students, albeit excising his claim as the theoretical inventor of the aerostatic globe.[91] Linguet's *Annales politiques* reprinted it as a personal journal potentially as valuable as

TRUE COLUMBUS | 97

that of Columbus's voyage that would have conveyed his thoughts, hopes, and fears in creating a new world! The public could at least read Charles's narrative, now an object of "universal curiosity." Linguet nevertheless qualified Charles's voyage as the second to that of "the true Columbus" Pilâtre and d'Arlandes, in tune with his royalist sympathies.[92] In doing so, he ignored the philosophical (scientific) requirement for the title, which was necessasry to secure public approbation.

More gratifying to Charles was the Academy's reception. On December 3, 1783, Sage introduced him to the general assembly where he was seated, along with Étienne Montgolfier, next to the president Bochart de Saron. The favorable reception, in part prepared by Meusnier's detailed calculations, gave Charles "the most complete satisfaction."[93] Although the motion to name the two inventors as extra associates went nowhere, Saron solemnly presented them with jettons to emphasize the extraordinary honor of the occasion. The Academy also made their respect for Charles's accomplishment known through the *Journal de Paris*: "His account finished, the Academy responded with the most vigorous applause, a testimony of the admiration that is rare to obtain in this sanctuary of sciences where one is enthusiastic only for the works of genius." The royal Academy of Sciences seemed to embrace this folk hero without reservation. The *Journal* also sought to "fix the opinion" on the mysterious royal prohibition on the day of ascent, attributing it to the "goodness, paternal tenderness of His Majesty" to prevent harm to his valuable subjects.[94]

Behind closed doors, however, the royal administration maneuvered to control public enthusiasm and the artifact that was overshadowing the peace celebrations. On December 6, 1783, Breteuil conveyed the express royal wish that further research by the Academy be conducted in collaboration with Montgolfier and Charles "to determine a better form of the machine, the most convenient material for its construction, and the means of producing the effect with as much simplicity, economy, and certainty as possible." The Academy might divert its annual budget of twelve thousand livres for new discoveries. Although this move would have brought the popular invention under their control, the Academicians who were engaged in various projects with this fund had to find other means of honoring the royal wish.[95] They decided to grant the prize of six hundred livres and the status of correspondents to the Montgolfiers, emphasizing that this special election was an exceptional distinction to honor the light (*lumière*) that their invention brought to their patrie.[96] They protested at the same time that including the inventors in their research would violate the statutes. The academic threshold proved difficult to cross. Breteuil had to modify his demand and promised on December 12 that the king would support with additional funds the Academy's zeal to render such an important discovery useful and to perfect it by consulting or supervising the inventors.[97] In their official report on December 23, the Academy's balloon commission issued a detailed report that recounted the Montgolfiers' and Charles's experiments without favoring either one.[98]

The heroic ascent set the public mood for the peace celebrations on December 14 with illumination, fireworks, and the distribution of food all over the city.[99] The "vivid sensation" among "all orders of citizens" made it necessary to honor the men who had offered "the most beautiful spectacle of the universe to the nation." Louis XVI was apprised that "the invention of the aerostatic machine was one of the most brilliant & useful discoveries that enlarged" human domain and that it suited the "grandeur & dignity of the Majesty who loves & protects the sciences" to reward the inventor and other emulators.[100] He dispensed honors and pensions—the *lettres patentes* conferring nobility on Pierre Montgolfier, the title of royal manufactory to his mill, the Cordon de Saint-Michel for Étienne Montgolfier, two thousand livres in pension to Charles, and one thousand each to the Robert brothers and Pilâtre (who apparently complained that he deserved at least as much as Charles).[101] The marquis d'Arlandes could only be compensated by a military rank.

Louis XVI refused the Cordon de St. Michel to Charles, ostensibly to distinguish the original inventor from the one who merely perfected the discovery.[102] Charles's rivalry against the presumed royal inventors must have played a role, even if the king could not publicly reprimand the folk hero. Projects for monuments and medals were afloat. An order to strike a medal went to Breteuil and another, to erect a monument at the Jardin des Tuileries, to the comte d'Angiviller.[103] Since the discovery marked the "highest honor to the human spirit," it was necessary "to preserve for posterity the memory of this discovery by a public monument."[104] Rumors flew around town on consultations and commissions involving various academies. Soon, the ideas of a column or an obelisk circulated in the public, along with inscriptions that celebrated "two mortals or two Gods."[105]

The apotheosis of Charles configured a popular philosopher-voyager with the "requisite personal qualities" of a scientific explorer—"a philosophic attitude, courage, and veracity," as Diderot had expected of Bougainville.[106] The aeronauts combined universal knowledge and superhuman courage to demonstrate an impossible hope for the populace—to rise above all constraints and obstacles and to open a new vista that promised infinite progress by means of science. High up above the atmosphere, as Diderot had hoped, a modest man could be free to enjoy the beautiful lands below where the arts and sciences were flourishing. Floating above the arrogant tyrant, he vowed to "shed tears for" persecuted genius, forgotten talent, and unrewarded virtue and "pour insult and shame on those who deceive and oppress men."[107]

Despite the initial treatment of Charles as a money-hungry and attention-grabbing imposter, poets rushed their pens to celebrate this human miracle: Charles and the Roberts were promised fame, immortality, and esteem rivaling those of the Montgolfiers, d'Arlandes, and Pilâtre, along with a "place in the heavens" or one "between men and gods."[108] Dubbed "the god of the air," each of these "great patriotic men" was celebrated for the glory their flight brought to the nation by opening the "azure empire" in their "car of god."[109] Adulation flowed unfiltered to Charles. Just as the works of Shakespeare, Pascal, Corneille, Racine, Montesquieu, Voltaire, and Raynal had been translated into nearly all languages, the architect Bélanger (working for the comte d'Artois) extolled, Charles's work would be imitated in all countries as a "scientific and ingenious discovery."[110] To some women, Charles became the object of all their thoughts and the hero of their heart.[111]

The memory of Charles's heroic voyage would linger (see figure 3.1), but the peace he desired did not come.[112] As the hundred-foot Lyon balloon began to circulate in rumors, the "great controversy" also continued, dividing all French amateurs into the "two sects" of Montgolfists and Charlists.[113] As if he wished to appropriate Charles's accomplishments for the queen, the comte d'Artois commissioned Léonard Alban and Jean Baptist Vallet at the Javel acid manufactory to make a hydrogen balloon that could be steered.[114] Charles returned to public lecturing, never to make a balloon or to fly again.

Ironically, Charles's heroic voyage enhanced the Montgolfiers' glory. Dubbed "the French Daedalus, the conqueror of the air," Étienne received a flood of letters from amateurs, imitators, inventors, and courtiers with their variegated schemes.[115] Many Parisian socialites and prestigious Masonic lodges sent invitations to Étienne, the inventor who made "all nations envious of France."[116] Others followed suit, wishing to use the Montgolfier name as their social capital.[117] Sincere homage came mostly from the underemployed professionals of intermediate ranks—military personnel, ship's captains, idle abbés, amateur scientists of indistinguishable heritage, and so on—who summoned their utmost *politesse* to compliment the inventor and to get into the circle of his glow. Étienne's glory offered deep consolation to their frustrations and a fleeting hope for an escape from their ordinary lives.

The balloon as a mobile theater opened an aerial vista free of physical, social, economic, and political constraints. Despite the hierarchical seating on the ground, its spectacular ascent ruptured the chain of gravity, highlighting the spectators' earthbound existence and servitude to the current order, perhaps to float an illusion of a technoscientific utopia that would disrupt their social bondage. By commanding the natural power of invisible fluids, the aerial Columbus conjured up the irrepressible power of the people as a mass collective. As Rivarol expressed in a sentimental mode, aerial travel broke the "imperious law of gravitation that pushes man ceaselessly to the surface of the earth, forces him to crawl on it, subjects him to such a painful life," and binds him as a "terrestrial animal." It offered to the sad, "weak and unfortunate creature" a "pleasant illusion," if only for a moment, that he could see once more "a lost patrie" and recover something "aerial and celestial."[118] Mass hope engendered a liminal realm of infinite possibilities, which could destabilize the society of orders.

More importantly, the balloon as a scientific artifact promised to erase the destructive past for a peaceful empire. Notwithstanding the British military interpretation, the French public transcript focused on the balloon's philosophical ambition and scientific capacity, which fostered the desire for an ideal polity governed by a philosophical majesty—be it an armchair advocate of liberal polity, a daring philosophical voyager who could traverse the universe and consolidate earthly compatriots, or a prince who might ameliorate the abusive system of privileges through the application of scientific principles. The imagined nation of universal citizens called for a philosophical majesty that would cater to their happiness above all else. The contestation to define who might become the most benevolent sovereign would shape competing visions of the imagined republican polity.

Figure II.1. Frontispiece to [Louis Guillaume de La Folie], *Le Philosophe sans prétention* (1775), depicts an imagined electrical machine that brought the visitor Ormasis from the planet Mercury. By turning the two glass globes of three feet in diameter with prodigious speed, while the percussion of light agitates the air underneath the machine, he could cancel out the atmospheric pressure to rise or descend at will.

PART II

PHILOSOPHICAL NATION

The aerostatic globe, as an "astonishing" scientific invention, honored the French nation.[1] This miraculous discovery excited restless minds chasing the vain phantom of immortality. It sensitized even those numbed by the political, moral, and scientific wonders of "the century of philosophy"—such as the deflation of prejudices and superstition, the abolition of slavery, the creation of the great American empire, the suppression of the Jesuits, Catholic divorce, and the rapid elevation of the Russian empire. No other period or nation could have nurtured such a marvelous invention, according to an enthusiastic pamphleteer, but for the century in which one made daily use of the pen and but for France where everybody hustled and bustled to invent something new.[2]

A philosophical nation would require noble citizens who, in their quest for an egalitarian polity, would inadvertently build a scientific empire. Equally astonishing to the vulgar and to the learned, this brilliant "flying observatory" opened a new epoch in "the empire of sciences."[3] The populist Mercier pontificated that the "Assumption" of a living being should command "universal admiration." France only needed more "citizens distinguished by their nobility and courage" and wishing to improve the balloon for their patrie and humanity.[4] If the balloon functioned as a national artifact, the imagined "nation" became a moving target with an elastic boundary.

In other words, the balloon opened a liminal moment of competing possibilities by demonstrating the power of (natural) philosophy in integrating the cultural nation-empire. The imagined philosophical nation became a powerful rhetorical resource for the politiques who sought to transform the populace into legitimate citizens. In contrast to the king-machine, a "terrible machine" Diderot wished to dismantle, the balloon ascent gathered citizens indiscriminately around a venerated scientific artifact and a philosophical voyager to constitute a people-machine that included the populace.[5]

An inclusive philosophical nation would push the limits of both the republican imaginaries and the absolutist polity. In his philosophical history of the Roman Empire, Montesquieu had attributed its decline to the geographical expansion that compromised republican practices, augmented military power, and implemented political slavery. The republic was lost in the empire. The absolutist polity, which excluded the populace and eliminated mediating powers, could enslave the citizens by introducing similar techniques of government. In *Lettres persanes* (1721), the harem functions as a metaphor of the despotic regime that reduces the citizens to comfortable slaves without any capacity for self-governance or development. Montesquieu warned explicitly in *The Spirit of the Laws* (1748) that republican principles were incompatible with despotic power. For the inheritors of classical republicanism, a tolerable path toward a republican empire was difficult to conceive, notwithstanding the variations on the theme.[6]

In articulating an ideal civil society in which the citizen would become "truly the master of himself," Rousseau illustrated the power of moral imagination to spur a broad spectrum of oppositional praxis under the republican banner. His republicanism offered a mélange of incompatible longings and prescriptions, but he saw public education, public spectacles, religion, and citizen militia as possible ways to institute virtue and justice.[7] The republican imaginaries of the Rousseauean generation insisted less on the egalitarian principle of popular sovereignty than on the moral integrity of sovereign magistrates.[8] A philosopher-king had to guarantee the citizens' happiness, a task traditionally allocated to the virtuous Christian sovereign, as seen in Fénelon's *Telemachus* (1699), which remained popular throughout the Enlightenment. This hybrid republicanism—one that melded the ancient republican polity and Christian values—produced diverse political dreams, fictions, and discourse in the public sphere.

The enlightened "cult of the nation" was nurtured by the illimitable trust in reason and Nature as well as republican sensibility, as can be discerned from the utopian fictions of the preceding decades.[9] Nature guaranteed common sense and reasoning capacity of the wise, just, powerful, enlightened, virtuous, and happy citizens in *La République des Philosophes* (1768), attributed to Bernard le Bovier de Fontenelle. This republic of the philosophes is a nation of purest mores, just laws, and perfect tranquility. The fertile island of Ajaoiens supplies all the needs of its inhabitants who enjoy the most pleasant and happy life imaginable. A "Sovereign Magistrate," who maintains a simple residence stripped of luxury, rules by just and wise law. The upward organization of republican villages ensures justice for all inhabitants, including the slaves. The Ajaoiens gather annually for a military exercise and festival at their champ de Mars and debate matters of shared concern in a general assembly.[10]

The "pure light of reason" would foster a free, egalitarian, and just society in Louis-Sébastien Mercier's *L'An 2440* (1771), an underground best seller almost immediately translated as *Memoirs of the Year 2500*. He called for a bloodless revolution by a philosopher-king, which would secure "the sacred character of a free citizen," equality, and justice.[11] Mercier's rational government is not a monarchy, an aristocracy, or a democracy. Seeing how contrary an absolute sovereignty is to the true interests of the nation, a philosopher-king leads a revolution to secure provincial autonomy, universal suffrage and jurisprudence, and the citizens' happiness. He restores the ancient estates for the legislative function and the senate for political and civil affairs. As a nominal king, he watches over the execution of the laws. He walks on foot among his people to observe natural equality and to learn from respected citizens. To ensure legitimate succession, the dauphin is brought up until age twenty as a commoner. All citizens are equal, except for their natural differences in virtue, genius, and industry. They endeavor to reconcile the good of the nation with that of individuals. Versailles lies in ruins.

Republican imaginaries in late Enlightenment deviated from the classical republicanism that idealized a small republic to accommodate the scale of modern politics that would include the populace. In an ideal polity where philosophers make the laws, the people's mores and happiness would become their central concern. The "invincible desire for happiness" was a part of the human soul, as the abbé de Mably asserted in *Des droits et des devoirs du citoyen* (1789), which justified a political revolution against an unjust government.[12] Mercier would insist that the spectacles "must be for the *people*, repeat ceaselessly this word *people*, reprimand us for scorning and ignoring the *people*." He wielded "the idiom of the populace," as the Academician Jean-François de La Harpe (1739–1803) worried, as if he wished to "turn the world upside down."[13]

Republican polity à la Rousseau demanded an entirely different form of spectacle from the royal theater. He preached that a republic did not need the theatrical performances (*spectacles*) that kept its citizens fearful and immobile in silence and inaction. Public festivals, "staged by and for the people themselves" in the open air, would combine pleasure and utility.[14] The flying machine would have been the perfect vehicle of mass Enlightenment in an effort to build a republican nation-empire, if we follow Rétif de la Bretonne's reverie in *La Découverte australe* (1781). A philosophical empire that would insist on the citizens' liberty and equality was a beautiful dream, which would entail in practice a strong monarchy and coercive moral compliance. The sanction against the enemies of the public good was a moral imperative, as Mercier envisioned for his utopian Paris. The monument of "sacred Humanity" would demand penitence from all nations that had committed crimes against humanity.[15]

As a people-machine, the balloon projected a nation of citizens that would build a moral empire. Its patchy body rising against severe natural elements floated a fragile dream of an egalitarian polity built with scientific prowess. Balloon gatherings could project an imagined republic stitched together by genuine sentiments à la Rousseau. Such hopeful citizens longed for a "philosophical majesty"—an enlightened prince who would cater to their happiness above all else—although who actually might qualify as one was a matter of contention.[16] Masonic princes such as the duc de Chartres dreamed of becoming a philosophical majesty that would tame the absolutist polity with republican principles and paternalistic care, but the malcontents claimed the moral liberty à la Rousseau to imagine alternative polities.

If the balloon became a national artifact by forging a broad consensus in public opinion, in other words, it helped change the scale of the imagined nation across the sociopolitical spectrum. It engendered a mass public that would populate an alternative nation-empire, which made it more urgent to define what the French "nation" was or could be.[17] The power elite would have hoped that the balloon would induce the citizens' voluntary submission to the royal authority and implement a regime of scientific hegemony. A repertoire of public and royal resources helped stabilize its public performance: royal emblems, parks, and police, along with standardized newspaper reports, forged a euphoric consensus on its scientific promise and public utility. The public complied with this hegemonic process out of their rational submission to science's authority and power.

A peaceful scientific nation-empire would require a philosophical majesty and virtuous citizens. The balloon's fall from grace in public opinion closed this promising path to a philosophical nation, however. The aeronauts' inability to steer the balloon against the wind exposed the limits of the royal administration and the scientific elite in controlling its political and material utility. As is well known, other events soon began to dominate the public sphere—such as the affaire Le Maitre in 1785, and the Diamond necklace affair in 1786.[18]

The fight for and against the revolution would turn on the cultural authority and memory of the philosophes. Would a "happy revolution," carried out according to their maxim, bring about the true happiness of citizens? Or would it lead to a reckless pursuit of passions and pleasures?[19] The revolutionary outcome would be shaped not just by philosophical ideas but also by a repertoire of social technologies that controlled the public sphere and popular culture. The revolutionary government would mobilize the taken-for-granted hostility toward the enemies of the public good to institute the Terror with a new scientific machine—the guillotine.[20] Insofar as historians limit their attention to the ideological, financial, and political causes of the French Revolution, the nature of this continuity in social engineering is likely to escape intelligible characterization.[21]

Figure 4.1. "Dedicated to the Amateurs of natural philosophy. Experiment of the aerostatic globe of M.M. Charles and Robert at the Tuileries, December 1, 1783." Gallica, BNF.

4

BALLOON SPECTATORS

The **aerostatic globe** as a national emblem was expected to forge a new theatrical relation—voluntary submission rather than formal domination—with the emergent citizenry. To this end, public newspapers reported the ballon ascent as an orderly gathering that focused the people's veneration on the scientific artifact. Such idealization served to establish the model *balloon spectator*. To be seen as a mass public, the journalistic public transcript seems to tell us, the balloon crowd had to behave according to the prescribed rules of the theater—offering silence or applause to express their rapture or approval, but no talking or crying out to voice their discontent. The balloon spectator would cultivate an aspiration for the universal values expressed in numbers and embodied in useful artifacts as well as an aesthetic appreciation of Nature's beauty and grandeur. He would be a "citizen of the universe."[1]

In order to fulfill its hegemonic function, the journalistic balloon spectator had to play a neat trick—assume the existing form of the spectator who could pass absolute judgment in the guise of public opinion, yet warp their critical voice into that of a royalist citizen who would voluntarily submit to reason and science for the public good.[2] This metamorphosis of the spectator performed serious political work by suppressing or concealing dissident subcultures and malcontents. The journalistic public transcript inscribed an emergent system of scientific hegemony—a cultural domain of material appropriation, public mastery, nonviolent subordination, and nonideological legitimation of the reigning polity in the name of science.[3] The rational, aesthetically disciplined balloon spectator, ever mindful of the public good, would populate a philosophical nation-empire that would include the entire humanity and, as a corollary, the populace.

A critical history of ballooning that probes its material politics has to reach beyond public newspapers, therefore, to uncover the hidden transcripts that addressed diverse audiences and to interpret the silence of the people in the public transcript.[4] By configuring a genealogy of the spectator in the Enlightenment public sphere, we can examine its relationship to the public and hidden transcripts on ballooning. For the historic event that triggered an avalanche of public and private records, the archive is overflowing with the documents that have shaped the triumphalist narrative of the aerial conquest. While it is not possible to recover the voices of the "people" from the archive, we can gauge the balloon's political agency as a people-machine from the dramatic disparity between the public transcript in newspapers and the hidden transcripts produced for diverse, less visible audiences (as illustrated, for example, in the stark contrast between the images of the Versailles ascent in figure I.1 and Charles's in figure 4.1).[5]

As French historians have noted astutely, pamphlet production in the 1780s was less an exclusive domain of malcontent writers than a contentious field of factional politics that aimed at controlling public opinion—absolutist politics by other means than outright suppression, as we have seen with the Montgolfists and the Charlists.[6] The fact that pamphlet literature has been written out of history affirms the strength of the public transcript on ballooning, especially since these pamphlets could have circulated more broadly than our trusted sources. While *Grimm's Correspondance* is a familiar sight in our libraries, it was produced for the exclusive elite in Enlightenment Europe and had only a tiny fraction of the subscribers of the *Journal de Paris* who numbered several thousands. We have no figures on how many pamphlets and loose sheets of balloon ephemera were sold, but we can make an educated guess. Sarah Maza estimates that the press-run of a popular judicial brief reached twenty thousand in the 1780s. Each copy would have been read by several persons and often to illiterate people. Mercier claimed that the advertisement of a famous street performer selling remedies would reach one hundred thousand people.[7] Enlightenment Paris had developed an intricate infrastructure of distributing news, including public readings for illiterate people.[8]

A GENEALOGY OF THE BALLOON SPECTATOR

A genealogy of the spectator, a textual figure that executed the critical gaze from an assumed distance, can help elucidate the manner in which this critical figure was modified to serve scientific hegemony. We have already seen how the figure of the cosmic spectator—an imagined critic of the home civilization from a cosmic distance—carried the Lucianic gaze down to the early modern period. In the Sun King's court, La Bruyère incarnated this critical gaze by offering a moralistic critique of the court from a deliberate distance (as discussed in the introduction). His function as a reflexive critic would be enhanced by the literary device of foreign spectators during the Regency. The "barbarian" Usbek in Montesquieu's *Lettres persanes* (1721) and a detached Swiss in Béat-Louis de Muralt's *Lettres sur les Anglois et les François* (1725) both could claim an authentic distance from the French civilization to expose the systematic corruption and its negative impact on citizens' happiness.[9]

Pierre de Marivaux (1688-1763) in his turn fashioned himself as a "modern Theophrastus"—a freethinker (*esprit libertin*) footloose in Paris. In his dispassionate observation and description, Parisians ruled their little corners of the world "sovereignly" with "despotic liberty" in language and action.[10] In *Le Spectateur français* (1721–1724), Marivaux cultivates the posture of a deliberate outsider who criticizes the artifice and false appearances demeaning to the natural man endowed with sincere sentiments. His posture of disinterested selections, random observations, and chance encounters underscores the prevalence of similar experiences. By presenting the miseries of the poor, dispossessed, and displaced along with the fantasies, dreams, and pleasures of the powerful and idle, Marivaux attuned his readers to the arbitrary and capricious injustice perpetrated with casual impunity by the political or commercial elite. The pompous entrée of the Infante prompts a poor cobbler, a "subaltern philosopher" or "brute Socrates," to pity the crowd's curiosity. The spectacle is "too beautiful for little men" and would only sadden them to lose interest in their vocation.[11]

The textual figure of the spectator continued to evolve in tandem with the critical public sphere during the Louis XV's reign, as Suzan Pucci has argued persuasively, which consolidated the public who could assert the individual's right to judge and thereby to function as an informed citizen.[12] It would be tedious to recount the contestatory political culture of the mid-eighteenth century, mostly around the parlement and its pamphlet warfare against the Crown. A major consequence by the 1770s was that the critical figure of the urban spectator was absorbed into the rhetorical figure of public opinion or an all-knowing critic in service of the public good. Public opinion as the absolute tribunal exceeded the authority of the king or ministers and changed how the absolutist politics was played. Pamphlets emerged as an important tool of opinion making in the incessant struggle to control public opinion.[13]

In this milieu, the balloon offered an almost miraculous tool of engineering consensual public opinion, which could be mobilized as a powerful advocate of the ancien régime technocracy. To this end, the journalistic balloon spectator had to fuse the absolutist "look of power" and public opinion and express the hybrid perspective in the voice of a neutral (journalistic) observer—one who stayed above the fray and thereby served the public good and humanity. The task of evaluating balloon documents and images becomes more complicated if the historian regards the balloon ascension as an event that served social imperialism by projecting a harmonious nation and a glorious empire.[14] Just as travel literature fashioned the narratives of anticonquest that naturalized Euro-imperialism, balloon literature generated public transcripts to facilitate internal colonization.[15] In turn, resistance was fashioned in the very Enlightenment ideals propagated by the educated elite—utility and the public good.

In other words, balloon ascension can be analyzed as a "contact zone" between the political, commercial, and educated elite and the lowly people (*menu peuple*), a space of inward colonization and integration that stabilized an outward discourse of the aerial empire. Contact zones engender not just contrasting discourses of the dominant and the subordinated, but many layers of discourse in between. Subordinated groups can "select and invent from materials transmitted to them by a dominant or metropolitan culture" to create the "autoethnographies" (in Louise Pratt's terminology) that reflect their standpoints vis-à-vis the dominant group.[16] Dominant groups could also modify their standpoints and praxis to win over the subordinated groups. Such hybridization of standpoints and perspectives, or a process of transculturation, makes it difficult for the historian to maintain a fixed analytic gaze to classify, objectify, and describe past events and peoples in the Rankean fashion that has dominated American historiography.[17]

Pamphlet literature functioned as a genre of autoethnography to highlight the elite colonizing outlook on ignorant people. In contrast to the number-ridden newspaper reports and the myth-invoking poetry that packaged the inventors as mythical geniuses, pamphlet literature often conjured up diverse characters to voice different standpoints and moral concerns.[18] Such a representational field configured a reflexive social gaze that destabilized the vision of a vast technological and cultural empire.[19] As literary performances, the pamphlets fractured the balloon's scientific identity and representative political function and, for this reason, offer us a glimpse of the imagined machine polity engendered by this ephemeral, yet monumental artifact. They harbor the memories, political claims, and identities of anti-balloonists. While most of these literary characters reflect the reading public and their concerns, the animal voyagers such as the sheep give voice to the people.

IMAGINED WINGS AND LIMINAL POLITIES

In order to fully appreciate the journalistic balloon transcript as a hegemonic discourse of integration, we must unearth the utopian dreams and critical voices concerning the flying machine before the Montgolfière appeared on the scene. The flying machine had long been perceived as a chimera, an impossible dream that nevertheless signified an intense longing for a better world, which fostered utopian reveries and dystopian satires. The quest for a happy nation found fictional wings, for example, in *La Découverte australe* (1781) by Rétif de la Bretonne, a nocturnal observer of Parisan life. The protagonist Victorin in this utopian fiction transforms his servile self into an ingenious inventor of artificial wings, which allows him to kidnap and marry his social superior Christine. He also rescues destitute men and women to establish a republican monarchy on an inaccessible mountaintop (see figure 1.4).

A republican monarchy would naturally expand into a benevolent empire, according to Rétif. When the population grows, the king Victorin transports his subjects to Australia, the long-imagined (and recently discovered) Southern Continent, where he finds many islands populated by diverse humanoid-beastly races. The Megapatagonians—the most perfect species probably named after the Patagonians described by Charles-Marie de La Condamine in *Relation abrégée d'un voyage fait dans l'intérieur de l'Amérique méridionale* (1749)—work equally for the common good, share the fruits of their labor, and strive for egalitarian justice. Victorin mixes these "races" to create "perfect men," classifies his people by their "degrees of perfection," and assigns them work matching their intelligence in order to solidify a republican polity whose constitution stipulates equality as the foundation of justice and "national happiness." Victorin's egalitarian monarchy expands through education and systematic interbreeding between French males and females of other races. It thereby offers a vision of a cooperative utopian polity in which "all individuals existed as much in the others as in themselves."[20] Artificial wings became a fictional transport to a republican nation-empire.

The flying machine also provided a means of criticizing unfit sovereigns, decadent aristocrats, and fashionable society and promoting moral rectitude, patriotic courage, and inventive genius. In *Réflexions amusantes et intéressantes sur le vaisseau volant* (1782), published soon after Blanchard's spectacular failure, the anonymous author introduces the contrarian (*l'homme au paradoxe*) into an imagined conversation among a good society of usual assortments—men of wit, power, means, talent, or knowledge and lovely women—to develop a reflexive gaze on French society.[21] A self-appointed representative of public opinion, somewhat like André Morellet (1727–1818) who regarded the Enlightenment as a reform project that would strengthen the absolutist regime, the contrarian scrutinizes the royal couple's theatricality, ministerial abuse, decadent sexual morals, intellectual fashions, and French national character—light and inconstant—to pass harsh judgments and outline the necessary moral reform.[22]

The royal couple came under irreverent scrutiny. The imagined transport would be useless to the king, the contrarian rebukes "the man of influence [*l'homme de poids*]," since a truly sensible prince should choose good ministers to administer justice. The king should work for his subjects' true happiness rather than displaying his heart only in the presence of spectators. The contrarian's admonition thus voiced the patriotic opposition to Maupeou's top-down revolution, which was often characterized as ministerial despotism.[23]

A sympathetic intervention by a duchess for Marie Antoinette, deemed a hostage princess imprisoned in her beautiful palaces run by an insufferable regime of etiquettes, elicited a stronger rebuke from the contrarian. Instead of seeking an escape, he admonishes, "Antoinette" should curb her sexual appetite and honor the men of merit. Why would a queen, "thoroughly French, lovable and loved," have anything to desire?[24] The public had no sympathy for a queen who sought pleasures in order to avoid the arduous career of a sovereign devoted to the happiness of her people, according to Pierre-Ambroise-François Choderlos de Laclos (1741–1803), the suddenly famous author of *Les Liaisons dangereuses* (1782) and future propagandist for the duc d'Orléans: ambitious courtiers obliged her wonton pursuit of pleasures with the design to dominate their sovereign and to oppress the people.[25]

Sexual decadence and frivolous fashion carry the main satirical load in the *Réflexions amusantes*. Criticisms of royal mistresses such as Madame de Pompadour and Madame du Barry had brewed deep resentments during the previous decades.[26] Libels against Marie Antoinette grew from the start of Louis XVI's reign in 1774, in large part thanks to the duc de Chartres's intrigues, although open attacks on the queen were still rare.[27] The rumors of her affair with the comte d'Artois produced a pornographic poem, *Les amours de Charlot et Toinette* (1779). Often repeated were the claims of her sexual excess, political ambition, and illegitimate space of personal leisure, in addition to her foreign identity, although the libels may not have circulated in print before 1789.[28] A new cluster of fictions depicting dystopias ruled by women reflected the dark political cloud gathering around the unpopular queen.[29]

Should the art of flying be carried to perfection, the contrarian predicts, French women with their light character and consummate art of dissimulation would travel all over the world and become more famous than Alexander or Caesar for their bloodless conquests of men. They are the "true Proteans" who could simulate any woman as needed: a serious Spaniard (who demands perfect love) with a migraine, a constant Englishwoman (who would poison a straying partner) with nonchalant demeanor and the words of honor and religion, and so on. The aerial empire would also allow women to collect new fashions from all over the world. The aerial carriage would itself become a new fashion statement. Such was its utility, according to this "pleasant" conversation on the frivolous expectations of the chimerical machine.

The *Réflexions amusantes* thus configures a field of representative characters whose standpoints are scrutinized by the contrarian—a textual figure that embodies incontestable public opinion, or what Mercier termed the "national spirit"—to articulate a collective judgment of abstract, objective, and universal authority.[30] By the 1780s, then, the rational sovereignty accorded to public opinion eclipsed the legitimacy of royal power, at least in imagination. Blanchard's failure offered an outlet for the rising critique of court frivolity and sexual decadence that had long plagued the Woman Question.[31]

The fiction of *Les hommes volants* (*Flying Men*), which follows the *Réflexions amusantes*, focuses on frivolity and other related vices that permeate the prosperous nation of flying men—an allegory of the French nation ruled by the fashion queen Marie Antoinette. It is a dystopia inhabited by colorfully made-up men, capricious women, and the Goddesses of Frivolity, Gallantry, and Fashion.[32] All men have wings and seem joyous, but they are cowards with a limited mental capacity. Women are light and inconstant because they never find men who would settle them down. Though naturally gentle, they use every opportunity to avenge male abuses. Unable to fly themselves, they exercise their right to choose men, guided only by caprice, which results in constantly changing sexual liaisons. In just a few pages, the author sketches out what is perceived as the main malady that ails French society. Effeminate men with privileges abuse their advantage over women and degrade themselves, while women seek revenge by playing sexual favors in the most capricious manner—a poignant situation brought into sharp focus by Choderlos de Laclos's *Les Liaisons dangereuses* (1782).

The nation of flying men and capricious women in *Les hommes volants* has only one cult and one divinity—fashion.[33] The Temple of Fashion stands on top of a majestic hill, flanked by the palaces of Gallantry and Frivolity where goddesses give out passports to the Temple, often capriciously. To reach them, one has to pass the Hôtel des Petits-Soins, which houses an entire arsenal of cosmetics (*toiletterie*). In other words, access to the Temple is controlled by illegitimate influence. The morning ritual of *toilette* in French society served to gather an assortment of friends and protégés who sustained the hostess's informal sphere of influence.[34] The Palace of Gallantry, built with flowers, is ruled by a vivacious and playful goddess who dispenses wings to her courtesans. Her sister, the Goddess of Insipidity (*Fadeurs*), competes for the incense the sisters receive. The Palace of Frivolity has an indented architecture and houses a library that contains all witty authors, which makes it an arsenal of conversation.

The vast Temple of Fashion has a brilliant exterior, which changes form every instant, each time more marvelous than before, according to the Goddess's caprice. Her messengers fly all over the world, collecting rare and beautiful things from all nations. She appears to the visitor as a large heap of garments in which he discerns only with effort a very small face encased in an enormous hairpiece. She pets a small monster, resembling the ancient Sphinx, called Pun (*Calembour*). Her advisors are Caprice and Coquetry, with an occasional intrusion from Simplicity (as would Marie Antoinette in playing a milkmaid at her farm Hameau that began construction in 1783). Ridicule, dressed in a grotesque manner, proposes to dress men as women and vice versa. Moral decadence reigns in this temple dedicated to fashion. The reference to "Antoinette," the fashion queen, is unmistakable.[35]

The king in *Les hommes volants*, who serves as the grand priest of the Temple, is the queen's favorite chosen not for his courage, strength, or merit but for his beautiful wings. Active and vivacious, he seems to appear everywhere at once, almost like an acrobat. He often advises the queen on matters of taste. The powerless king with his brilliant wings is the vainest of all men and easily replaceable. Women discard him as soon as he displeases them. The king who inscribes his power in fashionable displays has no claim to his subjects' loyalty—a radical political message embedded by the author in a rather straightforward caricature of French society and its theatrical polity. The juxtaposition of female sexual activity and social subversion would soon become "a central metaphor for political decay."[36] The French Revolution would bring about an avalanche of libels against the reputedly incestuous queen.[37]

MALCONTENTS AND THE PHILOSOPHER-VOYAGER

Critical reflections on the flying machine disappeared almost overnight with successful balloon ascents. Aeronautic historians have focused mostly on the chorus of balloon euphoria, which stifled dissenting voices, unaware of the existence of critical literature before the balloon appeared on the scene. Their disappearing act, when brought to light, can speak for the efficacy of social engineering achieved through a mass spectacle and journalistic control. In order to fully appreciate the balloon's historical agency in modulating the public sphere, we must pay close attention to the inchoate, hidden transcripts that often appeared as pamphlets. Exactly how the cracks appeared in the public transcript can offer critical insight on the power relations that constituted the propaganda market riding on the balloon's popularity.

With the duc de Chartres and the comte d'Artois on its side, the balloon was safe from the licentious libels produced in the two princes' privileged publishing enclaves at the Palais Royal and the Temple, except when their interests collided. An early exception to balloon euphoria was occasioned by the partisan spirit between the Montgolfists and the Charlists, which was probably fueled by the duc de Chartres's (or his coterie's) populist agitation. Antoine de Rivarol (1753–1801) anonymously published *Lettre à Monsieur le Président de* *** (1783) in order to defend the beleaguered Charlist camp.[38] In addition, Rivarol captured a range of hidden transcripts. When Paris fell into chaos from the "idolatry of novelties," the provinces anticipated a revolution. While some congratulated themselves on living in an "epoch of such a great revolution," others worried that everything in the civil, political, and moral world appeared upside down. They already saw an army swarm in the air, blood rain on the earth, and thieves descend by the chimney to carry away their treasures and daughters. The police would have to ride on globes to control contraband. People already imagined a lunar voyage, much as they had expected to see houses on other planets with telescopes.

Rivarol also registered the barely visible concern over the relative authority of science and religion that had been brewing for some time with Diderot's *Encyclopédie*, electrical experiments, and mesmerism.[39] The reasoners (*Raisonneurs*) mourned the loss of state and religion, calculating that religion lost a miracle for each discovery made in natural philosophy. The globe would undermine the significance of the Christ's Ascension as well as the Virgin's Assumption. Rivarol advised a measured philosophical stance—to cool the heads and to scale down the hopes pinned on this marvelous artifact—as philosophers meditated quietly on the means of improving and utilizing the globe for safe aerial navigation. His pessimistic objections to ballooning would induce lengthy rebuttals.[40]

Pamphlet literature on ballooning allows us to discern the traffic between public and hidden transcripts, which strengthened the rhetoric of utility as an inviolable component of public opinion. In *Lettre à Mr. M. de Saint-Just* (1784), we find a well-crafted satire that mocks balloon fever through a skillful mixture of rational and irrational anticipations. The aerial empire is in sight, the author proclaims, with a thoughtfully designed machine and the knowledge of chemistry, the queen of the sciences. If one could fly all over the earth regardless of the weather and the terrain, it would expedite foreign affairs and make wars less violent. A prompt, convenient, and cheap transport would also allow women to visit the four corners of the world in search of new fashions and lovers. The easy traffic of love—French courtesans meeting Asian Sultans and American millionaires, French chevaliers visiting harems, eloping lovers, and so on—forecast a rebirth of the "century of gold and pleasures." The aerostatic machine could presumably bring about "a happy revolution in politics, arts, commerce and the pursuit of pleasure."[41]

Utility emerged as an absolute imperative in engineering public opinion, but the meaning was open to interpretation. The more effusively the author extols the balloon's potential utility, for example, the more obvious becomes its potential to disrupt the existing social order. Unable to deter contraband at the city wall, the Tax Farmers (*fermiers généraux*) would have to patrol all quarters day and night and inspect all flying vessels at the moment of their arrival. This necessity would change the tax system to suppress all tolls and odious searches that discourage French commerce and industry. The transgressive potential of a free movement of persons and goods is in plain view, if not emphasized.

The most disturbing rumor on the street was the allusion to lunar voyages, often dismissed by the elite public as a chimera. In *Le Char volant ou voyage dans la lune* (1783), anonymously published by the impoverished baroness Cornélie de Vasse (1737–1802), the protagonist Eraste is characterized as a great philosophe, famous méchanicien, and intrepid voyager whose goal is pure, as is his soul—an ideal savant-inventor-voyager who would have captured the hearts of the French public, or an idealized version of Jacques-Alexandre-César Charles.[42] Polite, sociable, moralistic, and compassionate, he wants a suitable companion for his lunar voyage of learning. Not of high birth but of certain moral character (not too talkative, vicious, or pretentious), the ideal candidate should not export earthly ills to the new planet either intentionally or unwittingly. In Diderot's diagnosis, imperial voyages in the past had caused devastation across the globe by failing to send out "decent souls"—benign, humane, and compassionate.[43]

While endorsing balloon enthusiasm, the screening process for the lunar voyage serves as a clever foil to offer a panoramic view of the social ills that plagued the ancien régime. Eraste advertises the position in order to attract the most distinguished persons of all estates and professions to his residence—Hôtel du Bon Sens, rue de l'Enjouement. Common sense in Baron d'Holbach's *Le bon sens* (1772) would have obliged the cosmic voyager to reject the absurd tyranny of religion as well.[44] *Le Char volant* thus allows us a glimpse of the balloon spectators who were screened out by the public transcript. The candidates comprise men and women who circulate in high society, possess "the enthusiasm for novelty," and live "for the moment."[45] Their ambitions were shaped and ruined by the theatrical polity, which reveals systemic power dynamics and resultant sufferings in their society. The philosopher-voyager as a paragaon of virtue could diagnose their affliction and judge their potential to become virtuous citizens.

Undesirable citizens comprised not just the populace but also those controlled by unproductive passions. Decadent sexual mores within Parisian high society mark the most undesirable category of potential émigrés—the gallants who squander their fortunes on libertinage and the coquettes who prize and abuse their seductive traits. The baroness de Melencour (*la Beauté mécontente*) is a vain lady past her prime, but she still possesses seductive traits idolized by men and envied by women. Courted and disappointed by other European men—cool Englishmen, gravely respectful Germans with raucous table manners, barbaric Russians, and libertine Italians—she expects a "more absolute empire" over men in Paris, "the center of civility [*politesse*] & good taste." Instead, she encounters only lightness, indiscretion, and inconstancy, which fuels her "ardent desire" to leave the planet. Even the chevalier de Valmont (*l'Avantageux*), a character probably taken from Laclos's *Dangerous Liaisons* (1782) as a man "made for Paris," is bored and wants an adventure, especially now that he has been caught in a lie. He intends to be useful to his country by introducing French mores and sociability to the moon and by forcing the lunarians to admire him.[46]

Eraste refuses both characters for the honor of the Earth since their inconstancy, vanity, and checkered history would give a bad impression to the lunarians. He also declines the marquise de Losange (*la femme galante*) and the vicomtesse de Valmont (*le Scrupule*), both persecuted by their illicit love affairs, and the actress Sophie (*la Demoiselle de l'Opéra*). Born of the usual circumstances of pleasure-seeking nights and by age twelve accustomed to serving men, she had met a series of affluent patrons. After acquiring the furniture, jewels, and a carriage necessary to maintain her reputation, she was ruined by the need to keep up appearances when she lost favor with men. Actresses bear witness to this decadent society sustained by the pursuit of pleasure.[47]

No better suited are those accustomed to the power game. Eraste turns down Fierencour (*l'Ambitieux*), professedly languishing in humiliating oblivion or despicable obscurity, because of his intention to subjugate the lunarians. Apprenticed in the Parisian world of power laced with intrigues, Fierencour had built his career by flattering his social superiors and subjugating others. He studied their character, exploited their weaknesses, and deployed vices as powerful weapons to make them fear him. Love is a sentiment too gentle for the men in a position of power, he contends in the Machiavellian fashion. Eraste also rejects a group of men (*les reformés*) who were unwillingly divested of their fortunes by an enlightened prince who distributed their wealth to widows, orphans, defenders of the state, and teachers. Eraste advises them to lead an active life tending to the public good and patrie, which would atone for their previous parasitic lives.

Eraste the moral philosopher also refuses the fortune-seekers who set up the material condition of theatrical polity. The comte de Gerzac (*Le Joueur ruiné*)—who lost his fortune, distinguished family name, and reputation because of his cursed passion for gambling—illustrates another deeply rooted problem in French high society.[48] Thinking that the ordinances against gambling should not have reached the moon yet, he proposes a great fortune-making scheme—to set up a bank of Pharaon with the lunar emperor and divide up the profit. The former Tax Farmer Dryau (*Ex-Fermier général*) proposes a commerce of attic salt (*sel attique*) between the two planets by obtaining an exclusive privilege for thirty years. While Eraste shuns these men who intend to replicate earthly evils on the moon, he promises a place in the aerostat to an innocent victim who laments his foolishness in trading real happiness for an imaginary fortune. The chevalier de Velville had squandered all his assets enticed by a neighbor who constantly advertised the tremendous yield from the lottery. He now spends his days forming imaginary projects to make a fortune.

Men of false learning exhibit peculiar follies. Systematon (*Le Systémataire*), who sacrificed his life and fortune to amass a large number of honorary titles, claims the right to accompany Eraste based on his knowledge of all philosophical, astronomical, and geometrical systems. He could also turn ashes into glass, a project he deems wise and useful because it would purify the air. The process could also preserve genealogies better by turning one's ancestors into bottles, goblets, and chandeliers. Eraste mocks Systematon's "sublime genius" by suggesting that it would be better to vitrify the living instead. If Systematon knew all the violence in Paris, he would want to vitrify all of Paris. Why not the entire globe?[49]

Spleen (*Le faux Philosophe*) is an arrogant philosopher with unkempt appearances who feels persecuted. He imagines the moon as a place where natural innocence still exists to prevent corruption in the mores. Eraste advises a philosophical approach: let others amuse themselves at his expense. If his soul is pure, the rest is not worth bothering about. He should be optimistic and happy as a good citizen, loyal friend, and an honorable gentleman (*honnête homme*). His morality, as a rare novelty, would be worth a fortune in Paris but useless in a perfect society. A "true philosopher" should be a "friend of humanity."[50]

The candidates fall far short of the ideal philosophical voyger or the model balloon spectator. In desperation, Eraste settles on a few acceptable companions who possess some philosophical or moral virtues. The first is a curious Englishman named Travel with a broad trajectory of learning. He wishes to learn about the universe by making measurements with his loyal companion M. Hilly. In addition to this fellow philosopher in search of universal knowledge, Eraste chooses another hapless victim of the ancien régime society. Driven into the arms of a swindler by her husband's inconstancy, the vicomtesse de Villemant (*le fatal soufflet*) was stranded in a foreign country with her son. Although her husband forgave her and left her a fortune upon his death, she cannot live with herself out of remorse and wishes to leave with her son.

The six voyagers, including the chevalier de Velville, leave from the Tuileries Garden, which is packed by a crowd of all ages and estates. All Parisians debate Eraste's invention for two days but forget about it by the third day, as is customary—an observation that targeted the coquettish fascination with novelties, which defined the Parisian pursuit of pleasures. The parting look requested by Travel at the earth below, the theater of their misfortunes, also enacted a cultural memory dating back to Lucian—the cosmic spectator. The imagined distance from the familiar civilization allows the cosmic voyagers to articulate a reflexive gaze on the life they wilfully leave behind.[51]

COSMIC SPECTATORS AND EARTHLY COMPATRIOTS

Eraste and his crew travel to the moon, experiencing violent winds and difficulty in breathing, to find a paradise that subsists in eternal spring. An infinite number of flowers impart perfumed transpiration to the inhabitants who live for five centuries. Benevolent Nature with no tempestuous ocean, ferocious animal, or noisy bird also provides free nourishment, eliminating hunger and disease. Nature constitutes the only notion of divinity, the "supreme being," which unites the lunarians in language and worship. Not tyrannized by passions, they enjoy a tranquil and happy existence by utilizing their natural intelligence.

The moon consists of just five kingdoms, all governed by women on the same principles that constitute an egalitarian "universal monarchy," one diametrically opposed to the centralizing, hierarchical vision of the Catholic empire.[52] The kingdoms of love, fortune, justice, fame, and moderation serve as allegories of human fallibility and vices, personified in the caverns of black rocks that surround their kingdoms. The passions, vices, and violence represented and punished by the cavern monsters serve to cultivate noble sentiments—the love of duty, the horror of vice, and the taste for virtue.

The earthly voyagers in their grand tour of the moon develop a cosmic perspective and a compassion for their earthly "compatriots," but they all settle down on the moon. Eraste alone returns briefly to collect some necessary material for his next exploration. The story expresses the collective desire to escape from cruel rulers and unjust society. However improbable the lunar journey was, the balloon provided a conceivable means that could set the earthly inhabitants free from their oppressive homeland to find a "universal kingdom" of equality, justice, and love. Cosmic tales had long served to articulate universal polities for European malcontents, as we have seen in Cyrano de Bergerac's *Histoire comique* (1657), to configure earthly (rather than national) compatriots.

Even a gallant tale of lunar paradise could offer a few radical perspectives. In the *Histoire intéressante d'un nouveau voyage à la lune* (1784), the main character ("the disinterested") finds the moon inhabited by gentle creatures. The story is seriously warped by his lust for beautiful lunar women among whom he enjoys considerable advantage on account of their monstrous-looking men. Nevertheless, the lunar paradise contrasts sharply with his earthly homeland: calumny has no empire and life flows in gentle pleasures. Everybody enjoys liberty, simple taste rules, women are respected, and even actresses are virtuous. Attentively bathed, luxuriously robed, and ceremoniously entertained, the earthly voyager's thought runs to the "vicissitude of human affairs that makes a person one day a private (*particulier*) person and tomorrow a king." If such an astonishing change could happen in a day, who would attach such importance to social rank? "Man is everywhere man," he concludes in Rousseauean fashion. Having received such splendid hospitality, his companions refuse to leave: "Homeland is where one is treated well: here is ours."[53] While acknowledging this truth, "the disinterested" reminds them of the glory, honor, and their compatriots that all await their return. Finally driven out by the jealousy of lunar males, he returns with a celestial beauty as his trophy.

In *Lettre à Mr. M. de Saint-Just*, the imaginary liberation is completed in the region of fire, filled with the marvels of nature. Enchanted by energetic meteors, brilliant stars, auroras boreales, and other numerous celestial entities, all puns and slanders recede from the voyager's mind. Cleansed of all earthly ills, the cosmic voyager acquires new knowledge and reflection to become a "citizen of the universe." Nevertheless, he does not find a better landscape, perfect societies, or happier men but, instead, other "beings vicious by nature & virtuous by art" living in an environment similar to that on earth.

By drawing a parallel between the cosmic and earthly nations, the author can apply the well-worn climate theory of Montesquieu and Buffon—that the climate determined the characters and mores of nations—to cosmic peoples.[54] Since it has the most malign influences on Earth, the moon is likely to have volcanoes, seas, lakes, rivers, continents, and all kinds of peoples. Astronomers could build an observatory and form an academy of mathematicians to teach eternal truths to the lunarians. European powers eager to extend their commerce would send troops and artillery to conquer and divide up the moon. Afterwards, they would establish flying ferries between the two planets. The philosopher might still explore other planets, comets, and fixed stars in order to map a "cosmic empire (*ciel empire*)," if not deterred by the distance.[55]

The imagined cosmoscape would shape the "citizens of the universe" who could transcend unhealthy passions and petty interests to transform the earthly habitat. Lacking the "sublime virtues of these cosmopolites," the author would rather stay on his native planet, laugh at the follies of his contemporaries, complain about their delusions, and detest their real vices. He would not even make a short voyage to the terrestrial paradise or El Dorado since happiness should not derive from such extreme adventures, but from daily pleasures. Knowledge and reflections from the cosmic voyage would return the earthly traveler back home, secure and happy in his own patrie. Imagined interplanetary travels would implant in the voyager a deeper appreciation of the earth and its inhabitants—a "true path to supreme happiness" as Epicurus and Democritus had suggested.[56] By the same token, we can infer, European explorations and colonization would yield nothing of value. The illusion of the riches from terrestrial and cosmic empires would mean nothing to such a homebound soul.

THE PHILOSOPHER-KING: A Princely Vision

The pursuit of utility as a moral mandate, which was interactively stabilized between public and hidden transcripts, can help us appreciate the political imperative of a paternalistic, republican monarchy that would guarantee people's happiness. For Jean-Jacques Rousseau, a civil state had to compensate the loss of natural equality among citizens by ensuring their moral equality. Given the inevitable degeneration that accompanies the rise of particular interests, citizens had to remain vigilant while participating in politics to the fullest extent possible.[57] For Rousseauean writers, a republican polity based on philosophical principles would undermine the theatrical polity that promoted ostentatious displays, manipulated ignorant people through religious veneration and fanciful novelties, and forged an artificial consensus by silencing legitimate opposition.[58] Whether a republican polity should actually abolish the monarchy was not a subject of discussion.

This hybrid republicanism—an enlightened monarchy that would establish a just society based on egalitarian principles—had acquired broad support by the 1780s. A particularly interesting pamphlet in this regard is *Le Triomphe de la machine aérostatique, ou l'Anti-balloniste, converti par l'expérience* (1784), which requires a careful reading for a princely interpretation of the "philosophical majesty." Published anonymously soon after Charles's Tuileries ascension with approbation, the pamphlet was ostensibly designed to promote the utility and morality of ballooning.[59] A spirited debate between "the aerial philosopher" and "the envious" (or "the Anti-Balloonist") brings out not just their opposing views on ballooning but also their shared commitment to the public good, the nation's future, and even "the good of humanity." Their conversation thus configures an ideal polity based on the egalitarian principle of "philosophical tolerance" and participatory citizenship.

Even the pamphlets approved for publication can be analyzed, then, to discern how public opinion on ballooning was engineered to bolster a regime of scientific hegemony. The aerial philosopher as the voice of reason characterizes the balloon ascension as a marvelous spectacle that would "lead natural philosophy to its highest degree of perfection" and teach the populace its mysteries. Ignorant people can only be a threat to society, but once educated they would develop a sense of self-respect and an attachment to society to become virtuous citizens. As the balloon delivered humanity from the physical chain of gravity, the knowledge of natural philosophy could free the "people"—seen as "brutes" or "monsters"—from their social chains. This discovery, "a marvel more astonishing than the works of Hercules," thus belonged to the entire "French nation" that should be proud of its inventors and aeronauts. The critics who denigrated the balloon with outmoded word games (*plaisanteries, bon mots, calembours,* and so on) fell short of their duty. "True savants" should be devoid of envy or hatred. Nor should they condemn the credulity of the common people since everybody should have the "liberty to think in his own manner." The "principle of philosophical tolerance" consists in "condoning others' weaknesses and correcting their faults without irritating or mistreating them."[60]

The "Anti-Balloonist" claims to be "just as committed to the public good as any enlightened citizen of the Republic of Letters." He still regards the balloon as a useless and dangerous novelty, "proper to amusing children and surprising ignorant populace," one that would surely lead the aeronauts to their untimely deaths. While he professes reverence for the sciences and the savants, he mocks the "enthusiasm" of the "philosophical Majesty" (the aerial philosopher), the machine's shallow and familiar scientific basis, and particularly its theatricality designed to prey on the credulity of the vulgar. If he wished to make "experiments," he would do so in the presence of the savants capable of appreciating his discoveries, rather than staging a public spectacle for the "miserable applause of the people." An imposing apparatus belied philosophical intentions. Why else did the ancients, who certainly knew how to traverse the air, keep the knowledge from the vulgar? A flying machine would only augment the number of "infernal machines invented for the destruction of human kind." It could invert moral as well as physical order. Who would protect citizens' privacy from flying thieves? In war, just a few balloons could set cities ablaze, ravage the harvest, and destroy fortresses.[61]

Notwithstanding their divergent stances on ballooning, the aerial philosopher and the Anti-Balloonist share their commitment to the public good and the moral authority of science, or an enlightened outlook that scientific progress and technical innovation would contribute to public morality and social order. Such agreement opens the door for a legitimate polity governed by a philosopher-king who, in his effort to ensure citizens' welfare, could induce balloon enthusiasm.

The chevalier, upon his return from the site of ascension, vividly details the "majestic" spectacle the "new Argonauts" staged at the Tuileries Garden. Rising in the midst of an inner circle that represented the entire "nation," the aeronauts' car became their throne while their machine presented "the most astonishing spectacle" that had ever struck human imagination. Everybody succumbed to "astonishment, admiration, and fear, some kneeling, some trembling, and others crying out: What an astonishing machine! What kind of being is then man to invent such marvels! . . . what courage! What virtue! Love of the arts, this is stronger than Nature!" The chevalier contends that the balloon exalted everybody, especially the common people, while dismissing the rumors of a lunar voyage as "mere street talk [*Boulevard tout pur*]." He maligns "the envious" more colorfully as a satirist, demi-savant, insolent parasite, vampire, or Pyrrhonian.[62]

The baron, who returns from the site of descent, reveals the pamphlet's political agenda when he breaks into a serious panegyric. He reports that the Prince Louis-le-Grand had followed the balloon on horseback and personally greeted the "astonishing men" or "savant-voyagers" upon their descent. Animated by a noble zeal for new discoveries, which characterizes the greatest kings, the prince had followed balloon experiments with an enthusiasm dignified of his illustrious ancestors. The nation owed the most brilliant establishments and precious discoveries to this grand protector of talent. Like Alexander the Great, he welcomed the arts, encouraged savants through his generosity, and honored them with his friendship, while obliging the painful military duties.

The balloon seemed to offer a golden opportunity to endorse an alternative philosophical majesty for the French nation. Although the label of "Louis le grand" properly belonged to Louis XIV and to Louis XVI as a birthright, the prince who actually followed the balloon was Louis-Philippe, or the duc de Chartres (see figure 3.7). His family had a long tradition of patronizing scientists and artisans, from the chemist Wilhelm Homberg to the Périer brothers. His coterie worked to attract savants with pensions, encourage the arts, support inventors, and organize philanthropic societies. To win over the people tired of despotism, they ran a vigorous propaganda designed to camouflage the prince's short attention span.[63]

The baron's praise would have served as an endorsement of the duc de Chartres, heir to the duc d'Orléans, as a philosophical majesty. Whether the prince truly agitated to displace the king or not, his public support of the Charlière—more scientific and therefore potentially more useful than the Montgolfière—would not have pleased the queen's faction.[64] The vicious libel published in the same year, *Vie privée ou Apologie de très-sérénissime Prince Monseigneur le duc de Chartres* (1784), maligned his family while protecting the reputation of the king and his ministers, notwithstanding its title that advertised a defense of this "living example." Chartres's ambition and pretention as a philosophical prince would emerge more vigorously during the French Revolution when he sided with the radical nobility and the Third Estate to fashion himself as Philippe Égalité. Little did he know in 1784 that his coterie's overblown ambition to eclipse royal authority would end his own luxurious way of life.

ANIMAL METAMORPHOSES AND THE PEOPLE'S VOICE

The most subversive genre of satire characterized the animals in the Versailles flight as the first aerial voyagers in an effort to undermine the heroic narrative and to voice the people's suffering. Since La Fontaine and his fables, the animal kingdom had become a rich source of allegory and satire against the powerful. Just as Usbek, the "barbarian" observer in Montesquieu's *Persian Letters* (*Lettres persanes*, 1721), delivered a more potent critique of French civilization than a civilized Frenchman ever could, subhuman heroes could fundamentally erode the triumphant narrative of balloon ascents. By pairing the outward pursuit of glory and honor with the inward cowardice of calculating intelligence, the satires featuring heroic animal voyagers undercut the triumphalist rhetoric of intrepid aeronauts and mocked the theatrical polity that prized outward displays rather than inner character. When weak and vain humans hatched the "ridiculous and strange design" of traveling in the air, one poem went, they made a prudent decision not to risk their precious lives and sent up as emissaries a gentle and peaceful sheep, a proud and courageous rooster, and a noisy duck.[65]

In *Le mouton, le canard, et le coq. Fable dialoguée* (1783), the anonymous author mocks man's cowardice and vanity in pursuing immortal glory at the expense of innocent, powerless people. The rooster (*le coq*) represents an adventurous nobleman eager to undertake "the most audacious enterprise" for immortal glory. He relishes his superior position over the swarm of people and takes pride in the honor bestowed upon him as *"the bird of France"* to "announce her glory to the universe"—immortality, at least, like the mythological figures that shine brilliantly in constellations. As the heroic aeronaut, the rooster hopes that man would make a better use of flying than navigation. Rather than pursuing new territories, man might be guided by philosophical curiosity and humanitarian sentiments to uncover the secrets of nature in this "immense laboratory of the atmosphere where the most imposing wonders operate." If the rooster dies, he would then become a "glorious victim of the most astonishing, the most sublime discovery" that brought honor to France, long ashamed for having invented nothing of importance.[66]

The duck (*le canard*), characterized as a learned and philosophical "Doctor," explains the aerostatic machine and the human aspirations riding on it. As a calm, reasonable interlocutor, the duck queries if audacious humans would displace the birds' empire. In contrast to these winged creatures, the sheep represents a hapless victim. Trembling at the realization of the earth disappearing from beneath his feet, he queries: should a sheep be in a position to call men cowards? If the animals were chosen out of prudence, why choose the most innocent animal instead of criminals? Perhaps a rooster could pass for his audacity and pride, a duck for having nothing to fear in the air, but to send a sheep, such a simple and timid animal, for the most dangerous of all projects! He would rather stay on a beautiful pasture than fly to the most brilliant place of firmament among the stars.

The innocent sheep, a gentle and amicable creature, soon became a recurrent trope for reproaching humans for their chimerical pursuit of novelties, pleasures, immense treasures, and fortunes. The motif of the innocent victim to vainglorious philosophical pursuit takes on a much more somber tone in the *Réclamation du mouton, premier navigateur aërien* (1784). While taking a promenade in the park of Versailles, the anonymous author finds the aerial voyager-sheep in the royal menagerie full of exotic animals.[67] When he ponders aloud the injustice of making them prisoners of indolent human curiosity and placing them in slavery, the gentle and lovable sheep suddenly acquires a voice to indict the cruelty of unjust men. Wrenched from his mother in the countryside and destined for slaughter, the sheep survived by refusing to eat, which made him an ideal candidate for the balloon experiment. He experienced a total change in his organs as he rose in the air: his eyes shined with flames as vivid as those of the sun to his companions' fascination. He suddenly understood all of human nature.

The Sheep charges that humans ignore truths and unleash pitiless massacres against defenseless animals whose gentle nature only animates their furor and sharpens their dagger. While their pettiness made him laugh, the sheep also pitied the chimerical happiness of mortals and their caprice in chasing futile honors, novelties, and fortunes. They placed him in the aerial car not to honor his value but to sacrifice him for their glory. Most perniciously, humans appropriated the title of the first aerial navigator and planned to immortalize it by a public monument. This was the last straw for the hapless sheep whose kind had suffered for centuries in human hands. The sheep, perhaps representing the enslaved, vows revenge and vindication:

> Sooner or later, they would experience without delay the upheaval [*bouleversement*] we want to return magnified several times. Such has been our general resolution. We have supported for centuries and might still suffer more that they shear our coat, kill us, put our meat on the skew, and use our skin to protect their ostentatious [*fastueuses*] chimeras. But that they take our glory and insult the truth in a manner so shocking, we can no longer endure. It is necessary therefore that on a granite column of a hundred feet in height, an alabaster sheep may be placed with this inscription on the base: *ego sum primus navigator in auras* [I am the first navigator in the air]. The first humans who, after you, try this noble enterprise may be placed at your feet in good time. This would be sufficient for their arrogance, if they knew what we are.[68]

The monument of the miserably exploited sheep is reminiscent of the Negro figure in Mercier's utopian Paris—"the avenger of the New World" who delivered God's justice to the slaving nations and helped restore sacred humanity in its natural state of equilibrium.[69] The innocent sheep and his voice could stand for the powerless people and their deep-seated desire for justice.

In the politically static yet culturally dynamic milieu of the late Enlightenment, the balloon materialized a composite body of the French nation to engender a radically different sense of the citizen's place in the world. It visualized a new state-body that included the populace by mobilizing the existing material, institutional, and literary resources—royal police, transportation network, postal system, and newspapers. The balloon spectacle thereby constituted a liminal realm that encompassed all citizens and their inchoate dreams. A republican monarchy seemed within reach for the balloon audience. For Jacques-Pierre Brissot, science had to offer an emancipatory rationale for the "perfect republic" that would guarantee the "liberty of the human spirit."[70] The people seemed completely absorbed in a national celebration, in large part thanks to the balloon's extraordinary visibility and representativeness. Contemporary euphoria and centuries of celebration afterwards have sustained the balloon's universal and timeless appeal as a material fiction of infinite progress.[71] The asymmetric power relation between the educated "public" and the illiterate "people" in the case of ballooning is hidden under the euphoric public transcript that permeated newspapers, correspondence, and even personal diaries. In contrast to the model balloon spectator as portrayed through the newspapers, an assortment of balloon spectators in pamphlet literature represent diverse standpoints of the contemporary figures who constituted the composite audience for the balloon. Their desire for an escape and their critical gaze on social ills reveal to us what they perceived as irreparable problems of the theatrical polity that prioritized appearances over substance.

An archeology of mass silence must trace, therefore, how the balloon became a unifying spectacle and unearth the technologies of mass control and the concerns of the populace.[72] The eerie promise of revenge from the sheep in a temporary metamorphosis—that his oppressed brethren would destroy the privileges and pretenses of vainglorious humans—can be read as a vow to eliminate the oppression of the poor and the enslaved and to restore the "sacred humanity" in Mercier's vision. All nations would kneel before her and repent their bloody crimes against humanity.[73] Nobody was immune from the blame, except perhaps the people who were sacrificed for the sustenance and pleasure of their superiors.

Figure 5.1. "New Game of Aerostatic balloons for elevated minds." Gallica, BNF. The game represents all major balloons until the Roberts' September 1784 ascent.

1. Aug. 27, 1783. Champ de Mars
2. Sept. 12, 1783. Réveillon factory
3. Dec. 1, 1783. Tuileries
4. Nov. 21, 1783. Château de la Muette
5. Sept. 19, 1783. Versailles
6. Jan. 19, 1784. Lyon
7. March 2, 1784. Champ de Mars
8. July 11, 1784. Jardin du Luxembourg
9. July 18, 1784. Rouen
10. June 14, 1784. Nantes
11. July 15, 1784. St. Cloud
12. June 23, 1784. Versailles
13. Sept. 19, 1784. Jardin du Luxembourg

5

FERMENTATION AND DISCIPLINE

The year 1784 had an auspicious beginning for balloon enthusiasts. With the temperature rising over four successful ascents, balloon fever had reached its peak to "ferment all Parisian heads."[1] Citizens of the triumphant nation cheered for the audacious conquerors of the aerial ocean.[2] On January 19, Joseph Montgolfier pulled off a flight with six passengers at Lyon in the largest Montgolfière ever built. By early February, four balloons were under construction in Paris alone, and still more failed to secure funding.[3] Léonard Alban and Jean-Baptiste Vallet began experiments at their acid factory in Javel (with a hydrogen balloon named *Le comte d'Artois*) with the explicit aim of finding the means of steering it.[4] More than a hundred memoirs on the same subject poured into the Lyon Academy that had issued a prize competition in previous December to find "the most certain, the least expensive, and the most effective means of directing the aerostatic machine at will."[5] Some of them also found their way to the Paris Academy of Sciences and were often published.[6] Models of these devices, the most popular being the sail, were on display along the Seine and at the cafés.[7]

The French art of building a boundless empire made foreigners fear the coming of the most destructive weapon ever invented by humanity. Benjamin Franklin and Thomas Jefferson worried about an aerial invasion against which no nation could mount an effective defense. Which prince could afford enough troops to protect his country against the soldiers carried in five thousand balloons? They would still cost less than five naval ships. Inland countries would become as vulnerable as maritime states.[8]

The aerial empire that sprang up in French collective imagination nurtured a brilliant illusion of their global power (one that would see partial realization under Napoleon). It also stirred visions of a truly integrated nation that would include the populace. A system of scientific hegemony required an administrative control over the invisible fluids and insignificant bodies that filled the material public sphere.

A curious conjuncture nevertheless points to the subversive potential of invisible fluids that invited serious administrative efforts to tame it.[9] At the height of balloon mania, the long-brewing issue of mesmerism flared up and demanded public attention. Instead of welcoming these diversions from the looming financial crisis, the royal administration took decisive measures to curb both forms of science that could induce popular unrest. In March 1784, Breteuil set up two commissions to investigate Mesmer's claims of scientific authority and therapeutic efficacy, enlisting twelve Academicians and physicians from the Academy of Sciences, the Faculty of Medicine, and the Royal Society of Medicine.[10] A royal ordinance on April 23 banned the aerostatic machines with combustibles, which restricted Parisian ballooning to a privileged few—princes, savants, administrators, and their protégés.

While constricting these popular sciences that brought the elite public in contact with the populace, Breteuil also gave permission to stage Caron de Beaumarchais's *La Folle journée, ou le Mariage de Figaro* at the Comédie Française on April 27, 1784. He had persuaded the reluctant king by arranging a special reading, over the strong objection of the royal censor and the police chief Lenoir.[11] Why suppress a mute machine and an invisible fluid that promised much public good, while staging a torrent of anti-establishment declarations in a public theater? How was it possible, we may ask with Leo Weinstein, "that such an explosive play could be performed, against the king's will, under an absolute monarchy that had at its disposal every means of suppressing dangerous and subversive works and throwing their authors into jail?" Few episodes illustrate better, in his judgment, "the irresolution of Louis XVI, the disarray of his government, and the suicidal lightheartedness and lightheadedness of the nobility of the ancien régime."[12]

Are we to believe that Beaumarchais's long campaign and unparalleled cunning, helped by the aristocratic (especially the queen's) demand for novel entertainments, finally triumphed over the royal displeasure and frowning censors? The evidence for his courtly skills notwithstanding, were the administrators really so clueless as not to appreciate the danger the *Mariage* posed to the delicately balanced state-machine?[13] What if Breteuil took a calculated risk, especially since he probably did not wish to ignore the queen's desires? What if he wished for the *Mariage* to undermine aristocratic image and power, while suppressing ballooning and mesmerism that could pose a more serious threat to the absolutist polity? The theater crowd would have been much easier to control than private séances or the entire adult population pouring out into the street for the balloon ascent.

The play's longevity and popularity, as well as the enthusiastic commentaries from the conservative sector, make it plausible that it was a conventional masterpiece rather than a revolutionary one. The *Mariage* mocked aristocratic decadence without challenging royal authority, in contrast to the underground libels that had seriously eroded public respect for the sovereign.[14] Sarah Maza argues that malcontent writers showed no love for the "radical" Beaumarchais. The *Mariage* adopted the style of *parades*, or parodies of popular speech aimed at the fashionable society, to offer an entertainment that "reeked of moral corruption."[15]

The royal administration's effort to curb the theater of natural fluids —mesmeric séances in private quarters and balloon ascents in public gardens—evinces their potential threat to social order, as Simon Schaffer has argued convincingly, especially when they became associated with radicals and enthusiasts.[16] Although France had a much more controlled public sphere than Britain, or a system of scientific hegemony implemented through the symbolic hierarchy of royal institutions, the balloon opened an outlet for popular expression that was not otherwise allowed. Even if the raucous parterre disturbed the hegemonic function of royal theaters, the theater still presented a better-controlled public space where the administration could absorb dissident energy at the expense of the aristocracy.[17] When the Mesmerists refused to give in, the administration staged a play, *Les docteurs modernes*, in an effort to mock them into silence.

MATERIAL NATION AND ACADEMIC SCIENCE

Paris was the capital of European fashion. This imperial capital of taste fostered emulative consumption of populuxe goods (such as gold watches, formal clothing, fans, snuff boxes, and stockings) through fashionable shops, trade literature, and secondhand barter. Cissie Fairchilds has carefully documented this consumption pattern from the property records (*inventaires après décès*) of middling consumers such as small shopkeepers, master artisans, journeymen, day laborers, and domestic servants.[18] While the consumption of art and luxury goods such as silverware, furniture, soap, candles, sugar, and coffee had steadily eroded the distinction between the highborn and the merely wealthy, a process of social assimilation produced an evolving rank of nobility rather than its elimination, as exemplified in the case of patriotic art collectors.[19] The French civilizing mission depended on the geography of fashion, which homogenized the nation-empire through consumer desire and taste.[20]

Fashioned as a public artifact in the imperial center of novelties, the balloon became a ubiquitous consumer motif. Women wore their hair in globes, small societies were formed about globes, and small theaters played pieces about globes. Rose Bertin, the high-priced designer of Marie Antoinette's wardrobe, led the fashion on balloon hairstyle with large hats.[21] Innumerable prints, fans, mini-balloons, hair and clothing styles, faience, porcelain, wallpaper, fabric, jewelry, objets d'art, and furniture propagated the new fashion (see plate 8).[22] The balloon as a popular fashion motif disrupted the existing pattern of emulative consumption emanating from the court. It was highly visible yet cheap enough to reach down to the vulgar and to start a new fashion trend that enlisted the entire population except for the most destitute. Everybody wanted to launch his own balloon, albeit in varying sizes to match his fortune. Self-help manuals appeared on the market.[23]

The balloon as a material fashion propagated a different sense of the cultural nation than the traditional one controlled by academic networks. In order to appreciate this difference, we need to understand the actions of the Paris Academy of Sciences and its relationship with the provincial academies. Human ascents made Lavoisier—the rising star at the Paris Academy—worry that balloon fever might paralyze the projects that "tend truly toward the progress of sciences," or systematic research he sought to conduct within the Academy.[24] Nevertheless, the Academy did not entirely give in to popular furor or royal directives.[25] Although Lavoisier closely followed new developments for the balloon commission, he treaded carefully to protect the Academy's reputation as the highest court of European science without denigrating the Montgolfiers' public prestige. In their official report, dated December 23, 1783, the balloon commissioners gave approbation to the inventors while securing the authority of official science by rendering an objective judgment on the relative merits of the Montgolfière and the hydrogen balloon.[26]

Lavoisier performed a delicate balancing act in order to appropriate the popular invention for his theoretical reform in chemistry. Inflammable air, still a domain of experts, bolstered the balloon's and the Academy's scientific status. Lavoisier steered the balloon commission's research toward the hydrogen balloon, which allowed him to take advantage of its exceptional popularity for his ongoing research in pneumatic chemistry. In setting the direction of their research to perfect the aerostatic machine, the first balloon commission focused on two facets. The first consisted in finding a less cumbersome means of procuring inflammable air and containing it securely, which required chemical knowledge. The second involved steering the machine, which belonged to the mechanics.[27] Although the commissioners phrased their report carefully so as not to offend the presumed royal inventor, Lavoisier soon made it clear in forming the second balloon commission that a convenient method of producing light gases was a major focus, which meant that the Academy's research dealt mostly with the hydrogen balloon.[28] The research on inflammable air remained largely in the hands of a few qualified experts such as Lavoisier, Claude-Louis Berthollet (1749-1822), and Gaspard Monge, and well demarcated from the realm of amateurs.

The Paris Academy could not provide an effective leadership for national ballooning.[29] Royal institutions suffered from a repertoire of maladies that limited access and innovation, which allowed for a multiplication of the scholarly networks that often bypassed the capital. Notwithstanding repeated entreaties from the royal administration, it was the Lyon Academy that issued a special prize competition in December 1783. More than a hundred entries poured into this provincial institution during the following months, while less than a third made their way to the Paris Academy where research continued mostly behind closed doors. Enthusiastic amateurs not only penned these entries but also showcased their models on the banks of the Seine.[30]

The royal Academy's capacity to represent and control French science was seriously undermined by provincial amateurs. Even with the serious fire hazard, provincial amateurs clearly favored the Montgolfière over the hydrogen alternative, probably for its spectacular size and modest cost. Hailed as an "arch-patriotic" invention made by provincial papermakers, the Montgolfière made a convincing case against the exclusionist academic practice decried by Jean-Paul Marat.[31] Xavier de Maistre (1763–1852), brother of the future counterrevolutionary Joseph de Maistre from a noble family in the kingdom of Piedmont-Sardinia, declared in proposing a subscription at Chambéry that the balloon humiliated European savants. It proved that "men of all classes and of all countries" should engage in studying nature. Great inventions owed almost nothing to the titled savants gathered in large cities with all the instruction and resources. Academicians explained, corrected, analyzed, and perfected the arts without adding anything to human power. The journalist Linguet also asserted that academies were bad judges of new discoveries.[32]

In other words, the balloon fostered a more inclusive material geography of the nation, as in the case of the theater boom, linking urban centers rather than the ceremonial foci of Versailles and other royal palaces.[33] The aerostatic globe became a coveted repertoire in all fêtes, causing a provincial "sensation as vivid as in Paris" and creating a demand that exceeded Parisian supply.[34] As a fashionable commodity, the balloon also helped shape "citizen-consumers" to destabilize the social hierarchy. The "chain of buying" connected individuals more horizontally than vertically, Colin Jones argues, to undermine traditional status and prestige. Balloon consumption defined a material geography of the nation mostly determined by the wealth, rather than the status, of its consumers.[35]

Provincial replication induced a participatory national culture, much like singing.[36] Erasing the worry that it was just another Parisian folly that would "tire our provincial readers who perhaps do not share the enthusiasm of our Parisian heads," the balloon became a national pastime and fueled a sense of national pride.[37] Many provincials attempted to make their own balloons and, perhaps, their own majesties. Unlike the traditional entrée ceremonies that pulled the nation toward the royal body and its linear itinerary, balloons could multiply majestic bodies all over France.[38] Such multiplication in the absence of the royal body (or his substitute) did not always work in the king's favor.

Provincial ballooning required a complex mélange of resources, which modulated the process of nationalizing a scientific artifact.[39] The will to participate in the national celebration was often frustrated by financial and scientific shortfalls. A subscription issued at the Nancy Faculty of Medicine just after the Champ-de-Mars ascension took several months to fill.[40] The Rouen academician Scanégatty issued a subscription to make a fifteen-foot balloon in mid-September 1783, but it went nowhere. A rumor also circulated in early October that a Rouen nobleman had experimented with a small balloon (five feet in circumference) at his Chateau Drubec and that he intended to float a much larger one (forty-eight feet in circumference), second in size only to the Versailles balloon.[41]

Several attempts took place in Lille where the authorities printed an extract from the *Gazette de France* to inform the public of "a discovery of which the government deemed appropriate to publicize in order to prevent the terrors it could occasion among the People."[42] Nicolas Joseph Saladin (1733–1829), a physician and professor of mathematics at the Academy of Arts, and businessman Joseph Gosselin released a toy balloon made in goldbeater's skin on November 9, 1783. Town authorities also gave permission to Mr. Berschits to stage another on November 10. The project failed after selling numerous tickets at three livres each and caused a public grievance. On November 29, Saladin's students at the Academy launched a three-foot balloon made in goldbeater's skin at the Redoute de Wazemmes.[43] Lyon offers a special case because of its geographical proximity to Vidalon. A European center for the silk industry and Freemasonry, Lyon developed balloon fever early with Joseph Montgolfier's frequent visits. After a few trials during September 1783, royal intendant Jacques de Flesselles issued a public subscription on October 22.[44] Joseph floated another small balloon without public announcements on November 13, which nevertheless drew ten thousand spectators.[45] Five days later, local amateurs launched a balloon charged with fireworks. More than thirty thousand spectators watched, amidst "excellent music," when it rose at 8:45 PM and spread fireworks high in the sky. Illuminating the entire city, the firework balloon produced in the spectator "an effect . . . impossible to describe." The experiment proved "irrefutably," the *Journal de Paris* editorialized, that the balloon could send signals from a considerable distance.[46] Parisian balloon enthusiasts and journal editors kept track of Joseph's Lyon balloon.[47]

Replication of the hydrogen balloon was more troubled than the Montgolfière, which indicates a relatively lower level of pneumatic literacy and network in France than in Britain (see chapter 8). Of the fourteen cities that attempted human ascension during the year 1784, only three (Dijon, Nantes, and Bordeaux) tried their hands at making a hydrogen balloon, which were all made possible by the savants with a certain level of chemical knowledge and varying access to the Paris Academy. Only Dijon and Nantes succeeded, although financial woes soon marred their triumphs. The hydrogen balloon required a level of finance well beyond what a public subscription could raise in a provincial town, not to mention scientific expertise and logistical support such as the transportation of sulfuric acid. Exceptional research at a remote locale such as Leuven did not yield a working balloon.[48]

Unable to raise sufficient funds through a subscription even after five months, Nantes experimenters headed by Pierre Lévêque (1746–1814), a correspondent of the Paris Academy of Science, resorted to a number of associates willing to bear the financial burden. Nantes possessed a professor of natural philosophy Joseph Mouchet who had experimented extensively in pneumatic chemistry, retracing the experiments of Robert Boyle, Stephen Hales, Joseph Priestley, and Henry Cavendish. While attesting to this fact, the apothecary Jean Louvrier nevertheless disputed the professor's contribution to the balloon experiment.[49] Lévêque would organize a second experiment to settle the affairs and publish a definitive account.[50]

BLANCHARD'S RESURRECTION: Champ de Mars, March 2, 1784

The balloon's popularity prompted the discredited inventor Jean-Pierre Blanchard to announce an ascent, wishing to secure "the title of the sovereign of the air." He promised a show that would "merit universal suffrage." His pleading words nevertheless reveal a harsh reality for the aspiring inventor.[51] The Périers and the Montgolfiers were exceptions rather than the norm in the theatrical polity. If the "French Nation" were just, the pamphleteer Arnaud de Saint-Maurice could only hope, she would pay homage to her genius whose hard work fueled human hope.[52] Not to risk another blemish

Figure 5.2. Portrait of Blanchard with his balloon, printed for the Rotterdam ascent. Tissandier Collection.

on his reputation and credibility, Blanchard pulled off an astonishing task in harsh weather, finishing a twenty-six-foot balloon by February 21, 1784. He probably had help from the Robert brothers or their associates in coating the silk balloon.[53] He put it on display at the abbé de Viennay's residence (see figure 5.2). Amateurs could observe all facets of his "flying vessel" at three livres per entry.[54]

On March 2, 1784, the police secured the Champ de Mars for a free circulation of carriages. The launch site had to be enclosed to raise sufficient funds, yet visible from afar to attain universal approbation. A detailed instruction for entering the enclosure was printed in the *Journal de Paris*. Blanchard ran the operation like a clock, anxious to keep his word to the public. The balloon and the instruments were ready before noon, but a student named Dupont de Chambon, possibly a classmate of the future Napoleon Bonaparte at the École militaire de Brienne, jumped into the gondola with his sword drawn.[55] Blanchard's rejection led to Dupont's rampage that destroyed the parasol, the wings, the strings, and all instruments and left Dupont's hand bleeding.

Blanchard had to leave behind his companion, the Benedictine Dom Pech. Rising alone "stripped of everything," he was determined to fly for "the sole satisfaction of the Public." He had a single rudder to help him navigate against the wind. The spectators were gripped with fear, a sentiment rather different from what they had exhibited at the Tuileries. He flew for nearly two hours on a path to Versailles, reaching a "prodigious height" of about two thousand toises, rumored to be the highest ever. When he landed near the Sèvres porcelain factory, an officer from Lauzun and Billancourt's Legion invited him to their residence where he drafted the procès-verbal.

Religious reception of balloon mania is difficult to determine. A rumor circulated, perhaps to underline the perception of ecclesiastical oppression, that the Benedictine physicien Dom Pech had to overcome objections from his church at Saint-Martin des Champs. His church supposedly regarded the device as the devil's work, one that would diminish "the most extraordinary miracles such as the ascension of Jesus Christ."[56] Benedictines were the most scientifically curious branch of the Catholic Church, however, as seen from Viennay's patronage throughout Blanchard's career.[57] References to religious opposition rarely surfaced in the public literature on ballooning, which means that science had become a strong ally for the administration running on public opinion. A strong government fortified by consensual public opinion would have worked well for the Catholic Church as well, if not entirely in the Vatican version. Religious anxiety is understandable, but it remained mostly as a hidden transcript that rarely came to public attention, except for occasional surfacing as seen in the caricature that replaced the ascending Christ in Joseph Benoît Suvée's *Résurrection* with a balloon.[58]

Who could own the technical and moral qualification to represent the newly minted national artifact was a complicated matter. Blanchard's beautiful flight on March 2 failed to secure his place in Parisian aeronautics, although it garnered exuberant praise from the English for ushering in the "century of balloons."[59] Simultaneous reports in several journals indicate that his patrons circulated a favorable version, but it did not win over the establishment opinion.[60] *Grimm's Correspondance* mocked it as a mere repetition of Charles and the Roberts' experiment, but one that netted between forty and fifty thousand livres for the author.[61] With all the mechanisms and the instruments broken, Blanchard could not meet the advertised aim of steering his vessel at will and against the wind. His experiment became no more than a repetition of Charles's in public opinion, although broad commercial appropriations of his voyage tell a somewhat different story (see plate 8). Failing to gain sufficient recognition from the Parisian elite, Blanchard was soon on the move.

With the spectacular success of a hydrogen balloon in the capital and the proliferation of Montgolfières all over France, Étienne Montgolfier must have sensed that the great family dream was falling apart. Although Blanchard had not succeeded in steering his balloon, his long solo flight marked a sharp contrast with the botched Lyon flight (see chapter 6) and strengthened the scientists' belief in the superiority of the hydrogen balloon, now made self-evident to the public. Perhaps to claim his rightful place as the original inventor, Étienne read a memoir to the Academy on March 6, 1784, on the possibility of using oars to steer the balloon, advocated by many of the contestants for the Lyon competition.[62] He must have been aware that Lavoisier was steering the Academy's research toward the hydrogen balloon, which enlisted a number of talented chemists such as Guyton de Morveau at Dijon.

News also came from abroad that Paolo Andreani had piloted the first human flight outside of France from his estates near Milan on February 25, 1784. He made the envelope in the close-knit *toile de Rouen*, covering it with the finest Milanese paper both inside and outside, and coating it with layers of alum. Such up-to-date technical knowledge points to a communication channel with Paris.[63] The abbé Miolan and the engraver Jean-François Janinet (1752–1814) issued a public subscription in Paris for the largest Montgolfière with the help of the Academicians, while a rumor circulated that the duc de Chartres had commissioned the Robert brothers to build a sophisticated Charlière at the cost of forty thousand livres.[64] Another Montgolfière launched from Chambéry just outside the French border in the kingdom of Piedmont-Sardinia failed on April 22 and induced a barrage of criticisms, although the second trial on May 6 had a happy ending.[65]

The Montgolfière was beset with fire, which threatened civil order. When the administrators of Lille asked for instruction on how to deal with the fire hazard,[66] the police chief Lenoir began to talk with the principal magistrates of the Paris parlement to gauge their opinion.[67] The royal ordinance of April 23, 1784, banned all balloons with stoves, almost simultaneously with Lavoisier's highly publicized gun-barrel experiment, which promised to produce hydrogen in large quantity by decomposing steam. The process would have eliminated the need for iron filings and sulfuric acid, which took up a significant portion of the cost in producing hydrogen balloons.[68] All other balloon ascents had to obtain permission beforehand, reserved for those "with experience and well-organized capacity."[69] In other words, the ordinance brought the mushrooming balloon ascension under royal and academic control. Louis XVI talked personally to provincial intendants about the grave danger the balloon posed to public order.⁷⁰ Preventing riots was a prime directive in public policy, which translated into regulating grain prices and distribution, as the recurrent bread riots should have reminded them.⁷¹

The royal ordinance dimmed further prospects of the Montgolfière and Etienne's value as a scientific inventor. He did not have independent means like Lavoisier or Buffon to maintain a lifestyle ostensibly devoted to the pursuit of truth. From the royal perspective, the Montgolfiers were properly compensated with paper titles and honors whose vanity "only as solid as the smoke in the balloon" would not have been lost on Étienne. A modest royal pension could not compare with what he made in the paper business.⁷² In his absence, the Vidalon factory had lost a significant market share and the competitive edge in paper quality.⁷³ After processing the requisite paperwork for the Cordon de St. Michel and the title of Royal Manufactory through the bureaucratic maze, Étienne returned home at the end of May, leaving behind an unfinished balloon and the royal wish to stage it for the Swedish king soon due in Paris. Unable to let go of his beautiful dream, he would petition for funding to establish an aeronautic school at Vidalon, while nursing the paper business back to health. Bitterness would creep up when he realized the scale of his family's financial loss.⁷⁴

FLUID POLITICS: Mesmerism Controversy

A more worrisome turn of events for royal administrators in the spring of 1784 was the intensifying public debate over mesmerism. Franz Anton Mesmer (1734–1815) was a Viennese physician who circulated in high society, including the Masonic lodge La Verité et l'Union where he met Wolfgang Amadeus Mozart.⁷⁵ Vilified for his magnetic therapy, he moved to Paris in 1778, the year in which both Voltaire and Rousseau passed away and the philosophes' symbolic reign ended.⁷⁶ A man of broad learning, deliberate demeanor, and statuesque posture with powerful facial features, Mesmer cut a noticeable figure in the fashionable society to attract well-heeled female patients.

Mesmer's theory was simple, unified, and borderless: a universal fluid filled all space and acted on all bodies through flux and reflux. The accumulation and circulation of this "magnetic virtue" explained the origin, nature, and progress of all diseases.⁷⁷ A "truly universal fluid" would link electricity and magnetism in a way that could explain the human body, thought, and sentiments as integral parts of the regular workings of the universe.⁷⁸ While his practice of channeling this fluid for a cure incited criticism from the royal institutions of science, well-meaning philanthropists took up the cause with the aim to ameliorate human suffering. Controversy followed him around, however, compromising his effort to characterize animal magnetism as a scientific discovery with universal utility.

Mesmer failed to persuade the elite audience at the Academy of Sciences and the Royal Society of Medicine, although he gained a few disciples and many highborn patients along the way. After the Faculté publicly rebuked his earliest disciple Charles-Nicolas d'Eslon (1750–1786, physician to the comte d'Artois) on September 18, 1780, Mesmer announced his intention to abandon Paris.[79] To prevent this, his aristocratic female clients secured through the queen a royal pension of twenty thousand livres and an additional ten thousand a year to set up a Parisian clinic. Mesmer's public refusal of this offer negotiated by the chancellor Maurepas tarnished the queen's reputation, although Mesmer's target was less his compatriot queen than the Academicians and the physicians of the Faculté. He recounted their assaults on his credibility in *Précis historique des faits relatifs au magnétisme animal jusqu'en Avril 1781*, which began a war of words that transformed this elite fashion into a cause célèbre.[80]

As a self-styled benefactor of mankind in possession of "a truth essential to human wellbeing," Mesmer professed to be in search of a suitable government that would help him work for entire humanity without worrying about expense. For this cause, he had to mind the "judgment of nations and of posterity" and act courageously. What was, after all, "four or five hundred thousand francs more or less, well spent," to her majesty?[81] The happiness of her people was far more valuable. Mesmer's populist rhetoric appropriated Rousseauean idealism to his personal ends. His ability and willingness to do so indicate that such republican wrapping had already become a common rhetoric even among the self-serving elite circle.[82]

The literary public sphere had become scandalously politicized by the 1780s, in part thanks to the judicial briefs that manufactured causes célèbres for public consumption. The police chief Lenoir dated the Parisian "vogue of Mesmerism" to 1780, but the serious concern over "seditious speeches against religion and government" to 1781. Mesmerist debate attracted disillusioned scientists and writers such as Jacques-Pierre Brissot, Jean-Paul Marat, and Jean-Louis Carra, a self-styled "prophet philosopher" working as a royal librarian and much maligned by academic scientists. They had diverse reasons to take up the pen against what they perceived as academic despotism.[83]

The Lyon lawyer Nicolas Bergasse (1750–1832), one of Carra's patients, joined the fight with an anonymous *Lettre d'un médecin de la Faculté de Paris à l'un médecin de collège de Londre* (1781), which characterized Mesmer as a "true man" with revolutionary intentions whose "immense discovery" comprised all truths of nature. Academic societies, which judged him an imposter or an enthusiast, merely established an "empire of fashion" rather than fostering "advantageous revolutions." Bergasse charged that the institution of medicine was a political body "whose destiny is linked to that of the state, & whose existence is absolutely essential to the prosperity of the state." Outside of academies, people wished to extend the domain of sciences to improve their health. Shorn of harmful remedies, a new, vigorous generation would ensue to follow the laws of nature and thereby bring about a revolution in society.[84]

Mesmer's vainglory soon muddied the revolutionary consciousness. When d'Elson established a clinic in Paris and defended his independent authority to the Faculté during Mesmer's absence, Bergasse had to pen a reluctant defense of Mesmer's exclusive rights. He found it "impossible to speak and write for a man who, proclaiming everywhere that he possessed a discovery universally and essentially useful to humanity, made humanity's destiny subordinate to his own glory and, what was worse, his fortune."[85] Mesmer's greed must have been a factor in his scuffle with the Faculté since this prestigious institution examined and licensed in 1782 the electrical therapy of Nicolas-Philippe Le Dru, with the street name of Comus.[86]

Mesmeric threat to social order only intensified with the Société d'Harmonie universelle (Society of Universal Harmony), which joined the growing list of secret societies.[87] Nicolas Bergasse and Guillaume Kornmann, a Strasbourg financier, began to sell shares in this society (at one hundred louis d'or each) on March 10, 1783.[88] Headed by the chevalier de Chastellux (François-Jean, 1734–1788), subscribers included religious persona such as father Gérard (the superior of la Charité) and Dom Gentil (prior of Fontanet), notable public figures such as the baron de Montesquieu (Jean-Baptiste de Secondat, 1716–1796) and the marquis de Lafayette, and the chemist Claude-Louis Berthollet.[89] Mesmer was promised twenty-four hundred louis for his secrets.

The Société, modeled after Masonic lodges, gave a boost to provincial mystics and Rousseauean agitators alike—the marquis de Puységur (Armand-Marie-Jacques de Chastenet, 1751–1825) at Buzancy (Société harmonique de Bienfaisance des Amis réunis), the Freemason Jean-Baptiste Willermoz (1730–1824) at Lyon (La Concorde), and Joseph-Michel-Antoine Servan (1737–1807) at Grenoble, to name a few. The founding subscribers supplied public testimonials for the cause. When an illness intervened in publishing his monumental series, *Le monde primitif, analysé et comparé avec le monde moderne*, Antoine Court de Gébelin (1725–1784) sent out to his subscribers in July 1783 a substantial pamphlet that recounted his mesmeric cure.[90] As the founder of the Musée de Paris and a member of the Loge des

Neuf Soeurs, Gébelin's testimonial would have carried significant weight among the educated audience.

Notwithstanding an earlier start, then, mesmerism as a social movement exploded in sync with balloon mania during the spring of 1784. According to the *Mémoires secrets*, "men, women, children, everyone is involved, everyone mesmerizes." Magnetism occupied "all heads," including those whose opinion carried some weight.[91] A war of words broke out on the pages of the *Journal de Paris* between Mesmer and Galard de Montjoie (1746–1816), which attracted more public attention, vicious attacks, and vigorous defenses.[92] Father Charles Hervier (1743–1820) published *Lettre sur la découverte du magnétism animal à M. Court de Gébelin* (1784) with an introduction by Gébelin who was a royal censor.[93] The subscriber list and membership for the Société d'Harmonie grew rapidly. The marquis de Lafayette joined on April 5, 1784, before sailing to America as the ninety-first receiver of Mesmer's diploma in the Parisian Society. He would try in vain to convert George Washington.[94] Some sought to moderate the excessive claims of Gébelin and Hervier. Gébelin's pitiful death at Mesmer's quarters in May 1784, which labeled him a "martyr of magnetism," would prove their prudence.[95]

THE IMAGINED NATION: Republican Monarchy

Material politics of mesmerism and ballooning—invisible natural fluids gathering insignificant bodies around them—did not deter conservative rearticulation of paternalistic monarchy, judging from the *Grimm's Correspondance*, which catered to the European super-elite in status or education. On these pages, d'Alembert's death in October 1783 lingered on, as eulogies trickled into print to immortalize this distinguished mathematician who had attained "the most sublime heights in calculus." His preliminary discourse to the *Encyclopédie* was praised as "one of the most beautiful monuments that philosophical genius has raised to the glory of human knowledge."[96]

Other than such exceptional events, new publications and theatrical performances occupied the bulk of *Grimm's Correspondance*. The January issue of 1784 began with a discussion of Montesquieu's posthumously published tale *Arsace et Isménie*. This oriental tale of an "absolute master," endowed with all virtues and amiable qualities to rule his kingdom with felicity, was recognized as "the most useful and moving lesson for a despot who desires the happiness of his subjects and himself." Although the "natural government of the French" lay somewhere between despotism and moderate monarchy, the baron de Montesquieu (Charles-Louis de Secondat, 1689–1755) had to render despotism useful because a monarchy almost necessarily tended toward despotism. Montesquieu mixed the simplest amusement of imagination with the traits of "profound philosophy, useful views, and maxims dignified by the usual elegance [*hauteur*] of his thoughts."[97] The ghost of Montesquieu, the exemplary philosophe, had long haunted the literary public sphere. His legendary act of generosity in Marseilles had appeared on stage in various versions since the 1770s, most recently in Pilles's *Le Bienfait anonyme* at the Théâtre-Français in late December 1783 and later in Mercier's *Montesquieu à Marseilles* in October 1784.[98] While protecting the great philosophe Montesquieu's legend, *Grimm's Correspondance* was harsh as usual toward rebellious spirits such as Mercier and Rétif de la Bretonne who refused to stay within the carefully managed boundaries of respectable theater and society.[99]

For the most part, *Grimm's Correspondance* evaluated theatrical performances in terms of literary style, spectacular effect, musical quality, the performative efficacy of actors and actresses, and audience response. The theater as an institution of controlled publicity could nevertheless articulate an ideal polity. Shakespeare's plays provided superb occasions, for example, to criticize English mores and to project French ideals. When Jean-François Ducis (1733–1816) made an adaptation of *Macbeth*, which was staged for the first time on January 12, 1784, *Grimm's Correspondance* printed lengthy commentaries to chastise the English "nation whose mores and climates make its people indifferent to emotions." Their spectacles had been limited to cockfights and gladiators. For this primitive people, Shakespeare had to choose "somber and terrible subjects—atrocious crimes, extraordinary events that crush and degrade humanity." Instead of the principles and rules that formed eternal beauty in Corneille, Racine, and Voltaire, English spectators fell for rustic and savage nature, which shaped the character of their mores. "Romantic events, forced situations, [and] atrocious and nearly monstrous characters" conveyed terror, "the sensation that has the most *empire* over a people somber, melancholic and nurtured in the *revolutions*."[100] Cultural chauvinism served to camouflage the French administrators' long-standing fear of the English "philosophical wind of free and antimonarchical government" that would involve "the election of true tribunes of the people."[101]

What would be a political system without the reign of terror or incessant revolutions? The discussion of ideal polity continued in *Grimm's Correspondance* mostly along the line of a moderate monarchy that would incorporate republican virtues.[102] The collapse of the Caisse d'Escompte, exacerbated by the extremely cold winter that threatened the survival of the poor, called for a benevolent monarchy that would safeguard citizens' welfare. Revolution floated in and out of elite consciousness, completed at least in music by Gluck, Piccini, and Sacchini.[103] If the word "revolution" meant a sudden change, it did not always imply an irreversible or progressive change, especially in political systems. Swedish and Genevan revolutions in recent memory had toppled republican governments, for example. The "happy revolution" envisaged in *Grimm's Correspondance* was not for a democratic or a republican nation, similar to the physiocrats' or Mably's, but for a moderate monarchy ruled by a sensible, paternalistic, and heroic king.

The full translation of Charles-François Sheridan's *Histoire de la dernière révolution de Suède* (1783) provided an occasion to trumpet a "happy revolution" from above by a virtuous prince who became the "first citizen of his homeland." His "noble project" overturned the tyranny of the senate, as in the Roman Empire. Mercier had rehearsed the revolutionary role of a "philosopher-king" in *L'An 2440* (1770), although the Swedisg King Gustav's 1772 revolution was a poor instantiation at best. Sheridan demonstrated quite well, *Correspondance* nevertheless judged, how the abuse of unjust, tyrannical, and corrupt aristocracy had reduced Sweden to a deplorable state that worked against its people. The apparent system of liberty implemented by the "strange constitution" after the death of Charles XII precipitated a revolution by the young prince in an effort to approximate the character, disposition, and needs of that nation. He carried out this noble project as the first citizen of his country without losing his virtue or his humanity. This was a triumph of his just and firm will, great and sensible character, and gentle, yet powerful eloquence.[104]

THE MARRIAGE OF FIGARO

Breteuil commissioned official reports on mesmerism at the height of balloon fever when academic scientists could still claim a level of public trust. The first commission, appointed in March 1784, consisted of four physicians from the Faculté (Guillotin, Darcet, Sallin, and Majault) and five Academicians (Bailly, Lavoisier, Franklin, LeRoi, and Bory). The second commission in April comprised four members of the Royal Society of Medicine (Mauduyt, Andry, Caille, and Poissonnier).

The extraordinary intervention only inflamed in the Academician Jean-François de La Harpe's (1739–1803) opinion the "epidemic" that had already conquered all of France.[105] The "frenzy" over mesmerism had begun to eclipse balloon mania by May, according to the Parisian chronicler Siméon-Prosper Hardy, who anticipated its future as "the sole universal medicine."[106] Except for some notable defectors like the chemist Claude-Louis Berthollet, a founding subscriber who famously stormed out of a session after the government inquiry had begun, the fashionable society was concerned only with animal magnetism.[107] Unlike the balloon, which had quickly acquired a hegemonic public transcript and fostered a rather limited range of literary repertoire and hidden transcripts, mesmerism would attract a broad spectrum of patrons and writers who rose fiercely to its defense in the public sphere.

The scandalous success of Beaumarchais's *Le Mariage de Figaro* takes on a different meaning when assessed against popular fermentations of ballooning and government suppressions of mesmeric practices. Written between 1776 and 1778 when Beaumarchais was organizing the American war effort, the play had originally been accepted by the Comédie Française in 1781, with the royal censor Coquelay de Chaussepierre objecting only to a few minor aspects. Louis XVI refused to grant a public performance, however, citing the indiscriminate jabs at the government. Beaumarchais continued the campaign to put it on public stage, reading it in aristocratic circles. A Versailles performance scheduled on June 13, 1783, was canceled, but a private performance at the comte de Vaudreuil's Château de Gennevilliers on September 27 proceeded in the presence of prominent personages including the comte d'Artois. In other words, the queen's coteries favored the play. Long negotiations and tantalizing peeks in private theaters only intensified the anticipation for a public performance.

The play's emergence from the arena of semi-public to public performance marked the beginning of the balloon's demise as a public artifact. The first public performance of the *Mariage* at the Comédie Française on April 27, 1784, attracted a crowd never before seen at a theater.[108] The play enlisted the audience in mocking the arbitrary exercise of power with explicit symbolism.[109] All the intrigues—played "on the authority" of the government or public opinion—added to the fervor. Most aristocratic spectators judged the play quite immoral and absolutely inadmissible in a public theater. La Harpe, a member of the Academie Française and intent on monitoring public culture, cringed at the unrelenting tirade that threatened his raison d'être, staged in a sold-out theater with government permission.[110] The *Mariage* proved a "prodigious" success nonetheless, staged a total of sixty-eight times in a row until late in the year, dwarfing all other eighteenth-century plays in its box-office popularity.[111]

Enthusiastic endorsements in the conservative media, coupled with broad condemnation from the Grub Street writers such as Mercier, raise the possibility that the *Mariage* was not perceived as a revolutionary piece but as a conventional one that would entertain morally corrupt aristocrats and contain restless malcontents.[112] *Grimm's Correspondance* attributed the play's longevity to its intricate performative situations and a strangely complicated, unpredictable plot that had a foolish, yet original conception; pleasant yet unpredictable turns riddled with intrigues; and a clear, ingenious, comical, and natural end. In the reviewer's opinion, such a labyrinthine plot full of surprises was previously unknown in French, Spanish, or Italian theaters. The unforgettable power and resources of a genius shone through the play to sustain the spectators' attention for three and a half hours.[113] As for the charges of immorality, the reviewer interpreted that the play was a portrait of the actual mores and principles of the high society, drawn with boldness and naiveté, as if the author wished to correct his century's vices rather than contain them within the boundary of proper taste. Instead of pretending to present an austere or moralistic piece, Beaumarchais portrayed realistic situations with more licentious details in order to activate the audience's imagination, which would put together the portrait.

Public royal theaters were supposed to constitute the absolutist body politic by prescribing a well-defined space for the performance of power. The administration had long honed the art of policing the theater in order to discipline audience response and expectations, notwithstanding the increasingly raucous parterre. In this context, the *Mariage* would have channeled simmering discontent into a ritualized laughter, while driving a wedge between the literate "public" and the uneducated "people." The ritualized laughter served—be it in the theater or in the carnival—as an outlet for suppressed anger. If the government had the wit (*bon esprit*) to allow staging it so as not to suppress the "small pleasures that could never be too dangerous," the public could do nothing else but indulge in this inconceivable mixture of the most refined and the most vulgar. Notwithstanding the universal laughter and applause induced by truly comic situations, the parterre reportedly behaved with admirable tact. The explosive phrases did not threaten the fabric of theatrical polity that required amusing caricatures. Nor did anti-aristocratic or anti-ministerial sentiments always translate into a subversion of the monarchy, judging from their endorsement of the Swedish revolution. The "citizen Beaumarchais" generated even more publicity and stronger approbation by organizing the fiftieth staging as a benefit performance and by donating his profit, rumored to be more than thirty-six thousand livres, to helping poor nursing mothers.[114]

The interplay of popular science and mass politics in prerevolutionary France deserves our attention for the ways in which the process incubated diverse republican dreams and the mass public. Ballooning and mesmerism challenged the ancien régime theatrical polity built on public royal theaters. We are less familiar with the emotive efficacy of these material theaters than the intellectual articulation of mesmeric discontents well-documented by Robert Darnton's pioneering work.[115] Mesmeric practices formed hidden theaters of natural power to forge a strong bond across the existing social hierarchy. The natural flow of an invisible fluid, deemed universal, extended the community of proliferating secret societies and brought them in contact with the common people, projecting a world without boundaries or social distinction. However flawed Mesmer's personal ambitions were, his vision of a regenerated body politic required that the magnetic fluid circulate through all bodies without social barriers.[116] The authority of Nature provided an allure of the ultimate power that was just, egalitarian, and benign.

The balloon ascent visualized a unified nation by attracting an audience that comprised all orders of society. While the balloon carried royal emblems on its body, its ascent tended to fuel the emancipatory desire simmering below the hegemonic façade and the silent hope bubbling up with the rising balloon. In sharp contrast to public royal theaters, the balloon ascent constituted an open theater that fostered fermentation among the populace who could harbor emancipatory and revolutionary dreams. In success, the aerostatic machine opened the imagined vista of a new ocean in which one could travel with an exalted sense of "hilarity," free from the physical chain of gravity and the social chain of economic and political bondage. In failure, the crowd broke through the physical barrier, thereby explicitly violating the social stratification evident on the ground and displaying the irascible anger stemming from their bondage. The workers who forsook their daily wages to participate in an event that promised to give their miserable lives a historic meaning were difficult to appease once they were disappointed. The sense of betrayal had long defined the "culture of retribution" that incited popular violence.[117]

The convergent pattern of material politics in the 1780s through mesmerism and ballooning invites a close scrutiny of who their representatives were, how they forged their relationship with the natural fluids, and what these invisible fluids allowed them to advocate in the ancien régime society. Instead of emphasizing the differences between discursive mesmerism and material ballooning, we might focus on their shared capacity to multiply contact zones between the fashionable society and the populace. Both exhibited serious effects of the disequilibrium in nature and manipulated them to promise a superhuman ascent or a universal cure that was not under royal control. If the balloon ascension constituted a new body of the nation that would transcend existing social orders in its aerial expanse, the mesmeric fluid connected individual bodies into one seamless whole. The political imaginary of these two scientific spectacles would have been far more dangerous than any torrent of criticisms in the theater that often provide an outlet for bottled-up anger and thus serve an essential function in maintaining an oppressive regime. Scientific ideas and artifacts had become dangerous instruments in consolidating and mobilizing popular unrest.

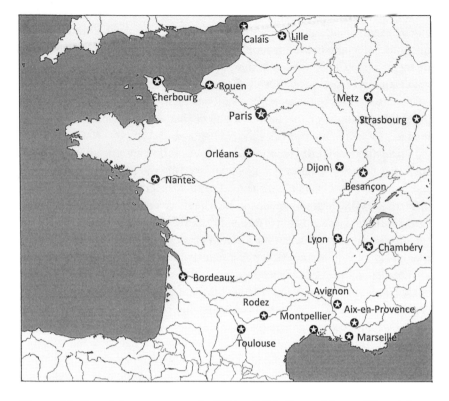

Figure 6.1. French human ascents in 1784—(M) indicates Montgolfière; (H) indicates hydrogen balloon. Dates in parenthesis indicate unsuccessful attempts.

Jan. 19. Lyon (M), Montgolfier
March 2. Paris (H), Blanchard
(April 4). Sorgues (M), Brantes
April 25. Dijon (H), Guyton
(May 3). Bordeaux (M), Grassi
May 6. Chambéry (M), Xavier de Maistre
May 8. Marseille (M), Bonin and Mazet
(May 15) Strasbourg (M), Adorne and Enslens
May 23. Rouen (H), Blanchard
May 29. Marseille (M), Mazet
May 31. Aix-en-Provence (M), Rambaud
June 4. Lyon (M), Laurencin
June 12. Dijon (H), Guyton
June 14. Nantes (H), Lévêque, Coustard, and Mouchet
June 15. Avignon (M), Scanégatty
June 16. Bordeaux (M), Darbelet
June 20. Avignon (M), Scanégatty
June 23. Versailles (M), Montgolfier
July 11. Paris (M), Miolan and Janinet
July 12. Besançon (M), Lepagnez
July 14. Strasbourg (M), Gabriel
July 15. St. Cloud, Paris (H), Chartres
July 18. Rouen (H), Blanchard
July 23. Bordeaux (M), Darbelet
Aug. 5. Toulouse (M), Duverney
Aug. 6. Rodez (M), Carnus
(Aug. 8). Bordeaux (H), Cazalet
Sept. 6. Nantes (H), Coustard de Massy
Sept. 19. Paris (H), Robert brothers

6

PROVINCIAL CITIZENS AND THEIR NATIONS

The "happy revolution in politics," which would undermine the king-machine as a national complex, seemed immanent in the proliferating provincial ascents that stirred intense regional pride and a sense of moral duty toward humanity.[1] A vast cloak of the immobile multitude surrounded the emergent balloon in Lyon, which made a critical observer "laugh and cry all at once" at the intrepidity of those who braved cold, rain, and hunger to stand shoulder to shoulder in a foot of mud, trying their best not to lose their shoes.[2] Spectators wrapped themselves physically around and virtually inside the patchy balloon to constitute a people-machine. Their collective yearning for a new machine polity—one that would allow the citizens to pursue happiness and contribute to the public good without being chained to specific places or functions in a fixed hierarchy—would find an expression, depending on the success or failure of the ascent, in a jubilant festival for all or a reckless riot by the dispossessed.

The absolutist theatrical polity was designed to spin a material geography of emulation centered on Versailles, but the spread of ballooning saw a process of variegated appropriations depending on local resources, technical expertise, and public composition.[3] Citizens with a strong regional identity and humanitarian ethos could not uphold the intricate, corrupt, and abusive system of privileges that turned the state into "a centralized instrument of private appropriation."[4] The monstrous overgrowth of Paris exhibited only too well the methodical extraction of wealth and resources by the central elite.[5] Such an exploitative system without virtue worked against the public good and the nation, as the proliferating discourse on political economy criticized.⁶ In other words, provincial ballooning does not reveal a uniform process of diffusion and integration emanating from Paris or Versailles, which complicates our assessment of science's role in the cultural construction of the French nation.

Instead, provincial ballooning depended on the complexion, motivation, and leadership of the local power elite whose mediating function between Versailles and their local public charted divergent pathways and identities for their balloons and publics. Instituting science's authority was an exercise in cultural translation that required "a realignment of power, a renegotiation of the unequal relationship" between the disparate languages of the colonizer and the colonized, as Gyan Prakash has discerned in colonial India.⁷ How a particular nexus of the local elite actualized the balloon ascent can help us understand the layered power relations that sought to integrate the nation as a cultural empire and thereby shaped the participatory experience of emergent nationalism in the provinces.⁸ The French nation remained a patchwork of asymmetric polities on the eve of the revolution. In this context, provincial balloon ascents configured a complicated political, commercial, and material geography of the French nation. Instead of mapping the multiplication of ballooning all over France as an automatic process guided by universal curiosity and commercial impetus, we can compare successful provincial ascents in an effort to probe how variegated provincial elites might have imagined their nations differently. Royal administrators, city officials, parlementaires, academicians, musée members, and the commercial elite, all wished to nativize the balloon as a point of regional pride, as they had done by building their own theaters.⁹ Early efforts at replication in Lyon and Dijon offer an illuminating contrast between their respective regional polities.¹⁰ Newspapers on cue localized the public transcript, which reflects diverse political agenda and strategies of legitimation while authenticating the balloon as a national artifact. Such mediated imperial use was contested, as we can see in the second Lyon balloon, which has been mostly hidden from our view. An archeology of hidden voices and artifacts can help us appreciate the limits of constructing a cultural nation in the theatrical polity.

LE FLESSELLES, JANUARY 19, 1784

The first Lyon experiment, organized by the royal intendant Jacques de Flesselles and funded by European Freemasons, offers us valuable clues to the political, cultural, and material work involved in staging a successful balloon ascent. The city possessed important assets—an original inventor, a vainglorious administrator, the commercial and Masonic elite, and a sufficient police force in the vicinity. A piece of underdeveloped real estate east of the Saône River offered an ideal site to launch a large Montgolfière. Lyon was a major hub for European Freemasons where Jean-Baptiste Willermoz (1730–1824), a silk manufacturer and amateur scientist, had consolidated an extensive network utilizing trade routes.[11] When Flesselles issued a public subscription on October 22, 1783, the prince de Ligne (Charles Joseph, 1735–1814), a Freemason and close friend of the duc de Chartres, paid for 100 subscriptions out of 360, issued at twelve livres each.[12] The project would attract a large gathering of Freemasons. Étienne's wife, Adélaïde, counted several lodges from Burgundy, Flanders, Grenoble, Annonay, and England. She also expected the princes of blood—the duc de Chartres and the comte d'Artois.[13] Aristocratic Freemasons traveled to witness the new scientific spectacle.

What seemed like a favorable ensemble of human and material resources to stage an impressive show failed to produce a matching result, due to the balloonists' worldly needs and desires, meager funding, severe weather, aristocratic pretensions, and the lack of cohesive leadership. Joseph Montgolfier faced a tough job in trying to make a large balloon under intense public and administrative scrutiny. His wayward imagination and social naïveté did not suit the massive project on a tight budget. Flesselles instructed the city architect Jean-Antoine Morand (1727–1794) to allow free passage to Les Brotteaux, a vast terrain east of the city developed by his company Saint-Clair. Flesselles also ordered guards to the construction site for a workshop and an enclosure, initially estimated at 190 feet in diameter and 570 feet in circumference.[14] All facets of supervision fell on Joseph, which included arranging accommodation for the carpenters through the police lieutenant.[15] Local and foreign dignitaries intervened frequently to demand updates and speedups. Lyon academicians had two meetings with Joseph to go over the size of the gallery, the weight it could carry, the nature of the combustibles, and the measures to ensure the voyagers' safety.[16]

Material constraints weighed down the leaky balloon. Poor fabric quality, made necessary by the low budget, plagued the whole endeavor. Solidity of the envelope was essential to ensure the success of the voyage and the safety of the aerial navigator.[17] Originally designed for an unmanned ascent to carry at most a large animal like a cow or a horse, Joseph made his balloon in paper and the coarse linen commonly used for wrapping postal packages. He could not afford to make the envelope in silk, which offered substantial advantages in weight and strength.[18] The subscription fell far short of the Parisian precedents. Stitching together a one hundred-foot balloon in severe winter weather was not easy, especially when one had to sew on layers of paper. Joseph had to coordinate sixteen teams (each consisting of a foreman, an assistant, three volunteers, and a workhand) to join 113 panels with 14,944 frogs.[19] Although the envelope was ready for assembly by mid-December, Parisian human ascents prompted Joseph to modify the design with a view to accommodate up to twelve passengers.[20]

A motley crew complicated the chain of command. Preliminary experiments began on December 26, 1783, after the Montgolfier family's arrival, to prepare for a public ascent on December 29. Much to their dismay, the balloon first inflated at the arsenal exposed many holes, in part due to the panels improperly sewn in different directions.[21] Then came Pilâtre and the marquis de Dampierre (Auguste Marie Henri Picot, 1756–1793), traveling under the names of Monsieurs Roland and Henri, on December 27. With his reputation at stake, Pilâtre delayed the departure, asking to strengthen the machine, increase its lifting power by reducing the weight, and make a few trial ascents. His "brilliant elocution" and "persuasive enthusiasm," not to mention the additional expense of almost three hundred louis, proved detrimental to the team spirit.[22] He demanded not just a stronger machine, the abbé complained to Jean-Pierre at Vidalon, but also "a thousand things that are more for looks than for utility" without even planning to stay long enough to remake the entire envelope in cotton, much stronger than coarse linen.[23]

Jean-Pierre agreed with Pilâtre, though, that the embellishments mattered more than the size since "it was necessary to speak to the eyes to cause sensations" among ignorant spectators. Above all, no accident should happen, and no funds should pass through Joseph's hands in order to avoid all suspicion of profiteering and thereby to protect the family's reputation. Jean-Pierre urged the abbé to conduct a few trials before the public demonstration.[24] Pilâtre's demands, as well as his decisive handling of idle spectators and thieves, probably saved the experiment. Joseph would later recall that the operation ran much more smoothly every time they checked entry cards and herded out useless people, apparently a problem Pilâtre was already familiar with.[25]

Pilâtre also worked tirelessly to publicize this "magnificent experiment," sometimes sending detailed news by a special courier, to secure his reputation as the nation's top aeronaut—Lyon brimmed with foreigners of the highest distinction; the entire city paid court to Pilâtre who commanded 150 workers; the list of amateurs wishing to join him in the aerial journey grew every day; the machine christened *Le Flesselles* could easily carry from thirty to forty men; Joseph hoped to fly with several friends to Paris or Marseilles, depending on the wind, to secure the glory of the first long aerial voyage, and so on.[26] Delays aside, the machine was worth gossiping about. It was well endowed with expensive instruments—nine thermometers by the Italian instrument maker Carlo Antonio Castel-Nuovo, three siphon barometers by Pierre Nicolas Changeux's method, and two hygrometers sent by Horace Benedict de Saussure (1740–1799) from Geneva, and so on.[27]

Family expectations and meddling did not help. Except for Étienne in Paris and Jean-Pierre at Vidalon with his father, the entire Montgolfier family had rushed to Lyon. The princess de Ligne received them politely, as her husband solicited a place in the aerostat for their eldest son, Charles (1759–1792). The abbé utilized this opportunity to canvass the luminaries' opinions on how his family should be compensated. Although he should have known better than to trust the easy promises of idle nobility, he wrote optimistically to his father that the comte de Dampierre and the marquis de Bellevûe had a plan to secure his brothers' fortune.[28] He favored the idea of obtaining a privilege to exploit the royal forest, an object of popular resentment, as it was a domain zealously protected from the general population. Pressed from all sides, Joseph must have been worried sick that the whole endeavor could end in a disaster and ruin the family's reputation and financial prospects.

More pressure came from the immobile crowd that stood watch every day in the frigid weather. As soon as the project settled in Les Brotteaux, persons of all estates flooded the place. The entire population stopped their routine affairs: all businesses, all shops, all offices, and even churches were closed, according to the pamphleteers. Throughout the construction period, which stretched over two weeks, an immense crowd stayed on the spot, save for nights, as the crew struggled with the balloon 300 feet in circumference and 545,000 cubic feet in volume, thawing the frozen bag, setting it accidentally on fire, mending the burnt cap, and learning to nurse the fire gradually to fill it up. Even the workers who lived from hand to mouth stopped working for fifteen days, one pamphleteer perhaps exaggerated. Nobles of the highest rank applied themselves day and night to the most menial tasks, another author perhaps idealized, dressed like workers and exposed to the rigors of the season.[29]

A humble paper-linen machine became a rallying point of Lyonnais pride and created a consensual public space of shared aspirations. The localized public transcript, idealized to the point of implausible exaggeration, conveys a sense of the regional patriotism that resonated far beyond the circle of overambitious balloonists and their vainglorious patrons. Everyone was supposed to regard the balloon as a "child of Lyon," abandon their routine and livelihood to follow its progress, and pitch in wherever they could. Les Brotteaux became an open site to patch together the Lyonnais patrie— perhaps wishful thinking by the city elite, a mixture of far-thinking aristocrats and merchants often linked through Masonic societies.[30] The Lyonnais elite had long cultivated the sciences for their bustling city, as can be seen in their spirited defense of the arts and sciences contra Rousseau's *First Discourse*.[31] The balloon would have offered the civic-minded leaders of this stagnant commercial and industrial city an opportunity to float a shared hope for a better future.[32]

The "enthusiasm was not universal."[33] The epic struggle against the inclement weather caused many more delays and fractured the fragile coalition. The harsh winter that did not allow the interment of bodies would be remembered for the misery it unleashed on the poor. Even after they carried the balloon material "triumphantly" to Les Brotteaux, incessant rain made a horrible working condition for the fragile linen-paper bag that was too large to be hauled inside.[34] The hope for "the great day of triumph" was repeatedly dashed for the prospective aeronauts and frosted spectators.[35] On January 10, the day began with half a dozen cannon shots at 6 AM, which called all Lyon militia to their stations. The whole town came out and lingered on throughout several futile efforts to raise the balloon. Women lost color and turned limpid to present "hideous sights."[36]

As days became weeks, pressure mounted on the intendant who had to accommodate foreign dignitaries at his residence, numbering more than thirty according to an unfriendly estimate. Prospective aeronauts grew anxious and pushy because Pilâtre talked of cutting their number to three—himself, Joseph, and the young prince. Flesselles also worried that the delay kept the workers away from their productive employment and the military commanders from their posts for far too long. Much rode on the patchy machine weighed down by numerous interests. Only its public success could confer glory to the nobility, the royal administrator, the inventor, and the collaborators. In failure, it could expose the problems Étienne and Réveillon had carefully hidden from public view—vainglorious ambitions of idle aristocrats and cunning projectors, uneven distribution of material resources, and inadequate scientific institutions to support technical innovation.

Le Flesselles finally broke free on January 19, 1784, only at the threat of serious violence. According to an unpublished source, the young prince Charles put a pistol to the collaborator Fontaine's neck and threatened to fire.[37] Although Joseph and Pilâtre repeatedly pointed out the machine's fragility and the need to reduce their number, the armed aristocrats refused to give up their seats at the center of mass adoration. The intendant would rather stage an inferior flight than force them out of the vessel. Confusion reigned over the ship's command, seriously compromising its mission. At the last minute, they threw out the ballast and three quarters of the fuel to lighten the load. Lyon academicians took up their stations at the Observatory, having measured the distance from the launch site, to calculate the height of the ascent.

A vast cloak of spectators covered Les Brotteaux, feasting their eyes on the imposing sight. A grey linen globe topped by a white cotton dome and tapering down to an inverse wool cone, *Le Flesselles* (see plate 6) displayed two medallions that represented history and fame. The pavilion carried the intendant's coat of arms along with a placard that read: "Le Flesselles. Madame Intendant, led by M. Montgolfier, attached this pavilion & declared herself godmother of the Balloon."[38] Tucked carefully inside were small pieces of oiled paper attached to lead weights, asking whoever discovered them to send a notice immediately to Mr. Intendant of Lyon, "noting the number, the place, the day, the hour, and if possible, the minute at which one sees the balloon," along with detailed observations on the height, the route, the winds, and so on.[39] A patriotic Lyonnais had to be a methodical observer.

The scene of mass absorption visualized a hopeful city. Never had anyone experienced "a mixture of such overwhelming, yet labored sentiments —a rapture of ecstasy that mixed joy, fear, and admiration that made the spectators voluntarily kneel on the ground."[40] Their arms stretched to the heavens, their spirit relished the inexpressible moment in a profound silence. In about thirteen minutes, the battered machine suddenly stopped rising even with strenuous efforts at feeding the fire. Pilâtre noticed a long vertical rent near the new dome where the fabric had been damaged by fire, possibly due to the difference in vertical and horizontal pressures.[41] Somebody instinctively poured water over the stove, nearly extinguishing the fire and thus imprudently putting the voyagers in peril.

Le Flesselles came crashing down from the height of 1,438 toises in three minutes, regardless of Pilâtre's desperate effort to slow it down by feeding the fire.[42] The sudden crash caused serious anxiety among the crowd, but the aeronauts emerged unhurt from under the massive envelope, except for the marquis d'Anglefort's broken tooth and the prince's bruised leg. Consternation gave way to a joyous parade to celebrate "the most astonishing spectacle" the city had ever seen.[43] When a chevalier draped Pilâtre in his cloak and put him on his horse, young men vying for the "glory of holding the bridle" created a furor. In the midst of incredible tumult and acclamations, others flagged down a carriage and ushered in Joseph and other voyagers.

There ensued a "triumphal march of a completely new genre," as if the Cyclops leaving the Vulcan's forge had come to life. An immense crowd followed the cortège to express their admiration, surprise, and joy for their safe return, shouting all around "Vive Montgolfier & Pilâtre, Vive le Prince de Ligne & all the courageous voyagers" who led this voyage "memorable" in size, weight, height, and the number of voyagers.⁴⁴ Once the voyagers had caught their breath from the traumatic fall, Flesselles led them to the theater for his moment of glory. The opera *Iphigénie en Aulis* was under way, but they stopped the performance and brought out seven garlands. Madame de Flesselles gave one to Pilâtre who relayed it to Joseph Montgolfier, inducing new applauses and bravos without end. The opera resumed, but the actress gestured homage toward the intendant's lodge when the scene suited the purpose, which renewed the applause. When the spectators found Augustin Montgolfier and the young collaborator-aeronaut Fontaine in the parterre, they pushed them onto the stage. When the performance ended, the crowd and musicians accompanied the voyagers to the military commandant's place where they dined.

Thus ended the story of *Le Flesselles*, an artifact that cost "24,000 livres and more than a million worker-days" according to an unpublished account. This sharp critic predicted that it would no longer be possible "to amuse the workers with ordinary entertainments."⁴⁵ He could not have foreseen the traveling guillotine as an extraordinary substitute. A few days later, young collaborators hosted a grand festival at the Hôtel de Vengeance.⁴⁶ The festive mood continued among the aristocratic inner circle as well. The abbé relayed home several grand projects under consideration by Prince Charles and the comte de Potocki (Stanislas-Félix, 1753–1805), a Polish aristocrat and Freemason.⁴⁷

The Montgolfier family gained few advantages. Flesselles pressed the city council to open a subscription for a pyramid and grant a suitable place for it, while asking the Academy to appoint a commission for the inscription.⁴⁸ The city council declined, noting that the experiment was not "an object of utility for the patrie," while granting Joseph Montgolfier the right of bourgeois in recognition of his talents and knowledge.⁴⁹ The refusal reflects their reservation against frivolous spending. They could not even count the Montgolfiers among their compatriots. No longer employed, Joseph returned home.

In contrast, Joseph made in early March a triumphant sweep of Montpellier, the seat of the Languedoc Estates, as the "author of a discovery that belongs particularly to this province." He gave a detailed report to the Société royale des Sciences, which made him an affiliated member (*associé*) and promised him the first vacant place. They praised this "modest inventor" and compatriot for having attracted at once the "rapture of the people and the admiration of the most thoughtful savants" who showered on him the honors reserved for "genius and distinguished talents."⁵⁰

Figure 6.2. An engraving to immortalize the Montgolfiers, approved by the Academy of Marseille. Underneath is written: *Montgolfier vole au Rang des dieux / et L'immortalité Ravie / Fiere d'un nom si glorieux / L'inscrit aux fastes du genie* (Montgolfier flies into the ranks of the gods, and enraptured immortality, proud of such a glorious name, inscribes it in the annals of genius). Gimbel Collection XL-2-1089.

Joseph was crowned at the theater and granted bourgeois status, complete with a gold bottle. He dined with local dignitaries such as the comte de Périgord, the bishop, and the royal intendant.[51] The festive occasion seared the Montgolfière in local memory (see figure 6.2).

Joseph Montgolfier was never asked to make another balloon, notwithstanding his ambitious plan for a three-hundred-foot silk balloon.[52] The only person who consulted him on ballooning was the marquis de Brantes (Marc Louis du Blanc, b. 1759) of Avignon whose wife had been receiving information from Versailles.[53] Their effort to make an elaborate Montgolfière seems to have begun with an order by the comte d'Artois for a hydrogen balloon, but the difficulty of procuring a large quantity of silk turned it into a paper balloon, while spinning off the design for a silk parachute.[54] The flight on April 4, 1784, was not successful.[55] Avignon would see other attempts by the Rouen Academician Scanégatty in June, but this balloon caught fire and crashed.[56]

Pilâtre's reputation also suffered significantly. Before the flight, the *Journal de Paris* reported on the balloon's physical characteristics, Pilâtre's status as the captain, and the reasons for the delay. After the flight, the Parisian press noted his initiatives to strengthen the balloon, lighten the load, and order precision instruments, initially ignoring (or perhaps uninformed of) the ignomious fall and Lyonnais enthusiasm. Even the gossipy *Mémoires secrets* relayed Pilâtre's last-minute objection to the excessive number of passengers, the intendant's failure to dissuade noble passengers in his eagerness to please such "illustrious voyagers," and the safe return of all voyagers. Still, the precipitous fall did not help the famous aeronaut's reputation. Pilâtre returned quietly to Paris, stopping at Dijon where Guyton de Morveau and Bertrand were constructing a rather sophisticated hydrogen balloon.[57] He had to shore up his lectures at the Musée de Monsieur, which had languished during his six-week absence. The "partisans of M. Cailhava" had established another public assembly on the same street, rue Sainte-Avoye, and began to publish their proceedings in public newspapers.

The *Mémoires secrets* mocked this self-fashioned god, feasted and crowned in Lyon, who came back to Paris as a simple mortal in an ordinary transport and began to make all kinds of excuses for the machine.[58] Reporting that the machine had descended "rudely," the *Mémoires secrets* pointed out Pilâtre's "false pride" (*amour-propre*) shown in his expressive disdain toward the provincials' exalted enthusiasm, aggrandized ideas, and feverish heads; in his "grand dispute" with Joseph over the honor of being the first captain; and in his incessant emphasis on the project's difficulty. The Lyon experiment was deemed "less than nothing."[59] Pilâtre himself would not have been satisfied with the flight, an astute critic noted. Pilâtre wrote to Faujas and made deferential gestures toward the Academy of Sciences to "merit" their esteem. Making a living on public opinion made him a puppet attached to their strings.[60]

Winning the Parisian approbation proved difficult for a provincial balloon, which also compromised its promise of cultural integration. There was no way to hide the inglorious fall. *Grimm's Correspondance* noted the "little success" of the Lyon experiment, while giving a high praise to the balloon scene in the opera *La Caravane du Caire*.[61] Skeptical from the beginning, as can be seen in Lalande's caution that such a large balloon could not be steered, the Parisian scientific establishment regarded the Lyon ascent as a failure. Ridicule circulated more virulently in Parisian gossip to denigrate the Lyon experiment in public opinion. Ever mindful of his symbolic capital in the court society, Flesselles commenced damage control to secure favorable press coverage. Lyon academicians also spent several months on a detailed report with calculations to vindicate their involvement.[62]

FOR THE PHILOSOPHER-KING: *Le Gustave*, June 4, 1784

More balloon trials followed in Lyon without much success until the comte de Laurencin (Jean-Baptiste, 1740–1812) built a sturdy machine.[63] Mostly forgotten in historical accounts, this second balloon represented the progressive Lyon elite's hope for scientific ballooning.[64] Laurencin, a member of the Masonic lodge La Parfaite Réunion and Lyon academician, was one of the seven voyagers in *Le Flesselles*.[65] His meticulous attention to the process —the solidity and impermeability of the envelope, its geometrical form, the arrangement of the cords, the design of the gallery and the stove, and the nature of combustibles—indicates that he had not been an idle spectator during the construction of Joseph's machine. He instructed the local artist Fleurant to build a new machine without issuing a subscription or applying decorations.

Making a sturdy balloon required material knowledge of the fabric, paper, coating material, and glue in addition to the Archimedean mechanics of equilibrium.[66] Laurencin wished to make them known for the public good and a compassionate, progressive polity that he envisaged. The details he printed in a public letter offer us a reliable source of what had gone wrong with Joseph's balloon. To make the envelope impermeable, he instructed the workers to properly layer and glue the paper on linen. To ensure its solidity, they used pieces of linen in their entire lengths and sewed them together horizontally for the body and perpendicularly for the other sections. Thirteen cords passed through the pole and were sewn onto the dome to make twenty-six cords for the gallery. Much calculation was done to strengthen the dome.

Staging the balloon required administrative coordination. While Laurencin's team was busy at work, the ordinance of April 23 came down. Town authorities distributed it in two parts, apparently repeating two earlier ones issued on November 20, 1783, and March 27, 1784. The royal prosecutor (*procureur du roi*) merely prohibited any aerostat undertaken by "persons without knowledge or capacity," asking the "learned and enlightened persons" to seek permission in advance. The lieutenant general of police Laurent Basset issued a stronger prohibition against all aerostatic machines with combustibles such as alcohol and fireworks.[67] When Vergennes heard of Laurencin's balloon, he gave permission expressly to stage it for the Swedish king Gustav III, traveling since September as the comte de Haga.[68] Vergennes had served as an ambassador to Sweden during the revolution of 1772, which was staged with a heavy French subsidy to create a tricky diplomatic situation.[69] Although Sweden lacked military power and Gustav fell short of complete confidence, Vergennes had to keep Russia from expanding, having spent the previous year managing the Crimean crisis. This daredevil king with a grandiose ambition and extravagant expenditure was an important diplomatic asset.[70]

Laurencin's team felt honored to entertain the Masonic "philosopher-king" revered in public opinion for his bloodless revolution. Charles Francis Sheridan's chronicle *A History of the Late Revolution in Sweden* (1778) had appeared in French translation first as an excerpt by Lescène-Desmaisons in 1781 with his discourses. The full translation soon followed as *Histoire de la dernière revolution de Suède* (1783), which characterized Gustav III as an enlightened despot who suppressed aristocratic feuds for the patrie.[71] For the ascent initially scheduled for June 8, 1784, the balloon was transported to a new enclosure that Morand had built in Les Brotteaux along the contour of the city.[72] When Gustav suddenly moved up his schedule, the entire crew doubled their zeal to meet the date.

On June 4, the team began preparations at 3 AM and finished by 4 PM. Gustav arrived at 6 PM, examined the balloon and asked questions, indicating that this was a new spectacle for him. He took his seat at the residence of Antonio Spreafico where he had the vast enclosure in full view (see figure 6.3). Women seated in the arrangement of an amphitheater presented a colorful parterre with their ribbons and parasols. Beyond the enclosure, an immense crowd lined the Rhone River against the city's most imposing neoclassical architecture designed by Jacques-Germain Soufflot.[73] At 6:10 PM, Laurencin asked Gustav to signal for departure. When *Le Gustave* got off the ground, a number of collaborators held the gallery and presented it before the king, each wearing a medal that had the arms of France and Sweden on opposite sides and a white handkerchief on his left arm. This was the sign by which the king had "distinguished his friends on . . . the day of the revolution that without shedding a drop of blood . . . substituted order for anarchy, tranquility for trouble, celebrations for public calamities, [and] laws for factions." As an "enlightened & just monarch wrapped in the power

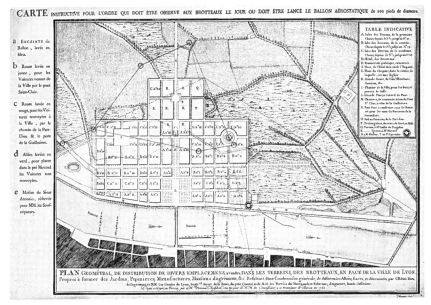

Figure 6.3. Morand's map of instruction to maintain order at Les Brotteaux on the day of the balloon launch, drawn on the plan of the parcels of real estate to be sold. It was meant to be color-coded to designate the place of ascent, the routes of the carriage traffic coming in (yellow) and going out (red), the alleys for parking carriages (green), and the house of Antonio reserved for the subscribers. AM Lyon, 14II-422P.

his people bestowed on him" with love and veneration, he had delivered them from a group of tyrants.[74]

After a sequence of precise maneuvers, the majestic machine left the earth "disdainfully . . . to establish its empire in the celestial region."[75] Laurencin was not on the machine he had built in such a painstaking fashion with serious investment, having yielded his place to Marie Elisabeth Thible, a Lyon native to whom life had not been so kind. Married without love at age twelve to an ignorant man and chained to this condition, she probably wished to console herself. Laurencin sympathized in praising her courage, composure, and determination. Upon descent, the first female aeronaut in history and her companion were honored at the Lyon Academy with a medal.

The Lyon press propagated an enthusiastic public transcript not on Laurencin's balloon but on Gustav III, the protector of Swedish Freemasons. A sensible and patriotic philosopher-king, they extolled, he had established a just political system to safeguard his subjects' happiness against the forever-feuding aristocrats who had ruined Swedish patriotism through a series of frequent revolutions. The irony of an absolutist revolution against the republic was lost on the progressive Lyonnais in search of a pragmatic, benevolent monarchy. The *Journal de Lyon* published everything they could get their hands on, ranging from Gustav's portrait to his speech to the Swedish Estates, which condemned the "aristocratic despotism" that bred hatred, vengeance, and persecution.

Public adulation of Gustav III as a philosopher-king speaks for the progressive milieu of the Lyonnais elite who yearned for a reformed monarchy. As a patriotic king who wished to govern a "free people," the *Journal de Lyon* editorialized, Gustav advocated a wise, law-bound government that would guarantee true liberty for his people. Invariable laws would secure their properties, reward their industry, establish impartial justice, expand general opulence throughout the kingdom, and bring back pure piety devoid of all hypocrisy and superstition. To attain such "happiness," it was necessary to govern the kingdom by a clear and precise legal code that would not allow false interpretations. With a solid and unshakable foundation of public order, which would guarantee true liberty, the king and the estates could work for the common good of the kingdom.[76] Laurencin's balloon, staged for Gustav III, allowed the progressive Lyon elite to articulate their desire for a just political system under an enlightened king instead of despotic aristocrats, which echoed the theme in the *Mariage de Figaro*. Such liberal royalist sympathies would resurface during the Federation movements.

ACADEMIC BALLOON: Dijon, April 25, 1784

Not far away in Dijon, the administrative capital of Burgundy proud of its literary and scientific prowess, events unfolded in a more orderly fashion under the leadership of the local Academy, which deployed a well-coordinated campaign to sell the balloon experiment as a patriotic science. Established in 1740 to counter the judicial culture of literary erudition, the Dijon Academy had maintained a strong commitment to useful knowledge. While it gained notoriety through the prize competitions that produced Jean-Jacques Rousseau's two discourses, its routine business centered more on natural philosophy and medicine. Outside of Paris, Dijon became a rare exception to the rule by successfully launching an expensive hydrogen balloon that acquired Parisian endorsement. The success depended critically on the existing institutional support for academic science, especially by the Burgundy Estates.[77]

The Dijon experimenters orchestrated a strong academic, administrative, and media control of public opinion, which reflected a well-governed paternalistic polity. After the physician Hugues Maret (1726–1786) and the lawyer-chemist Louis-Bernard Guyton de Morveau (1737–1816) assumed leadership in the 1770s, the Academy acquired substantial symbolic capital in the European Republic of Science. As a member of the Burgundy parlement, Guyton worked as a skillful advocate for the Academy. Under his vigorous promotion and Maret's internal management, the Academy expanded its library, installed a chemical laboratory, acquired a new building, and set up a series of public lectures, which contributed to its local reputation and European visibility.[78] Guyton and Maret routinely advertised the Academy's activities in the local and Parisian newspapers.[79] They could also mobilize their network of correspondents to gather information and build a superb machine.

On December 2, 1783, the recently resurrected *Affiches, Annonces et Avis divers de Dijon, ou Journal de la Bourgogne* printed a detailed account of Pilâtre and d'Arlandes's ascent on November 21, praising their courage to rise up between three and four thousand feet in the air "next to a burning furnace" in "a gallery full of the most combustible matters."[80] At the Academy's rentrée on December 4, Guyton the skilled lawyer made an urgent plea for replication: it must be astonishing for the public living "in a city where sciences and the arts flourish" that the Academy appeared indifferent to the Montgolfiers' beautiful experiment. His plan, immediately endorsed by the Academy, fixed the subscription at twenty-four livres each. If the number did not reach a hundred by December 21, the plan would be abandoned. There is no mention here that a subscription of this size could not cover the actual cost, indicating that Guyton probably expected a subsidy from the Burgundy Estates. He also wrote to local newspapers for public support, again feigning astonishment.

Public enthusiasm was immediate, judging from the odes and verses that conjured up a "charming nation, flying and light" on her path to build the "empire of the air" and renounce the empire of the sea.[81] The Academy's mixed composition of poets and physicians worked well to promote the project. The *Affiches de Dijon* helped with a long article on a bird-shaped balloon that could rise to a prodigious height, descend gently, and travel everywhere. The *Affiches* also announced the Lyon Academy's prize competition on steering the balloon to support Guyton's scientific crusade. It suppressed failures, which implies a strong hand in controlling the news that might adversely impact the Academy's experiment. Chaussiers and Dubard, students of natural philosophy, apparently tried to float a barrel-sized paper balloon in vain on December 21, 1783. A prodigious number of spectators packed the small garden of Capucins, but the affair did not make the news.[82]

Dijon academicians carefully circumscribed vulgar enthusiasm and idle curiosity in addressing the public, whose upper crust consisted of solemn magistrates, lawyers, and administrators, by emphasizing the balloon's potential utility. Their public transcript represented a well-governed province (*pays d'états*) to consolidate Burgundy patriotism, which left their balloon unnamed. None of the players involved in the process dared to put his name on it, lest it be perceived as a vainglorious pursuit of fame. The subscription filled quickly enough, or so it was claimed, to begin work in less than two weeks. By December 21, the experimenters received permission to make the machine in the garden of the intendant's old residence, although the subscription still fell short.[83] Curious notables invited Guyon to their homes for an explanation, dangling the subscription as a carrot.[84]

The endeavor had to be perceived as a scientific experiment, if the experimenters wished to obtain public funding in Dijon and public approbation in Paris. In a letter artfully crafted for the *Journal de Paris* and simultaneously released to the *Affiches de Dijon*, advertising an ascent scheduled between January 20 and 25, Guyton again emphasized the scientific character of his balloon. The twenty-seven-foot balloon would utilize new, cheaper material for the envelope, the inflammable air obtained in a more convenient manner without using any metal or acid, and the gondola designed to "steer its progress at will."[85] Parisian newspapers acknowledged Guyton as the "famous chemist, of the Academy of Dijon," which intensified their anticipation.[86]

The experimenters' need to publicize their expertise for public approbation provides us with rare information on how the hydrogen balloon was made, in sharp contrast to the Parisian precedents. They researched the subject methodically to perfect the invention, while contemplating its novel use as a motor to pump water from the mines. Guyton was in contact with other experts, including the Parisian Academician Le Roy who was also working on the means of steering the balloon.[87] Taking the cue from the Paris Academy, the Dijon academicians focused on four aspects—the envelope, coating material, inflammable gas, and the means of steering the machine. Unlike the Paris Academy, they publicly favored the hydrogen balloon over the Montgolfière whose large volume posed a singular obstacle to steering it, notwithstanding other advantages such as the low cost of the envelope and the facility of making the gas quickly and anywhere.

To reduce the cost, the envelope had to remain airtight for a long period. The Dijon academicians experimented and deliberated extensively to find a fabric that would provide a combination of solidity and levity and a varnish that would make it truly impermeable. All varieties of linen and cotton proved too heavy or too coarse. Nor could they be varnished evenly. The experimenters eventually settled on varnished taffeta, as Charles and Robert had done with their second balloon. Instead of ordering special fabric, they gathered the best available on the market—strong silk called *gros de Florence* (or *taffetas d'Italie à trois bouts*)—regardless of color.

To find the right varnish, the experimenters tested more than thirty different compositions in different proportions before settling on *la glu* (which, according to Crell's *Annalen*, had properties quite similar to caoutchouc or the elastic resin from Cayenne), a commonly sold substance drawn from holly bark. They found that a direct combination of la glu with linseed oil yielded a simple, adherent, flexible, and economic composition better than others. It dried slowly, however, causing an impossible logistical problem in the cold winter. In the end, they chose a composition of linseed oil and Copal resin that dried faster.

The team began to cut the fabric on December 28, 1783, for a flight in late January, but severe weather forced the balloon inside.[88] They commandeered a forty-five-foot hall at a great expense and transformed it into a veritable sweatshop. Making an airtight envelope was a laborious process. They cut the fabric in two-feet-by-two-feet *fuseaux* and spread them out on a large table for three coats (two outside and one inside) of hot varnish made in terebenthene. Multiple coating dispensed with strips, a difficult operation that could be accomplished only imperfectly. The pieces were assembled first two by two, then four by four, always sewn in the same manner, to make the entire envelope in twenty-one parts. The finished envelope withstood many tests to conserve its force and quality.

Research on inflammable gas, "an object of quite considerable expense," ran into many obstacles. The order went out in early January 1784 for large iron retorts to distill Solanum roots.[89] Since the unrelenting snow did not allow a flight, they continued to search systematically for a gas as light as the one made from metals and acids but at a lower cost, and made more conveniently. It remained to "determine if there is really only one inflammable gas, always identical when it is pure."[90] Notwithstanding the substantial advancements in pneumatic chemistry, the techniques of identifying and differentiating new gases were limited. The identity of inflammable air depended primarily on its specific levity for which different authorities provided diverse numbers, depending on the acid and the metal used in its production—according to Cavendish, 1/12; Fontana, 1/15; Priestley and Kirwan, 1/11; Achard, 1/6 or 1/12; and Faujas, 5/53—in comparison to the atmospheric air.[91] The search continued with many different substances, using an air pump to weigh the gas exactly, until the experimenters settled on the inflammable gas made from zinc and vitriolic acid.

Although the academicians also pondered on the pressing question of the day, or whether one could steer [*diriger*] the machine, they had a more pragmatic attitude—it depended on what one meant by steering. If one wished to transport a body aerially at all times in all winds, such perfection would require centuries of experiment and practice, a goal even maritime navigation had not yet met. In the meantime, it was necessary to develop a theory of winds, a subject incidentally of the Academy's prize competition in 1780–1783. One should at least give the aerostat a form that would conserve the direction it received, for which the sphere seemed best by occupying the greatest volume under the smallest surface.

Guyton also took care to protect the Dijon academy's precious symbolic capital that he had built up with diligence to sustain funding from the Burgundy Estates. When the *Mercure de France* attributed the delay to the machines' unfinished state, he sent a modified announcement to the *Journal de Paris*.[92] When the *Gazette de Leyde* reported troubles with the preliminary trials, attributing the delay to the gas extracted from charcoal, Dijon Academicians made a swift resolution to disabuse the public from the "news as false as [it was] ridiculous." They spent time reading the memoirs that came their way, be it on the nature of the gas, on the form of the machine, or on the means of steering it.[93] On January 22, 1784, Guyton presented a memoir sent by the engineer Émiland-Marie Gauthey (1732-1806), director of the Burgundy canal project, on constructing an aerial vessel that could be directed vertically and against the wind.[94] Maret as the secretary also channeled various balloon news.[95]

Dijon academicians had a tight leash on public opinion throughout the repeated delays, thanks to their reputation and insignificant opposition. With the snow continuing, they displayed the balloon for the subscribers in the great hall of the Academy, scarcely larger than the balloon. The public experiment was set for February 18, weather permitting.[96] Public enthusiasm in Dijon did not wither, characterizing the prospective aeronauts as classical heroes.[97] Town administrators commandeered gunpowder for the cannon from a nearby château, while banning all other balloons that carried a stove, ostensibly for public safety.[98] The retorts arrived on February 22. Realizing that most of the produced gas leaked, the experimenters abandoned distillation on February 28 and began to mix vitriolic acid and metals. Quickly exhausting the supply, they scoured neighboring cities at a great expense. When the balloon began to bounce off the ground, it was transported to the garden of the Benedictine abbey St. Benigne.[99]

A trial commenced on February 29 with the balloon only two-thirds full in the midst of numerous spectators including many foreigners. Carriages numbering more than thirty blocked the gate. The balloon rose to sixty feet, but it could not lift the gondola. Reports of these preliminary experiments were promptly sent to the local and Parisian newspapers whose editors had grown impatient as days stretched into weeks. Strong winds forced the team to fold the balloon and put it back in the grand hall of the Academy on March 15. Although this suspension prompted satiric verses and pamphlets, they were kept out of public newspapers. The rumor spread nonetheless that the balloon had cost 25,500 livres of which the public's share amounted to 9,000.[100] To fill the long wait, Guyton delivered a lecture to the Academy on constructing and steering the aerostatic machine. He also published a proper excuse in the *Affiches de Dijon*.[101]

The flight finally took place on April 25, 1784. The operation began anew at 6:15 AM to fill the twenty-seven-foot balloon and was finished by 11 AM. They attached the lightwood gondola, five feet and nine inches in length and equipped with oars for steering. Although the departure was announced only in the morning, a crowd numbering about eighteen thousand gathered to fill the places all around the garden by 1 PM. The city of twenty thousand residents had swollen with outsiders. At 4:48 PM when the cannon and the drums signaled the takeoff, "a vast silence reigned." The balloon rose with six ropes still attached since the strong wind threatened to smash it against the wall. The aeronauts, Guyton de Morveau and Claude Philippe Bertrand (1755–1792), had to throw out nearly eighty pounds of ballast to rise above the church steeple and begin a rapid ascent. After disppearing two or three times behind the clouds, they drifted out of sight.

The aeronauts landed at Magny-les-Auxonne, six leagues away from Dijon, at 6:25 PM where they dined with M. Buvee, a civil and criminal lieutenant of the bailiwick. A spontaneous festival broke out, encouraged by the artillery officers, but the aeronauts decided to return home at the late hour. They traveled in an open carriage, accompanied by many Dijonnais who had followed the balloon. On the way, they were greeted by a group of balloon workers armed with trumpets and timbals. Several highborn ladies also joined the growing cortège of at least twenty-five horses and carriages. The word of their safe landing prompted nearly three-quarters of the city's population to pour out into the streets, lining the road from Auxonne to Dijon. The path from the port Saint-Pierre to Cremolois filled with carriages, horses, and women in colorful dresses. Most returned to the city with the approach of the night. Even with the hurried pace, the aeronauts missed this spectacle, which, albeit in a spontaneous disarray, simulated the medieval entrée ceremony—a traditional rite of passage for a new ruler in consolidating his domain, which also served to affirm the rights of urban residents. No longer in use, the ceremony would acquire a radically different meaning during the revolution when Louis XVI returned to Paris on July 17, 1789.[102]

The success brought exuberant acclamations to the aeronauts-turned-heroes. When they entered the illuminated city at 8:30 PM preceded by the cavalcade, they were greeted by renewed applause from the remaining crowd and crowned with a bouquet of flowers. Several young men of distinction accompanied the chaise on horseback, carrying torches. Such a "triumphal entrée" expressed, according to the journalistic public transcript, the esteem and "sentiment of merit" that the citizens in their hearts and minds unanimously felt toward the aeronauts, and their wish to "distribute glory with justice."[103] Such a public attribution of glory to the men of science in the form of the entrée ceremony, which was traditionally reserved for new rulers in an effort to display their divine rights, would have fundamentally changed the meaning of the ceremony. Letters of felicitation arrived from Parisian and other European scientists.[104]

Parisian newspapers reported the Dijon flight as a success, but without much enthusiasm. They did not trust the provincials' claim to have steered the vessel with oars "at will" or "against the wind."[105] To address this concern, Guyton announced a second flight, apparently supported by another subscription to better equip the original balloon.[106] On May 29, 1784, the balloon was filled with common air to dry the new coat of varnish applied in several places. The following day, they tested a theory by the Dijon academician Charles André Hector Grossart de Virly, a président at the Chambre des Comptes of the Burgundy parlement, that solar heating would contribute to raising the balloon, but this experiment only damaged the balloon and delayed their departure.[107] The rumor on the street, however, was that they were waiting for Gustav III, who was scheduled to arrive on June 7 or 8.[108] When Gustav passed up Dijon after the Lyon ascent, the academicians staged their second ascent on June 12 and claimed to have "really steered" their machine "despite the sarcasms of unbelievers."[109] The Parisian press was not overly impressed by an aerial journey that lasted only an hour to land on Etevaux on the way to Pontaille (twenty-three kilometers from Dijon)—why didn't they descend in Paris instead of stopping at a poor village?[110]

The balloon developed an unprecedented capacity to congregate and impress a mass audience, which might have integrated the cultural nation by subjugating diverse provincial cultures and local patriotism under the banner of science. Despite the "universal" balloon enthusiasm, however, provincial replication depended on the complex topography of power and resources that allowed for regional appropriation. Invariably paternalistic and often republican in principle, regional power elites sought an alliance with their people while serving the royal administration. Although the complicated infrastructure of urban commerce and governance made it difficult for the regional elite to plan adequately for a mass gathering, they sought to tame the co-emergent, mutually penetrating body of the machine, the populace, and natural elements for an illusion of a harmonious patrie.

In order to maintain their public image as the paternalistic, benevolent, and intelligent elite who catered to the public good, provincial magistrates sought to control and discipline a whirlwind of bodies thrown their way. In successes and failures alike, we can discern the regional elite's serious effort to appease and educate the public. Their mediating function partially explains the early bans on the Montgolfière in Dijon and Lyon as well as the lethargic progression of the royal ban issued on April 23, 1784. While the ban immediately limited Parisian ballooning to the well-connected few, twelve provincial cities attempted to stage human ascents between April and August notwithstanding the fear of public safety, an "absolute necessity in such a considerable crowd this operation always produces."[111]

Provincial ballooning illuminates the asymmetric nation as seen from Paris and the diverse provinces in their differential capacity for scientific hegemony. The Dijon Academy pulled off two beautiful flights with an expensive hydrogen balloon by aligning the local power elite and their purses with public desire. Even so, the Academy had to control the public transcript and dictate what the balloon should mean to the regional pubic. Only with media complicity could the Dijon academicians utilize their pneumatic capacity and network to promote a scientific experiment that would serve both humanity and regional patriotism. Their philosophical nation could only be imagined as a body of citizens under a benevolent administration that would serve their interests and desires. Such an ideal entity was difficult to reconcile with the Versailles court and its aristocrats.

In large commercial cities like Lyon and Bordeaux, the fragmented civil authority and funding venues proved less conducive to a successful venture. A failure would expose the fault line under the fragile surface, which threatened to delegitimize their carefully crafted illusion. The economically stagnant and politically fragmented city of Lyon barely managed to float an inexpensive Montgolfière.[112] The second balloon accentuated the asymmetric nation by exposing the inadequacy of the royal intendant and the royal inventor in representing the public good. If Bordeaux comes across as a uniquely troubled spot (see the next chapter), its troubles speak even more dramatically for the spontaneous initiatives from diverse sectors that hampered the elite effort to maintain the illusion of a well-governed patrie.

Figure 7.1. "A cat is a cat and Jeannot is a rascal [*Fripon*]." Below is a "Dialogue of two famous physiciens Miolan and Janinet." Gimbel Collection XL-5-1181. They talk about the large number of satires, which made them the most persecuted martyrs.

7

THE FALL OF A NATIONAL ARTIFACT

The **aerostatic machine** became a national artifact by staging a majestic spectacle for a composite audience, ranging from prince to pauper. Drawing on a repertoire of public and royal resources, it forged a broad consensus on its moral value and thereby floated a vision of the philosophical nation that would utilize science for the happiness of its citizens. As a technology of crowd mobilization, inspiration, and integration, it seemed to promise a system of scientific hegemony that would include the populace. The balloon facilitated a "spontaneous distillation" of various historical forces, one could argue, to create the modern French nation out of religious and monarchial sentiments—a process that has been primarily attributed to modern newspapers.[1]

If the balloon weakened the king-machine, it failed to implement an alternative machine polity or the people-machine. Although the rhetoric of utility rallied a mixed crowd from all orders of society, the spontaneous coalition became fractured when the promise of immediate utility vanished. The "philosophical majesty" and his nation of virtuous citizens would disappear without a trace as the liminal moment of uncertain promises devolved into a revolutionary crisis. The rise and fall of ballooning in public opinion captures the fragile space of mass politics in prerevolutionary France. Just as its meteoric rise proved the capacity of ancien régime institutions to muster requisite resources for spectacular entertainments, its precipitous fall exposed the limitations in channeling those resources to productive technical, social, or political processes.

The balloon fell from public grace within a year of its debut. In addition to provincial troubles, three "degenerated" Parisian ascents during the summer of 1784 failed to maintain the allure of scientific progress and national regeneration. Once hopeful Linguet characterized them as useless, ridiculous, or nearly disastrous—"useless" because the *Marie Antoinette* flown at Versailles on June 23 merely served to entertain a foreign king and the royal court; "ridiculous" because the failure of the Luxembourg Montgolfière on July 11 induced a riot and a flood of satires; and "nearly disastrous" because the exquisite hydrogen balloon rising from St. Cloud on July 15 almost killed the duc de Chartres, the populist prince of blood and the grand master of French Masonic lodges.[2]

After the mid-year ascents, the balloon faded from Parisian public attention (if not popular affection), notwithstanding the Robert brothers' beautiful flight on September 19, 1784, and a somewhat longer entrenchment at popular entertainment venues such as the Pavillon Chinois and the St. Laurent fair.[3] The intrigue-ridden and spectacle-driven culture of power at Versailles played a large role in the Montgolfière's rise as a public artifact. Its fall may be seen as a natural process in the absolutist polity, which ran on personal favors, political factions, and endless intrigues. The "fall of the favorite" was a routine mechanism that sustained the prince's arbitrary power over his subjects.[4] One could argue that the Parisian public, in following the fashion of curiosity with their coquettish attitude, simply extended the court culture. Mercier thus blamed the balloon's undeserved demise on fashion-mongering Parisians.[5]

The balloon was not just another fashion, but a scientific artifact that exuded virtue and utility to forecast a "regime of publicity" which, in Jeremy Bentham's projection, would require an active public and transparent communication.[6] In the context of intensifying financial woes and headline scandals, however, nothing short of a balloon ship that could transport soldiers and goods against the wind would have satisfied public opinion.[7] When the prospect of the balloon's material utility dimmed, its political utility waned as well. Exactly why and how the balloon lost the ability to represent and promote the imagined philosophical nation of active citizens deserves our serious attention, especially because this national artifact of mass hope would be replaced by one of aristocratic despair—the guillotine.

BORDEAUX: A Global Province

This opulent Atlantic port of uncertain identity made it self-evident that public ballooning could become a problematic political experiment. With its rising merchant aristocracy and ingrained judicial opposition to the Crown, the city possessed a unique polity that utilized a range of royal institutions to build a strong regional pride and a global trading presence.[8] This hybrid identity as a global province—a French province that functioned as an economic unit in global commerce—made its relations to the Versailles-centered nation relatively unstable, somewhat akin to the southern plantations of America.[9] In other words, Bordeaux had developed as a self-consciously global economic unit rather than an isolated province that developed personal relations of exchange with the rest of the world.[10]

As the capital of Guyenne, Bordeaux was endowed with a parlement, a governor and his military commander, a royal intendant, and the city administration (*Jurade*).[11] The daily administration of this dynamic city, bustling with immigrants and transients, fell mostly on the Jurade, which consisted of a mayor, a lieutenant mayor, six *Jurats*, a syndic, a secretary, and a treasurer. Inconsistent directives stemming from this overlapping administrative hierarchy made it difficult "to maintain good order and public tranquility" in a boom city where scandalous riches from the wine and transoceanic trade could not eliminate bottomless poverty.[12] Bordeaux pride is easily discerned from the "theater-temple," a scandalously expensive building that opened in 1780, designed by Victor Louis (1731–1800).[13]

Regulations for the new theater reflected the city's complex polity that mixed royal institutions, regional administration, and a transient population.[14] The existing royal ordinance of 1769 had stipulated a free circulation of carriages "without any distinction" and entrance strictly by the ticket to each section (loges, balcony, and parterre), while prohibiting any disorderly conduct or interruption of actors during the performance with an extraordinary penalty. The Bordeaux ordinance of 1780 added that an officer should arrive with his troops an hour before the performance to control admission and any potential trouble, to monitor large gatherings within the theater, and to ensure a smooth circulation of spectators.[15]

The military commandant, the duc de Mouchy (Philippe Noailles, 1715–1794) still sought to block nighttime performances because they tended to create large gatherings. Even with these precautions, a major disturbance broke out in May 1783, which exposed the inchoate chain of command in policing the city. The spectators' demand for their favorite actor, the Jurats' dismissive handling, the soldiers' violent policing, and the military commandant's inability to calm the situation resulted in a public protest at the Jardin public and further military action.[16]

Mushrooming balloon trials seriously challenged the Jurats' capacity to maintain order, as James Martin Hunn has shown in his pioneering study of failed Bordeaux ascents, alerting the royal administrators to the dangers posed by popular enthusiasm and mass gatherings.[17] The trouble began with André Belleville, proprietor of the new entertainment complex, the Colisée.

With the support of the vicomte mayor Duhamel, Belleville purchased two toy balloons in Paris for a public trial on December 3, 1783.[18] The spacious Colisée, walled on three sides, seemed like an ideal locale but a crowd numbering forty thousand easily spilled over the walls. When Belleville sent up two toy balloons, a sense of betrayal gripped the crowd. The nonpaying spectators outside the privileged enclosure broke through the barrier and the guards. They advanced to the terrace and then to the café. A culinary pause there gave the guards a chance to disperse the crowd by nightfall.[19] The Jurats promptly arrested Belleville while looking for the troublemakers.

An intense power play ensued between the Jurats and the royal administrators over their respective authority in policing the city. Relieved that the incident did not result in disastrous consequences, the Jurats reported it to the marshal Mouchy, the duc de Richelieu (the governor), and the comte de Vergennes. In response, Mouchy conveyed his extreme distress and Vergennes's dismay over what was portrayed as a festival, blaming the Jurats for their lack of foresight. It was a particularly "bad lesson to allow the people to be masters of the battlefield." Mouchy seized this opportunity to demand that the Jurats follow his instructions and the governor's rules on policing spectacles. While defending their preparation, the Jurats promised more serious measures to avoid gatherings of the people.[20] They also heeded Vergennes's instructions to keep Belleville in prison for three weeks, while promising a statue of Louis XVI in celebration of the peace to appease the minister.[21] Their correspondence evinces long-standing frustration on the part of the royal administrators and the aristocratic governor with what they perceived as the Jurats' inadequate maintenance of public order.

Notwithstanding the ominous prelude, Bordeaux officials still wished to nativize the marvelous spectacle, as they had done with their splendid theater. Balloon fever spread quickly across the city. Three middling ascendants in their early thirties took the lead by launching a barrel-sized Montgolfière on December 8, 1783. It traveled eight kilometers at a height of about six hundred meters. Antoine Darbelet (a lawyer from a merchant family), Jean Chalifourt (an architect), and Jean-Baptiste-Alexandre Désgranges (merchant and the son of a prosecutor at the Bordeaux parlement) then made an eight-meter Montgolfière in coarse linen, covered inside and out with wallpaper, for an ascent on January 12, 1784, from a rope manufactory bordering the Garonne River. Unfortunately, the impatient crowd accidentally released the balloon, which came down in one minute, causing a torrent of epithets and ridicule on its makers.

After another failure on January 17, the Montgolfière finally "rose majestically" the following day amidst "the acclamations of all residents of Bordeaux," reaching the height of about two hundred meters in four minutes.²² The trio immediately began to work on the means to steer it, probably responding to the Lyon Academy competition.²³ A local captain sent the news to Étienne Montgolfier, suggesting a sail, a rudder, and parallelogram-shaped wings as the means of steering.²⁴ The stream of letters that reached Étienne reveals the provincial pride in and veneration for the modest inventor who symbolized the chimerical hope of the Rousseauean generation.

A more ambitious project took shape around Jean-André Cazalet (1753–1825) whose failure provides a striking contrast with the Dijon Academy. A member of the Bordeaux Academy and pharmacist, he taught a public chemistry course from 1780 and served as a tutor to the royal intendant Nicolas Dupré de Saint-Maur's children.²⁵ On December 14, 1783, Cazalet proposed a subscription (numbering 318) for a thirty-foot hydrogen balloon to raise twelve thousand livres. The Academy published the prospectus under its publishing privilege and contributed six hundred livres, appointing six commissaires.²⁶

The Bordeaux Musée also became involved when its secretary Chauvet (1746–ca. 1841), a founding member of the Parisian Loge des Neuf Soeurs, brought the matter to the members' attention.²⁷ Established only a few months earlier by the abbé Dupont de Jumeaux, a correspondent of the Paris Academy of Sciences and professor of mathematics, the Bordeaux Musée professedly aimed at promoting useful and amusing knowledge under the royal intendant's protection. With the motto of liberty and equality, often used in Masonic societies, it comprised diverse members who would later become revolutionaries.²⁸ The members allocated six hundred livres from the general fund and solicited additional contributions from those "animated by a true zeal" for the arts.²⁹

The beginning of Bordeaux's hydrogen balloon seemed as promising as Dijon's, then, with the two rival institutions pulling in the entire spectrum of the local scientific and power elite.³⁰ Cazalet also published an *Avis* with the Academy's permission to discourage public anticipation of a spectacular amusement that could "no longer interest the enlightened public." While he was occupied with the construction, the Bordeaux Academy publicly defended the Montgolfiers' originality, kept abreast with the Lyon Academy's prize competition, worked on the theory of aerostatics, and entertained local proposals on steering the machine.³¹ The slow subscription rate postponed the flight originally scheduled on February 18. Unlike the Dijon experimenters, Cazalet could not rely on institutional patronage to begin work early or to stop other competitors, which reflects the inchoate administrative layers.

More serious trouble came with the first attempt at human ascension with a large Montgolfière by the physician Candide Frédéric Antoine de Grassi (1753–1815) and a perfumer named Périer.[32] Having obtained permission through the comte de Fumel, military commandant of the Château de Trompette, they fixed the date for April 17. The Jurats promptly published instructions for carriage traffic, setting up barricades on the boulevards.[33] Postponed until May 3 because of bad weather and other accidents, the launch took place at the Jardin public, which was ordinarily reserved for genteel society. An enclosed space with six gates, over five hundred meters long and three hundred meters wide, it offered an ideal site with a basin at the center, albeit divided by four flowerbeds, two antique colonnades, and a café.[34] The Jurats took "all possible precautions to avoid even the idea of disorder," dispatching royal guards to circulate an ordinance around the town and posting a brigade of mounted guards at each door of the Jardin and foot patrols inside. They deployed their entire police force, insufficient as they were (at most one hundred on foot and sixty mounted), and mobilized royal guards.[35] They did not call in soldiers from the neighboring regions.[36]

Le Bordelois, the sixty-foot Montgolfière named after the city and decorated with a view of Bordeaux enthroned on clouds, receiving a globe from Zephyr, promised a glorious celebration of the prosperous city. On May 3, 1784, according to the Bordeaux Academy president François de Lamontaigne, "the entire city was put into a motion to witness this experiment." The crowd filled the Jardin, the surrounding avenues, the château's lawn and ramparts, and the windows and roofs of neighboring houses. Grassi and Périer placed their balloon (see figure 7.2) on a high platform built over the basin by utilizing a system of pulleys operating on two poles.[37] Strong winds frustrated their repeated attempts to fill it. When the experimenters decided to postpone the flight, those standing outside the garden broke through the wooden barricade and the gate. The guards drew their swords and fired into the crowd but a shower of planks and stones pushed them back. Fumel had to deploy two brigades before he could get a handle on the situation.

The riot lasted three hours during which the guards killed two, wounded a dozen, and arrested eleven. Most were craftsmen employed in various métiers—a coppersmith, a baker-carpenter, a drawer, a cobbler, a cook, a porter, a mason, a painter, a hairdresser, and a barge-master. Placing the blame instead on "foreigners, vagabonds, seamen and porters," the Jurats meted out severe punishments to set "a frightening example" for the "turbulent and seditious spirits of this sort of men who own nothing and have nothing to lose, and are free to be carried away by fermentation and the spirit of revolt."[38] On May 6, 1784, two were sentenced to death—Guillaume Gautier (a coppersmith from Villedieu-des-Poêles in Normandy) and Louis Caduc

Figure 7.2. *Le Bordelois* scheduled for ascent on May 3, 1784. Gimbel Collection XL-2-1087.

(an apprentice baker from Monpezat in the Benauge, near Bordeaux). Two others—Pierre Pompart (a hairdresser from Bazas in the Gironde) and Jean Jeanpierre (an apprentice carpenter from Saint-Mathurie in Anjou)—were sentenced to follow the tipcart, witness the execution, be branded, and serve six years on the king's galleys. The parlement reduced the latter sentence to merely witnessing the execution and released seven others.

The severe punishment for a relatively minor incident speaks for the administration's fear of public disturbance. The condemned were taken the next day in a tipcart to the Cathedral of Saint-André at the town center. They had to kneel, holding a burning torch of wax, to ask for the pardon of God, the king, and justice. They were hanged at the garden gate, while soldiers stood ready to handle any potential unrest. An exemplary execution was deemed necessary in "such an immense city as Bordeaux, which abounds in foreigners and the *menu peuple*."[39] Only the *Courier de l'Europe*, published in London outside the reach of French censorship control, reported the incident in any detail. How this terrible turn affected the population is not known.[40]

The orderly staging of a philosophical spectacle mattered to the Jurats, if they wished to retain their authority to govern the troubled city. Determined "to satisfy the engagement . . . with the public," the Jurats paid no attention to the royal ban and gave permission to the same Périer for another ascent on May 14, which did not materialize. Fortunately, Darbelet, Desgranges, and Chalifour were ready to try their luck.[41] They requested in March the use of the poorhouse, Hôpital de la Manufacture, located on the western bank of the Garonne River. Although they offered to donate the net profit to the inmates, the administrators took over the ticket sales. After the riot, the neighboring manufacturers at the faubourg Saint-Croix objected that a failure could ignite the combustibles in their establishments. Hospital administrators asked the Jurats for several fire pumps and to close the nearby grounds of Fort Louis and a Benedictine convent.

For the ascension on June 16, 1784, the trio built a sixty-six-foot Montgolfière in light fabric painted black on the outside to absorb solar energy and white inside to reflect it. It took only seven to eight minutes to inflate and the flight went off beautifully.[42] The crowd became silent with concern but turned jubilant when the aeronauts threw out their hats and leaned out of the balloon to reassure them. After an hour of flight, the aeronauts landed in a nearby field. Back in town, they dined with the lawyers who were hosting a fête to celebrate the election of their syndic.

The successful flight was a long-awaited affair that filled the town elite with a sense of pride, as the trio boasted to the Montgolfiers: "At last, Sirs, at last, Sirs, we have seen a superb aerostat flying in the sky with the majesty of an eagle."[43] The trio planned another flight, this time at the Jardin public. Although the Jurats implemented a ban on all unmanned ascents on July 6 and restricted manned flights to the persons of proven ability, this did not apply to the successful trio. With an enlarged balloon (seventy-eight feet in diameter), their second flight on July 26 lasted two hours.[44] After much celebration, they were inducted to the Musée and granted the rights of bourgeois along with their family.

The Bordeaux Academy's hydrogen balloon did not fare as well, which indicates a sizable gap in pneumatic literacy and technology between Paris and Bordeaux. Delayed significantly by the slow subscription rate and other technical issues, Cazalet finally set the date for August 8 at the Place des Armes on the grounds of the Château de Trompette for a smaller, better-heeled audience. Unfortunately, the balloon escaped in the process of filling it. Ridicule poured out and Cazalet lost his teaching post at the Musée to Darbelet. Public support for ballooning dried up after this incident. When the trio opened a new subscription a few months later, they could not find "one écu among our citizens who have been pushed to the limits of credulity."[45]

The financial requirement of material ballooning, in addition to its inherent danger to public order and property, became a serious obstacle even in this prosperous trading city that thrived on the tales of adventure. The failure of a hydrogen balloon in Bordeaux, a city of fabulous wealth and egalitarian drive, offers an instructive contrast with its success in Dijon, a city of conservative judicial wealth with stable social stratification. Expert science required disciplined personnel, intricate knowledge infrastructure, and stable institutional support. Public spectacles demanded the administrative know-how in mobilizing and controlling the public—a lesson that would not have been lost on the royal administrators.

USELESS: *Marie Antoinette,* June 23, 1784

Parisian balloon mania peaked in late spring, as seen in a celebration (probably hosted by the Loge des Neuf Soeurs) at the Waux-Hall of the Foire St. Germain on May 6. In the illuminated oval room near the entry was placed a large allegorical painting by the royal painter Charles Monnet. It represented nine muses sculpting a rock for a temple with the motto: *Their glory will be born from their works.*[46] The members scheduled to deliver lectures sat behind a table. Ladies were seated in the form of an amphitheater surrounded by male members. A military orchestra opened the academic session at 8 PM, which included a discourse by the president, the comte de Milly (Nicolas-Christiern de Thy, 1728–1784), fables by Le Changeux, and an ode by Rocher to the reestablishment of the French Marine under Louis XVI. After a concert intermission by famous musicians, the session continued with literary works by Imbert and Nogaret and ended with Rocher's ode to Pilâtre, d'Arlandes, and the Montgolfiers. Afterwards, the attendees viewed in the upstairs gallery the works of famous artists such as Claude Joseph Vernet, Jean-Baptiste Greuze, Jean-Antoine Houdon, Jean Pierre Houël, Charles Monnet, Donat Nonnotte, Étienne Fessard, Charles Étienne Gaucher, François Godefroy, and Jean-Louis Couasnon. Music resumed for the subsequent banquet, singing, and a ball.[47] The balloon and its projectors had become honored guests in this elite society of intellect.

With the mesmerists growing louder, it must have been a reprieve from the troublesome fluids for the royal administration to welcome Gustavus III on June 7.[48] The fashionable society immediately focused their attention on this self-fashioned philosopher-king. His movements created a visible track of credible propaganda for an enlightened monarchy. He attended a performance of the *Mariage de Figaro* on June 9 and the public séance of the Académie Française on June 15, which attracted "the most numerous and brilliant audience" who desired to see the Francophile sovereign renowned for "a great revolution." As soon as the heroic prince appeared in "the tribunal," *Grimm's Correspondance* extolled, this "public assembly" of literary talents in thunderous applause became "the organ of the entire nation." The marquis de Montesquiou (1739–1798) ended his reception speech by praising Gustav's "love of the public good." The enlightened despot surrounded by a learned company supposedly represented the ideal nation to the privileged audience.[49]

The itinerary of royal pleasures—including a ball hosted by the duc de Chartres at the newly finished Palais-Royal on June 18 and a balloon ascent at Versailles on June 23—only raised a specter of frivolous consumption.[50] Étienne Montgolfier had completed a new balloon before leaving Paris, although he could not secure an adequate sum (between forty and fifty thousand livres) from the newly appointed controller general Calonne to make it useful for practical ends. He used a low-quality material too heavy to carry much load, but he strengthened the dome with 1,540 pieces of sheepskin to help it withstand the upward thrust.[51] Although Étienne left instructions for Réveillon to send the balloon after him, Pilâtre commandeered it with Calonne's letter and moved it to a new hall at the Place de la Porte, faubourg St. Antoine.[52] Réveillon had to promise his full cooperation to bring it back to his factory.

The situation did not bode well for Étienne: If the flight succeeded, Pilâtre the "zealous collaborator to Étienne in appearance" would take all credit. If it failed, Étienne the designer would be blamed.[53] He had no choice but to ask Réveillon to ensure its success by making it more seductive.[54] Slightly enlarged to eighty-six feet in height, the *Marie Antoinette* performed well for the diplomatic occasion. On June 23, Pilâtre took off from Versailles at 4:45 PM with the chemist Joseph Louis Proust (1754–1826) teaching at his Musée, avowedly chosen "to avoid all rivalry" from fifty-four vainglorious applicants.[55] They traveled for forty-five minutes and descended at 5:30 PM in the forest of Chantilly, about twelve leagues from Versailles. Although the aeronauts managed to land it gently, the wind tipped over the gallery to start a fire. They barely saved the dome and the cylinder with the help brought by the prince de Condé. After refreshments, the aeronauts left for Versailles the same evening. The place of landing was named Pilâtre de Rozier.[56]

Symbolism mattered in staging French royal power for the European courts. *Grimm's Correspondance* worked as an organ of royal propaganda by characterizing the *Marie Antoinette* as the most "magnificent and imposing" spectacle: an enormous mass emerged like "a powerful magic" in just a few minutes to dominate the entire palace with a body painted in azure blue, gold royal monograms and arms entwined with the arms of French nobility. The intervals between the cords were covered by the drapery that bore twelve crosses, each eleven feet tall and seven feet wide. An arm wrapped in a white scarf, with the hand receiving the crown of laurels, signified the newly resurrected Swedish kingdom. The gallery, painted with royal monograms and fleurs de lis, had a yellow pedestal. After receiving the two "intrepid mortals," this "astonishing machine" rose "majestically" to open a new route in the air. Struck by such a great spectacle, the editors posed rhetorically, could they not imagine that they were witnessing "the creation of a new world?"[57] The imagined globe in *Grimm's Correspondance* projected a French cultural empire.

The fortunes of the royal inventor and the royal aeronaut changed quickly. Pilâtre claimed the glory as "the true Columbus" who served royal desires despite the inventor's imperfect method that caused the landing accident.[58] He also printed a separate pamphlet ostensibly to immortalize the "first experiment of the Montgolfière constructed by the order of the king." By presenting himself as a humble aeronaut who willingly went into the "battlefield" for the king, Pilâtre rallied powerful figures to his side—the king ordered the flight; the controller general and the queen assured him that he was only charged with a new experiment regardless of the outcome; the comte de Vergennes lit the fire in the stove; the prince de Condé extended his hospitality at descent to show his "sensible and charitable heart"; a prodigious concourse of persons of all ranks gathered at the Musée to greet his co-pilot Proust; the royal family, Vergennes, De Castries, Monsieur, and Calonne all extended their grace.[59] A royal aeronaut had to be a spineless courtier.

Pilâtre's supple maneuvers alarmed Étienne's friends who sensed a complete takeover in progress. Réveillon assured Étienne in vain that "the entire court knows that this is your machine."[60] Left with a dilapidated machine that Réveillon managed to salvage and send to Vidalon,[61] Étienne sought a financial assistance of twelve thousand livres from Calonne and De Castries.[62] He should not have bothered. Calonne had already promised forty thousand livres to Pilâtre for a new balloon to cross the English Channel. Étienne's renewed request was met in early October by a short rejection letter from the inspector of finance De la Roche.[63] The royal inventor was no more.[64]

RIDICULOUS: Luxembourg Riot, July 11, 1784

The fiery landing of a skilled aeronaut highlighted the inherent danger and commercial uselessness of the Montgolfière, a fact well known to the royal administration from provincial failures. Its threat to public order became obvious on July 11 when the total failure of a large Montgolfière caused a riot at the Jardin du Luxembourg. A "seductive" prospectus had appeared toward the end of February, promising "a more complete spectacle."[65] The abbé Miolan, a lecturer of physics, and Jean-François Janinet (1752–1814), an engraver of considerable merit, issued a subscription to make a seventy-foot Montgolfière that could be steered at will.

The Academician Le Roy, a member of the balloon commission, played a critical role in publicizing the project as a scientific experiment that would contribute to the progress of the sciences and the nation. Public announcements portrayed the experimenters as humble, honorable physiciens devoted to truth and utility. Bound by the duties of friendship and eager to please the public, they were motivated "above all by the zeal with which the entire Nation is animated for the progress of sciences & for the perfection of a discovery" that honors the French. They would share the responsibility equally, appreciate the opinions of savants and amateurs, and publicize their contributions. They hoped to stage a public trial before the end of April.[66]

Despite their rhetoric of patrie and philanthropy and their homage to the respectable and enlightened benefactors of science, Miolan and Janinet were motivated by profit, judging from the ticket sales that added up to more than fifteen thousand livres. Étienne the honorable royal inventor was excluded from this commercial venture even as he struggled to obtain funds. The patrons "distinguished by birth and enlightenment" subscribed at five louis each for twenty tickets. They could follow all phases of the experiment, while each ticket would allow an entry to the innermost enclosure for three experiments performed on different days. In order "to place the greatest number of persons in view of the progress of this new branch of natural philosophy," the second tier of subscription at six livres would allow the ticket holder into the second layer of the enclosure.[67]

By March 29, an eight-foot model balloon went on display at the Observatoire for publicity. It was attached to a gallery, three feet and ten inches in diameter, with its appendage—wings designed by Palmer, paddle wheels to help maneuver landing in strong winds, a stove designed after Langue and Quinquet's lamp to produce vigorous fire and positioned to move vertically at will, and a small boat containing water and suspended under the stove to collect ashes, flames, and charcoal. By modifying its distance from the stove, one could adjust the current of air that would reach the stove. The gallery was designed by Montfort, a Neapolitan engineer working for the duc d'Orléans and operating a carriage shop, with a completely new material that was intended for making carriages and bathtubs.[68] Much stronger and lighter than ordinary carton, the new material would also turn the gallery into a floating device should it fall into a river. Two trials took place at St. Cloud "for the convenience of the largest number of amateurs" in the presence of the duc d'Orléans on April 10, before they began to construct the seventy-five-foot balloon at the Observatoire.[69]

By all measures, the project advanced smoothly. On June 19, a trial with the "enormous machine" was conducted successfully in the presence of numerous spectators including the Academicians Jean-Dominique Cassini, Edme-Sébastien Jeaurat, and Pierre François André Méchain. Taking off without prolonged efforts, the balloon assumed "the most beautiful form" long enough to test all necessary operations, its force of ascent, and the weight it could carry. Confident that it would satisfy the public, they closed the prior subscription and began to sell individual tickets at three livres at the cafés around the royal theaters, the Palais Royal, and the Observatoire.

The experiment changed course at this juncture with the marquis d'Arlandes's command. Miolan and Janinet had planned to use two small balloons to detect air currents—a hydrogen balloon elevated at 150 feet above and another balloon filled with the atmospheric air and suspended at 150 feet below the large aerostat.[70] D'Arlandes advised the team to adopt a new means of steering the aerostat based on Joseph Montgolfier's principle, which consisted of a lateral opening through the balloon. They also decided to enlarge the balloon to 112 feet in height and 84 feet in diameter, perhaps to outperform the Versailles show. The last trial on June 30 was successful nonetheless, carrying up nine passengers in addition to nine hundred pounds of ballast. The trial was witnessed by a company of distinguished scientists including the duc de Chaulnes, Cassini, Jeaurat, Méchain, and the comte de Milly.

The public flight was scheduled for July 11 with four aeronauts—Miolan, Janinet, the marquis d'Arlandes, and the mécanicien Bredin. All of Paris was "avid for the day of their triumph" at the Jardin du Luxembourg, the "theater" of the experimenters' choice.[71] Provincial and foreign visitors such as Madame Cradock abandoned their plans in favor of this fashionable event.[72] The crowd was the largest ever, according to Réveillon, which did not please the elite clientele accustomed to the usually quiet garden.[73] While the Palais Royal "swarmed with courtesans and washed-out libertines" seeking the most indecent entertainments to create "a noisy tumult . . . of confused and free-roaming voices," the Luxembourg Garden usually attracted more serious visitors for a "wise, tranquil, solitary and philosophical" promenade in silence.[74]

Nearly six hours passed after the scheduled time of noon, but the gigantic balloon refused to inflate in the summer heat. When they announced that the balloon would not take off, a riot broke out. Angry people forced the barriers past the exhausted guards, destroyed the balloon, chased the culprits, and scattered the balloon pieces all over Paris. The crowd had exercised their right of retribution to punish those who betrayed their trust.[75] Barely having escaped the scene with their lives, Miolan and Janinet pleaded with their patrons for a financial rescue.[76] D'Arlandes found himself at the Café du Caveau defending the monstrous machine. This "lamentable history" continued on the stages of the *Foire* and through all kinds of songs and caricatures.[77]

The failure could have been predicted, if the experimenters had paid any attention to provincial developments. As Laurencin noted, the Montgolfière worked much better when the surrounding air temperature was significantly lower than the hot air inside. Miolan would realize only too late that both of his successful experiments had taken place at dawn with the thermometer at 9R (52 F).[78] The size of the aerostat, its permeable envelope, and the inconvenient place further compromised their endeavor. The Luxembourg disaster turned off even "the most enthusiastic partisans of the aerostatic science," allowing the anti-balloonists to ridicule the balloon fashion and its projectors who had seemed invincible just a day earlier.[79]

It was perhaps for the better that Étienne avoided the debacle of a large Montgolfière burnt and torn to pieces at the heart of Paris (see plate 9). The social fallout from such a public failure would have been more than he could bear. A plethora of visual satires that suddenly burst into the public sphere (see figure 7.1 and plate 10) could mean that the Charlists and their patron the duc de Chartres took up this opportunity to avenge their earlier sufferings at the hands of the Montgolfists and their patrons—the queen's faction.

NEARLY DISASTROUS: A Princely Fall

The all-too-public humiliation of a Montgolfière would not have been good for the royal image. To make matters worse, the entire court society soon attended a grand fête staged by the duc de Chartres, the prince of blood who loved to compete with the royal couple for the public spotlight.[80] He had commissioned the Robert brothers for a hydrogen balloon at the cost of forty thousand livres back in March for a flight on April 15. According to the hopeful rumors, he would depart from St. Cloud with six aeronauts and descend on St. James's Park in London "to merit the promised compensation . . . by George III for the daring French" who would travel by air from Calais to Dover.[81]

The Roberts built an elaborate machine, taking advantage of the Paris Academy's research and advice.[82] They settled on the cylindrical form, fifty-two feet long and terminated by two hemispheres (each thirty feet in diameter), which doubled the balloon's solidity while diminishing its surface and aerial resistance by a quarter. The gallery was a rectangle with a blue pavilion topped with a golden eagle. Instead of a sail, the most common choice in steering the balloon, they decided to use an immense rudder and two six-foot circular oars made in taffeta and attached to a lever sixteen feet and six inches long. To regulate vertical movement, they suspended a smaller balloon filled with ordinary air within the larger hydrogen balloon, like a fish bladder. They could change its weight by adding or subtracting the exterior air, thereby moving the machine vertically without disturbing the equilibrium.[83] The idea of using an ordinary air balloon as a variable ballast had been known for nearly a year.[84]

The prince's balloon had begun to circulate in Parisian gossip by July.[85] Just two days after the Miolan fiasco, the entire court society descended upon St. Cloud, the site that Henri III had once designated as his capital city but saw his assassination instead.[86] Nothing—uncertain date, inconvenient hour, or the trouble of dressing up—could stop fashionable women from converging on this beautiful spot. Ordinary people walked to the site and spent several nights. The flight finally took place on July 15 with the duc de Chartres, the Robert brothers, and their brother-in-law Hullin. For the early morning spectacle postponed from the previous day, "an infinite number of carriages" (about two thousand) drove through the night to arrive by 4 AM. The entire "beau-monde de Paris" gathered around the Orangerie before 6 AM.

The crowd could enjoy a full view of the picturesque site around a superb basin because the surrounding lawn was gradually elevated like an amphitheater, curtained by beautiful greenery. This "magnificent amphitheater," ready to stage one of the "greatest marvels of human industry," recalled for some the ancient Olympic games. The balloon in its cylindrical form offered a "new spectacle." Nothing was more magical, elegant, and noble than the form of this new "aerial chariot."[87]

A "singular" scene unfolded when the duc de Chartres, the grand master of the Grand Orient, entered the car and saluted the crowd. Responding with acclamations, those in the immediate vicinity kneeled on the ground, crouched as if "in adoration before the machine and the prince." When he gave the signal, the machine took off promptly in the midst of "the assembly thus graduated, quivering in fear and pleasure" to be lost from spectators' view in three minutes. The kneeling gesture, apparently induced by the complaints from behind that they could not view the scene of ascension, had been used only in the army's "murderous discipline" according to the *Mémoires secrets*.[88] A troubling vision of princely Assumption from the royal viewpoint, the spontaneous scene of adoration was not reported in the censored newspapers.

Public newspapers focused instead on the precipitous fall that happened about twenty minutes later. Once the balloon rose above the clouds, traversing from twelve to fifteen hundred toises vertically in about seven minutes, winds tossed it about, rendering all of its instruments useless, breaking the silk cords that sustained the bladder balloon, and blocking the valve designed to release the enclosed hydrogen. The balloon continued to rise unchecked, threatening to explode with the sun heating the gas inside. Sensing the impending peril, the duke poked a hole in the envelope, causing the balloon to tumble down in a few minutes.

Contestations over scripting the princely fall underline the symbolic meaning of his public ascension. At stake were the duc de Chartres's honor and political capital, with which to counter scandalous libels as in *Vie privée ou apologie des Très-Sérénissime Prince, Monseigneur le duc de Chartres* (1784).[89] Ridicules and explanations began to circulate immediately, although many spectators noted that the fall of this "new phaeton" did not diminish its merit or their pleasure. The *Mémoires secrets* mocked the prince for his cowardice; Linguet's *Annales* mounted an eloquent defense of his intrepidity; the *Gazette de Leyde* endorsed his quick action that saved all the passengers' lives. In addition to contributing "his purse and his own person to the progress of an art yet little known," the prince earned public gratitude by donating his aerostat to the Academy of Sciences.[90] The Roberts would petition to the Academy, fix the balloon through a public subscription, and stage a beautiful flight at the Tuileries on September 19, 1784, the anniversary of the first Versailles experiment (see figure 7.3).[91]

Parisian balloon ascents in the summer of 1784 took place amid the increasing royal apprehension over the balloon's potential threat to public safety and political order. As the consensus slowly began to emerge that steering the balloon against the wind was impossible, public opinion began to sour definitively. The balloon's inevitable visibility and powerful mass agency made it all the more urgent for the politiques of various persuasions to demonstrate that it served the public good, lest it should become another genre of useless spectacle that merely distracted the people from

Figure 7.3. The St. Cloud balloon by the Robert Brothers that was used for the September 19 ascent from the Jardin du Luxembourg. Gimbel Collection XL-4-1156.

their productive activities and moral thoughts. It did not help that the Montgolfière—a royal object that should represent the French nation—was "infinitely dangerous" because of fire hazard.

It seems that the public transcript on ballooning was consciously dismantled. Parisian newspapers duly noted the mid-year ascents, but without enthusiasm. They showed much pomp but no progress, *Grimm's Correspondance* concluded, even while relishing the flight of the *Marie Antoinette*.[92] Now that the floating pyramids at the Cherbourg harbor provided a "new way of bounding the sea," had the art of penetrating the air made a similar progress? Linguet asked rhetorically, to answer decisively in the negative, notwithstanding his earlier defense of the duc de Chartres.

The millions spent on these pyramids seem to have escaped his notice or knowledge. He judged harshly that new balloon trials produced nothing to strengthen the earlier hopes of easy transport and future uses. If the aeronauts could accomplish nothing more than penetrating the veil of clouds behind which nature hid her "vast & efficient laboratory," it would truly be a children's toy. To become more than a puerile amusement or a chimera, the flying carriages had to show a steady march toward routine air travel. Their slow progress justified the English reception with disdainful silence and cool curiosity. Pilâtre's and Charles's "quite flowery" narrations only exhibited literary pretensions without value.[93]

ACADEMIC JUDGMENT

The Paris Academy of Sciences spent the year of 1784 investigating and researching two popular sciences, only to condemn mesmerism as quackery in early August and ballooning as impracticable in mid-November. Their report on August 9 charged that mesmerism deceived only the susceptible, the poor, and the ignorant.[94] Having designed experiments to separate the effects of animal magnetism from those of "imagination alone," the commissioners judged that the mesmeric touch was useless and only induced dangerous crises. They implored the public to take the report seriously as "an excellent model" of the scientific method in drawing conclusions from experiments via "the chain of reasoning," or a calculation of probabilities.[95] In conjunction with the report by the Royal Society of Medicine, official reports sought to accomplish "a great revolution in public opinion" against the mesmerists and the hypothetical existence of animal magnetism.[96] Such a concerted effort by the royal administration, its scientists, and the conservative media reveal their increasing anxiety over the invisible fluid and its peopling power.

Official censure only inflamed the public debate over mesmerism, which began to dominate the newspapers. Unlike ballooning, which attracted a mixed bag of opportunistic courtiers, commercial projectors, science brokers, career-seekers, and vainglorious daredevils, mesmerists mostly comprised wealthy idealists and their advocates who pursued science for the public good. Not only did the mesmerists rise up against "a cabal of self-interested academicians" by publishing a flood of pamphlets, they also wrote a petition to the parlement calling for the destruction of tyrannical medicine, "the oldest superstition of the universe" that seized man in the cradle and burdened him with the force of religious prejudice.

Nothing seemed to work as an antidote against mesmerism. The government apparently printed and distributed twelve thousand copies of the

Academy's report, trying to float the idea of an edict outlawing mesmerism. Nicolas Bergasse (1750–1832) argued that this administrative move violated the first principles of natural law that should safeguard justice and morality. Upon his request for an honest investigation, the parlement appointed a commission on September 6, which nevertheless failed to act decisively, mired in their usual circumspection, even with some partisans and protectors of mesmerism in its midst.[97] Other anti-establishment radicals joined in by publishing a stream of pamphlets that tune us to the subterranean mass politics. The "great subject of all conversations in the capital" was still animal magnetism in early October.[98]

When the mesmerism controversy refused to die out, the royal administration staged an anonymous play *Les Docteurs modernes* at the Comédie Italienne on November 16. It was written by Jean-Baptiste Radet (1752–1830), a Burgundian writer working as a librarian to the duchesse de Villeroy. The "government had the wisdom to sense," according to *Grimm's Correspondance*, that "the weapon of ridicule would be more powerful than all the decrees and prohibitions," which were "not only useless, but sometimes even dangerous." The play generated resounding laughter, which promised to deliver a "more certain death" to the new sect of mesmerism than any academic report or legal maneuver could.[99] The mockery incensed the mesmerists: the conseiller au parlement Jean-Jacques Duval d'Éprémesnil (1745–1794) distributed a pamphlet during a performance to denounce the "new genre of despotism, that of ridicule with which the authorities are armed to stifle the truths they do not want to recognize." Unable to get a serious response from the official sector, he denounced the play's authors as "the lackeys who ridicule, under the protection of the authority, a man of genius quite superior to Newton."[100]

Mesmerism nevertheless withstood "even the most biting shafts of ridicule," especially in the provinces, which worried the royal administrators.[101] The Société d'Harmonie multiplied like Masonic lodges, netting over 340,000 livres for Mesmer by June 1785.[102] Utilizing a theory of "universal fluid" and thereby usurping the public appeal of science for the medical practice that stood in desperate need of a fundamental reform, mesmerist organizations tapped long-suppressed spiritual elements and provided an outlet for the emergent radical ideologies of the 1780s. Almost all cities (Strasbourg, Bordeaux, Lyon, Grenoble, Dijon, Montpellier, Marseilles, Douai, Nîmes, and so on) in the kingdom offered Mesmerist cures by 1785.[103]

The steady spread of mesmerism over a series of official condemnation begs the question: why did the balloon fall so quickly in public opinion, especially after forging a hegemonic consensus across the social hierarchy on its potential utility and morality? One obvious answer is that the balloon was an expensive commodity that required a total coordination of resources including royal parks and police. It was not an object that could survive underground, but it became a problematic political experiment because of its profound mass appeal. It formed a mass public to visualize the emancipatory potential of a fluid that broke the chain of gravity. The balloon crowd was too large to be tamed within the ancien régime.

Neither could the balloon prove its immediate utility to warrant continuous support. In comparison, mesmerism promised medical utility to attract public-minded writers and elite patronage. It failed to develop a hegemonic discourse in large part because of its unorthodox methods and the profit margin that threatened medical professionals. Such a failure seems to have produced, paradoxically, many erstwhile representatives who voiced diverse hopes and frustrations, which provide us with a range of critical discourse that are not available in the case of ballooning. Although the balloon quickly lost its public representatives after the mid-year ascents, radical malcontents and hopeful benefactors of humanity continued to fuel the mesmerist movement.

In addition to the cost and distant utility, the balloon did not fit the literary public sphere that was shaped by the theatrical polity—one that thrived on factions and controversies. The slow pace of the improvements made on the balloon fell far short of this fast-turning world built on literary propaganda. In other words, its material power in enlisting the populace did not translate into an ideological propaganda that could mobilize the progressive public. The balloon embodied a dream too beautiful, perhaps, to survive in the culture of calumny that governed the public sphere.

The interplay of public and hidden transcripts on popular science can help us discern how political maneuvers shaped these transcripts to forestall a potentially transparent "regime of publicity," as Jeremy Bentham would demand of the national assembly during the French Revolution.[104] The suppression of popular sciences and their representatives may have been a rational power play on the part of the royal administration in an attempt to control the emergent mass public that challenged the administrative technologies of social control. If the royal administrators succeeded, their vision of benevolent monarchy would have steered the populace away from the path of violent revolution.

We can identify a path of reform designed to regenerate the monarchy based on republican principles by means of scientific progress, then, if we move away from the teleology of violence in the French Revolution.

However, the philosophical nation failed to materialize. The "philosophical majesty" and his nation of virtuous citizens would disappear into the revolutionary turmoil. The national artifact of mass hope would be displaced by one of death—the guillotine.

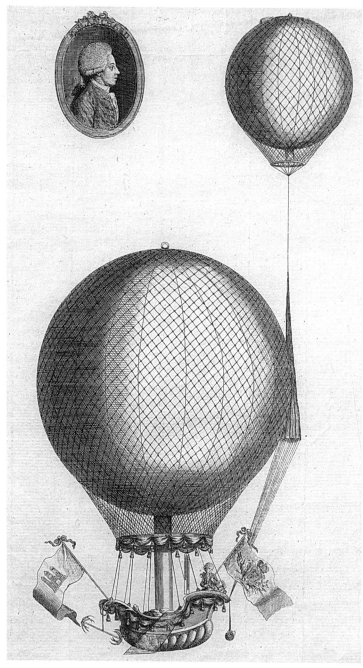

Figure III.1. Jean-Pierre Blanchard's balloon with the flags representing Hamburg (left) and the French Crown (right). Staatsarchiv Hamburg 720-1_284-13.

PART III

MATERIAL EMPIRE

The balloon became a new fashion that swept over the European continent and beyond, subject to cultural translations. As a French spectacle par excellence, it focused princely desire, aristocratic attention, patriotic resistance, and popular enthusiasm and fostered a patchwork modernity as fragile as its body. Masonic aristocrats who could afford to travel and patronize various forms of entertainment, local administrators eager to appropriate fashionable pleasures for political stability, politiques intent on expanding the enlightened public, and the populace who did not meet their expectations, all formed a composite audience that changed complexion at each locale. Balloon translations reveal diverse polities, publics, and levels of scientific and material knowledge.

The French scientific artifact did not automatically engender a uniform cultural empire. Notwithstanding its promise of universal wonder, the balloon's journey across linguistic, cultural, and political borders took effort in securing materials, sites, transport, patronage, and sometimes political will. The enlightened despots—Frederick the Great of Prussia, Joseph II of the Holy Roman Empire, and Catherine the Great of Russia—quickly banned ballooning, citing its potential danger to public safety.[1] The geography of European ballooning maps the technologies of communication and transport that built the Republic of Letters as well as the various polities that fragmented this imagined republic of universal citizens.[2]

Balloon literature proliferated in several languages but material ballooning developed divergently, limited to prosperous urban sites. Some

Italian states began to stage balloon trials and human ascents soon after the French.[3] Balloon news and pamphlets, including Faujas's semiofficial history, appeared in Italian as soon as they became available.[4] Dutch scientists took up ballooning early and eagerly translated balloon literature, which indicates their robust pneumatic literacy, but the revolutionary events swept away their efforts.[5] British ballooning took root in a couple of years despite a slow takeoff, but few translations from French literature appeared on the market. Faujas's history was never translated into English, except for long excerpts in various magazines.[6] Instead, John Southern and Tiberius Cavallo published English treatises, which strengthened British priority claims for the theoretical invention of the aerostatic machine.[7]

We can infer from these developments an equivalent level of material knowledge in France and Britain and, to a lesser degree, in some Italian states and the Dutch Republic. In comparison, German scientists translated Faujas's history in various versions, while making no attempt at human ascension, which hints at the lack of material resources and cultural motivation.[8] Pneumatic literacy and network played a vital role in the European reception of ballooning.

Complex balloon geography reflects diverse cultural and political appropriations. The balloon did not unite Britons, except for their claim to its theoretical invention.[9] The London media remained diverse in their opinions and coverage, commercially motivated in advertising balloon ascents, and indiscriminating in their choice of heroes. They reveal a public sphere infused with political debates, commercial concerns, and individual interests rather than one disciplined by administrative control. The imperial metropolis possessed an expansive infrastructure of commerce, transport, communication, and knowledge that allowed commercial projectors to replicate the new French artifact, but the media and the gentlemanly elite did not forge a system of scientific hegemony or envisage an alternative machine polity based on ballooning. London ballooning began initially in the commercial sector and was undertaken by foreign adventurers. In contrast, English provinces and the Irish capital generated balloon enthusiasm far more pronounced than in London.

The itinerant career of Jean-Pierre Blanchard and his boundary identity offer a rare measure of cultural translation across Europe. Blanchard moved to London because he could not make a decent living in Paris as a balloonist. French theatrical polity required that entertainers acquire symbolic distinction at their patron's expense and maintain a lifestyle proximate, yet subservient, to that of the aristocratic.[10] As an experienced French aeronaut in London, Blanchard helped create a hybrid zone that mixed the Francophile aristocracy, gentlemanly scientific community, commercial sector, and the populace to activate the balloon's extraordinary promise as a crowd-forming spectacle. In this hybrid zone that mixed diverse social elements for unstable representations of nature, he strove to outshine the other adventurers that the imperial capital attracted in droves.

Cultural translations often turn on misrecognition. The prestige of French science and, ironically, Blanchard's marginal status in French ballooning helped him to become a respected aeronaut in London. Few British notables patronized his flights. The American-born physician John Jeffries seized this unique opportunity to cement his royalist identity, having served British troops during the American War.[11] Their collaboration outpaced Italian and English competitors, which revealed London's complex metropolitan culture that offered multifarious opportunities for self-fashioning foreigners. Nevertheless, Blanchard's expensive shows did not fit snugly in London's entertainment sector, which catered primarily to the middling audience. His failure in negotiating the commercial topography of London launched his career as the most famous balloonist in continental Europe.

Blanchard cut a noticeable figure in revolutionary Europe thanks to his itinerant ballooning, although his fame has been much obscured in history. His balloon served as a brilliant example of French civilization that would build a cultural empire to induce, paradoxically, a reflexive need for distinctly national cultures and polities. Material necessity associated Blanchard with the local aristocracy and administrators who commanded his marvelous artifact at their pleasure in order to solidify their public authority. He decorated his balloon with a local flag during his aerial voyage and presented it to the local authorities (princely patron, titular ruler, or city magistrates) upon his return (see figure III.1). This routine ceremony transformed Blanchard's nondescript balloon into a malleable emblem of regional polity, which was then left behind at the city hall or in the palace.

While the continental aristocracy welcomed the balloon as a symbol of progress that vindicated their rule, German patriots opposed its frivolity that created only an illusion of unity to exploit the populace. Scandals followed him around, partly because the local power elite compensated his show extravagantly while paying scant attention to the poor among them.[12] His ascents in the German-speaking states became mostly associated with the reigning polity and its administrators, which aggravated the patriots who were plotting a moral and educational enlightenment of the populace. Blanchard's balloon itinerary charted a contentious pathway of German patriotism that would solidify as enlightened nationalism after the Napoleonic invasion.[13] Modern European nations came into being under French cultural dominance and military provocation.[14]

Material artifacts and spectacles often flow over local structures and integrate diverse communities into a cultural empire well before attempts at political integration are in operation. Their facility and efficacy in crossing political and cultural boundaries pose troubling questions on their imperialist uses, notwithstanding the simultaneously evolving nationalist rhetoric.[15] The travel and contact of peoples and things helped forge modern nation-states and empires by constituting and maintaining their boundaries, often with violence. Border zones gained a "paradoxical centrality" in mapping the geography of mass Enlightenment and European modernity, that is, by facilitating the movement of mediating bodies and artifacts across cultural boundaries and thereby shoring up the possibility of modern polities and cultures. The identities of European nations contra the French were shaped by a series of "trans-local" actions that produced a "heterogeneous modernity."[16]

The balloon mapped a liminal geography of mass Enlightenment, revolutionary ferment, and modernity that laid the groundwork for the Napoleonic Empire. Where Blanchard went with his balloon can help illuminate the relationship between Enlightenment cultures and political revolutions at the turn of the nineteenth century. Wherever the balloon traveled, crowds constituted a liminal zone in which the existing order and the uncertain future intersected to visualize an emergent nation of citizens. For this reason, we must pay attention to the balloon ascent as a technology of crowd mobilization that crisscrossed the continent, creating an illusion of a shared material empire, which ironically focused the patriotic desire for a distinct nation.

Plate 9. Miolan and Janinet's burnt balloon. Gimbel Collection XL-5-1185. Notice the animal figures on the left running away from the scene.

Plate 10. Definitive judgment in favor of Miolan and Janinet. Gimbel Collection XL-5-1186. They are again depicted as an ass and a cat.

Plate 11. *The East India Air Balloon*, December 30, 1783. BM J,2.122 © The Trustees of the British Museum. Fox hangs from a globe that depicts the East India House, holding a paper inscribed "[Bill of] Pains and Penalties." A large fox's brush hangs from his person, inscribed "The Man of the People." George III (left) holds a pair of scales balancing half a royal crown and "America." North (right), given the "Letter of Dismission," begs for the office.

Plate 12. The reception of British diplomats at the Court of Pekin, September 14, 1792. NASM A19680064000cp03.

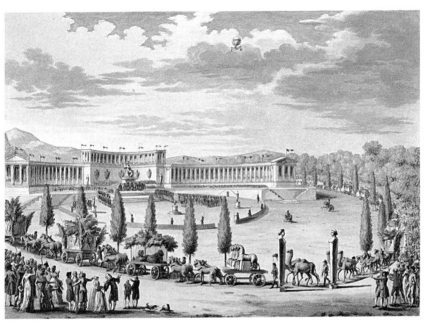

Plate 13. The triumphal entry of the monuments of sciences and arts in France, the festival on 9 and 10 Thermidor, An VI of the Republic. NASM T20140072272.

Plate 14. *Current Fashion. Ascension of Madame Garnerin*, March 28, 1802. Gimbel Collection XL-5-1399.

Plate 15. Fête du sacre et couronnement de leurs majesties imperials on December 2, 1804, seen from the Place de la Concorde. NASM A19780300000cp02.

Plate 16. Entrée of Louis XVIII on May 4, 1814. NASM A19780300000cp02.

Figure 8.1. "Political Balloon, or the Fall of East India Stock," December 4, 1783. BM J,2.125 © The Trustees of the British Museum. A satire of Fox's momentary triumph with the Bill that appropriated mining rights from the directors of the East India Company. They are shown falling.

8

MODERN ATLANTIS

Balloon news reached London soon after it began to circulate in Paris, but the gentry and the respectable media remained noticeably cool toward the new French invention. Spectacles fulfilled a different sociopolitical function in London than in Paris, especially after the unpopular American War. Royal splendor and emulative consumption did not dominate London's entertainment culture as they did in Paris. Although London's fashionable elite (*beau monde*) had developed a lifestyle much like the French aristocracy, a measure of public Francophobia tamed the aristocratic taste in all things French.[1] Neither were the gentry as overtly seduced by any dizzying array of novelties.

If the British public warmed up to the notion of royal grandeur when their empire seemed invincible, they now faced radical politics and a constitutional crisis. While the parliamentary radicalism of John Wilkes had turned a corner with the Gordon riots in June 1780, the "images of smoke and fire" lingered as a specter of mass violence. A mass petition (headed by Lord George Gordon) for the repeal of the Catholic Relief Act had resulted in the widespread destruction and killing of several hundred protesters.[2] Peace negotiations in 1783 brought the sobering loss home to spark an acute political crisis.

The balloon became a timely tool of visual satire during the parliamentary debate over the East India Bill. The *Political Balloon, or the Fall of East India Stock*, printed on December 4, 1783, caricatures Charles James Fox in his soaring success to appropriate Indian gold and silver from the directors of the East India Company (see figure 8.1). George III dismissed Fox and appointed the younger William Pitt on December 22 to end the Fox-North coalition, however, which precipitated a parliamentary confrontation with the Crown that threatened to destabilize the constitutional monarchy established after the Glorious Revolution of 1688.[3] The *Aerostatick Stage Balloon*, printed on December 23, presents three tiers of passengers ready for a departure perhaps to the moon. In addition to the morally suspect ladies and quacks on the top and bottom layers, the middle layer is occupied by political outcasts—North, Fox, the Duke of Portland (Lord-Lieutenant of Ireland), and Edmund Burke. Irish volunteers and their push for independence had become an issue as thorny as the parliamentary opposition to the American War. In *Original Air Balloon*, printed on December 29, the balloon labeled "America" is freed by Benjamin Franklin and escapes the contending forces of Spain, France, and Holland, which serves to marginalize the Fox-North coalition.[4] The *East India Air Balloon*, printed on December 30, 1783, announces the coalition's demise in vivid color (see plate 11).

A sensational French invention that posed a potential military threat could hardly have been cause for celebration when the loss of the American colonies forecast an imperial decline. As the astronomical debt began to surface in public consciousness, George III desired alternate paths that could "repair the mischief." He turned his attention more to domestic policies, recognizing that the outward stream of capital, industry, or population challenged the "preservation of the British power and consequence."[5] Although he gained the upper hand over the Fox-North coalition in the constitutional struggle, his young minister William Pitt sought to rehabilitate the royal image as "a genial, homespun farmer" rather than a "drain on Britannia" or an "oriental tyrant."[6] Agrarianism became the banner of national integration and patriotism in the effort to piece together the nation as a "moral community."[7] "As imposing and luxurious in appearance as they were hollow," Paul Keen notes perceptively, balloons could evoke a modern commercial culture "pried free from the foundational certainty of landed wealth and subject to a fluctuating network of exchanges."[8]

The general election in the spring of 1784 stirred strong sentiments against illicit aristocratic influence, which focused on the Duchess of Devonshire who canvassed votes for Fox.[9] Painful recriminations of imperial humiliation and dismemberment rang across Britain to shape the somber public mood that manifested in the commemorative concerts of George Frederick Handel's death.[10] Influential opinion makers such as Joseph Banks, Samuel Johnson, and Horace Walpole watched the balloon development with caution, worried that a useless amusement might spark something far more dangerous.[11] Satirical uses of the balloon, as in the anonymously published

Modern Atlantis (1784) and Elizabeth Inchbald's *Mogul Tale* (1784), indicate a robust political culture that could wield the critique for a meaningful reform. Those who characterized London (or Britain) as a "modern Atlantis" in the process of self-destruction could also configure a transcultural framework that would refashion British attitude toward the world.[12]

Reservations on the part of leading scientists, respectable media, and economy-minded administrators left London's ballooning largely in the hands of foreign adventurers and the enthusiastic populace. Merchant adventurers—figures of popular fascination and elite contempt—were regarded as a necessary evil that threatened moral fabric and social order at home and imperial interests abroad.[13] Even less welcome were the foreign adventurers who "infested" the capital to "assume every false appearance that can deceive the credulous and unwary" and to artfully insinuate themselves into the fashionable society.[14] Strong reactions against foreign adventurers reflected a metropolitan culture that fostered an intricate geography of science, pleasure, and patronage and thereby provided multifarious opportunities for a variety of projectors.[15]

The slow uptake of ballooning in London reveals the importance of gentlemanly scientific culture in moderating royal power, Francophile aristocratic taste, and vulgar enthusiasm. Regulated less by royal institutions and official titles than by peer reputation, London gentry had developed a more discerning taste for spectacle and an intricate hierarchy of refined sentiment.[16] The French patriotic invention initially became no more than a curiosity for the vulgar multitude in London.[17] The balloon failed to acquire a hegemonic public transcript as a virtuous scientific artifact that would forge a harmonious nation of citizens. Instead, it floated the specter of French invasion, which made British visitors in Paris follow the balloon experiments there "with great assiduity."[18]

THE ROYAL SOCIETY OF LONDON AND GENTLEMANLY SCIENCE

Scientific institutions modulated the development of material ballooning in competing imperial capitals.[19] The Paris Academy of Sciences had outwardly embraced balloon projectors and performers per royal request, setting off a wave of emulation nationwide. French Academicians worked for the Crown, notwithstanding their citizenship and symbolic capital in the Republic of Letters. The Royal Society of London did not have similar prestige in the Republic or obligation to the Crown. Fellows relished their independence as the gentlemen who would define British national interests on their terms. In its capacity to bring together Francophile aristocracy and plebian populace through commercial staging, the balloon posed a unique challenge to the gentlemanly scientists who claimed their moral authority over the nation-empire.

London science depended on power brokers like Joseph Banks (1743–1820) and his gentlemanly friends who maintained a loose liaison with the state. Having inherited the family fortune at age twenty-one, Banks joined the Royal Society in 1766, when it had lost much of its scientific luster since Isaac Newton's passing in 1727. Banks acquired instant fame as the traveling botanist on James Cook's Pacific voyage on the *Endeavour*. Caricatured on his return as a "botanic Macaroni," or fashionable botanist, he lacked serious scientific reputation even after he became president of the Royal Society in 1778. His enemies considered him deficient in scientific credentials and sought to undermine his charismatic presence and powerful network.[20] In order to gain control, he began to mobilize his allies to appoint Charles Blagden (1748–1820) as the Society's secretary.[21] The turmoil within the Society would last until the end of 1784.[22]

Encumbered by his botanical network that spanned the globe and a rebellion within the Society, Banks had to tread carefully on balloon news. Although he hoped privately that the discovery would rapidly increase the "stock of real knowledge" and improve transports, publicly he did not profess a serious interest in the French novelty.[23] Banks employed a dual strategy: he collected balloon intelligence judiciously through his scientific network and then peddled it among his gentlemanly friends, while feigning disinterestedness in public. Visiting Paris to restore communication with the Academy of Sciences, Blagden sent balloon news as early as July 4, 1783.[24] After Blagden's return, news continued from Benjamin Franklin, Faujas, and the naturalist Pierre Marie Auguste Broussonet (1761–1807). In sending concise and intelligent balloon reports, Franklin hoped that men as reasonable creatures would "settle their differences without cutting throats" and steer military expenses to the works of public utility that would have turned England into "a compleat Paradise."[25]

Summer recess prevented Banks from presenting Franklin's letter to the Society.[26] Blagden read it instead to the Thursday dining club, an informal gathering of Fellows. On this occasion, the group included William Watson, Sir George Baker, Joseph Planta, Alexander Aubert, John Hunter, John Smeaton, Henry Cavendish, and Samuel Hemming, a friend of Banks "who by his silly questions, & impertinent interruptions, made himself a greater nuisance than ever."[27] Banks's well-heeled friends who feigned interest in science annoyed Blagden who relayed Society's communication regularly to Banks residing at the Soho Square. The Society's librarian Jonas Carlsson Dryander also sent letters, newspaper clippings, pamphlets, and miniature balloons that conveyed Parisian development blow by blow.

London newspapers began to fill with balloon intelligence from early September. Commercial newspapers such as the *London Chronicle*, the *Morning Herald and Daily Advertiser*, the *Whitehall Evening Post*, the *Morning Chronicle and London Advertiser*, and the *Public Advertiser* began to report regularly on developments in Paris.[28] Hoping that English industriousness and inventiveness would turn it into "the newest, cheapest and most expeditious" transport, the *Morning Herald* even published a technical letter of Antoine De Parcieux, the nephew of the Academician Antoine Deparcieux (1703–1768), with detailed calculations on hydrogen balloons.[29] In order to satisfy public curiosity, the *St. James's Chronicle or the British Evening Post* translated a long article from the *Journal de Paris*.[30] The *European Magazine* included a plate in its October issue to give "some idea of this new mode of traveling," presumably for Charles's ascent from the Champ de Mars on August 27, 1783, but identified as the Montgolfiers'.[31]

As the president of the Royal Society, Banks had to oblige all sectors and their numerous questions with precision and speed. At the rentrée of the Society, Banks read Franklin's letters, a pamphlet, and some prints including an "exact representation" of the Versailles balloon, but he did not press the Fellows for replication.[32] Blagden combed all available sources to stay ahead of London newspapers, searching through Dutch newspapers for technical details. Parisian newspapers arrived in packets, leaving London scientists "entirely ignorant" in between. When a packet of the *Journal de Paris* (covering from August 1 to September 17) arrived, Blagden quickly turned the pages "to hunt for information about the great aerostatic experiment" at Versailles on September 19.[33] A traveling antiquary supplied Banks with a copy of De Parcieux's *Dissertation sur les Globes aerostatiques* and drawings.[34] News of Étienne Montgolfier's tethered experiments at the Réveillon factory in mid-October reached London only in early November.[35]

The Royal Society maintained a public posture of indifference, however, which discouraged the balloon's image as a scientific invention. When mini-balloons arrived in mid-October, one intended for Lord Bute (tutor to George III), Blagden showed it only to a limited circle of gentlemen. He admired the goldbeater's "most delicately thin" skin but found it "so new, so subversive of our established habits" as to prevent its assimilation. Blagden warned the Society's dining club members to steer clear of the potential projectors such as the artist John Russell (1745–1806) who almost "jumped out of his skin . . . at an opportunity of beholding the wonderful sight." Blagden continued the conversation instead with Watson, Richard Kirwan, Paradise, Cavendish, and Aubert, while waiting for Aimé Argand. Soon, the toy balloons sent by Franklin also arrived via Dr. Price. Blagden still preached against "the folly of simple imitation," although he was willing to participate in "some capital experiment" that would flatter his identity as a gentleman philosopher. For the moment, he would retain "true philosophical tranquility" rather than be seduced by "Montgolfiers Flying Medusa," as Broussonet named it.[36]

The rumor that George III asked the Royal Society in vain for a balloon ascent probably indicates these gentlemen's deliberate public stance on the vulgar enterprise masquerading as science.[37] Such feigned indifference also masked the internal division between gentlemanly and aristocratic Fellows. It would be a secretary to the Neapolitan ambassador—the prince of Caramanico (Francesco d'Aquino, 1738–1795), a Freemason and a member of the Royal Society since his arrival in 1780—who would stage the first human ascent in London. Parisian rumors of partisan bickering and aristocratic frivolity did not help the gentlemanly attitude. The duc de Chaulnes, who visited the Society's dinner club frequently, relayed the vicious partisan spirit between the Montgolfists and the Charlists that deployed ridicule and invective without mercy. Frivolous parties by the duc de Crillon (Louis des Balbes de Berton, 1717–1796) and the duc de Chartres (who apparently served up a balloon pie) seriously undermined the respectability of ballooning and its advocates in London.

FOREIGN ADVENTURERS AT THE STRAND

London newspapers filtered Parisian balloon news to denigrate French national character, a strategy often used to prop up British national character, which discouraged a public transcript on the balloon's scientific and moral identity.[38] The press initially named the new artifact the "Air Balloon," a term previously used in pyrotechnic entertainments to describe the canvas or pasteboard container for the rockets that burst in the air.[39] The editorial tone also fell far short of Parisian enthusiasm.[40] The *European Magazine* reported it only with the French government's apprehensive warning of potential terrors. The *Gentleman's Magazine* highlighted the "most ridiculous light" shed by French wits, while acknowledging the contribution of British pneumatic philosophers such as Robert Boyle, Joseph Priestley, Henry Cavendish, and Richard Kirwan.[41] The difference between "a dull Englishman" and "a lively Frenchman" who talks "with equal volubility on fashions or philosophy, especially when it is *the fashion* to talk on philosophy" was quite evident, the *Morning Herald and Daily Advertiser* editorialized, in the lively conjectures and animadversions on various airs. Considering "how *very little*" French interlocutors generally knew of the subject, "a person who had never considered French character would really be astonished to hear how *much* they have to say upon the subject."[42]

Unable to secure royal protection, scientific cooperation, or media enthusiasm, the initial batch of London trials fell squarely within the genre of novelty shows that spiced the consumer culture of the imperial metropolis. London entrepreneurs favored the hydrogen balloon, which was more durable and required less space than the Montgolfière. Francesco Zambeccari (1752–1812), formerly a mercenary Italian soldier in the Spanish army, launched a five-foot balloon on November 4, 1783, from Cheapside, sponsored by the Italian artificial-flower-maker Michael Biaggini. This stirred some public interest, but "the more respectable part" of the Royal Society became more guarded against "Ballomania."[43] Undeterred, the Italian pair made a ten-foot hydrogen balloon in oiled silk, tarred at the seams and covered with gilt, designed to carry about sixteen pounds. They exhibited it for a shilling per entry at the Lyceum, a building used for the annual exposition of paintings and located in the Strand, London's vibrant, sex-ridden shopping district.[44] Released on November 25 at the Artillery Ground for half a crown per entry, it descended two and a half hours later at Graffham, a village located about forty-eight miles from London.[45]

London normally outpaced Paris in news coverage, but the lackluster press response produced a pamphlet to peddle the balloon. William Cooke anonymously published *The Air Balloon* (1783) in late November to render balloon experiments "familiar to the plainest capacity" by explaining the properties of air, the methods of filling the balloon, and its possible uses "for the Benefit of Mankind."[46] By addressing the productive class long primed for useful inventions, he wished that their *"strong but unlettered minds"* would turn the balloon into "one of the most novel and serviceable discoveries" of the century by discovering how to steer it. As the inventor was a papermaker, Cooke wished, an English miller might "add wings, or some aerial rudder to guide it through those regions with certainty and precision." An improved machine could dispatch war intelligence, make observations at sea, send messages during sieges as in Gibraltar, and extinguish fire. English philosophers and artists must "erase the ravages of war, by the cultivation of useful and ornamental science." The balloon experiment would allow the relatively slow progress of natural philosophy to keep up with the moral philosophy that spread "the spirit of toleration" even to those countries where "fanaticism and bigotry seemed to have taken up their eternal residence."

Cooke's short tract of about forty pages had gone through four editions by early 1784, while Faujas's lengthy *Description* was never translated even in part, which made it difficult for the British public to appreciate French scientific pretensions. The balloon's sensational impact can nevertheless be gauged from other uses. A pantomime, *Lord Mayor's Day, or a Flight from Lapland in an Air Balloon*, made a debut in November and ran for over a year.[47]

Figure 8.2. George III and royal family viewing a toy balloon in the garden of Windsor Castle. Gimbel Collection XL 5-1178.

Royal curiosity had to be satisfied by foreigners and in the seclusion of Windsor Castle. Upon an invitation by Jean André De Luc (1727–1817), Aimé Argand took an apparatus to make pure inflammable air for a thirty-inch balloon made with goldbeater's skin. He entertained the royal family and about two thousand spectators on November 24, 1783.[48] He had just arrived in London armed with introductions from Faujas and Broussonet to promote his new lamp, unaware that elite British scientists had been primed "to explode" his adventurous scheme.[49] Notwithstanding the gentlemanly censure, he was elected on October 29 to the Coffee House Philosophical Society, which included Magellan, Josiah Wedgwood, William Nicholson, James Watt, Matthew Bolton, Richard Price, and Joseph Priestley.[50] The modesty of the Windsor experiment with a toy balloon (see figure 8.2), which contrasts sharply with French royal balloons, would have served to distance George III from the frivolous French invention. He had extensive education (with Lord Bute) and long-standing interest in exquisite scientific instruments, which would induce him to sponsor the largest telescope in Europe by William Herschel (1738–1822) at the cost of four thousand pounds sterling.[51]

To make a profit in a tight urban space, London balloonists displayed the balloon for a prolonged period in a busy commercial district before the launch. Biaggini soon made a sixteen-foot balloon "of force sufficient to carry up a child of eight years old."[52] Exhibited at the Pantheon, an elegantly decorated building normally used for masquerades, concerts, operas, and fashionable assemblies, Biaggini's balloon drew a diverse crowd that included the inventor Aimé Argand accompanied by the duc de Chaulnes.[53] The poet-tourist Anna Laetitia Barbauld went to see the balloon first and then the famous novelist Frances Burney, following the reigning order of "public curiosity."[54] The "vast concourse of people," promptly reported in newspapers, impressed and worried newsmakers and royals alike.[55] Extensively advertised in all newspapers throughout December, the balloon was set for release on January 13, 1784, from Mackenzie's Rhedarium. The admission fee was fixed at two and a half shillings.[56]

TRANSLATING HUMAN ASCENTS

The sharp division between popular enthusiasm and gentlemanly disdain began to erode noticeably with the manned ascensions in Paris, especially among the aristocratic Freemasons. Notwithstanding the parliamentary standoff over the East India Bill, successive grandmasters of the Premier Grand Lodge of England—the Duke of Manchester and the Duke of Cumberland—traveled to France with their wives for balloon shows. Their public trail would have undermined the British government's effort to maintain a somber façade. An anecdote circulated that the Duke of Cumberland nearly perished on the crowded Pont Royal but for a French soldier who rescued "the brother to the king of England."[57] The Prince of Wales, who had recently acquired his own household with parliamentary aid, raised many eyebrows when he appeared at the Queen's birthday celebration dressed in "air-balloon satin embroidered down the seams with silver" to upstage the Majesty.[58] A new trade of balloon makers emerged to satisfy market demand. The discerning public could only hope that the whimsical Parisian fashion of balloon hats would not invade England and cause still more inconvenience at the theater.[59]

Private correspondence and public newspapers began bursting with balloon news. The balloon engaged "all mankind" as a "wonderful and unexpected addition to human knowledge," according to Samuel Johnson who heard that a daring projector had raised eight hundred pounds sterling to make iron wings "better than Daedalean wings."[60] The balloon had "already invaded the whole Earth," according to Giovanni Valentino Mattia Fabbroni, a Florentine correspondent of Joseph Banks. John Strange also reported from Venice of a Milan trial.[61] Banks acknowledged that the balloon experiment now warranted full replication: the French could receive the credit for "Practical Flying," while Britons deserved the credit for "Theoretical Flying," citing Bishop Wilkins's precedent and Henry Cavendish's soap bubbles filled with inflammable air.[62] The *European Magazine* supplied relatively accurate information on the first two Parisian human ascensions.[63] The *Annual Register*, edited by Edmund Burke (and carrying articles on history, politics, and literature), began to report major balloon events in detail.[64]

The imperial rivalry persisted in muted forms, downplaying the French royal Montgolfière in favor of the public hydrogen balloon. The *Gentleman's Magazine* carried only a brief report of d'Arlandes and Rozier's ascent in the section on foreign affairs, while publishing Charles's narrative of the "memorable aerial journey" on the front page of the December issue, professedly to "gratify, as early as possible, the public curiosity."[65] *The Air Balloon* (1784) reported that Charles and Robert's ascension drew a crowd of between three and four hundred thousand, adding maliciously that on their arrival, "they were arrested by order of the King, who, as *father of his people*, was advised by some bigoted Ecclesiastics to prevent the father endangering the lives of his subjects." Such distortions served to ridicule French royal pretensions. Expressing sympathy for the aeronauts' plight, the author hoped they would be speedily discharged since Charles's next experiment was designed to fly from Calais to Dover, accompanied by Bougainville, the erstwhile French competitor to Captain Cook. While a design to cross the channel circulated in French court circles (as we have seen in Deslile's failed subscription), this was the first public exposure in print, which would have raised the specter of an aerial competition to supplant the naval one.

The balloon's destructive potential worried many. Horace Walpole hoped that these "new mechanic meteors will prove only playthings for the learned and the idle, and not be converted into *new engines of destruction to human race, as is so often the case of refinements or discoveries in science. The wicked wit of man always studies to apply the result of talents to enslaving, destroying, or cheating his fellow creatures.* Could we reach the moon, we should think of reducing it to a province of some European kingdom."[66] A judgment day seemed near, in the poet William Cowper's opinion. He could easily conceive "a thousand evils which the project must necessarily bring after it; amounting at last to the confusion of all order, the annihilation of all authority, with dangers both to property and person, and impunity to the offenders." If he were "an *absolute* legislator," he would order instant death for a man convicted of flying, notwithstanding potential blame from the philosophers, scholars, and historians who would reproach him of stupidity and oppression. The world would go on quietly, however, perhaps with less liberty and more security.[67]

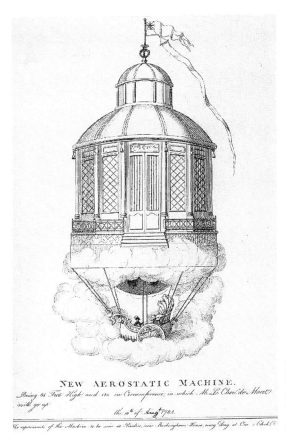

NEW AEROSTATIC MACHINE.

Figure 8.3. Moret's balloon that failed on August 10, 1784. Upcott scrapbook 2:175, Smithsonian Libraries.

Without firm endorsements from the scientific elite or conspicuous material support from the fashionable society, balloon experiments in London multiplied in the rather haphazard manner typical of British science.[68] Surging public interest only invited more entrepreneurial endeavors as well as frequent failures. Zambeccari issued a subscription on January 3, 1784, addressed to "the nobility and gentry," to raise eight hundred pounds for a fifty-foot aerostatic globe. He wished to steer it, make various meteorological experiments, try a new odometer of his own invention, and measure the speed of vertical movements.[69] By early March, Zambeccari forecast a flight "with his majesty's permission" from Hyde Park, but it did not materialize. He had willing collaborators—the silk manufacturer Frederick Breillat at Spital Square (for eleven hundred yards of silk, oiled at Saxay and Goldin's warehouse, St. Martins le Grand), the carpenter Gillow at Oxford Street, and the watchmaker Simpson at Petty-France, Westminster—who attest to the strength of London's commercial sector.[70]

Another prospect, the *Chinese Temple* (60 feet high and 95 feet in circumference with the capacity of 33,400 cubic feet; see figure 8.3), circulated in the press by the "learned Chevalier de Moret, a genius for distinguished discoveries."[71] Denied the use of the Artillery Ground, the project dragged on until August. Allen Keegan of Oiled Silk and Lawn Manufactory, a merchant of umbrellas and other waterproof articles in the Strand, began to make balloons for the Covent Garden Theater and other customers.[72] The fourth edition of *The Air Balloon*, published in early 1784, reported some gain in public opinion, although the author claimed that the indolent, skeptical, and unlettered part of humanity would stifle any new discoveries by modern philosophers such as Columbus, Galileo, and the Montgolfiers.[73]

Respectable London magazines and the gentlemanly public still looked down upon the populace attracted to the marvelous spectacle. In reporting the large number of spectators gathered for Astley's launch in St. George's Fields on March 12, the *Gentleman's Magazine* noted an "ample harvest for the pickpockets." The magazine also sought to undermine the Montgolfiers' reputation by identifying the antiquity of the principle. By quoting from Dr. Powel's *Human Industry* (1661), it pointed out that Pythagoras's student Archytas had made a wooden pigeon that could fly by means of the rarefied air enclosed within. It also mocked the accidental origin of the Montgolfiers' ideas, "the conical paper cover of a sugar-loaf thrown into the chimney which remained suspended by the smoke."[74] The "Grand Aerostatic Pyramid," scheduled to launch from the Ranelagh Gardens in May, ended in complete failure to raise further concerns.

The gentry found another reason to shun ballooning in the French royal ordinance of April 23. The *European Magazine* noted that "these precautions are not intended . . . to let this sublime discovery fall into neglect, but only that the experiments should be confined to the direction of intelligent persons." The necessity of a similar ordinance, or a restriction "on the madness of fashion," derived from the fire hazard. Such was "the *uncontrouled freedom of Englishmen in their philosophical as well as every other pursuit*," the *Gentleman's Magazine* lamented.[75] Even so, these skeptics could not ignore extraordinary French accomplishments, especially by notable scientists such as Guyton de Morveau in Dijon.

NATIONS WITHIN: Irish Patriots and Scottish Britons

Provincial enthusiasm easily outpaced London, as can be discerned from the newspapers, which indicates a high level of pneumatic awareness and literacy among the British provincial elite. The pneumatic network that was centered on Edinburgh and Birmingham defined the frontier of British natural philosophy, which allowed provincials to strive for balloon distinction.[76] Itinerant bodies helped the cause. James Dinwiddie (1746–1815), a Scottish lecturer of experimental philosophy then staying in London, released a small hydrogen balloon "much resembling that of a pocket of hops" on December 18, 1783, from the Bowling Green Tavern near the Buckingham Gate. He then traveled with it to Bath, Bristol, Waterford, Salisbury, Exeter, Portsmouth, Southampton, and Winchester.[77]

Members of the Lunar Society of Birmingham received balloon intelligence early through visiting luminaries such as De Luc, but they regarded it mostly as an amusement.[78] They were more interested in the properties of gases than in the show of human ascents, true to their philosophical and gentlemanly aspirations. As Joseph Priestley put it, they were aerial philosophers not aerial navigators.[79] They also lacked hands-on knowledge of coating materials. Erasmus Darwin, newly established in Derby, launched a four-foot hydrogen balloon on December 26, 1783, as an inaugural event for the Derby Philosophical Society.[80] The leaky silk envelope, though painted and varnished, could not hold inflammable air long enough for a flight to Soho, Birmingham, or a public demonstration. Josiah Wedgwood tried to make an air-balloon with goldbeater's skin to obtain impermeability.[81]

Matthew Boulton amused himself with paper balloons. A Parisian correspondent named Pradeaux sent him a packet of information in July 1784, which contained Faujas's two-volume accounts, Bertholon's pamphlet, and copper prints of Pilâtre's and the Roberts' ascents.[82] Boulton's "explosive balloon," as James Watt recounted to James Lind at Windsor, was about five feet in diameter and filled with a mixture of one part common air and two parts inflammable air. He used a match about two feet long to cause an explosion when the balloon was high up in the air to simulate thunder.[83] Watt's assistant John Southern would publish *A Treatise upon aerostatic machines* (1785), which contained calculations and tables of the balloon's ascending power and other instructions on making various balloons.[84] Watts sent an advance copy to Banks.

Human ascension was not a mere entertainment, but a collective performance that stirred a utopian desire. If such intense sentiment was muted in London by its culture of commodified pleasures and imperial competition, it mobilized the Irish patriots contending with radicalism.[85] The project of consolidating Britain as an imperial state had not been successful in Ireland where economic integration did not produce a broad distribution of benefits or impede peasant revolts. Building on the military volunteer movement during the American War, the Irish Parliament led by Henry Grattan (1746–1820) secured legislative autonomy in 1782.[86] This effort by Irish Protestant leaders to nativize Anglican power did not win wholehearted support from the Irish Catholics, judging from the *Volunteers Journal*, "a new patriotic paper" committed to fighting against "corrupt" parliamentary intrigues and "baleful aristocratic influence."[87] The balloon rode on the patriotic maneuver that sought to patch together nativized Protestant leaders and native Catholic peasants.

Balloon news reached the Irish public from September 1783 through newspapers.[88] The *Hibernian Magazine, or Compendium of Entertaining Knowledge* even printed a representation of Charles's hydrogen balloon the following January.[89] Richard Crosbie (1755–1800) began to work on an "Aerial Chariot" soon after the Parisian human ascension. He chose to make a hydrogen balloon, which means that he had a prior knowledge of new gases. Descended from a distinguished family and educated at Dublin Trinity College, he cultivated catholic interest in sciences, arts, and mechanical inventions.

In publishing Crosbie's subscription on February 7, 1784, the Irish press emphasized that he was not a foreigner trying to exploit the multitude's ignorance and curiosity with a "child's bauble," but an Irish gentleman who ardently wished to perfect the art for his native kingdom.[90] By identifying his social standing and patriotism, they conveyed a sense of national pride invested in his scientific attempt. His supporters included prominent Irish Protestant patriots—John Hely-Hutchinson (University Provost), Charles Manners Rutland (Lord lieutenant), the Duke of Leinster, and so on.[91] When the subscription did not fill on schedule, the astronomy professor Henry Usher recommended it to Lord Charlemont, commander in chief of the Irish Volunteers. Although Charlemont regarded balloons as "silly inventions of a trifling age," his "patriotic character" inclined him to promote Irish ingenuity.[92]

Crosbie advertised his intention to solve the outstanding riddle of steering the balloon with his "Grand Aeronautic Chariot," which resembled "in some respect a boat or wherry with two masts" with a light wooden frame covered with thin silk or linen.[93] Exhibiting his machine at the fashionable Ranelagh Garden during August 1784, Crosbie sought to shore up the subscription by floating a twelve-foot model balloon—with press coverage and also conducting a public dispute with the popular lecturer James Dinwiddie.[94] It took several months before he could stage a successful flight in January 1785.[95]

During the intervening months, factional struggle within the Irish Parliament intensified, and there began a permanent decline of the Volunteer movement.[96] Peter Wilkins took aim at the inflated political rhetoric and philosophical goals that delivered no real practical benefits to the public: How about transporting their "thoughtless and distracted politicians," who were neglecting trades, manufactures, and harvests while trying to patch the constitution, to the moon whence they should bring back an adequate plan of reform that might answer all grievances and eradicate all corruption?[97] In the visual satire, *Grand Irish Air Balloon* dated October 7, 1784, Edmund Burke is sandwiched between Fox and North on a boat attached to the balloon filled with "patriotic gas" and set free by the Duke of Portland. In the balloon are two figures representing Britannia and Hibernia engaged in a sword fight.[98]

Irish mobilization drew a sharp contrast from Scottish inactivity.[99] Since the Act of Union in 1707, the Scottish landowning and commercial elite had been successfully integrated into the British Empire. The ascendancy of Lord Bute, tutor to George III, exemplified this process of elite assimilation, although the Scottish Enlightenment developed a moralistic tone.[100] Edinburgh's fashionable elite did not mobilize until after the imperial aristocracy caught on to balloon fever, except for a lone "madman" who managed to pull off something like the first aerial jump in Britain.

James Tytler (1745–1804), a former medical student of Joseph Black at Edinburgh University, was meagerly employed from 1776 as an editor of the *Encyclopaedia Britannica*.[101] In writing articles, he kept up with the emergent field of pneumatic studies and air balloons, perhaps through his professor Black's contacts.[102] Tytler soon made a thirteen-foot model Montgolfière that was in the news by June 19, 1784. Although it was exhibited at the Comely Gardens, a rundown pleasure garden modeled after the Vauxhall in London, it did not spark a successful subscription campaign. The undecorated Montgolfière Tytler completed, measuring about forty feet in height and thirty feet in diameter, was in the news from mid-July. He scheduled an ascent during the race week in August, wishing to take advantage of the crowd.

The "Edinburgh fire balloon" struggled hard against the wind to make its public appearance, losing its gallery to a fire and a mob. Unable to carry the stove, Tytler resolved to project himself into the air "like a log or piece of ballast" by inflating the balloon to the maximum, the resolution of a "madman" as he would later admit. The balloon without the ballast was nothing but a projectile that would fall as fast as it went up. On August 25, 1784, Tytler's balloon took off "with the swiftness of an arrow," but the spectators dragged it down by the rope, giving him "scarce time to taste the pleasures of an aerial journey." Subsequent experiments on August 27 and 31 were successful enough for him to claim the title of the first British aeronaut. He also raised enough money to rebuild a new gallery. The experiment on September 29 went miserably. On October 11, another attempt failed because of the small stove, which invited much public abuse. Identified as a "public enemy" in the newspapers, he was called a "Cheat, Rascal, Coward and Scoundrel by those who had neither courage, honesty, nor honour."[103]

The cold reception of Tytler's modest balloon contrasted sharply with one that would be accorded to Lunardi's, when the latter traveled to Edinburgh from London with aristocratic support. Well integrated with the imperial elite, the Scottish elite followed the balloon fashion from London, rather than trying to develop a regional tradition. Tytler continued with more unsuccessful attempts for nearly a year, a lawsuit taking his balloon away for a period, until it burned to pieces on July 26, 1785. He sank into obscurity until he was prosecuted for sedition in 1792. He fled to Ireland and then to Salem, Massachusetts, where he built a respectable life as a physician-chemist and hatched an ambitious publishing project.[104]

A HANDSOME NEAPOLITAN: Vincenzo Lunardi

A flamboyant Neapolitan secretary aged twenty-two, Vincenzo Lunardi (1759–1806) became the first visitor of the English atmosphere. Although Naples had become an independent kingdom only in 1734, the capital had developed a vibrant Enlightenment culture with complex theatrical outlay.[105] The Neapolitan ambassador Prince of Caramanico, Lunardi's employer and a member of the Royal Society, was a Freemason who would have socialized with British Masonic aristocracy. They probably lamented, as Lunardi parroted, that English philosophers observed the French invention "with a silence, and apparent indifference," notwithstanding its potential utility. Lunardi must have sensed that a balloon project would introduce him to "persons of merit and consequence."[106]

The timing was right, judging from the success of Elizabeth Inchbald's play *The Mogul Tale*, which was staged at the Haymarket Theater ten times during the summer of 1784 and scored forty-seven more performances subsequently. In the aftermath of the Indian affair that shook the foundation of the British Empire, Inchbald presented a reflexive comedy on British arrogance (as exhibited in their balloon travel) and immorality (as seen in their willful deception) in sharp contrast to the Indian mogul's knowledge and generosity.[107]

Lunardi had followed Zambeccari's experiments and failures with interests. By July 1784, he formed a more ambitious project of ascending from the military hospital grounds in Chelsea. He obtained permission to use this picturesque spot near the Thames, an "object of national attention" and princely care, by promising to donate his profit to the patients.[108] Although he wished to publicize his ascent as one commanded by the king, such an explicit honor was not forthcoming for the insignificant foreigner.[109] Lunardi then appealed to the "liberal promoters of ingenuity," which gathered allies such as the wealthy gentleman George Biggin and the physician-chemist George Fordyce (1736–1802), a Fellow of the Royal Society and a member of the Royal College of Physicians. For a guinea, subscribers could visit the experimental site and sit in the enclosure on the day of ascension. Money trickled in slowly, even with Joseph Banks and the Prince of Wales on the subscription list.

Lunardi's "stupendous Aerostatic Machine" was on exhibit at the Lyceum by early August.[110] Made of oiled silk in alternate stripes of red and blue, the perfectly spherical balloon measured 33 feet in diameter, 102 feet in circumference, and 18,200 cubic feet in volume. The *Universal Magazine* printed an exact representation of the machine, which included a pair of wings meant for horizontal movements in calm weather. An additional pair of oars, differing only in size, were designed for vertical movements without releasing inflammable air. The machine did not have valves to release the gas, which was construed as an evidence of its "rudimentary or amateur science." It was covered with a strong net with forty-five cords to suspend the gallery. The Lyceum saw numerous and splendid visitors numbering more than twenty thousand.[111]

The difficulties Lunardi experienced in coordinating his ascent, even with broad coverage in the press, highlight the lack of official patronage. His balloon came in competition with the chevalier Moret's plan for "the first English aerial journey," scheduled for August 10 in the Bowling Green, Chelsea. Moret's forty-foot Montgolfière (see figure 8.3), decorated with a representation of constellations, failed to take off and caused a riot. Breaking down the fences, "the multitude rushed in, tore it in a thousand pieces," and robbed many spectators.[112] The incident occasioned unkind reflections on "the unfortunate coalition" of Fox and North, some reporting that Moret prostrated himself at the feet of Lord North, which made a "Representation of the whole British Nation."[113] This "terror" alarmed George Howard, the governor of the Chelsea Hospital, who absolutely forbade all balloons including Lunardi's, scheduled for August 14.

Apologies and satires streamed out in the public media, characterizing Lunardi's balloon as one "constructed on a plan and principle totally different from that of Moret's," assisted by the reputable chemist George Fordyce and to be accompanied by "an English gentleman of character and fortune, whose sole motive" was to perfect a useful experiment.[114] The prohibition was "a poignant mortification" to Lunardi, "reduced to the most complicated circumstances of . . . desponding anxiety."[115]

Public frustration built up. Almost twelve months after the first French ascent, "England with all its learning, ingenuity and encouragement of the arts" had not produced "one John Bull who wishes to take an airing this way."[116] Although it would undermine the "character of an enlightened kingdom to pre-judge an experiment," Lunardi observed, the English distrusted the "Gallic vanity" that had produced such pompous accounts of aerial voyages and regarded their potential advantages as no more than "romantic visions." He could only hope that the ladies, who "regulate the opinions and manners of a nation at pleasure," could "smile into acquiescence that uncouth monster, public prejudice."[117]

Lunardi had to fall back on the aristocratic circle around the Neapolitan ambassador. He applied to the Artillery Company, an independent assemblage of officers under the command of the Prince of Wales. Reverend Walter Blake Kirwan (1754–1805), chaplain to the Neapolitan ambassador, pleaded his case with Sir Watkin Lewes, a member of the Parliament and Freemason, who promised the Artillery Ground in exchange for one hundred guineas for Sir Bernard Turner's family. They began to sell additional tickets at one guinea, half a guinea, and five shillings.

As one crisis passed, another began. The Lyceum owner, who had already seen more than twenty thousand visitors, refused to release the balloon. When Lunardi relayed the situation to the Artillery Company, their committee conceived this as a deception and threatened to rescind their earlier resolution to guard the avenues with armed men and to stock the materials necessary for fixing and preparing the balloon, unless Lunardi paid them the promised sum and an additional security of five hundred guineas. In turn, the Lyceum demanded "a part of all possible advantages that should now and in future be produced by it."

Lunardi was rescued from this tangled web of commercial interests once more by "the generous and humane"—Sir Watkin Lewes and Reverend Kirwan, who offered themselves as securities to the Artillery Company. The police under the orders of Sampson Wright and William Addington extracted the balloon from the Lyceum and provided an escort to the Artillery Ground the night before the scheduled ascension. George Fordyce, a Fellow of the Royal Society advertised as "the first chemist in Britain," set upon filling the balloon overnight "with a readiness that does honour to his character and ingenuity."[118]

On September 15, 1784, luminaries including the Prince of Wales, Fox, Edmund Burke, and Lord North took up their seats near the globe. The Duke and Duchess of Richmond, along with William Pitt, occupied an apartment at a nearby floor-cloth manufactory. Banks was absent, apparently unable to find any "species of traffic" to attempt the journey from Soho.[119] The price of admission ranged from five shillings to one guinea, the latter for a chair near the balloon. Lunardi had planned to rise with George Biggin, but the slow production of inflammable air agitated the enormous crowd, ranging in estimates from 150,000 to 350,000. The "mob was on uptoe, ready to mount the cock-horse of their own sagacity, and to trample the imposter with all his apparatus into the earth" for his *"outlandish* impudence to tell them he could fly through the air."[120] Fearing the populace, "an impetuous, impatient and cruel tyrant" that dispensed violence to any disappointment, he had to leave alone at 2 PM in a flaccid balloon that looked more like "a pear, with its stalk downwards."[121] When the balloon was actually seen to rise "with all the majesty that heart could wish," "the most profound silence prevailed." In astonishment, the multitude "beheld it with a kind of awful terror, which rather closed their lips in *stupid silence*, than prompted them to rend the air . . . with joyful acclamations."[122]

British newspapers did not develop a hegemonic public transcript on balloon spectatorship, even though every newspaper published a detailed account of the ascent.[123] Some delivered favorable accounts like the *Oxford Journal*, which reported that "the immense Multitude who attended this sublime Spectacle, conducted themselves with the most respectful Decorum, and manifested an Enthusiasm in his Favour, and a Solicitude for his Safety, that did Honour to the Character of Englishmen."[124] Others chided the "motley multitude" that murmured different sentiments ranging from incredulity and concern to callous indifference, as the *Gentleman's Magazine* carefully recorded. The variegated typology of balloon spectators offered a sharp contrast to the French press, which insisted on the model spectator:

> The notions and opinions of this *motley multitude* were certainly as various as their situations in life were different. The *populace*, who composed the far greater part of the company, were sure the *thing* could not be done by day-light, for *no Christian could fly through the air*, and Goblins and Sprites were not permitted to ramble abroad till the dead hour of night. The *next class* to these had very little more faith than their fellows: "they could not think *as how it could be* that a bubble could carry a man, and they feared the whole story was but a bubble," and so they *divin'd* it would prove in the end. The *middle ranks* were doubtful, but not without hope. The *more enlightened* were anxious for the event, and were not without sharing in that concern which every sensible mind could not but feel for the issue of so hazardous an enterprise. *Men of real science* were otherwise affected; they were at rest as to the practicability of the expedition; but they could not help expressing, by their looks, the sympathetic concern they entertained, lest some untoward circumstance should intervene, to defeat, or even to delay the execution, either of which would have been equally fatal to the adventurer. While the thoughts and apprehensions of the subordinate classes were thus occupied, those of *elevated rank, who look upon the life or death of an individual, and the good or ill success of an experiment, with equal indifference, and calculated only to kill time, and contribute to their amusement*,

diverted themselves with the bustle among the *canaille*, and laughed and talked of *Lunardi, as another Bottle Conjurer*, whom all men went to see, though no man thought possible what he pretended to perform.[125]

Financial remuneration began to crystallize only after the successful ascent. Although the exhibition at the Lyceum yielded nine hundred pounds at one shilling per head, the crowd numbering a hundred thousand brought in only about two hundred pounds sterling.[126] When it became known that the subscription did not meet the cost, another opened to net over two thousand pounds sterling.[127] In addition, Lunardi's "authentic account" soon fetched a crown (equal to five shillings) with his portrait (by Bartolozzi after Cosway), and half a crown without, to generate more financial reward. In his narrative, Lunardi delivered his experience of "sublime pleasure" without terror, contrary to Edmund Burke's imagined sublime.[128] His account went through three editions within the year.[129] Newspapers including the *Gentleman's Magazine* had refrained from printing a full account in order to protect the aeronaut's profit margins.[130]

The popularity of Lunardi's ascent, reported in all the provincial newspapers, quickened the pace of British ballooning just as the Parisian fever began to cool.[131] His balloon was on display at the Pantheon, together with the dog and the cat that had traveled with him. Lunardi spent much of his time there receiving the compliments of women. A silver medal was struck. He issued a subscription for a larger (by 5,000 cubic feet) balloon to be painted in the colors of the national flag out of "his regard and attachment to every thing that is English." The ornamented gallery, weighing seventy-six pounds, was meant to carry ten people.[132] Allen Keegan seized the moment to make an eighty-foot cylindrical Montgolfière, made with nearly thirty-five hundred yards of coarse linen and varnished with boiled oil. He installed an elaborate stove equipped with an extinguisher and a gallery of about seventy feet in circumference to accommodate four aeronauts and the fuel. After several delays and changes of site, this was destroyed on September 30, costing him nearly six hundred pounds sterling.[133] Discerning gentry remained apprehensive about the "Balloon madness" that did not accomplish any real scientific benefits.[134]

THE AERIAL SPECTATOR

The popular artifact also unleashed the Lucianic gaze on the ailing British Empire. In *The Modern Atlantis* (1784), the sage Urgando residing on a utopian island is carried by an air-balloon conjured up by the devil Uriel to the island of Libertusia.[135] Hovering over a circus where "the rich, the beautiful and the voluptuous . . . vie with each other in a pompous display of dress and equipage," Uriel presents the entire spectrum of the fashionable society in order to unveil their lifestyle of dissipation and debauchery that shattered their mighty empire.

A long list of shameful examples with concrete biographical details provides the key to the contemporary notables—the inheritors of noble family names who squandered their fortunes and reputation on horse races, dice boxes, charlatans, and women; noble women who fell from grace through illicit affairs; the adventurers who made their fortunes abroad and displayed them in gaudy manner; the adventuresses and actresses of humble origin who bedded gullible men and achieved higher status and fortune; the highborn women ruined by amorous affairs, and so on. Their shared lifestyle of extravagance, folly, and debauchery shook the foundation of the nation-empire. Lacking proper education and moral principles, the aristocracy led a decadent lifestyle focused on ruinous pleasures such as gambling, horse racing, and libertine nights. The metropolis catered to such indiscreet and immoral pleasures in the "temples sacred to prostitution."[136]

Not so subtly dedicated to the irreverent populist John Wilkes (1726–1797), characterized as a patriotic senator and religious paragon, the political message is clear.[137] Factionalism fractured the "unrivalled mistress of the world" to see "a melancholy decline" of imperial power.[138] The consummate politician (Charles Fox), the "bulwark of people's liberty," did not enjoy the favor of his sovereign, who was unable to "distinguish between the false lustre of affected patriotism, and real worth."[139] Neither did the popular senator (Edmund Burke)—suddenly seized with a reforming zeal and advocating an "imaginary system of economy"—last long. Other senators cultivated populist idolatry by preaching the *"glorious majesty of the people."* In addition, Fanaticus (Lord George Gordon) in all his apparent innocence proved to be a "shocking compound of hypocrisy, cunning, vain-glory, and deceit." Albeit young, he could hide his pride and lust under that exquisite disguise of meekness and humility to be seen as a "modern saint." His "deep-laid stratagems" shook the "tottering empire to its foundation" by working up a religious frenzy that burned the metropolis. The fanatic with such blind enthusiasm and polished hypocrisy escaped without punishment.[140]

The only hope of preserving the "tottering empire from impending ruin" lay with the senators who exemplified their "patriotic glory," exhibited "great abilities more than Roman integrity" as citizens, and provided an "uncommon example of conjugal fidelity and of paternal tenderness" in domestic life. Their leadership alone could safeguard the modern Atlantis. The unhappy king had "to distinguish between the virtuous zeal of those

Figure 8.4. Human ascents in Britain, 1784–1785.

	Others	Sadler	Lunardi
1784			Sept. 15 (London)
	Oct. 16 (London) Blanchard	Oct. 4 (Oxford)	
	Nov. 30 (London) Blanchard	Nov. 12 (Oxford)	
1785	Jan. 7 (Dover) Blanchard	Jan. 4 (Birmingham) Harper	
	Jan. 9 (Dublin) Crosbie	Jan. 31 (Birmingham) Harper	
	March 23 (London) Zambeccari & Vernon		
	May 3 and 7 (London) Blanchard	May 5 (London) Wyndham	
	May 21 (London) Blanchard	May 16 & 19 (Manchester)	May 13 (London)
	June 3 (London) Money, Lockwood	June 1 (Norwich) Deeker	
	June 15 (London) Blanchard	June 23 (Norwich) Deeker	
	June 17 (Dublin) Potain	June 24 (Oxford) Fitzpatrick	June 29 (London)
	July 25 (London) Money		July 20 (Liverpool)
		Aug. 25 (Worcester)	Aug. 9 (Liverpool)
		Sept. 10 (Worcester)	Sept. 8 (Chester) Baldwin
			Oct. 5 (Edinburgh)
			Oct. 21 (Kelso)
			Nov. 23 (Glasgow)
			Dec. 5 (Glasgow)
			Dec. 20 (Edinburgh)

independent senators, whose object is their peoples' good, and the false patriotism of those ambitious spirits, who would gladly lay the foundation of their greatness on the ruin of their king and country," like the conceited and avaricious senator Thomas Onslow.[141]

Material ballooning diverged significantly in patronage, spectatorship, political use, and commercial development in the two imperial capitals of Paris and London. Some attributed the differences to their projected national characters: the balloon was an invention perfectly suited to French lightness prone to diversions, as was the steam engine to British seriousness.[142] In Mercier's imagination, the London public competently "policed spectacles" instead of the royal police who often brutalized the parterre audience in Paris.[143] Parisians also followed the objects of fashion in a concerted manner: they got excited with frenzy over a new amusement only to ridicule it the next day. The French living in the theatrical polity "knew only one kind of pleasure, that of seeing and being seen," while the English had "more active, varied and profound tastes," not determined by vanity, and enjoyed "more substantial entertainments."[144] Mercier's judgment derived in part from his utopian longings and served more as a critique rather than an accurate description of Parisian manners. He stood in the venerable tradition of Voltaire's *Letters on the English* and Montesquieu's *Persian Letters*. His caricature nevertheless pointed to the veneer of civility that modulated Parisian existence and stifled overt expressions of plebian political consciousness.[145]

The well-demarcated national characters reflected distinct histories and political cultures that had produced different publics for entertainment and different populaces for political action. The form of scientific hegemony that the French royal administration pursued with a set of royal institutions, pensioned scientists, material resources, police, and censorship was unattainable in London. Gentlemanly Fellows of the Royal Society considered it unthinkable to make observations on rooftops or to follow the balloon on horseback, as Blagden noted in shock, much less to "let the world know" through the public newspapers that they "employed themselves seriously in observing the globe."[146]

The two imperial governments also approached their fiscal insolvency divergently. The seventy-foot balloon flown at Versailles in full public view and a toy balloon (less than three feet in diameter) floating in the seclusion of Windsor Castle provide a visual contrast of the desired public images of the sovereign, functions of spectacles, polities, and fiscal policies in the dueling nation-empires after the American War.[147] If the French sought to celebrate their military victory with a glorious scientific artifact, Britons saw their nation as a modern Atlantis that desperately needed serious reflection and a new moral foundation. Handel's somber music served this purpose better than the frivolous French balloon.

Figure 9.1. *An Air Balloon Engagement for the Empire of the Sky*, December 1783. Gallica, BNF.

9

CROSSING THE CHANNEL

Thus Britannia—I sustain'd Blanchard for his glory gain'd,
When mighty *Jove* gave me the crown, proclaiming to the Gods around,
Immortal Blanchard should receive the Laurels, which the Gods did give,
Then He gave commands to *Fame* to found *Immortal Blanchard*'s name
And to *Great Britain's Isle* repair to fix *Minerva's Standard* there.[1]

The American War did not end the imperial competition between Britain and France (see figure 9.1), judging from the tens of millions the French government spent on building a seawall at the Cherbourg harbor by immersing massive truncated cones. Despite the enormous cost, the construction began in 1783 even as the royal household began to tighten its spending. Louis XVI would pay a royal visit in 1786 to publicize its importance, which endeared him to the local public.[2] The government also funded Pilâtre de Rozier's ill-fated attempt to cross the English Channel from Calais to Dover, a project that had surfaced in court circles immediately after the first human ascent.[3] His success in commanding the *Marie Antoinette* in June 1784 consolidated court patronage around his channel-crossing endeavor.

Jean-Pierre Blanchard was not privy to such privilege. Upstaged by the Montgolfiers' originality, Charles's scientific appropriation, and Pilâtre's aeronautic and courtly skills, in addition to his own notorious failure in 1782, Blanchard missed too many opportunities to become a Parisian balloon notable. His beautiful ascent on March 2, 1784, though captured in numerous consumer items (see plate 8), did not bring him sufficient acclaim to rehabilitate his reputation.[4] His inability to handle inflammable air also raised serious doubts about his qualification as a *physicien* (if not as a *mécanicien*). Only a complete mastery in steering the machine would have secured universal acclaim.[5] Then came the royal ban. Even Étienne Montgolfier, who enjoyed a national reputation as an original inventor, failed to obtain a lucrative position in Paris and returned home by the end of May. Charles reverted to public lectures, which gained in students and remuneration thanks to his aeronautical renown.

Unable "to obtain in Paris the permission to repeat his experiment" and no doubt frustrated by the lack of recognition, Blanchard moved northward to Rouen in his home province of Normandy.[6] This industrial city had kept up with balloon news without notable enthusiasm.[7] Blanchard must have had a strong patron, perhaps through the Benedictine connection of Viennay, since the judicial elite and city officials persuaded Vergennes and obtained permission for an ascent by their compatriot.[8] After a successful ascent on May 23, 1784, Normandy newspapers bristled with hyperboles, but some objected to Blanchard's excessive claims for speed and manipulation.[9] He responded with a second flight on July 18, which attracted a larger crowd, including many members of the Normandy parlement.[10] His patron the abbé de Viennay traveled from Paris to witness the ascent on a Sunday, the day of the fête de Sainte Céleste. The flight, accompanied by Dominique Bertrand Joseph Boby (of the Normandy parlement), went beautifully and sparked a public celebration. Blanchard was crowned at the new two-thousand-seat theater. A portrait, a medal, and engravings were commissioned to celebrate this compatriot.[11]

Provincial success did not warrant Parisian approbation. The Parisian press remained skeptical of Blanchard's capability and accomplishments and allocated their pages grudgingly.[12] His soaring ambition eventually took him to London, apparently to fly back and secure the glory and fame that nobody could dispute.[13] The rival imperial city brimmed with foreign projectors who crisscrossed its social boundaries to enrich the empire and themselves.[14] Skilled artisans of stately spectacles such as fireworks had become mobile cultural agents in early modern Europe who transposed and translated the illusion of power.[15]

Blanchard's career provides a convenient window to London's entertainment culture that sat uneasily with gentlemanly sensibility. Commercial balloonists had little difficulty in commandeering the requisite material technologies—the fabric, coating materials, and the production of hydrogen. Their material technology soon outpaced the French in cutting and coating the fabric, adding floating devices made necessary by the geography. The lack of elite and royal patronage entailed serious constraints, however, especially in securing the sites of ascension and policing the crowd. More often than not, funds were raised by a long preview period indoors, which limited the choice mostly to hydrogen balloons, in addition to selling the tickets to the experimental site. London's thriving commercial culture with a layered public for entertainment did not exactly work better than the more homogeneous, fashion-driven one in Paris that supplied a long list of affluent subscribers in addition to royal gardens and police.

The way in which the balloon was appropriated in Britain, as was the steam engine in France, shows that the ongoing competition between the two world empires required a steady exchange of bodies and things, which formed a shared material culture that mapped the civilized globe. Scientific communication resumed before the peace treaty was ratified to tackle the composition of water—a frontier topic in pneumatic chemistry. British steam engines and French fashion spread over the European continent and beyond. Strong nationalism in the modern world depends on a uniform material empire, as the Olympic and World Cup games illustrate in their standardized facilities, equipment, and uniforms.

A FRENCH EXPERT IN LONDON

Blanchard crossed the Channel on August 14, 1784. His balloon followed a few days later, detained by the Rouen mayor's attempt to levy taxes.[16] He arrived in London at an opportune moment, a month before Lunardi's first ascent, and joined the informal Balloon Club. He announced a "Grand Aerostatic Experiment" on September 1 to spike balloon fever. His balloon was exhibited at Christie's Great Room in Pall Mall, a vibrant art district around the Royal Academy.[17] In sharp contrast to their haughty attitude toward Lunardi, several Fellows of the Royal Society took interest in Blanchard's balloon. As a renowned French aeronaut, Blanchard carried more weight with the Fellows than the flamboyant Neapolitan Lunardi in translating the balloon as a potentially useful artifact that would contribute to the public good.

Lunardi's successful flight on September 15 also brought media attention to Blanchard's credentials. The *Universal Magazine* printed an "exact narrative" of his second Rouen ascent, which marked a drastic change from their earlier hesitant and sketchy reporting.[18] It introduced the British public, now more attentive, to the scientific claims of French aeronautics. In a judicious account with instrumental readings, Blanchard had emphasized his ability to steer the machine vertically and "against the wind."[19] In a separate pamphlet, sold at a shilling and six pence with a balloon engraving, the translator deployed Baconian rhetoric—that he wished to publicize the balloon's importance, "its rapid and unexampled progress, the numberless improvements" already made, and its potential to benefit mankind, as was already acknowledged by "the more enlightened part of the world." The greatest obstacle to the progress of sciences lay in "the despair of mankind, and the supposition of impossibility," as the "prophetic philosopher Lord Bacon" should remind them.[20]

Blanchard's aeronautic renown and the prestige of French science helped mediate between London's scientific elite and commercial sectors. John Sheldon, a Fellow of the Royal Society, signed up to become the first English aeronaut, paying for the scientific instruments to be taken up. His participation advertised Blanchard's endeavor as a respectable scientific experiment. On October 16, 1784, "a considerable concourse of people" converged on the Military Academy at Chelsea. Fellows of the Royal Society—Charles Blagden, Henry Cavendish, Alexander Dalrymple, and the scientific instrument maker Jesse Ramsden—got ready to take measurements at nearby windows.[21] Aimé Argand supervised the filling, which took about an hour and a half. Although the site was deemed easier to negotiate than Lunardi's Moorfields, the ascent took some trouble. Blanchard had to throw out most of the instruments and provisions except for a barometer, a compass, a telescope, a flageolet, and a bottle of wine. Once the balloon cleared the wall, the aeronauts saluted the crowd with English and French flags. When they reached a desired height for the advertised maneuvers, they found the left wing missing its handle, which apparently prevented their aerial show over the launch site. Instead, they saluted the health of English and French kings, the Prince of Wales, and the royal family.

Remarkably, there exists no trace of public resentment against the French aeronaut in the London press. When the aeronauts descended at 12:50 PM on a meadow near the village of Sunbury, about fourteen miles from London, the inhabitants swarmed around the balloon and broke the remaining wing with their "awkward eagerness . . . to make themselves useful." Securing the fly and the helm "from the violent effect of their zeal," Blanchard took off alone. After many near stops over various towns, he came down at 4:30 PM on the meadow called Goosey. With Blanchard still standing in the boat, "his balloon was conveyed through Lord Palmerston's park into the middle of the market-place, amidst the acclamations of a vast concourse of people." As soon as Blanchard alighted, Mr. Penton greeted him in French and took him to his house for refreshments. When Blanchard and Sheldon arrived back at the Military Academy, "a numerous retinue" obliged them to keep their seats in the boat and conducted them, with their respective flags in their hands, in a "splendid cavalcade."[22] They entered London in this manner, accompanied by a band playing military music and a great number of carriages and people. The French aeronaut's triumphal procession through London must have made a peculiar impression on the discerning public emerging from the summer's somber commemoration of Handel.[23]

With two successful human ascents, London finally came under the siege of balloon mania, awash in songs, plays, pamphlets, and merchandise.[24] A

French poem appeared in translation by "a friend to mankind" to encourage prospective *"Aerial Beings"* to "evade the laws of their King or Country" instead of creeping on the earth "like the brute beasts which perish."[25] At the Covent Garden Theater, Frederick Pilon (1750–1788) staged a farce, *Aerostation, or the Templar's Stratagem* on October 29. The actors tossed up the ideas of bombarding Algiers with a "fleet of 14 balloons, each carrying 12 cannonades, 10 bombs, and 4 twenty-pounders," traveling to the moon, and stealing the Saturn's ring for the British Museum. The farce would be published the following year with a plea for the "Legislation of the upper Regions" to curb "the Excursions of celestial Travellers" and to ensure security against "the bold Aeronaut" who may "rob us of our Daughters, and our Wives from our highest Windows."[26] A tale of lunar voyage emerged to "extort a smile at the expence of philosophic pride," which often trampled upon "the modest pretensions of humbler claimants, perhaps of greater merit."[27] William Blake featured "Inflammable Gass the Wind" or "the Glory of France" in his unpublished piece, *An Island in the Moon*.[28]

Popular and aristocratic enthusiasm trumped gentlemanly caution and imperial rivalry. The *Gentleman's Magazine* belatedly printed the narrative of Blanchard's second Rouen ascension, adding that he had contemplated a flight to England and was now planning to fly back to France "by a passage that birds only have made before." The November issue printed the Academician Lalande's list of balloon ascents, augmented by recent British ascents, while judging the Montgolfière too unwieldy and accident-prone.[29] The "Dress of the Month" column in the *European Magazine* reported current court fashion as "plain frocks, dark brown, blue, or Lunardi's maroon." The *Critical Review* apologized for the earlier caution in reporting French balloon news. It had arrived in the form of the highest panegyric, decorated with so many ornaments that "the cautious philosopher hesitated in silence."[30] As a gentlemanly scientist unwilling to fashion himself as a flashy aeronaut, Blagden could only grumble that Sheldon's aerial voyage was "foolish enough, or rather absolutely good for nothing." He judged Sheldon ridiculous, Blanchard unworthy of his reputation, and Lunardi ignorant and vain.[31]

FIRST ENGLISH BALLOONIST: James Sadler at Oxford

The list of remarkable events for the year 1784 included three balloon ascents—Lunardi's, Blanchard's, and Sadler's—among the notable political, military, and natural upheavals.[32] James Sadler (1751–1828) was working as an assistant at the refurbished chemical laboratory of Oxford University, then under the direction of Martin Wall. Educated under Joseph Black before becoming the Reader of chemistry at Oxford in 1781, Wall sought to rally support for chemistry by enumerating its relevance to medicine and the arts.[33] His knowledge of pneumatic chemistry, conveyed in public lectures, must have played a role in Oxford balloon experiments and their reception but Wall stayed away from public ballooning.[34] Sadler enjoyed the patronage of the botany professor John Sibthorp, which allowed him to launch a Montgolfière in the garden of Queen's College on February 9, 1784.[35]

After a few successive projects, Sadler completed by September a Montgolfière measuring 170 feet in circumference and 38,792 cubic feet in volume, equipped with wings, oars, and scientific instruments.[36] He proceeded to stage a human ascent on October 4, 1784, sandwiched between Lunardi's and Blanchard's ascents, from the physic garden at Oxford.[37] His second ascent in a large hydrogen balloon on November 12 scored a resounding success that brought together a "surprising concourse of people of all ranks" who filled the adjacent "roads, streets, fields, trees, buildings and towers . . . beyond description." Identical reports in several newspapers characterized him as the first "English adventurer" who became his "own architect, engineer, chemist, and projector." For his display of "genius, intrepidity, and cool resolution," he deserved the patronage and liberality of a generous public.[38]

Native English ballooning emerged not in London but in the provinces, then, to rouse public enthusiasm. The Englishman Sadler's ascent inspired Henry James Pye's poem *Aerophorion* (1784). The "bold ambition . . . urged by the love of manly enterprize" could not stop at the shore. The balloon adventurer had to "leave the beaten paths of life behind" and "in Freedom's holy cause to brave the adverse legion and the hostile wave." He would earn "from Virtue's breath a purer fame than all the poet or the sage can claim."[39] News circulated that this "Ingenious English Aerial Navigator" planned more ascents in Dublin. His balloon was sold for 120 pounds sterling to Birmingham where Mr. Harper announced an ascent on December 27, which attracted a large crowd numbering sixty thousand. The attempt failed, causing much damage to the balloon.[40] While hoping that an Englishman would win the windy competition, the press taunted Sadler, "the little Oxford pastry cook," and told him to return to his famous cheesecakes and apple puffs.[41] The first English aeronaut would enjoy a distinguished ballooning career before he began to work for Thomas Beddoes (1760–1808), who assumed chemistry lectureship at Oxford in 1788.[42]

A HERO OF TWO EMPIRES

As a foreign adventurer, Blanchard was fortunate to find a steadfast patron in John Jeffries (1745–1819) who contributed a hundred guineas toward the second London ascent with the Channel crossing in view. Without elite or royal patronage, Jeffries encountered many difficulties in securing a site, ostensibly stemming from the disorder, mischief, and property damage occasioned by previous ascents. He settled on Mackenzie's Rhedarium and purchased a long list of reputable instruments—a compass, a pocket thermometer graduated by both Fahrenheit's and Réaumur's scales, a barometer tailor-made by William Jones and graduated down to eighteen inches, a pocket electrometer by Edward Nairn and Thomas Blunt, a hydrometer, a timepiece by John Arnold, a mariner's compass, and a good telescope. He also packed several yards of very thin, light-colored ribbons to make short strips for judging the vertical motion. Henry Cavendish supplied several numbered phials filled with distilled water to collect air samples at different altitudes. A silver pen and a notebook in quarto were meant to note the times and readings of the thermometer, barometer, and electrometer, along with "transient remarks of what passed." Warm clothing and a detailed map secured on a thick pasteboard (to keep it open during the flight) rounded up the preparation.[43] Jeffries probably named his instrument makers and scientists to publicize the ascent as a scientific experiment.

Blanchard's second London ascent took place on November 30, 1784, in the presence of many dignitaries, including the Prince of Wales and the French ambassador, the comte d'Adhémar (Jean-Balthazar, 1736–1790). The immense body of the people that gradually diminished into "mere pigmies" magnified Jeffries's "joy and exultation, at having at last overcome every obstacle." The clouds denied him the magnificent "bird's-eye view" of London, "the greatest and most opulent city in the world" with thousands of ships nearly covering the River Thames. After traveling through "a boundless expanse," the aeronauts' descent was an oddly humble affair, witnessed by a few locals including a farmer, a printer, a draper, and two surgeons. Unlike in Paris, noble patrons did not follow the balloon on horseback. The aeronauts returned to London the following day and sent the phials of air to Blagden, leaving scientific inferences to their "superiors in philosophical enquiries and reasoning."

No financial remuneration came from the fashionable London elite, although Blanchard had expected the cost to be shouldered by "the wealthy and public-spirited people" as in Paris.[44] The mayor apparently paid "for a voyage undertaken just for him" with a large bottle of fruit brandy.[45] The British press still classified the balloon as a novelty rather than a useful invention, judging that its advantages were "remote, uncertain and even improbable." Given the "distant chance of attaining an adequate compensation," English philosophers would be justified to view it with "cautious silence" rather than "the ardour of curiosity."[46]

The Channel crossing became a mad dash, with the anticipation of French and British competitors—Pilâtre and Sadler. Jeffries supplied the capital of more than seven hundred pounds sterling.[47] After sending the balloon in advance, the aeronauts left for Dover on December 17. Fortunately for Blanchard, Sadler had been delayed in his competitive bid when the fresh varnish on his new balloon damaged the fabric during its boat ride. If Sadler had succeeded, he would have won British public sentiments and pushed Blanchard into oblivion.[48] Pilâtre's visit only aroused a bitter sense of rivalry, although Blanchard had the presence of mind not to express it publicly. "Everything conspires against me," Blanchard wrote to his parents at Les Andelys, characterizing Pilâtre as a Judas who came merely to mock his battered machine. Blanchard vowed to make the French government regret their mistake in making him an antagonist of their famous charlatan. He would vanquish his enemies or die trying.[49] The daring competition between the two pioneering French aeronauts, mostly hidden from public view, attests to their marginal existence in the theatrical polity that dangled meager compensation for entertainers.

Tempestuous weather delayed the impatient aeronauts. Worried about the weight, Blanchard apparently schemed to leave Jeffries behind. Slim built at 110 pounds, any companion would have been a heavy responsibility for Blanchard. Jeffries had to present their written agreement, enumerate the cost he had borne, and even promise he would jump out of the car if his weight should ever threaten Blanchard's survival. British newspapers reported satirically that one of their principal ports was under a Frenchman's (Blanchard's) control with an American (Jeffries) commanding their sailors.[50]

Dover Castle, designed as a lookout post with four towers and a high castellated wall, stood on top of the chalk cliffs curving in a crescent. On January 7, 1785, Blanchard collected Jeffries at 6 AM and went to the pilot's lookout with the experienced Captain Hugget to judge the wind direction. Shortly after 8 AM, they fired the signal gun and hoisted a flag to announce the impending ascent. As James Deeker (a balloon maker from Soho) worked on filling the balloon, people began to line the shore and the cliffs. The aeronauts gauged the wind direction many times with Hugget, a paper kite, a paper Montgolfière, and a small Devonshire gas balloon. They took off at 1:15 PM with only a few instruments (a barometer and a mariner's compass) to reduce the weight. Blanchard wore his great coat but Jeffries was in light sailor's dress.[51] They also had oars, a fly, letters to the French nobility enclosed in an inflated bladder, a bottle of brandy, a copy of the French edition of Blanchard's voyage with Sheldon, two silk ensigns, a large parcel of pamphlets, a number of other inflated bladders, two small anchors or grapnels, and two cork jackets. They had to throw out most of the ballast, leaving only three ten-pound sacks of sand.

The flight lasted an hour and a quarter. The aeronauts enjoyed "enchanting views of England and France being alternately presented" by the rotary motion of the balloon, along with greetings from the vessels below. When the balloon began to descend well in advance of the French coast, they threw out the remaining ballasts, pamphlets, and anything dispensable. At about five to six miles from the coast, they stripped the balloon and themselves—silk and finery, the wings, and their own coats.[52] After discharging five to six pounds of urine as the last measure, they were ready to put on cork jackets when the balloon rose again to bring them twelve miles inland to the woods of Guines.

The French welcoming party ushered the aeronauts through the night, first on horseback and then in a postal carriage, to Calais. When they arrived past 1 AM, the commandant ordered the gate to open, having waited with his wife in a tent pavilion. The following day, the aeronauts dined at the London Hotel with "all titled and principal" persons of the region. Town authorities presented Blanchard with the right of bourgeois in a gold box, while making repeated apologies to Jeffries since they could not offer it to a foreigner without explicit permission from Versailles. As a newly minted "Citizen of Calais," Blanchard belonged to "the heroes of the French nation."[53] Calais authorities asked to hang the boat in the local church, while petitioning Versailles authorities to erect a monument at the landing site through a subscription, which was readily granted.[54]

Thus began a whirlwind tour of the desperados-turned-heroes in the public spotlight, much like a royal visit, which lasted two months. Jeffries recorded it carefully, wishing to secure his glory in the eyes of the British public and posterity.[55] They arrived at Paris on January 11, where Pilâtre and the abbé de Viennay (Blanchard's "foster father") greeted them. Armed with letters from the English nobility to "the principal personages of the French court," they visited Versailles where they met Vergennes, the Polignacs, Blanchard's uncle (apparently charged with the royal Ménagerie), and the extended royal family—the comte d'Artois, the comte de Provence, Louis XVI, and the duc de Chartres. They attended the royal theater and the chapel before returning to Paris on January 13. Elated, Jeffries wrote a long letter to Joseph Banks to repay his patience, candor, and favor earlier at Soho. It seems quite likely that Banks recommended Deeker for filling the balloon.[56]

The aeronauts became the men of the hour. M. Hirschberg paraded them around Parisian fashionable society and public space. At the Café du Caveau, they were informed that their portraits would be placed among the busts of the greatest men of wit. At Pilâtre's Musée de Monsieur, they received a billet d'entrée reserved for the founding members and enjoyed a long conversation with the duc de Chartres. Afterwards, they met with the British ambassador, the Duke of Dorset, dined with Benjamin Franklin at Passy and then with Madame Hirschberg at the Hôtel de Calais, and spent the evening at the Opéra. On January 16, they went back to Versailles to dine with the royal couple where they met the comte d'Artois, the Polignacs, the baron de Breteuil and his daughter, and other courtiers numbering fifty. The marquis de Laroche du Maine then took them back to his Parisian residence. In the evening, they were invited to a grand ball at the residence of the duc d'Orléans where they were "received by universal and continuous shouts and claps of applause, embraced and complimented by hundreds of the first ladies and gents in Paris."[57] Festivities continued until February 27 when Jeffries left for Dover. Back in London, he published an account of the voyage that would immortalize his name and provide some financial compensation.

Notwithstanding the hero's welcome, the triumphant return fetched only a small pension for Blanchard. When he requested permission to deliver a lecture at the concert hall of the Tuileries (where Charles had previously made his balloon), charging three livres per person for a charitable cause, Breteuil refused and demanded that Blanchard quit flying.[58] Nor would he have benefited from the sale of Jeffries's travelogue, the truth of which he bitterly disputed. He protested that he did not deceive the doctor by wearing a lead jacket, that he took care of all necessary tasks including instrumental recordings, and that he alone handled an impeding peril while the doctor fortified with brandy remained oblivious. The doctor was a bag of weight that Captain Blanchard could not abandon without cruelty. He vowed to engage English and French papers to refute Jeffries's account and to show, in addition, how the "English nation" did not compensate him for practicing this "sublime art."[59]

Strangely, Blanchard's flight was embraced as God's blessing to Britannia. The "immortal Blanchard" in his "triumphal entry into the ethereal world," crowned with "Godlike Bays," delivered the standard of wisdom to Britannia in his ultimate descent to the British Isle—the blessed island where Gods reside, or the "Paradise of happiness" where "a *noble-virtuous King*" reigned.[60] Blanchard had become a shining star for both the British and French Empires. Neither compensated for his troubles.

IRISH PATRIOTS: Dublin Ascents

The Channel crossing opened a busy year for British ballooning, along with Harper's Birmingham trials on January 4 and 31, 1785, in Sadler's balloon.[61] The most notable ascent was Richard Crosbie's from Dublin, which made a serious statement for Irish independence. With no standing police at hand, the entire city bustled with preparations for a patriotic festival led by a committee of prominent Irish Protestants.[62] On January 9, "undescribable crowds" of horses, carriages, and pedestrians "hastened to the curious and awful scene" by 11 AM. The balloon—measuring 34 to 40 feet in diameter and 20,579 cubic feet in volume, with the lifting force of 1,286 pounds— was adorned with paintings of Minerva (goddess of wisdom) and Mercury (messenger of the gods), enveloping the arms of Ireland and inscribed with a phrase from Virgil's Aeneid, *Prapetibus pennis acerus se credere caelo* (He dared to trust his life to the sky, floating off on swiftly driving wings), which referred to the story of Daedalus.[63] They had to remove the windmill device and sails at the last minute in order to reduce the weight.

Uniformly positive descriptions of the spectators indicate that the Irish patriots paid attention to how their nation might be portrayed in British newspapers.[64] Although the filling was much delayed by heavy rain, the spectators numbering twenty thousand "waited with the utmost good nature, candour, and patience" until 2:35 PM when their "kind and polite patience" was amply rewarded by "a most beautiful, solemn, and wonderful sight."[65] Crosbie cut an impressive figure with his six-foot-three-inch frame, dressed in a robe of oiled silk lined with white fur, Morocco boots, and a Montero cap of leopard skin. When he landed at the North Strand near the island of Clontarf, a large crowd carried him shoulder high in his aerial chariot to the home of Lord Charlemont. The brave aerial adventurer became a patriotic philosopher who could stir "the most grateful, benevolent, and sublime sensations" in the Irish audience. Newspapers praised his noble ancestry, mechanical genius, perseverance, and industry. The "grateful nation" would expect "discoveries of the highest import" from Mr. Crosbie, this "great ornament of our university, our country, of science and of human nature."

How the newly independent Irish Parliament would function as a patron of science and offer public munificence was yet to be seen. Should he cross St. George's Channel to England, Crosbie would "prove to the World that Ireland in scientific knowledge is not inferior to any part of" England.[66] Science seemed to promise a new vehicle of Irish independence. Irish nationalists turned the balloon ascent into a meaningful discourse of utility and public good for their cause. Although Crosbie's later ascents proved disastrous, this fervent rhetoric of patriotic glory probably helped the founding of the Royal Irish Academy in April 1785. Lord Charlemont became a founding member. Richard Kirwan would return to Dublin in 1787 to assume its presidency.

The balloon's immediate political value consisted less in its scientific merit than in its capacity to mobilize a crowd, judging from the fact that Irish patriots sponsored another ascent by Dr. Potain, a French surgeon residing in Dublin. Potain announced a balloon crossing to England by advertising his credentials of working with Blanchard, including at the Parisian ascent. An illustrious committee headed by the Earl of Moira (Francis Rawdon, 1754–1826), facilitated his endeavor.[67] Throngs of nobility and gentry went to view his balloon during the month of May.[68] At a meeting on May 26, 1785, the committee appointed Thomas Bell as treasurer. Sir Edward Newenham (1732–1814), the ebullient Protestant radical nicknamed "the Fox of Ireland," apparently announced the subscription in a parliamentary session testifyng that he had received letters of support from Benjamin Franklin and the marquis de Lafayette.[69] He planned to accompany Potain on the voyage with his eldest son, having failed at his own endeavor to make a *"Lillypution* air Balloon," but was dissuaded by his wife.[70]

The price of admission to nonsubscribers was set at five shillings and five pence. The Irish banker Latouche sent two hundred pounds sterling. General Pitt, commander of the English troops, provided security. The residents along the canal Saint-George were promised a reward of twenty-five pounds should they come to rescue the voyagers in need. With the parliament closed for ten days, the departure was delayed until June 15. Potain's ascent on June 17, 1785, just two days after Pilâtre's accident, brought the entire town to Malborough. The balloon took off at 12:30 PM and went straight up to about four hundred toises, hovered over the site for some time, and then shot up to a prodigious height before floating away.

The attempts to mobilize Irish patriotism through ballooning did not have a happy ending, however. Potain's balloon ruptured suddenly midair, admitting extremely cold vapors and condensing the gas inside, which caused a rapid descent into the mountains. Potain was thrown from the boat while trying to cut the cords and was carried, hanging upside down, for more than two thousand feet. Freed from this disposition, he was dropped in the mountains of Mullinaveugue, near the house of M. Price. He sent the news of his safe return to the committee, barely communicating in English. The Irish patriots (the Duke of Leinster, Lord Charlement, Sir Frederick Flood, Mr. Hutchinson, Sir Henry Cavendish, Mr. Latouche, and so on), wishing to show the world that they were "not inferior to any other nation, for enterprize, genius, and generosity," also funded another ascent by Crosbie on July 19, 1785, to cross the Irish Sea. With the failure of this endeavor, all of Crosbie's "castles" built on "frail foundations" disappeared "in one capricious moment."[71]

FOREIGN ADVENTURERS AND ENGLISH NOTABLES

London skepticism persisted. The press continued to mock the balloon enterprise, often citing "many bands of pickpockets" that claimed Edmund Burke among their victims (see figure 9.2).[72] The enterprise still remained mostly in the hands of business establishments like the Lyceum with reluctant cooperation from a few administrators who provided the sites of ascension only with significant financial concessions. More commodious sites such as Hyde Park or the Kensington Gardens were not available to London balloonists. Lacking wealthy and public-spirited patrons, they had to bear the expense upfront in the hopes of being reimbursed by the curious public. While the lack of central planning or control had long been the signature advantage of London market, ballooning required substantial capital and could not thrive in the patchwork environment of London's entertainment industry.[73] Even after the Channel crossing promised scientific and commercial utility and fostered more favorable public opinion, Jeffries could only hope that conditions might improve in the future.[74]

As in France, the discussion turned increasingly to steering the balloon so that it could be more than just a fashionable pastime.[75] Unless these "adventurers can acquire the power of steering their buoyant bark," Anna Seward summarized elite sentiments, "the experience is as idle as it is dangerous." The enormous disproportion between the boat and the balloon made it difficult to conceive how any device could steer it.[76] It did not bode well that leading aeronauts were all foreigners.

> Nothing can more fully demonstrate the folly of the age, than their rage for ballooning. Foreigners of every description have their day. Witness the Lunardi, Blanchard, Zembeccarri, &c. who live on the folly of our countrymen. But to speak seriously on the subject, it is high time that some restriction should be laid on the madness of their frequent trips to the air, without one single good purpose being produced. If a calculation was made of the number of manufacturers it takes from their employment, a very alarming loss to the nation would appear to the reader, which all the money spent by foreigners in this nation will not compensate for; besides the dissipation, idleness, and mischief, which are the natural consequences of workingmen making holidays.[77]

A notable change of attitude came from the military personnel and their influential political friends, which ameliorated the press snobbery. In reporting a new endeavor by Count Zambeccari, who had enlisted Sir Edward Vernon (1723–1794) as a passenger, newspapers characterized Zambeccari as "an Italian nobleman with great scientific reputation" whose

Figure 9.2. "The New Mode of Picking Pockets. Frenchmen, Dutchmen, Italians, Swedes, and Hungarians. If you have any dancing bears, monkies, camels, butterfies, beetles, lap dogs, or balloons or any other whims—bring them to England and by gar you will be loved and well paid for your pains—For de English be one great pack of fools—beside john de Britain is very good temper'd if you can tell him one very good storie he will belief you and his pocket is yours." The pickpocket is diverting the gentleman with one of the greatest curiosities that the world had ever seen, one that would fly 3,000 miles in a day. The gentleman, admiring, concludes that "the French people are the most ingenious hands living" (September 14, 1784). Gimbel Collection XL-5-3431.

"philosophical zeal, and enterprising spirit" promised great discoveries and improvements in aerostatics.[78] Vernon, who had recently returned from distinguished service in the Navy, carried a proud family name earlier borne by the Admiral Edward Vernon (1684–1757), a national hero of the mid-century British Empire.[79]

Zambeccari finished a thirty-four-foot hydrogen balloon by December 1784, funded by the Lyceum proprietor Lockwood.[80] Advertised as the "largest, and most magnificent" hydrogen balloon ever constructed, it was exhibited at the Lyceum for one shilling per entry. Notwithstanding the unfavorable weather, "a great number of respectable and scientific persons" paid a visit. The "public were never before entertained with such a happy combination of grandeur, ingenuity, and elegance." The "noble constructor" received "numerous and flattering" compliments for his affable and attentive explanations of the "assemblage of mechanical excellence."[81] As Zambeccari became a public persona, divergent portraits of his character emerged in newspapers. The *Morning Post* accused him of forcing out the third passenger, Miss Grice, for the three hundred guineas paid by the admiral. In contrast, the *Public Advertiser* carried a flattering story of his noble ancestry in Bologna where his father had served as minister plenipotentiary.[82]

Zambeccari's balloon was indeed a beautiful artifact. Composed of thirty-two pieces of "particularly delicate, elastic and transparent" oil silk "manufactured on purpose," each measuring eighteen yards in length and forty-two inches in width, it was sewn together without the equatorial seam to enhance its beauty. A netting of "the finest *India* Silk," formed upon mathematical principles, encompassed the envelope to suspend a gondola of "unparalleled Elegance," fifteen feet in length and four feet in width, to accommodate three or four persons. The gondola, about eleven feet in length and five and a half in width, was shaped exactly like regular boats on the Thames. It was made in linen cloth painted green and decorated with crimson silk and gold fringe.[83]

Notwithstanding sleet, snow, and rain, the ascent on March 23, 1785, at Tottenham Court Road was attended by "most of the principal people of fashion."[84] The hydrogen balloon was filled in two and a half hours (in comparison to Lunardi's sixteen hours), but a tear caused it to fall precipitously in the King's Fields near Horsham in Sussex at about 5 PM.[85] Another ascent on May 3, scheduled with the Lockwoods, failed to take off.[86] After this public failure, which offered a clear point of comparison with Blanchard's entertaining show on the same day, Zambeccari was on the way out.

London ballooning stagnated by the summer 1785, in part because of the rivalry among the entrepreneurial showmen who had no desire for scientific emulation and discovery.[87] Lunardi's strategy of catering to the aristocratic

Figure 9.3. Lunardi's balloon representing the Great Union (England and Scotland).

audience did not endear him to the public press. His new balloon representing the Great Union (England and Scotland), "exceedingly large, and more beautiful than" any other, was on exhibit at the Pantheon (see figure 9.3). The ascent on May 13 from the Artillery Ground drew a greater concourse of people than ever before. Unable to fill the balloon properly even at the expense of four hundred pounds sterling, Lunardi had to abandon the other voyagers, Mr. Biggin and Mrs. Sage, "to the disappointment of all present." The crowd broke down the paling between five-shilling and one-shilling sections and filled the ground with "one shilling gentry."[88] Many "illiberal reflections" on his intentions for Mrs. Sage and his "ingratitude to a gentleman" emerged.[89] The *Gentleman's Magazine* scolded that those "who furnish *entertainment for the people*" should avoid this kind of "deception" that broke an engagement with the public. His "feeble attempt to ascend" after the intense publicity forecasting a miraculous performance made for a most "contemptible exhibition."[90] After staging another ascent on June 29 at St. George's Field to remedy his reputation, Lunardi traveled north to Liverpool, Chester (by Baldwin), Edinburgh, Kelso, Glasgow, York, and Newcastle (see figure 9.4), before returning to Italy.[91]

THE BALLOON AND PARACHUTE AEROSTATIC ACADEMY

Blanchard's solution to the growing competition and persistent elite disdain was rather typical of a Frenchman who regarded the school as a safe and respectable path of social mobility.[92] His Balloon and Parachute Aerostatic Academy opened on a square plot of grassy field, conveniently enclosed by high brick walls to collect fees, on the Stockwell Road near the Vauxhall, a twelve-acre pleasure garden on the southern banks of the Thames River. Rehabilitated in 1732 by Jonathan Tylers, the garden opened during the summer for supper, promenades, orchestra, and other performances. Frequented by the Prince of Wales and his entourage, actors, and visiting dignitaries, it staged for some "a spectacle of cohesion" in its mingling of ranks.[93] The educated gentry dutifully distanced themselves from such vulgar pleasures and sexually charged atmosphere, however, to demarcate their self-conscious elitism.[94]

The site of Blanchard's Academy thus betrays his understanding of public entertainment as a Frenchman accustomed to the trickle-down economy of pleasure. His choice of site, theatrics, and admission price catered to the affluent rather than to the middling public. While his shows attracted the fashionable society inside and the populace outside the walls, he did not win over the discerning gentry despite his repeated efforts to this end. Nor could he attract enough funding to make a great balloon that would allow him to carry more passengers.[95]

Blanchard varied the repertoire for "the most distinguished of the nobility and gentry of both sexes," rising with Miss Simonet, the fifteen-year-old daughter of a French dancer on May 3 and 7, 1785. He performed well, ascending and descending at will and crossing the yard from one end to the other "with infinite ease."[96] Another ascent on May 21 was performed with

Figure 9.4. Satire of Lunardi. "Aerostation out at Elbows or the Itinerant Aeronaut." Gimbel Collection XL-7-1264.

"Behold an Hero comely tall and fair!
His only Food Phlogisticated air!
Now on the Wings of Mighty Winds he rides!
Now torn thro' Hedges—Dash'd in Oceans tides!
Now drooping rooms about from Town to Town
Collecting Pence to inflate his poor Balloon,
Pity the Wight and something to him give,
To purchase gas to keep his Frame alive."

the still younger Simonet in the Calais balloon, which drew an immense crowd to the vicinity. Blanchard was dressed for the theatrical occasion in a green frock, a white dimity waistcoat, nankeen breeches, white silk stockings, and shoes fastened with black silk ribbons, drawn neatly into the form of cockades. His hat was covered with a green japan, to which was affixed a green cockade ornamented with a small ostrich feather. His designated companion, the colonel Thomas Thornton (1757–1823), was dressed in the same costume, except that the ribbons on his shoes were decorated with a pearl-color edging. The younger Simonet, weighing exactly eight and a half stone (a stone being fourteen pounds), wore a cotton gown of Devonshire brown mixed with white, a light green silk petticoat, a small black cloak, and a black hat. Finding the ascending power of the balloon insufficient, Blanchard rose with Simonet. Descending after two hours of flight, they were led back to the place of ascension in their car, "preceded by a band of music, and accompanied by a vast concourse of applauding spectators" reportedly including the nobility. A French showgirl could not claim the title of the first female aeronaut in Britain, notwithstanding her "*star-gazing tour*." Pilâtre and a large party of gentlemen went to the Vauxhall in the evening.[97]

In order to sustain the Academy, Blanchard had to vary the repertoire and work the media. He scheduled a parachute experiment for June 3, advertising it through newspapers, wall postings, and handbills.[98] Priced at five shillings when the entry to the Vauxhall Gardens cost one shilling, this novelty show was designed for the affluent audience. In a prelude, Thornton rose in a tethered balloon with a terrier. Dropped from about two hundred feet, the dog was found "without a whole bone in his body."[99] Blanchard boarded the car at 4 PM with a twelve-pound black cat, placed in a net and connected to the parachute by a long cord. It was dropped from a considerable height and found intact in the net caught between the two branches of a tree with the umbrella completely spread. Blanchard then flew across the Thames, descended near Woolwich "amidst the shouts and acclamations of the delighted multitude," and dined with officers of the Royal Regiment of Artillery who signed the minutes "describing the evolutions and maneuvers" he had made. They sent him off ceremoniously with the firing of guns. Blanchard arrived in London at 8 PM "accompanied by his friends, and a numerous train of people, who carried the flags before him all the way from Woolwich to Mr. Sheldon's, Great Queen-Street."[100]

Blanchard's fame guaranteed press coverage. His parachute experiment was hailed as the "climax" of British balloon madness—"miraculous to common capacities, and gratifying in the highest degree to the scientific."[101] Still unable to turn a profit, he advertised a sequel.[102] He would drop a sheep from a considerable height (at least a mile) via "a peculiar mechanism"—a box of fireworks designed to explode ten minutes after the ascent in order to release the parachute and open the valve of the balloon. With broad advertising, combined with specific measures against counterfeit bills, "a very numerous, and indeed a very genteel company" turned up on June 15, 1785, despite the hot weather. Unfortunately, the show plan was derailed by a lady who entangled herself in the parachute while tending to the animal and released the balloon, which then exploded in mid-air. The newspapers were relatively sympathetic, but the failure must have caused a serious strain on the business since the first parachute experiment had not been a financial success.

Blanchard's London career came to a screeching halt by this incident. He advertised a substitute experiment for June 18, but the balloon was too damaged to serve the occasion.[103] "Extremely concerned," he advertised yet another experiment with the Calais balloon. Although the admission price was adjusted upward to half a crown, those who had purchased a ticket on June 15 were promised free entry. The public had become wary of human ascents by this time because of Pilâtre's death in trying to cross the channel from Calais to Dover on the same day. Since the "silken vehicles in the atmosphere" were found to break human bones as well as those of other animals, many were determined to "remain on *terra firma*."[104] In life, Pilâtre had been an enemy that Blanchard resolved to fight to the death. In death, Pilâtre dissolved Blanchard's plans to settle in England. Blanchard left for the continent with his trusted Calais balloon, vowing to return and reclaim his honor. The British press were surprisingly sympathetic: although he had much skill and more modesty, "poor Blanchard" had "French notions" of addressing his experiments to the aristocracy and setting his admission prices high enough to keep away "that rank of the community, by which all public spectacles in this country are cherished and maintained."[105]

MODERN ICARUS: Death of Pilâtre

Pilâtre had received forty thousand livres for what he billed as a "national enterprise"—crossing the English Channel from Calais to Dover. As a self-fashioned royal aeronaut, he had to prove his worth to the court. His new Aero-Montgolfière was contracted to Pierre-Ange Romain (1751–1785) on September 17, 1784. It had a thirty-seven-foot hydrogen globe on top of a cylindrical Montgolfière (ten feet in diameter and twenty feet in height). The cords made in very light silk and lined with "silver paper" formed a moveable curtain around the Montolfière.[106] The design represented a "happy accord of the two procedures," which would allow Pilâtre to take advantage of the hydrogen balloon's maneuverability without undermining the royal image (see plate 7).

Pilâtre arrived in Boulogne-sur-Mer on December 21, 1784. The wait for favorable winds stretched out to several months, with the increasing realization that the winds would almost always work against him. He often visited England on account of a lady named Dyer, frequently taunting Blanchard along the way.[107] He and Romain finally took off in a much-battered balloon on June 15, 1785, under a favorable wind. Although they had deployed three test balloons beforehand, the wind changed quickly to push the aerostat back to the coast. To the spectators' horror, the balloon fell from a considerable height, estimated at seventeen hundred feet. The aeronauts were found broken to pieces in the gallery near the place of their departure.

Exactly what caused the crash became a prolonged dispute. According to the marquis de La Maisonfort (Antoine-François-Philippe Du Bois Des Cours, 1763–1827) who delivered the news to the royal administration, Pilâtre tried to pull the valve (that would have released hydrogen gas) to land and wait for another day and different winds, but the hydrogen balloon collapsed on top of the Mongolfière. In short, Maisonfort attributed the accident to a design flaw in the hydrogen balloon and insisted that the Montgolfière remained intact and its stove closed. In contrast, another brief report in the *Affiches de France* relayed that a column of fire rose from below. They tried to determine, in other words, whether the accident was caused by a defect in the hydrogen balloon or in the combination design that would implicate the Montgolfière.

British newspapers almost unanimously blamed "the fire balloon," indicating their aversion to the French royal artifact. They were gentler with the fallen Icarus, "the first martyr to the new science . . . distinguished for" an acute mind, warm sentiments, gentle manners, and a disinterested devotion to science and philanthropy. Blanchard's parachute would have saved Pilâtre's life. The loss of this "intrepid philosopher" would be regretted by "all friends of science and lovers of genius throughout Europe."[108]

The sorrow that engulfed the underprivileged balloon public is palpable in the anonymously published letters attributed to Jean-Paul Marat. One can sense that Pilâtre's death delivered a shock wave that was absorbed only with difficulty, especially by those wishing for progressive science, society, and polity. The balloon had created a bright spot in public mood that rallied the politiques of diverse dispositions. If Mercier's moral valuation of the aeronauts expressed a sense of popular hope, Marat's deep pain conveyed that of popular despair and a new resolve. On hearing the news of the fatal catastrophe, Marat relayed, all hearts were gripped with pain over the cruel deaths of "the two men interesting for their personal qualities, love of the sciences, and thirst for fame & dear to the public by their devotion" to the most exquisite spectacle of the age.[109]

According to Marat, everybody admired the intrepid men carried by this frail machine to float in the ethereal plain—like a superb eagle circling above the mountains, rivers, and seas—and to invade the empire of Aeolus. What a precious gift it was for the unfortunate inhabitants of the earth—where disorder, trouble, and confusion reigned—to have an asylum for indigent and virtuous beauty, to have the means of taming Nature and protecting the vineyard, to have at their disposal a vast theater of war and festivals in the air, and even to dream of a new habitat or a "new universe of which man would become the master." The expensive, "pompous excursions" had only resulted in "vain promenades."[110]

After examining some twenty letters from Boulogne, Marat attributed the accident to an inherent design flaw. Given the considerable distance of the balloon from the shore, he opined, Maisonfort's report contained little more than "improbable conjectures." If there was a tear in the hydrogen balloon, it would have released the gas to stabilize it and descended more slowly than it did. The observations—the column of fire rising from below, the detonation (rather than the explosion) of inflammable air, the considerable charring around the top calotte, and the severely burnt appendix at the bottom—made it plausible that the inflammable air escaping from the appendix, mixed with the atmospheric air, exploded thanks to the fire from the stove. In his rousing critique of such a "revolting" and dangerous combination, Marat targeted royal symbolism and the savants who had neglected such a promising invention.[111]

Although Charles had earlier used a small balloon filled with the atmospheric air as adjustable ballast, placing a live Montgolfière with combustibles under a hydrogen balloon (so that it could also provide additional lifting power when needed) was like putting a stove under a pile of gunpowder.[112] After his later visit on January 5, 1786, Blanchard would conjecture that an accelerated ascent with the help of a live Montgolfière caused the hydrogen gas to escape from all possible places such as the appendices and the valve. When this cloud of inflammable air descended on the foyer, or when the foyer rose due to the vehement ascension, fire broke out and was communicated through the open valve to cause an explosion.[113]

Marat could not let go of the balloon's promise. Notwithstanding the two failures by Blanchard and Pilâtre on June 15, 1785, he insisted that the art could be perfected with knowledge [lumières], genius, and systematic support. The aeronaut's risk was less than the navigator's. An aerostat was a "precious instrument in the hands of a skilled physicien," which could serve useful purposes if constructed in the most certain, advantageous, and convenient manner and launched with all the precautions that prudence dictated. Marat judged that a hydrogen balloon was incomparably less risky than the Montgolfière, which required a large quantity of combustibles and continuous attention and made it difficult for the aeronauts to make observations. In contrast, a hydrogen balloon could be maintained with a simple operation of the valve. All trials for "pure amusement" or "pure ostentation" had to stop. If they exposed the aeronauts to peril, they at least had to have a justifiable goal—the "progress of sciences" for the good of humanity, happiness of the state, and welfare (*salut*) of the patrie. To this end, he wished to transform the balloon's gallery into an observatory that offered more stability and less danger. A properly built balloon could be commanded by military chiefs to establish a correspondence between combined armies. Was this just an illusion spawned by his wishful, patriotic heart?[114] Marat's position would change radically after the Revolution began, as seen in his diatribe against the "modern charlatans."[115]

London proved a difficult conquest for the French aeronaut Blanchard, but he received a measure of esteem from the British press and scientific establishment as a skilled aeronaut. Although he could not attract much investment from the fashionable elite, his Channel crossing helped domesticate the invention. British aeronautics was no longer a field dominated by foreigners by the summer of 1785. James Sadler made well-publicized flights from Moulsey-Hurst (with William Wyndham), Manchester, Oxford (by Colonel Fitzpatrick), and Worcester.[116] Major Money, Lockwood, and George Blake (naval officer) took off from Tottenham Court Road on June 3 in Zambeccari's balloon.[117] Afterwards, they dined at Lord Orford's (George Walpole) with Pilâtre and Blanchard who had also staged a flight in his new Aerostatic Academy. James Deeker made two flights from Norwich (on June 1 and 23) in Sadler's balloon.[118] Major Money's flight on July 25 dropped him into the sea, but Lord Orford continued to support the balloon endeavor.[119] Since they promised "a prodigious navy in the air," Horace Walpole mused on such proliferation, "what signifies having lost the empire of the ocean?"[120] Various treatises appeared in the market on the theory and practice of aerostation.[121] British ballooning continued, with a variety of makers, experimenters, and adventurers, to squeeze out the foreign adventurers. It dropped off dramatically after the hectic year in 1785, however, as its novelty status wore off and accidents marred the celebrations.

Blanchard's London career highlights divergent cultural topographies and polities in London and Paris. The two imperial capitals shared material and scientific resources, popular enthusiasm, and media infrastructure, but they differed noticeably in the elite response to and finance of entertainments. While London's robust political culture quickly appropriated balloon images to run a political commentary on current affairs, the London press did not forge a consensual script on the balloon's scientific identity. Public press assumed a substantially different function in Britain and did not allow for a dominant public transcript that would support scientific hegemony. Public spectacles played a more commercial function in London, while they helped consolidate the balloon's scientific identity in Paris. The interactive stabilization of British and French "national characters" through the cultural translation of ballooning illustrates how national identities were maintained through the transcultural actions at contact zones, or "along the policed and transgressive intercultural frontiers," which were mapped by material artifacts. The nation as a cultural empire requires "constant, often violent, maintenance."[122]

Figure 10.1. Jean-Pierre Blanchard's continental ascents, 1785–1791.

1785 12. The Hague, July 12
13. Rotterdam, July 30
14. Lille, Aug. 26
15. Frankfurt, Oct. 3
16. Gand, Nov. 20
1786 17. Douai, Apr. 21
18-9. Brussels, June 10, 16
20. Hamburg, Aug. 23
21. Aix-la-Chapelle, Oct. 9
22. Liège, Dec. 27
1787 23. Valenciennes, Mar. 27
24–25. Nancy, July 1, 2
26. Strasbourg, Aug. 26
27–28. Leipzig, Sept. 29
29. Nürnberg, Nov. 12

1788 30. Basel, May 5
31. Metz, June 2
32. Brunswick, Aug. 10
33. Berlin, Sept. 27
1789 34. Warsaw, May 10
35. Breslaw, May 27
1790 36. Warsaw, May 14
37. Prague, Oct. 31
1791 38. Wien, July 6
39-41. Wien, Aug. 2, 14, 15
42. Prague, Sept. 11
43. Hanover, Nov. 8

10

A LIMINAL GEOGRAPHY

Balloon travels mapped an emergent geography of French cultural empire through textual, material, cultural, and political translations. In Britain, a competing empire that possessed an equivalent level of material and scientific knowledge, translation was minimal except for the newspaper reports that sought to rehabilitate the balloon as a British artifact. Translations in the Dutch Republic point to an active community of scientists who kept up with scientific news from Britain and France, developed their own expertise and societies, and tried their best to keep up with scientific fashions. They could not muster enough resources to stage human ascents in the fractured republic tangled in diplomatic maneuvers and a patriotic revolution.

Blanchard's first continental stop at The Hague in 1785 sent a mixed message. The French "aerial prince" in staging a scientific spectacle seemed to affirm the authority of the hereditary ruler facing down a patriotic revolution. British ambassador Sir James Harris in his effort to support the territorial prince did not object to a stately spectacle that could help the prince's public image.[1] In pursuing complicated maneuvers to keep the Dutch Republic under French control, Versailles administrators and their diplomatic and military officers juggled for influence with the Dutch Patriots while trying to pacify the hereditary ruler, the Prince of Orange. Blanchard's binary identity as a citizen of Calais and a British hero of the Channel crossing helped satisfy all parties.

The reception in the German-speaking states presents a peculiar mix of proliferating balloon literature and few human ascents. While German scientists kept up with the news, tried some experiments, and extracted the most useful part of French balloon literature for their reading public, they seldom engaged in the more dramatic attempt at human ascension.[2] Although this could be attributed to the scarcity of material resources, given the small size of German-speaking cities and the relative poverty of their aristocratic layer, the libels that followed Blanchard around hint at more complex factors at work. The Prussian debate over "What is Enlightenment?" began almost simultaneously with the balloon invasion to spotlight the problems of popular Enlightenment and legitimate censorship.[3]

Enacted by Jean-Pierre Blanchard, a pensioner of "the most Christian king" often patronized by French diplomatic and military personnel, human ascents in central European cities became a material spectacle divorced from the balloon's scientific promise. Hamburg patriots littered the streets with libels. They saw Blanchard as a pensioner of Louis XVI intent on exporting French civilization to German territory. Suspicions of financial greed followed him around: since he was a pensioner of the French king, richly compensated to the tune of fifty louis or twelve hundred livres in annual pension, why should he be further rewarded by the expensive jewelry drawn from the German purse?[4] If Blanchard's ascents illustrated aristocratic power and authority, they also galvanized patriotic resistance to the Francophile aristocracy and sharpened "German" consciousness. His travels across cultural and linguistic boundaries, often determined by a confluence of aristocratic patronage and commercial prosperity, mapped an evolving geography of patriotic Enlightenment in the German-speaking states, which would later foster enlightened nationalism in response to the Napoleonic Empire.[5]

DUTCH REVOLUTIONARY MIX: Summer 1785

A complicated diplomatic situation that focused the attention of the major European powers (Britain, France, Prussia, Austria, and Russia) on the United Provinces seems to have facilitated Blanchard's escape from London.[6] He arrived at The Hague on June 24, 1785, accompanied by the marquis de Breuilpont (1763–1836), a French dragoon stationed there.[7] French ambassador, the marquis de Vérac (Charles Olivier de Saint-Georges, 1743–1828) presented him to the Stadtholder William V, the Prince of Orange, two days later. A subscription at one ducat per ticket was advertised in Dutch newspapers from July 7.[8]

The hereditary ruler agreed to stage a French balloon spectacle, then, in the midst of a revolution that threatened his status.[9] Dutch Patriots had issued a call to arms in 1782 when the republic was mired in peace negotiations between Britain and France. The national assembly of militia began in

Utrecht from December 1784. The hereditary prince saw nothing but misery and disgrace in his family's future and was contemplating voluntary abdication when the British ambassador Sir James Harris arrived with the intention of scrambling the Patriots' alliance with France and Prussia.

The British bid for Holland made French diplomatic maneuvers more difficult. Sensing that the French government had promised the Patriots "a pure republic independent of the Stadtholder," which would in effect render Holland a French province, Harris schemed to create, foster, and mature "a counter-revolution" that would "fortify the national independence of Holland under its ancient constitution, and recover her friendship and alliance."[10] Vergennes dispatched the comte de Maillebois (Yves-Marie Desmarets, 1715–1791) in March 1785, with Frederick the Great's blessing, to replace the Austrian general Brunswick as the commander of the prince's army.[11] At the same time, Vérac was trying to forge an alliance with the Patriots to keep Britain, Russia, and Austria in check.[12] The third assembly of the Patriots took place just before Blanchard's arrival and required all participants to swear allegiance to the "true Republican constitution" and "People's government by representation."[13]

Intense diplomatic negotiations must have facilitated the flow of balloon news to Holland. In addition to an early report in the French language *Journal d'Amsterdam* on August 27, 1783, Dutch newspapers such as *Oprechte Haarlemsche Courant, Middelburgsche Courant, Nederlandsche Courant, Hollandsche Historische Courant, Amsterdamse Courant*, and *Leeuwarder Courant* followed Parisian developments closely from September.[14] Trials began in November.[15] Jan Daniël Huichelbos van Liender (of the Batavian Society of Experimental Philosophy) at Rotterdam wrote to Étienne Montgolfier for guidance.[16] Johannes van Noorden, a physician and a member of the Rotterdam Physical Society (*Natuurkundig Gezelschap*) successfully launched a hydrogen balloon on November 26 for a small private audience. Although his second attempt on December 1 did not succeed, his detailed knowledge of Parisian developments hints at a channel of communication with Parisian scientists. Not only was he cognizant of the relative merits of the Montgolfière versus the Charlière, but he was also able to calculate the cost of a thirty-foot hydrogen balloon that included the fabric, the varnish, and labor.[17] Charles Diller, a notable instrument maker associated with the Leiden professor Jean Nicolas Sébastien Allamand, a correspondent of Joseph Banks, launched a balloon at The Hague on December 11, 1783, which was reported in all major newspapers (see figure 10.2).[18]

Dutch newspapers reflected French influence. Their reports on the Montgolfière were uniformly positive and often mentioned the French king's kind regard for his subjects' safety. Patriotic criticism against paying for the

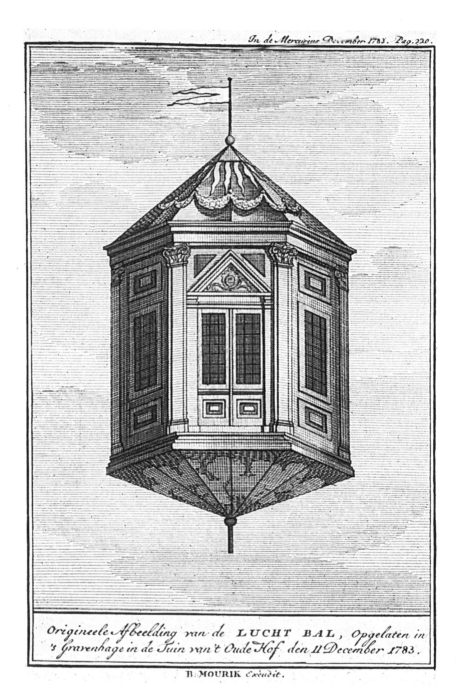

Figure 10.2. Charles Diller's balloon flown from the Hague on December 11, 1783. Gimbel Collection XC-10-2C-3394. It looks similar to Moret's failed balloon.

"French wind" as well as warnings against the fire hazard appeared in pamphlet literature, but these were not printed in major newspapers.[19] Although Dutch balloonists conducted their own debate over the respective merits of the Montgolfière versus the Charlière, the public transcript favored the Montgolfière and marginalized Noorden's open criticism of its danger. A letter of the Montgolfiers (dated November 18), probably Étienne's response to Liender's letter of November 3, was published in the *Groninger Courant* on December 12. Maarten Willemsz Houttuyn (1720–1798) in translating Faujas's *Description* also defended the Montgolfière, citing the cost and technical expertise required for making a hydrogen balloon. As a physician with considerable reputation as a naturalist and close ties to the family's publishing enterprise in Amsterdam, his voice would have added significant weight.[20]

Who controlled Nature mattered in a nation whose landscape was constructed by windmills and canals.[21] The Patriots had already staged a Montgolfière on January 11, 1784.[22] More than a dozen trials took place during the following spring. Tens of thousands turned up for the Amsterdam ascent of a Montgolfière on May 5, 1784, but the balloon crashed and burned after a short flight, unleashing much ridicule. Rising concerns over the fire hazard and the crowd control, probably worsened by the Patriots' mobilization, began to curb Dutch ballooning afterwards. Henri-Louis de Bosset nevertheless organized an illuminated ascent at The Hague on August 9, 1784, as a part of the birthday celebrations for the Princess of Orange—wife to the ruler and niece to Frederick the Great—who commanded much diplomatic effort behind the scene.[23] It could not have been Vérac's earnest desire to undermine the Patriots' political authority by staging a public alliance between the Dutch territorial prince and the French aerial prince, unless it was meant as a camouflage of his real intentions.

On July 12, 1785, the hastily constructed apparatus did not allow Blanchard to fill the balloon sufficiently. He had to leave behind one of the prospective aeronauts—his travel companion Breuilpont or the captain d'Honinethun serving in Maillebois's legion. A lottery between the Oranges and Vérac, who presided over the scene, settled on Honinethun. After traveling through some thunder-producing clouds, the aeronauts landed at 9 PM on a meadow near the village of Zevenhuizen, two leagues from Rotterdam, where they encountered peasants wielding stakes and pitchforks. Riders from The Hague could not follow the balloon on horseback due to the numerous canals that intersected the roads. The landowner even demanded ten ducats as a compensation for the damage to his field, but Blanchard only gave him a promissory note to be redeemed at The Hague.[24] In his futile effort to collect, the farmer argued that whatever fell from the clouds legally belonged to the landowner.²⁵ The lack of pompous celebrations or financial remuneration reflects the difficult political situation.

Ironically, a French aeronautic show bolstered the British effort to prop up the sagging Orangist party. Although Dutch newspapers were somewhat critical, Blanchard's flight was hailed as a success in the British newspapers, sometimes before they could verify the facts of the actual ascent.²⁶ Broad coverage advertised their claim to Blanchard's identity as well as their diplomatic need to recover continental relations.²⁷ They emphasized the occasion as the first aerostatic spectacle in the Netherlands that summoned the entire court and noble personage, which showcased the prince's public authority.²⁸ The "Prince of Aerostatic Adventurers" rose with the flags of Holland and Maillebois's Legion, the *General Evening Post* extolled, to induce a "universal applause from all ranks of people, both the learned and the ignorant, all joining with wonder and admiration at this new and brilliant experiment." Despite their generally positive reports, the British press also asserted a sense of national superiority in reporting the landing. The *Morning Herald* poked fun, for example, at the "uninformed people" of "the *Low* Countries" who imagined Blanchard's balloon to be "nothing more than a prodigious large bladder, filled with an infusion of quicksilver."²⁹

The accident of landing near Rotterdam occasioned an ascent there, which must have pleased resident balloonists such as Noorden. Having lost substantial capital at The Hague, Blanchard demanded at least two thousand subscriptions at two and a half guilder each.³⁰ On July 30, 1785, he had to wait out a downpour until he managed to leave between bouts of violent wind. The turbulent flight navigating through the windmills prevented him from consuming provisions or writing down observations. Maneuvering to avoid landing in a swamp, Blanchard began to descend when he felt the freeze and noticed the sea at a distance. With the barometer broken, he relied on air-filled bladders as well as ground noise to judge vertical distance, while looking for a dry spot. He landed near the town of Ysselstein, about nine leagues from Rotterdam, where an enthusiastic crowd greeted him. When he returned to Rotterdam around noon, spectators filled the streets and the palace. According to a local wit named Klairwal, this intrepid aeronaut's glory surpassed even that of Hercules. Engraved on marble and bronze, his name would last longer than a sovereign's, which proved rather prophetic.³¹ The Stadtholder would be forced out of The Hague in September when the Patriots began to prepare for a war with Joseph II.

BORDER ZONE: Lille and Flanders

Blanchard's Dutch experience must have been traumatic, what with the stormy weather, watery landscape, and revolutionary turmoil. It also cost him a good deal of capital, which means that neither the Dutch prince nor the French ambassador compensated for his show.[32] Somewhat disheartened, he visited with a grand cavalcade the Forest of Guines, "the place named by the King, *Canton Blanchard*." As Breteuil had informed him back in January, a majestic marble column stood at the site of his descent after the Channel crossing. Blanchard wrote to the newspapers that it would require fifty thousand reams of libels, heaped one on top of another, to cover the column all over.[33] The immoderate boasting masked his sense of alienation as a French adventurer traveling along the border zone, greeted by fashionable aristocrats and harrassed by serious patriots. As a lone traveler of the uncharted air and foreign lands, he must have struggled with extreme anxiety and, sometimes, a sense of aimlessness.

An invitation from Lille must have been a welcome respite. Lille was the seventh largest city in France with a population hovering a little over sixty thousand. Well connected by rivers and canals, it served as a commercial hub for the tiny province of Walloon Flanders, which possessed only two other notable towns—Douai and Orchies.[34] Located outside the invisible customs boundary (Five Great Farms), the region had developed a vibrant textile industry. Lille was also a hot spot for ballooning. Early endeavors (see chapter 5) continued with larger Montgolfières, ranging from thirty to eighty feet in circumference. As early as in 1784, some began to dream of crossing the Channel from Lille to England with a waterproof aerostat, even after several failures in trying to stage human ascension.[35]

Vigorous ballooning worried Lille authorities who had experienced serious grain riots during the previous decade. Integrated to France through Louis XIV's conquest a century earlier, Flanders maintained a tiered administration of local and royal magistrates whose political allegiances diverged somewhat.[36] In their turn, merchants and guild members sought to control textile production and distribution. Notwithstanding this complicated patchwork of administration, daily governance of Lille fell largely on town authorities.

After a balloon launch caused a fire on March 13, 1784, Lille magistrates asked Paris police for instruction, which prompted Lenoir to work the Paris parlement for a broad ban that was issued on April 23. Lille authorities issued a local ordinance on May 12 after receiving explicit orders from the royal intendant Charles François Hyacinthe d'Esmangart who had an audience with the king. While an experiment to perfect the discovery could immortalize its authors, the century, and the nation, they had a duty to maintain public order and protect the citizens.[37] Although the ordinances made it more difficult to launch a Montgolfière, balloon fever continued unabated. Joseph Pâris de l'Épinard (1744–?), publisher of the *Feuilles de*

Flandres, caught the "epidemic malady" early on. He had planned to invite Pilâtre, but this famous aeronaut's death in June 1785 directed his attention to Blanchard.[38]

In advertising the upcoming ascent in the *Feuilles*, Blanchard embellished his aeronautic skill by imparting his affective experience of the magnificent scenery in gliding over an "immense sea of vapors." High above the clouds at sixteen to nineteen thousand feet, he found himself in the "vast solitude of the heavens" while variously shaped clouds floated around, beneath, and above him. The diversity of clouds, some azure bordering on purple and others in more vivid colors, formed a thousand pictures.[39] The communication of such a sublime experience shaped a rhetorical topography that interwove space, time, the aerial navigator, and the balloon spectator. Printed in a newspaper, it enticed the balloon public to participate in a rare moment of heightened awareness as the emergent mass subject.[40]

Blanchard had become a seasoned entrepreneur mindful of financial loss. Upon his arrival on August 9, he petitioned Lille administrators directly, overriding l'Épinard's earlier announcement for subscription, to secure their official patronage.[41] They approved it on August 12, hoping that it could attract a great number of "foreigners" to serve as a public fête of general utility. They promised to contribute fifty louis and to subscribe individually at six livres, higher than the normal rate of three livres.[42] In the prospectus published as a supplement to the *Feuilles de Flandres,* Blanchard promised the most brilliant ascent that would build a "human empire of the air."[43] With the date fixed on August 25, 1785, the day of the festival of Saint-Louis, he made two preliminary shows with a parachute, first at the Hôpital général, and then at the grand Magasin near the port St. André in the presence of the prince de Robecq, chief magistrates, and military officers.[44]

Blanchard was a practiced showman with the capacity to satisfy the local power elite's need to exhibit a unified ruling structure for the region—a countryside dotted with densely populated textile villages. The decision to invite outsiders marked a sharp contrast to Bordeaux's approach and evinces an integrated regional economy and social order that depended on coordinating the neighboring regions. Lille was less a global province than a provincial center. A magnificent public festival would provide an occasion for the Lille magistrates to showcase their governing authority. The magistrates decided to send the prospectus to neighboring towns "as was customary in hosting public fêtes." Five hundred copies were sent to forty-two neighboring cities and villages, each receiving from two to twenty copies.[45]

Even for a town that loved spectacles, the magnificent balloon ascent became a public festival of unusual magnitude, which served to integrate the region. As French and Austrian "foreigners" began to flow in, Lille authorities deployed all of their administrative resources to maintain public order. The police issued a detailed instruction on August 23, which was read in public, posted around town, and published in newspapers, "to prevent accidents too large a crowd could occasion" on the Esplanade. They set up a linear barrier to form an enclosure around the Citadel, designated points of entry, issued instructions for the carriage traffic, and specified launch signals. The level of specificity in the thirteen-article public notice reveals the magistrates' justified fear of potential disturbance. Two years earlier, they had to face off an organized opposition by the guild masters who demanded a more equitable distribution of royal military requisition.[46] The wishful image of a well-ordered polis can be discerned from the scene of ascension painted by Louis-Joseph Watteau (1731–1798), professor at the local École de dessin (cover illustration), and engraved by Isidore-Stanislas Helman (1743–ca. 1806) for broad distribution.[47]

Heavy rain delayed the launch until August 26 when they began to fill the balloon at 6:30 AM. With three cannon shots from the citadel ramparts, the balloon rose "majestically" at 11 AM in the midst of military music and applause. The aeronauts waved a flag that had the arms of the king and the city on different sides.[48] After dropping a dog strapped to a parachute just outside of the city, they traversed many towns, alternately rising to 13,104 feet and coming close enough to the ground to converse with the inhabitants. The aerial voyage lasted seven hours and displayed royal and city emblems to the countryside. They landed at Sarvon, sixty-three leagues away from Lille, where they could not dispatch the news.

Watteau also painted the scene of Blanchard's return, which simulated the entrée ceremony. When the aeronauts sent news of their landing while approaching the city on August 30, the magistrates sent the carriage of the colonel marquis de Wignacourt (Antoine-Louis, 1753–?) with two valets, escorted by dragoons and a military band, to greet the aeronauts at the city gate. The cortège proceeded to the city hall surrounded by numerous "citizens of all classes" and presented the flag from the voyage to the mayor amidst public acclamations. The cortège then moved on to the residence of the marshal Charles François de Virot de Sombreuil. In the evening, the aeronauts attended the theater where they were crowned with laurels. *La Belle Arsène* was staged by design.[49] They supped at Wignacourt's with a firework display in their honor.

The magistrates also decided to register the event for posterity; to compensate the aeronauts financially; to offer Blanchard a gold box inscribed with the arms of the city and inscription, valued at twelve hundred livres; to present l'Épinard with a *caffetiere d'argent* with the arms of the city, valued at three hundred livres; and to present regional wine as a special gift. When the magistrates offered twelve hundred livres in cash or a gold box decorated with the arms of the city, Blanchard chose the latter, perhaps to demonstrate his honor. During the festivities, two official ushers (*huissiers à verges en robe*) took a box containing twenty-five bottles of wine from Champagne mousseau covered with a tapestry embroidered with the arms of the city to l'Épinard's house where Blanchard stayed. He was deeply gratified by these signs of appreciation.[50]

What a grandiose balloon festival should mean was subject to interpretations. An alternative story of princely reception (rather than city magistrates) points to multiple appropriations of its symbolic value.[51] Even more troubled was its reception among the religiously inclined, which was mostly suppressed in the public accounts. An anonymous pamphlet offers a rare glimpse of the balloon's disruptive impact on the Christian worldview. The author summoned his utmost wit to appropriate the human ascent as a part of Christian narrative, only then to reverse his position. In his first letter he adopted the aeronauts as the "Apostles of Christianity" and complimented these "new Coelicoles" for exalting his spirit. Their "triumphant ascension in the air" visualized "the resurrection of the bodies of the just which, united to their souls, leap triumphantly" to the highest heaven to receive "immortal glory" among the saints. He interpreted the spectacle as real proof of the human soul—"an active spirit, an intelligent motor," or "an immaterial, spiritual, immortal substance" that "thinks and reflects on his thoughts and his proper reflections"—and the aeronauts as "Christian Heroes." Their "obviously supernatural & entirely divine courage" confirmed the existence of "God the supreme author of all things . . . this being eternal and infinitely perfect."[52]

This religious pamphleteer would soon raise "the most cruel alarm," pitching that the ascent posed a certain peril without "just necessity" that risked the precious gift of life from "the supreme author of all beings." Blanchard's dangerous experiments only excited curiosity, this "passion so natural but so harmful to man in nurturing his laziness, inspiring distaste for work and taste for frivolity, and making him forget, neglect and abandon the essential duties of his estate, to be occupied with bagatelles and to chase after vain amusements and useless spectacles."[53] The flip-flop speaks for the deep anxiety the religious public must have felt in trying to negotiate such a public triumph of science.

Lille became a hub for Blanchard's continental journey by supplying material and human resources, while nurturing other balloon trials.[54] Well connected through rivers, canals, and highways and enjoying tax relief in international trade (*droit de transit*), this northern border city offered an ideal locale from which Blanchard could negotiate the outlying regions.[55] For a planned ascension in Gand (nowadays Ghent, Belgium), Lille administrators authorized the transport of sulfuric acid by boat. The apothecary Carette made several balloons for Blanchard to be used later at Douai and Hamburg.[56] Blanchard would publicly acknowledge his sincere attachment and "eternal esteem" for this master apothecary's probity, consideration, and talents.[57] The remarkably favorable reception of Blanchard at this border city perhaps compensated for his sense of marginality in French theatrical polity. His continental journey hovered over the northeastern border zone of France with occasional excursions to German-speaking cities.

IMPERIAL CITY: Frankfurt

Frankfurt showcased the new French scientific fashion for the continental aristocracy. A majestic balloon ascent was a fitting spectacle for this ceremonial city where the Holy Roman Emperor was elected and crowned. Frankfurt was also an international center of trade and finance, thanks to its location on the Main River, which merged with the Rhine River downstream, and its proximity to the major roads connecting central European cities. The city's fortune derived from the influx of Protestant refugees from the Spanish Netherlands in the late sixteenth century. The Thirty Years' War brought immigrants from Switzerland, the Palatinate, eastern France, Italy, and the Netherlands. Although its prosperity depended critically on the fairs and banking business that brought visitors from all over Europe, the city (with a population of about thirty thousand in the 1780s) was ruled by an oligarchy that formed a distinct rank through their inherited wealth and service on the city council. The constitutional conflict during the early eighteenth century highlighted the city's complex polity mediated by imperial authority.[58]

Initially scheduled for September 25, 1785, the ascent was delayed until October 3 due to violent weather.[59] Blanchard had to abandon the first balloon, torn from top to bottom in the storm, and its prospective passengers —Prince Frederick of Hesse-Darmstadt and Mr. Schweitzer, an officer in Schomberg's regiment. All seemed lost, but he managed to take off alone in his trusted Calais balloon.[60] After traveling for about forty-eight minutes, he landed on the borders of the Lahn River, about a quarter of a league from Weilburg.

The imperial aristocracy treated the aeronaut in style. When Blanchard returned to Frankfurt, a festival broke out. Two actresses crowned his bust on stage. He appeared on the balcony of the Russian ambassador the comte de Romanzow's residence. The next morning, Blanchard traveled to the theater in a carriage borne by a number of men who took the place of horses. In the midst of applause, he visited the princes in their boxes to receive compliments. Afterwards, he dined with city notables who gave him a generous gift. The following morning, he presented to the Senate the flag that had traveled in the air bearing the arms of the city. City magistrates rewarded him with fifty gold coins (a hundred ducats in value) struck for the coronation of Joseph II and promised to compensate him for the entire cost of experiments.[61]

Blanchard's impending departure occasioned a pompous farewell at the theater on October 8. After the first piece, the scene changed to a superb palace in which his bust was elevated on a magnificent throne, the foundation representing the Temple of Memory with nine muses guarding the door. Three actresses representing Grace, richly decorated with flower garlands, addressed several couplets and came to the box where he sat with the Russian ambassador. The princes and princesses, numbering 122 altogether, resolved to issue a subscription for a balloon capable of lifting fifty persons under Blanchard's direction.

When Blanchard sent a flag from the voyage to Frederick William, the Prince of Nassau-Weilbourg (1768–1816) at Kirchheim, the prince invited him to his castle (on October 13) and presented him with a gold box. The princess gave him a watch to mark the exact time of his descent there. The flag was deposited in their archive. Blanchard staged more trials of the parachute on October 17, dropping a dog from half the height of the town church. After the magnificent reception at Frankfurt, Blanchard must have appreciated the opportunities that prosperous German cities could present. He hoped to stage future ascents in Hamburg, Vienna, Warsaw, St. Petersburg, Rome, Milan, Naples, and Spain.[62]

Balloon fever was already in decline, however. Blanchard went back to Gand for a promised ascent, which proved both perilous and without profit.[63] During the winter, he visited the site of Pilâtre's accident, perhaps wishing to expel the lingering shadow of death. In the spring of 1786, he staged an ascent at Douai (April 21) and two at Brussels (June 10 and 16) where an imperial order had forbidden the Montgolfière.[64] Blanchard experienced many logistical and financial difficulties, which points to his borderline identity.[65]

In order to entertain princes, Blanchard had to sustain his image as a gentleman balloonist, which made him vulnerable to attacks on his character. A few days before the first Brussels ascent, a female merchant solicited him for 150 florins. When he refused, a rumor spread that he would pay any bill presented to his secretary. The female merchant appeared again with a bill and two other "rascals." He had to oblige them not to cause a scandal that might delay his forthcoming engagement with a hundred thousand spectators and the royal couple. The ascent did not even turn a profit, which forced him to stage another. The wealthy did not pay to sit in the enclosure, indicating their indifference to its scientific value. Blanchard also had to pay the innkeeper thirty-two hundred livres for six weeks of residence, notwithstanding the loss of a valuable diamond buckle and other possessions. Although he wished to maintain his reputation and lifestyle of a gentlemanly balloonist, the experience pushed his veneer of civility beyond limit. He threatened to publish a complete and honest chronicle of his voyages to vindicate himself.[66]

INTERNATIONAL PORT: Hamburg

French diplomatic personnel, including the ambassador then staying at Brussels, facilitated a call to Hamburg.[67] Claude Antoine Viviers, Vergennes's brother-in-law who was serving as minister plenipotentiary, also extended his patronage. Hamburg was a major gateway to central Europe. With a large stock exchange and a population of nearly ninety thousand, it was the third largest German-speaking city next to Vienna and Berlin. Ships from all over Europe blanketed the canals and the Elbe River that connected many German cities. Visitors found "a perpetual motion of all nations and peoples caught up in the business of money-making," although its governing elite struggled to keep up with the influx of immigrants and rising poverty.[68]

The road to Hamburg was treacherous, which put Blanchard in a chauvinistic mood. Traveling via Aix-la-Chapelle (nowadays Aachen in Germany), he found Westphalia dreadful where one was lucky to get a piece of black bread before joining "the poor beasts" for the night, much like the stable of Bethlehem. Language barriers and depreciating currency compounded the situation. Blanchard had to pass "an infinite number" of customs barriers and endure a long wait (from ten to twelve hours) between stagecoaches. Normally, one could travel six leagues a day, divided into two three-hour journeys with a three-to-four-hour wait in between. Postal coaches on this route moved slowly with no set schedule, stopping at any time for the driver to take a sip of brandy.[69]

Blanchard arrived at Hamburg behind schedule, probably in mid-July, and went on complaining.[70] Although the owner of the London Hotel greeted him in French, Blanchard found the city ill-paved, uncomfortably configured, and shoddily constructed. He received an enthusiastic applause at the theater and a civil welcome from the local society in the company of the abbé Kentzinger, deputy French ambassador who had negotiated the deal. He declined invitations from many notables of the boomtown—a volatile mix of merchants, bankers, lawyers, magistrates, intelligentsia, aristocrats, and diplomats.[71] In sharp contrast, Hamburg patriots welcomed Blanchard with a deluge of libels. Clueless to local politics, he resolved to dine with the malevolent journalist Jonas Ludwig von Hess in full view of the spectators just before the ascent.[72]

Blanchard's haughty treatment of local notables, combined with his magnanimous gesture toward their patriotic opposition, reveals his misguided sense of superiority as a French gentleman. In reality, he was nothing but a hired entertainer—an expensive one at that. A short prospectus printed before his arrival stipulated the subscription price at one ducat per person and three ducats per family. His secretary would collect the subscription, print the list of subscribers on July 15, and begin distributing the tickets the following day. Ground preparations took extensive negotiations with the Senate over the site and necessary construction, which required bilingual personnel. A renowned lawyer Misler apparently offered his services without compensation.[73] A full prospectus was published on August 7 in French and German (side-by-side) with a brief history of Blanchard's path to flying.[74]

Blanchard's ascent provided a splendid occasion for the Hamburg magistrates and aristocracy to display their authority in public. On August 23, fourteen hundred troops guarded the enclosure at the Sternschanze (the star-shaped fort west of Hamburg) against 160,000 spectators forming a superb amphitheater. The city population had nearly doubled by the convergence of foreigners from Holland, England, Denmark, Russia, Hannover, Brunswick, Saxony, Prussia, Poland, Westphalia, Lübeck, Mecklenburg, and so on. Blanchard had ordered three balloons from the Lille apothecary Pierre Louis Joseph Carette—one of 5,577 cubic feet for the ascent, another of 900 cubic feet to raise a sheep and a parachute of his own invention, and a smaller one to be released by the Princesses of Holstein and Mecklenburg. They could not fill the third one after an accident that wasted sulfuric acid.

Blanchard left at 4:30 PM, saluting with the flag that bore the arms of Hamburg. At about nine hundred toises, he cut the cord (from a distance of sixty feet) that tied the animal to the small balloon. The sheep descended with the parachute slowly to arrive gently on the ground. Several dragoons surrounded the parachute to protect it. When Blanchard landed on a plain near Altona, a dozen dragoons with their swords drawn had to extract him from the overzealous populace. They conducted him, hovering from six to eight feet above the crowd, back to the place of departure. Blanchard arrived thus in the midst of acclamations at the Sternschanze where the august audience still remained in their seats. The comtesse de Bentinck offered her carriage and presented a crown.

In the evening, Blanchard attended the theater and then a fireworks display. The town was packed beyond its capacity with many foreigners having to spend the night, unable to find horses for the return trip. Balls and fêtes took place at all corners. Girandolini staged in the fireworks a balloon with a boat, flags, a voyager, and all accessories. Toward the equator of the firework balloon were written in the letters of fire, "Vivat Blanchard." Afterwards, Blanchard had supper at the comtesse de Bentinck's residence where the hereditary Prince of Brunswick and other nobility gathered. He was crowned by the countess and received an exquisite gold box from the prince. He then stopped at the town resident Doorman's for another gathering.[75]

On August 24, Blanchard presented the flag with the arms of France to the chevalier de Viviers who also took charge of the sheep and ordered a beautiful collar engraved with the words: "On August 23, 1786, at 4 hours 35 minutes, this sheep was thrown from about 900 toises by M. Blanchard, citizen of Calais, Pensioner of S.M.T.C., correspondent of several academies." Blanchard also presented the other flag with the arms of Hamburg to M. A. Von de Sienek who received it for the Senate and promised to keep it in the city hall. The Duchess of Mecklenburg sent him a watch encrusted with diamonds and a gold chain, as did the Princess of Holstein. Blanchard left the city on August 27, confident that he had made a favorable impression.

Hamburg patriots were not happy, however, with the useless French frivolity that center-staged Francophile aristocrats.[76] They covered the streets with libels and wrote to the newspapers. Not speaking German, Blanchard had little comprehension of their discontent or empathy for their oppositional patriotism. When he reached Aix-la-Chapelle, he received several letters that informed him of a Hamburg libel, advising him to respond. In asking for Blanchard's word to protect his reputation, the Hamburg elite would have been just as anxious to protect their own. He dismissed its significance, while inserting a response in his narrative of the Hamburg ascension. Simply asserting that his conduct had been irreproachable, he vowed to act honorably and to maintain his tranquility of mind. If he wrote against all newspapers that attacked him, he would have to write all his life, which was not his métier.

Blanchard's refusal to defend his reputation in German newspapers indicates his distance as a French entertainer from the German reading public. He attributed the libels that dogged his sojourn to the ignorant individuals jealous of his glory, akin to French courtiers. Unable to attain such status on their own talent or courage, he surmised, they sent their "sad productions" to the German press that would publish anything for payment. Confident of his success and virtue, he even printed one of critical epigrams in his account of the voyage. Stylish epigrams and press articles tell us a story of the French aeronaut who filled his silk balloon by emptying meager "German" purses.

The libel Blanchard printed in an attempt to "give an idea of the atrocities" of German newsmakers actually reveals how far he had strayed from any scientific pretensions of ballooning. Although he wished to prove that the calumniator was a miserable "Anti-French" seduced by cheap compensation (two ducats), the German author actually pointed out that the balloon was supposed to be a scientific machine. Natural philosophy did not gain anything from Blanchard's show—there was no observation, not even with the barometer or the thermometer, or any effort at steering the machine, contrary to the promise. It had become a *faut perilleux*, a dangerous amusement that served no public purpose. Except for its visual effect and cost, it was "exactly the same thing as that of ropewalkers" and other acrobatics. Blanchard received more than 11,400 Marks in ticket sales plus 3,000 for his "useless *Balonnerie*," drawn from the German purse.

In response, Blanchard merely printed an epigram by the Hamburg physician Wittenberg who compared the satirists to children crying for milk. Blanchard meant to ridicule German critics into silence by invoking elite patrons, without appreciating the legitimacy and public resonance of German patriotic sentiments. He mocked German journalists as "stupid beasts [*lourdes bêtes*]" who, without censors or merit, published worthless things for a small fee. He would ensure publicity for all critics and their criticisms, provided that they were well written.

Not quite finished with his rant, Blanchard then plunged into a line-by-line response, which provides us with a summary of German criticism. Other than denigrating Blanchard's performance and relevance to the progress of natural philosophy, the author had also pointed out that Blanchard agitated the populace with repeated shouts of "Vivat Hamburg"; that he was a shady imposter who hid the gifts he had received; and that he was a great gambler and had lost a considerable sum in Frankfurt. Blanchard acknowledged that he made a large sum, twice what the author mentioned, for which he gratefully shouted "Vivat Hamburg"; that he kept the jewelry with a grand seigneur at Boulogne-sur-mer to make "the most rare and precious cabinet" when he eventually returned home; and that he never gambled. Some had solicited the bankers at Aix to raise a subscription, between four and five hundred louis, saying that they would win the money in one session with him.

Blanchard insisted that he had always conducted himself as an honorable gentleman (*honnête homme*) far above these calumniators below "the class of respectable citizens." He could not appreciate the fact that his idea of the *honnête homme*—a civilized socialite in his understanding and aspiration—was not what German patriots desired from a virtuous scientific aeronaut. Misrecognition and ignorance would only fuel anti-French sentiments among the German reading public, although defensive publications did also appear in German.[77]

French prospects disintegrated noticeably as well. After Hamburg, Blanchard went back to Aix-la-Chapelle, lured by an innkeeper. When he was passing through the city on the way from Frankfurt to Lille, he received a bill from the London Hotel. This was a scheme by the innkeeper Rouisse to meet Blanchard, but he agreed to stage an ascent. He arrived on September 5, only to discover that the subscription by amateurs came to less than a quarter of the agreed-upon sum. He opened a new subscription, but there were few takers because most residents chose to watch the show outside the enclosure. For the audience who did not care for the bragging rights of knowing science, their proximity to the machine was not worth the admission price.

The ascent on October 9, 1786, more than a month after his arrival, proved successful. On his return, a performance at the theater began and ended with the crowning of his bust on stage. The pedestal was inscribed by Marc Doberny. He sat next to Prince Gagarin, a Russian Freemason from St. Petersburg with ties to Berlin lodges.[78] Afterwards, he dined festively with Princess Gagarin at the old Redoute. Although Blanchard received public recognition, he did not make a profit because he had to pay the innkeeper's exorbitant bill that came to two thousand ecus. The experience made Blanchard somewhat despondent of his prospect as an itinerant balloonist, as is evident in his published accounts.

POWERFUL ABSENCE: Vienna and Berlin

Vienna and Berlin would have been obvious destinations in Blanchard's itinerary after Hamburg, but their enlightened despots—Joseph II and Frederick the Great's successor—declined his advances, citing the lack of utility and potential danger. Hamburg libels must have worried these powerful rulers and their administrators. Vienna, the largest city in central Europe, had not yet seen a human ascent. Blanchard wrote to the comte d'Escherny (1733–1815), a Swiss diplomat circulating in European courts, who had assured him of the emperor's permission. He wished to schedule an ascent in December 1786 since the prince de Ligne was scheduled to visit Vienna then, or in April when the comte de Paar would be there.[79] Having attended the Hamburg ascent, the comte de Paar had promised a hundred ducats in subscription. His father, the prince de Paar, was a Viennese aristocrat who frequented Versailles.

The Habsburg emperor declined the proposal, writing that a useless spectacle would not afford him pleasure.[80] After a spectacular debut among the imperial aristocracy in Frankfurt, this polite refusal from Marie Antoinette's brother must have come as a surprise to Blanchard. The Habsburg Court was not known for shunning useless novelties, as can be seen in the case of Wolfgang von Kempeln's (1734–1804) chess-playing automaton that began a European tour in the summer of 1783.[81] However, the balloon as a people-machine would have induced an unsettling mass spectacle, especially when diplomatic relations with France were severely strained. Blanchard had to wait until after Joseph II's death to stage a series of ascents in Vienna.

In Berlin, balloon fever had started early through the *Gazette littéraire de Berlin*, a French-language journal initially edited by Joseph Dufresne de Francheville (1704–1781) and sponsored by Frederick the Great from 1764. Balloon trials began cautiously by the academician Franz Karl Achard (1753–1821), mostly for the court elite and Freemasons. After a fatal accident on February 18, 1784, Frederick the Great made a sharp turn and forbade all aerostatic experiments in his dominion, citing their danger and inutility.[82] The *Gazette* stopped publishing balloon news altogether, instead publishing a polemical article against useless amusements. It also began a serialized article on proper forms of government by the powerful pro-English statesman Ewald Friedrich von Hertzberg (1725–1795).[83] The *Gazette* did not even carry the news of Blanchard's Channel crossing or Pilâtre's tragic death. Strong censorship of balloon news would not ease until after Frederick William II was securely in power.

Prussian balloon fever among the court circle of Freemasons was overshadowed by the more serious debate on "What Is Enlightenment?" On December 17, 1783, the royal physician Johann Karl Wilhelm Möhsen presented his essay on "What is to be done towards the enlightenment of fellow citizens?" to the well-intentioned patriots of the Wednesday Society (*Mittwochsgesellschaft*) that had started two months earlier. He defined their patriotic duty as the enlightenment of themselves and their fellow citizens in the entire Germany, their "shared fatherland." While Möhsen's lecture called for a broad understanding and definition of the Enlightenment that would allow Berlin's literary, scientific, religious, and administrative elite to craft strategies of progress for the imagined nation in step with their king, the debate within the Society quickly came to focus on press and print freedom. Most members, especially jurists, bureaucrats, and religious educators, shared the concern that a free and unrestricted discussion of religious, moral, and political issues might undermine the conventional mores and beliefs that sustained the social order.[84] A flurry of reflections appeared on the slow progression of the popular Enlightenment and the censorship of popular literature.

Whether the balloon spectacle could or should enlighten the populace would have been a serious concern for the Prussian intellectual and administrative elite wishing to form a national culture distinct from French civilization. In other words, the balloon as a spectacular artifact that valorized French theatrical polity did not fit the Prussian vision of enlightened absolutism. Blanchard seems to have contemplated an excursion to Berlin after Hamburg. Frederick the Great's death in the summer of 1786 created a sensitive diplomatic situation for France, which prompted Vergennes to dispatch the comte de Mirabeau (Honoré-Gabriel Riqueti, 1749–1791) on a secret intelligence mission.[85] A balloon ascent could have passed as a stately celebration for the new king, Frederick William II, but he declined Blanchard's advance, citing its danger.[86] The "philosophic Mirabeau" apparently persuaded Berliners that the famous French aeronaut knew little science, that he gambled, and that he had three wives.[87] Blanchard had to wait until William was securely in power before he could stage his balloon ascent in Berlin from the Tiergarten on September 27, 1788.[88]

A FADING DREAM

Unable to persuade the powerful German princes, Blanchard went back to the French border zone where he staged an ascent from Liège on December 27.[89] He resumed his travels in the spring of 1787 and went to Valenciennes, the hometown of Joseph Watteau who painted the Lille ascents, and then moved on to Nancy and Strasbourg.[90] Since the balloon was no longer a novelty, he had trouble meeting the cost as political events came to occupy the attention of local elites. For those without scientific pretensions, an expensive seat around the balloon seemed pointless. At Nancy, a notary named Pierre invited his extended family, friends, and town notables to his house directly facing the place of experiment, not all of them gratis, to avoid paying the subscription fee. In a small town, this must have put a serious dent in Blanchard's purse and future plans. He publicized Pierre's action as a dishonorable act in the Strasbourg prospectus.

Blanchard's desperate attempt to fend off common objections provides us with a useful gauge of the public mood. Their objections were reasonable. Since the balloon could not be viewed for a long time, blocked by the surrounding trees and houses, was it not better to stage it in an open field? Would not one get exhausted in the citadel while waiting for the departure? Would the place be large enough to accommodate all amateurs? Wouldn't it be risky to take children to such a crowded place, especially at the entrée? To the charge that he forced people to enter the citadel for a profit, Blanchard claimed that it was a just reward for an author who incurs great expenses and causes gold to spread in the town.[91]

Strasbourg had been receptive to ballooning from early on, judging from the publications of Christian Kramp (1760–1826). The town elite had subscribed to a failed human ascent organized by the mécaniciens Gabriel and Pierre in May 1784, which enlisted at least 254 subscribers.[92] A Montgolfière made by the Adorne and Enslen brothers took off on May 15, 1787, from the Citadel. Unable to economize on gas production, Blanchard had to open a subscription.[93] To justify expenses, he claimed to have opened "a vast field for physics" and garnered much knowledge. He urged amateurs to place themselves in the enclosure of the Citadel, which could contain more than a thousand persons, rather than losing time and suffering pain by trying to follow the balloon. He claimed that preliminary preparations and the moment of ascension were the most interesting part of the event. He offered to show the preparations to the amateurs who could not afford to pay six livres by personally applying to him. He also put the aerostatic apparatus on display in a large wooden hall constructed for the purpose at the Citadel from 9 AM to 7 PM, at twelve sols per entry for the guards. The Calais balloon became an object of particular interest. With the permission by the marquis de La Salle, military commandant of Alsace, the Strasbourg ascent took place on August 26, 1787. The event was not popular, judging from the fact that a third-class ticket went on sale for thirty sols at the gate on the day of ascension. Since the events of the French Revolution had begun to reach this border town, which set up a provincial assembly on August 18 according to Calonne's proposal, the royal commandant's wish to stage a public spectacle must have been somewhat contentious (see figure 10.1).[94]

Blanchard weathered the early phase of the French Revolution in foreign lands, perhaps fearing retribution against his aristocratic associations. From Strasbourg he moved on to German-speaking territory transporting all necessary apparatus on a specially constructed carriage.[95] Economic necessity probably drove him outward to find places where his ascension would be compensated. Except for a brief return to Metz in 1788, his itinerary included Leipzig, Nürnberg, Basel, Brunswick, Berlin, Warsaw, Breslaw, Prague, Vienna, Hanover, Lübeck, and Philadelphia.[96]

The jewel of Enlightenment sciences came to the central European states as a French royal artifact devoid of scientific virtue to shape contested Enlightenments, modern aspirations, and national identities. Its displacements engendered a geography of the French cultural empire that excited aristocratic desire and popular enthusiasm to invite patriotic resistance. Blanchard's continental itinerary illuminates how material ballooning constituted a boundary genre of performance between universal science and stately spectacle. As a malleable cultural performance, the balloon spectacle enacted variable networks of power at different locales, which projected diverse cultural visions and polities. Because he strove to fashion a career by pulling together scattered resources and by transporting his balloon for most journeys, Blanchard's itinerary can also highlight where the opportunities of and the hindrances to mass Enlightenment lay. Staging a public ascension depended on the availability of materials such as sulfuric acid, the existence of advertising venues for a subscription, and the enthusiasm and hospitality of local princes, notables, and administrators who depended in turn on their relationship with the larger populace.

Blanchard's struggles in America show his determination and the unfavorable sociocultural conditions, notwithstanding the eager audience of a new republic. Balloon news had arrived there early through the American travelers in Paris and fostered various attempts without a successful human ascent. The hydrogen balloon was not a reasonable choice in the former colonies lacking an industrial infrastructure. For a single ascent in Philadelphia, Blanchard had to transport forty-two hundred pounds of acid from London, along with the balloon, and settle for the jail ground in order to accommodate the paying audience. After the successful launch on January 9, 1793, he built an "aerostatical laboratory" as a business venture. Vandals damaged his balloon to make further human ascents impossible. He continued with smaller balloons made in the city but the yellow fever epidemic drove him out. Further ventures in Charleston, South Carolina, and Boston faltered when Jeffries filed a lawsuit to recoup the cost of the Calais balloon. Losing the legal battle, Blanchard sought a reprieve in New York, but fate would strike it down again. His "Balloon House" was destroyed by a tornado on September 14, 1796, even as he struggled to gather enough subscriptions. Blanchard left New York in May 1797.[97]

Blanchard resumed a ballooning career in France with the ascent in Rouen on August 12, 1798, and his travels lasted until 1809 when he fell to his death. His new wife, Marie-Madeleine-Sophie Armant (1778–1819) picked up the trail to stage a magnificent show for Napoleon's marriage. She would also fall to her death on July 6, 1819.

Figure E.1. *Fête de la Fédération at the Champ de Mars, July 14, 1790*, painted by Charles Monnet and engraved by Isidore Stanislas Helman.

EPILOGUE

Revolutionary Metamorphoses

The aerostatic machine embodied the promises and paradoxes of French Enlightenment sciences under the absolutist monarchy. The enlightened trust in the authority of science to interpret and control Nature had fostered, on the one hand, state sciences embodied in royal institutions and, on the other hand, popular artifacts and practices like the balloon and mesmerism. One cannot understand Jean-Paul Marat's radicalism and his desire to rupture the "chains of slavery" without unearthing his aspirations for and frustrations with the French regime of science.[1] The Périers and the Montgolfiers became national heroes without cultivating a broader foundation of scientific innovation.

Born in the victorious aftermath of the American War, the balloon seemed to promise a system of scientific hegemony that would integrate a nation of citizens under the royal authority through the conquest of Nature. The reformed nation in this elite imagination would be a scientific imperium that would tame arbitrary power with the authority and reason of Nature. If the invention of the aerostatic machine required the ancien régime technologies that aimed at integrating the cultural nation via theatrical performance and sociability, its performance nevertheless ruptured the "chain of gravity" and intimated a comparable degree of social emancipation. In other words, the balloon opened a liminal period of uncertain possibilities when utopian dreams and dystopian anxieties stirred indiscriminately.[2]

Dreams and realities would clash during the French Revolution.[3] Balloon experience helped shape the revolutionary rhetoric against the collective memory of exclusion and injustice, as seen in abbé Sieyès's "What Is the Third Estate?" He identified exclusion as "a social crime, a veritable act of war against the Third Estate" and argued for a common system of law and representation that would constitute a nation of citizens. The principles of reason and justice, he preached, would abolish privileged orders, cultivate virtuous citizens, and achieve social unity. The law in his imagined republic would occupy the center of an immense globe and would place every citizen at an equal distance on its circumference. Every citizen would find "an equal place" in his new republic.[4]

Translating the imagined philosophical nation into a working republican monarchy was not easy. If the balloon offered a plausible image of the just nation, motley balloon publics did not automatically constitute an electoral polity. For the nation whose simple citizens had long been excluded from public functions and discussions of administrative principles, Condorcet in his proposal for provincial assemblies and a "truly representative" national assembly called for citizens who possessed necessary knowledge and public spirit.[5] The middling order of underemployed lawyers like Robespierre and educated manufacturers such as the Montgolfiers easily qualified, but the populace remained outside of the republican nation.

In a telling testimonial to this continuing exclusion, French balloonists and their patrons met divergent fates during the revolution. In April 1789, Réveillon narrowly escaped the crowd that condemned him "to be hanged and burned in a public square."[6] Although the Parisian assembly of electors and the duc d'Orléans (formerly Chartres) intervened, the crowd broke through the barrier of soldiers to loot and burn the endless array of luxuries at his residential factory—fifty thousand books, expensive furniture, thousands of bottles of wine, and so on. The crowd must have been hungry to mobilize over a rumor of a wage cut amid rising bread prices. They must have been angrier about the exclusion signified by Réveillon's *folie* that stood like a castle in the artisanal district. During the balloon years, the factory had served as a displaced court, which attracted the fashionable society in their reckless carriages and became a tantalizing site of distinction and exclusion. Having entered the premises in part out of curiosity, the rioting workers (not Réveillon's, as Rudé pointed out) would have been reminded of the artful distinction that had sustained the theatrical polity.

The boundary character of the Réveillon riot—seen either as the last outbreak of the ancien régime or as the first popular violence of the revolution—points to the transitional identity of the ancien régime entrepreneurs who became the representatives of the Third Estate. The identity of royal administrators was not as malleable. On July 14, 1789, the crowd stormed the

Bastille, the infamous locus of persecution and injustice in public opinion.[7] They killed its commander the marquis de Launay (Bernard René Jourdan, 1740–1789) and then, at the city hall, Jacques de Flesselles. Their severed heads were paraded through the streets of Paris on pikes to signal an ominous turn.[8] When a mob rushed into the Louvre three years later, in contrast, they stopped at the sight of Charles's balloon.[9] The Montgolfiers also found their place in the revolutionary process as respected local notables. Joseph Montgolfier would be appointed to the new Conservatoire national des arts et métiers and inducted to the Légion d'honneur by Napoleon.

An archeology of public space would exhibit similar shadings of continuity and discontinuity. When the call for the Assembly of Notables triggered an avalanche of pamphlet literature, the Champ de Mars became the authenticating site of the Estates General.[10] As the call for a free nation based on a new constitution began to accrue momentum, the anonymous author of *Clovis au premier Champ de Mars* (1789) offered a vivid mental image of "the first assembly of the nation" that had supposedly gathered bishops, high nobility, all able-bodied Frenchmen, and numerous deputies of Gallic villages. While innumerable people in the vast enclosure contemplated the majesty of the assembly, Clovis in imperial garments appeared in the midst of his nation and imposed a respectful silence to lay down fundamental laws of the new "empire." Skillfully mixing Rousseau and Montesquieu, the author sought to envisage a constitutional monarchy for the "great nation." Only a monarchy supported by a free assembly of the Estates General could secure liberty, eliminate fear, and instill hope, which would induce patriotism (*l'amour de la patrie*) among the citizens. This moderate solution was remarkably in tune with the Orleanist vision. The duc d'Orléans would fashion himself as Philippe Égalité, mounting a vigorous defense of his revolutionary identity. The intensifying royalist slander, as in the *Vie de Louis-Philippe-Joseph, duc d'Orléans* (1789), indicates the strength of the Orleanist conspiracy as a political specter, if not as a real intention.[11]

The Champ de Mars also hosted the Festival of Federation, which staged a vision of republican monarchy.[12] Federation movement began in late November 1789 on the banks of the Rhône River with twelve thousand National Guardsmen from Dauphiné and Vivarais (where the Montgolfiers lived) and then spread to Lyon, La Rochelle, Troyes, Anjou, Brittany, and Strasbourg. The Parisian celebration, imagined as a public festival like the Roman Saturnal when all distinctions of rank supposedly disappeared and slaves talked freely to their masters, was meant to visualize a popular revolution and dramatize the spirit of liberty and patriotic enthusiasm.[13] Massive construction drawing on voluntary labor prepared an elaborate stage for this patriotic festival (*fête de la patrie*) where the king in his royal costume shared the spotlight with the president of the National Assembly (see figure E.1). The reciprocal oath between the National Guardsmen and Louis XVI to uphold the constitution might have ended the revolution for a regime of republican monarchy.[14] According to the plan, the commander general would summon Parisian guards at the Champ de Mars, review them, and make a show of military art as the Romans had done in their own champ de Mars.

A balloon ascent would have marked a beautiful ending to the national festival that brought together liberal aristocrats such as the marquis de Lafayette, constitutionalist Jacobins (see figure E.2), their patron the duc d'Orléans, and the royalists, while excluding the populace.[15] A republican balloon, stripped of royal emblems, was scheduled to rise at the end of the weeklong carnival on July 18, 1790. As the troops made their maneuvers, they would prepare a "superb balloon in the colors of the nation" and walk it around at a fixed height so that all spectators could pay homage to the patrie. Two or three brave aeronauts would then take up the aerial voyage. One could not better celebrate a "festival of the nation," as the rhetoric went, than by perfecting an art whose discovery brought honor to France and reduced the limits of human spirit.[16] The tricolor balloon appeared after Lafayette, suspended at a height of three to four meters. It caught fire and crashed, injuring six soldiers engaged in their military maneuvers.[17]

The utopian hope for a peaceful revolution proved just as fragile. In 1791 the clergy's refusal to swear an oath to the civil constitution, the royal family's flight to Varennes in June, and the Champ de Mars massacre in July quickly changed the course of the French Revolution.[18] On September 18, the Festival of Constitution took place at the Champ de Mars, renamed Champ de Fédération, as a celebration of the king's acceptance of the constitution. The expensive balloon (estimated at fifteen thousand livres) that took off from the Champs Elysées afterward carried the fading hope for a republican monarchy. Decorated with four constitutional emblems (liberty, l'amour de la patrie, France, and law) and carrying a rooster-shaped boat, the thirty-foot hydrogen balloon allowed the aeronaut to spread exemplars of the constitution from the air.[19] Another balloon festival offered by Louis XVI on September 25 to ameliorate the public opinion was in vain. With the beginning of the revolutionary wars in 1792, subsequent balloon efforts quickly turned into a military project. Guyton de Morveau, who moved to Paris and became the citizen Guyton-Morveau, took the lead in the military mobilization. A reconnaissance balloon would float with much acclaim in the victorious battle of Fleurus in June 1794 (see figure E.3).[20]

The balloon's powerful, yet indeterminate agency complicates the existing interpretations of the French Revolution—Marxist, revisionist,

Figure E.2. *The Jacobins are going to revolutionize the moon with balloons.* Gimbel Collection

Figure E.3. The Battle of Fleurus, June 26, 1794. NASM A19680192000cp04.

Tocquevillian, or performative.[21] Although it was an eminently material object that functioned like a national artifact, the balloon did not exactly induce bourgeois or plebian political consciousness, much less determine the onset or progress of the revolution, as a Marxist interpretation would require. It was a majestic artifact that catered to the fashionable society except for its crowd-gathering capacity, which explains its resurrection under each successive regime (see plates 13–16). Neither did its hegemonic representation in the public transcript lend itself to a discordant repertoire of oppositional discourse, as a revisionist interpretation would demand. Unlike mesmerism, the public transcript on ballooning screened out malcontent voices and forged a consensus.

By the time the French Revolution began, the balloon had lost much of its allure as a national artifact that could potentially bring about a scientific utopia. Its fall in the public opinion made room for an alternative scientific artifact just as spectacular and crowd-engendering—the guillotine. The scientific killing machine was conceived by the physician Joseph Ignace Guillotin (1738–1814), formerly a member of the Mesmer commission (and of the Loge des Neuf Soeurs) and now a delegate of the Third Estate of Paris, as a simple mechanism that would carry out democratic justice. As a form of capital punishment traditionally reserved for the nobility, universal

Figure E.4. Execution of Louis XVI at the Place de la Révolution on January 21, 1793.

beheading would eliminate the consideration of rank and social stigma as well as spectacular pain. In other words, the revolutionaries envisioned a regime of scientific justice—or a democratic form of scientific hegemony— that would legitimize their reign.[22]

The efficient killing machine also fulfilled the unspoken need for a mass spectacle that would take over the public spotlight, if not public affection, from the balloon.[23] After the first execution on April 25, 1792, the guillotine took up residence at the Place de la Révolution (now Place de la Concorde). In the following year, the march to the guillotine of Louis XVI on January 21 (see figure E.4), and of Marie Antoinette on October 16, was soon followed by their cousin the duc d'Orléans (Philippe Égalité) on November 6. Regardless of his effort to sell a populist vision of republican monarchy and his vote to execute Louis XVI, the prince was guillotined as one "unfaithful to tyrants and traitor to his patrie" as a prelude to Robespierre's regime.[24] The scientific spectacle of death would continue with a band of Tax Farmers (including Lavoisier), the deputies of the National Assembly that had approved the guillotine, the Jacobins who implemented it (including Robespierre), and so on, which marked the Reign of Terror that "trampled upon" humanity.[25]

The balloon and the guillotine as successive, overlapping national artifacts in their contrary movements highlight the power of their material agency that intersected with human agency. The royal administration welcomed the balloon as an instrument of scientific hegemony that would strengthen the absolutist theatrical polity, but in performance it came to embody the people's hope for a republican polity. In contrast, the guillotine was designed as a symbol of democratic justice—a humane machine that would execute nobles and commoners in the same manner. In practice, it demonstrated the horrors of mechanical justice that proved too efficient to control. The guillotine did not unite the republican nation but split it asunder and invited a military dictatorship and royal regimes in succession. The fact that these monumental scientific machines did not perform as intended speaks for their potent historical agency in opening a liminal period when "almost anything may happen" for the society caught "betwixt and between" the normative structures and scripts that regulate the citizens' aspirations and conducts.[26]

The descent of the national artifact from the balloon to the guillotine, from one of public hope to one of aristocratic fear, inaugurated a republic of the Terror that would find an outlet in Napoleon's brilliant war games.[27] By accommodating these machines within the historical account, we can develop a genealogy of the French Revolution that weaves together divergent historiographical explanations of the crowd's agency.[28] The search for the "popular" culture of Enlightenment has not produced a robust explanation of the crowd's revolutionary agency, much less the utopian longings Mona Ozouf has identified in the revolutionary festivals.[29] If we wish to piece together the utopian longings and the violent actions of the revolutionary crowd, we must link individual intentions, rhetoric, and discourse to the technologies of crowd mobilization that defined the sites of collective memory, mythic heroes, and the rituals of collective action.

An archeology of the guillotine should direct our attention less to its singular revolutionary status in representing the Terror than to the role of spectacular scientific artifacts in crafting and sustaining a system of scientific hegemony, a legacy bestowed upon and nurtured by Napoleon.[30] Instead of a radical ideological break deriving from philosophical ideas and their downward flow to the populace, we can discern the material, social, and literary technologies that packaged political power as concrete operational units such as a mass spectacle.[31] Such operational units worked across the revolutionary divide to sustain successive power regimes and their variegated ideological agenda. Contingent revolutionary actions and their divergent representations both drew on these stabilized technologies of publicity, mass control, and state maintenance. Before we can assess the crowd's political agency, we must develop a better understanding of the kinds and degrees of freedom the populace could exercise under the given regimes of power. Fascist mass spectacles offer a poignant example of the totalizing spectacles that were designed to mobilize and discipline the illiterate population.[32]

A genealogy of scientific-national artifacts across the revolutionary divide can help us to address one of the most difficult problems in modeling the French Revolution—the conjuncture of radical discontinuity in revolutionary rhetoric and substantial continuity in state technologies.[33] As I have shown in the limited affair of the Chemical Revolution, a radical break in representation was born of the long-term changes in the analytic techniques that burst the limits of conventional representations. It was the strength, not the weakness, of conventional practice (the chemistry of salts) that made the existing theoretical framework (the affinity table) impractical. In order to discern this simultaneous yet contrary dynamics of chemical theory and practice, the historian of chemistry has to understand the symbiotic relationship between theory and practice across the revolutionary divide. Chemical theory was not meant to serve the internal logic of natural philosophy but the laboratory practice of humble empirics.[34]

Alexis de Tocqueville has similarly observed that it was the advancement, not the backwardness, of French administrative technologies that led to the revolution.[35] In order to put this twisted causal claim in context, we should add that the revolutionary rhetoric—based on Enlightenment principles of reason, equality, and justice—had to be responsive to the emergent, murky political practice.[36] Concrete revolutionary actions were constrained by the same set of state technologies as before, except for the guillotine. The revolutionaries' capacity or inability to deploy these technologies with efficacious supervision charted the actual course of the revolution. Freedom of speech was a noble cause, for example, which in the sudden vacuum of censorship and other disciplinary measures irrevocably corrupted the revolutionary culture. The "culture of calumny" that had been nurtured under the absolutist regime required a system of internalized checks and balances, which could uphold the moral parameters of a legitimate public language and thereby lessen the corrupting power of a poisonous language.[37]

A Foucauldian archeology and genealogy of scientific artifacts and practices can sharpen our understanding of the relationship between the Enlightenment, the French Revolution, and the Napoleonic Empire.[38] Ken Alder has shown how the "technological form of life" in the artillery corps and the scientific effort to standardize the metric system persisted through the revolution.[39] These continuities can be seen as a part of the emergent system of scientific hegemony that helped build the modern nation-empire.

An archeology of the guillotine would unearth an alternative scientific artifact of mass mobilization—the balloon—and an ensemble of public, administrative, and communicative technologies that stabilized its scien-

Figure E.5. Étienne Gaspard Robertson, *La Minerve* (1804). Gimbel Collection XL-16-1627. This imagined hydrogen balloon ship, 150 feet in diameter, was designed to carry a load of 161,000 pounds including 60 scientists. It was equipped with a dinghy, an anchor, and living space that featured a church, a library, a gymnasium, separate quarter for female observers suspended at the back, cannons, gas lamp, and soldiers. Clive Hart, "Printed Books, 1489–1850," 84.

tific status and philosophical virtue to engender a national artifact. Such excavation (as is shown in this book) will lead to a genealogy of the French Revolution that configures the strength and stability of ancien régime technologies in fostering a republican dream, a modern nation, and a material empire. A republican monarchy was supposed to guarantee the people's happiness by cultivating useful sciences, which would build a scientific empire without domination. Despite their divergent characters in public function, in other words, the balloon and the guillotine worked as state technologies of mass mobilization and affective discipline.

Although the balloon served as a technology of mass mobilization across the revolutionary divide by materializing the vision of a unified nation, it remained a malleable artifact whose symbolic meaning kept changing (see plates 13–16).[40] When the winds of the revolution calmed down, the second international science exposition was staged at the Champ de Mars in 1798 to carry on the balloon's imperial legacy. One may argue that the balloon inaugurated the science carnival, a collective celebration of the imagined, imperial nation based on scientific prowess.[41] Napoleon would consolidate a vision of the republican empire that would enlist the sciences for the military might. This vision of the moral empire would sustain Napoleonic scientists and future republican regimes.[42]

NOTES

ABBREVIATIONS

AD	Archives départementales
AM	Archives municipales
AN	Archives nationales de France, Paris
BIF	Bibliothèque de l'Institut de France
BM	Bibliothèque municipal
BNF	Bibliothèque nationale de France
ECCO	Eighteenth Century Collections Online
FM	Fonds Montgolfier, Musée de l'Air et de l'Espace, Bourget, France
Grimm's Corr.	Correspondance littéraire, philosophique et critique par Grimm, Diderot, Raynal, Meister, etc. Edited by Maurice Tourneux. 16 vols. Paris: Garnier Frères, 1877–1782
NASM	National Air and Space Museum, Washington, DC
RAD	Registre de l'Académie de Dijon, 15 vol. BM Dijon
SCBanks	The Scientific Correspondence of Sir Joseph Banks. 6 vols. London: Pickering and Chatto, 2007
TC	Tissandier Collection, Library of Congress

PROLOGUE

1. Sannazaro's (1458–1530) sonnet, translated in Martindale, *Ovid Renewed*, 40.
2. Ovid, *Art of Love*, 20.
3. Lucian, "Icaromenippus."
4. Barkan, *Gods Made Flesh*; Morton, *English Enlightenment*; Haskell, *Prescribing Ovid*; Ziolkowski, *Ovid and the Moderns*.
5. See the different editions of Ovid at http://ovid.lib.virginia.edu/.
6. Xavier de Maistre, *Les premiers essais*, 26.
7. Bynum, *Metamorphosis*, 77–111; Yates, *Astraea*, 2–12.
8. Bartlett, *The Natural and the Supernatural*, 111–48; Lindberg, "Science as Handmaiden"; Hackett, *Roger Bacon*; Clegg, *First Scientist*; Thorndike, "True Roger Bacon"; White, "Eilmer of Malmesbury."
9. Brownlee, "Phaeton's Fall"; Hawkins, "Metamorphosis of Ovid."
10. De Boer, *Ovide moralisé*, 229.
11. Black, *Humanism and Education*; Moss, *Latin Commentaries*; Moss, *Ovid in Renaissance France*; Barolsky, "As in Ovid"; Greenblatt, *Renaissance Self-Fashioning*.
12. Goldstein, *Flying Machine*, 14–40; Laurenza, *Leonardo on Flight*.
13. Martines, *Power and Imagination*; Jones, *Italian City-State;* Tabacco, *Struggle for Power*.

14. Strathern, *Artist;* Hart, *Prehistory of Flight,* 94–115.
15. Nicholl, *Leonardo da Vinci,* 203–11, 394–99.
16. Duhem, *Musée aéronautique,* 48–75.
17. Greenblatt, *Marvelous Possessions;* Knapp, *Empire Nowhere;* Wegner, *Imaginary Communities.*
18. Burton, *Anatomy of Melancholy,* 34–69.
19. Lepenies, *Melancholy;* Gowland, "Early Modern Melancholy."
20. Lewis, *Galileo in France;* Brockliss, "Scientific Revolution in France."
21. Alcover, *La pensée;* Harth, *Cyrano de Bergerac.*
22. Aravamudan, *Enlightenment Orientalism;* Said, *Orientalism;* Elliott, *Atlantic Empires.*
23. Lucian, *Certaine select dialogues* (1634); Nicolson, *Voyages to the Moon.*
24. Gove, *Imaginary Voyage;* Dejean, *Libertine Strategies;* Elliott, *Shape of Utopia;* Ouvard, *Libertinage et utopies;* Yardeni, *Utopie et révolte;* Merrick, "The Cardinal and the Queen."
25. Yates, *Astraea;* Masuzawa, *Invention of World Religions.*
26. The story of Icarus and Daedalus was the subject of major artists in the seventeenth and eighteenth centuries, including Peter Bruegel, Jacob Peter Gowry (see plate 1), Joos de Momper, Ludovico Lana, Charles Le Brun, Joseph Marie Vien, Pyotr Ivanovich Sokolov, and Antonio Canova; Jones, *Antonio Canova.*
27. Roche, *France in the Enlightenment,* 11–74.
28. Crouch, *Genesis of Flight,* 12–13, 60. The *Passarola* was featured, for example, in Pierjacopo Martello's poem *Versi, e prose* (Roma: Francesco Gonzaga, 1710).
29. Hart, *Prehistory of Flight,* 164.
30. Rousseau, *Nouveau Dédale* (1910). He would later develop a blistering critique of the arts and sciences for their role in the corruption of mores; Rex, "Rousseau's First Discourse."
31. Argenson, "Invention des Ballons."
32. Bourdieu, *Field of Cultural Production;* Saisselin, *Enlightenment against the Baroque.*
33. Restif de la Bretonne, *Monsieur Nicolas,* 360.
34. Lynch, *Economy of Character;* Guha, *Dominance without Hegemony;* Bourdieu, *Outline.*
35. Debord, *Society of the Spectacle,* 150–54.
36. Baecque, *Body Politic.*
37. Adas, *Machines.*

INTRODUCTION: A PEOPLE-MACHINE

1. Thirteen years before, in 1770, a stampede had killed hundreds during the marriage celebration of Louis XVI, then the dauphin, to Marie Antoinette; Gruber, *Les grandes fêtes,* 74–87.
2. Crouch, *Eagle Aloft,* 13. They would become a revolutionary prince (Philippe Égalité) and successive rulers after Napoleon who was still at the École militaire de Brienne (he transferred to the Parisian school in October 1784 and was crowned emperor in 1804).
3. Hardy, "Mes Loisirs," Aug. 27, 1783. For contemporary fashion, see Cornu, *Galerie des modes.*

4. *Correspondance littéraire, philosophique et critique par Grimm, Diderot, Raynal, Meister* 13 (Aug. 1783): 348 (hereafter cited as *Grimm's Corr.* with volume number, date, and page).
5. Benjamin Franklin to Joseph Banks, Aug. 30, 1783 (*SCBanks*, 2:127); *Journal encyclopédique* 56 (Oct. 1783): 126.
6. *Mémoires secrets*, Aug. 28, 1783.
7. *Courier d'Europe*, Sept. 9, 1783.
8. Debord, *Society of the Spectacle*, 12–13; Linger, "Hegemony of Discontent."
9. Rousseau, *Discourses*, 6.
10. Kantorowicz, *King's Two Bodies*; Apostolidès, *Le roi-machine*; Burke, *Fabrication of Louis XIV*; Marin, *Portrait of the King*; Melzer and Norberg, *From the Royal to the Republican Body*; Merrick, *Descralization*; Outram, *Body and the French Revolution*.
11. Greenblatt, *Renaissance Self-Fashioning*, 29.
12. Bourdieu, *Outline*; Bourdieu, *Distinction*; Dejean, *Essence of Style*.
13. Geertz, *Negara*, 13.
14. La Bruyère, "Of the Court," in *Characters* (1702).
15. Cowan, *Triumph of Pleasure*.
16. La Bruyère, "Of the Court," in *Characters* (1702); Koppisch, *Dissolution of Character*; Elias, *Court Society*.
17. Biagioli, *Galileo*, 313–52.
18. Schaffer, *Wonders of the Clockwork World* (BBC documentary).
19. The historiography is extensive and can be traced from recent works; Kaiser and Van Kley, *From Deficit to Deluge*; Chartier, *Cultural Origins*.
20. Hunt, *Politics*, 1–16; Jones, *Great Nation*, xviii–xxvi.
21. Habermas, *Structural Transformation*; Calhoun, *Habermas*; Jacob, "Mental Lanscape"; Nathans, "Habermas's 'Public Sphere'"; Lilti, *Le monde des salons*; Spary, *Eating the Enlightenment*; Brooke, "Reason and Passion"; Cowans, *To Speak for the People*.
22. Mah, "Phantasies of the Public Sphere."
23. Foucault, *Power/Knowledge*, 55; Rousseau, *Social Contract*.
24. Kaiser, "Strange Offspring"; Smith, "Between Discourse and Experience"; Sonenscher, "Enlightenment and Revolution."
25. Osborne, "Science and the French Empire"; McClellan and Regourd, "Colonial Machine"; Bret, *L'état*; Fox, *The Savant and the State*.
26. Gillispie, *Science and Polity*; Baker, *Inventing the French Revolution*, 153–66.
27. Alder, *Engineering the Revolution*.
28. Cooter and Pumfrey, "Separate Spheres."
29. Schaffer, "Natural Philosophy."
30. Darnton, *Forbidden Best-Sellers*, 240; see the critique in Burrows, *Blackmail*.
31. Gillispie, *Montgolfier Brothers*; Thébaud-Sorger, *L'aérostation*; Lynn, *Sublime Invention*.
32. Carlson, "Theater Audiences"; Carlson, *Places of Performance*; Case and Reinelt, *Performance of Power*; Reinelt and Roach, *Critical Theory and Performance*; Keen, "Balloonomania."
33. As for later revolutionary festivals, see Ozouf, *Festivals*.
34. Pratt, *Imperial Eyes*, 4.

35. On the role of historiography, see Ranum, *Artisans of Glory*; Leffler, "French Historians"; O'Regan, "Myth of Place"; Nora, *Lieux de mémoires*. On the historical imaginary, see Fogu, *Historical Imaginary*; Bishop, "Myth Turned Monument."
36. D'Argenson, *Journal d'Argenson*, 8:152–53. See also Major, *Representative Institutions*; Major, *Representative Government*; Margerison, *Pamphlets*, 1–31.
37. Ellis, *Boulainvilliers*, 76; Furet, *Workshop*, 125–39; Stone, *French Parlements*.
38. Margerison, "History."
39. Franklin arrived in Paris at the end of 1776 to coordinate a public campaign as well as diplomatic maneuvers. The peace treaty was signed on September 3, 1783; *Grimm's Corr.* 13 (Jan. 1783): 264; Wright, *Classical Republican*, 176–87.
40. Furet, *Workshop*, 157. I am using the phrase "ancien régime" mostly as a period designation for the lack of a better term, but the concern about its negative reference should be taken seriously; Gordon, *Citizens without Sovereignty*, 24–25; Jones, *Great Nation*, xx.
41. This was put in the mouth of Fontaine, a collaborator in the Lyon balloon project. He jumped into the balloon filled with illustrious aristocrats and responded to the reprimand by saying; "on earth I would respect you, but here we are equal"; *Mercure de France*, January 24, 1784.
42. Diderot, *Political Writings*, 198.
43. *Histoire du Ballon de Lyon* (1784), 41. The term "bourgeois" in eighteenth-century usage referred to the urban resident with legal rights; Maza, *Myth of the French Bourgeoisie*. On carnivalesque critics who utilized a diverse repertoire of popular genres (rather than radical criticism) to undermine official culture, see Fort, "Voice of the Public."
44. Bakhtin, *Rabelais*, 7–10.
45. Bristol, *Carnival and Theater*, 5.
46. Farge, *Subversive Words*, 36; Chartier, "Culture as Appropriation."
47. MacAloon, "Olympic Games," 250.
48. Linguet, *Mémoires sur la Bastille* (1783); Lüsebrink and Reichardt, "La 'Bastille'"; Schama, *Citizens*, 165–68, 394–99.
49. Mercier, *Memoirs* (1772), 1:37; see also Darnton, *Forbidden Best-Sellers*, 115–36.
50. Rousseau, *La nouvelle Heloïse*, 588–98 (book 5, letter 7); Marshall, "Rousseau."
51. Mercier, *Tableau*, 2:207–10, translated in Popkin, *Panorama of Paris*, 158.
52. *Lettre à Messieurs Blanchard* (1784); *Seconde Lettre* (1784).
53. Fried, *Absorption and Theatricality*.
54. Balloon ascension has received only passing mentions in Schama, *Citizens*; Darnton, *Mesmerism*. In the revisionist historiography, the emphasis remains on textual politics, see Chartier, *Cultural Origins*.
55. Roche, *People of Paris*, 37. See also Birn, "Reinventing le Peuple in 1789"; Lucas, "Crowd and Politics"; Cobb, *Police and the People*.
56. Schama, *Citizens*, xv; Tocqueville, *Old Regime*; Darnton, *Mesmerism*.
57. Alpaugh, *Non-Violence*.
58. Montesquieu, *Reflections* (1734), 90.
59. Debord, *Society of the Spectacle*, 13. See also Turner, *Dramas*.
60. Edelstein, *Terror of Natural Right*; Israel, *Radical Enlightenment*.
61. Cooper, *Colonialism*; McClellan and Regourd, "Colonial Machine"; Osborne, *Science of French Colonialism*.

62. Brey, "Artifacts as Social Agents."
63. Carroll, *Science, Culture*; Melzer, *Colonizer or Colonized*; Anderson, *Imagined Communities*.
64. Kim, "Archeology"; Kim, "Material Enlightenments."
65. Newtonian mathematics as a source of absolute oppression is seen in Swift, *Gulliver's Travels*.
66. Foucault, *Power/Knowledge*, 80–81.
67. Secord, "Knowledge in Transit"; Tresch, *Romantic Machine*.
68. Foucault, *Madness and Civilization*, ix–xii; Foucault, *Order of Things*, xi, xx–xxi.
69. Kümin, "Popular Culture," 204. See also Krantz, *History from Below*; Guha, "Prose of Counter-Insurgency."
70. Scott, *Domination*, 18, 55.
71. Nietzsche, "Use and Abuse of History"; Taylor, *The Archive and the Repertoire*.
72. Haraway, "Situated Knowledges."
73. Miller and Reill, *Visions of Empire*.
74. Scott, *Domination*, 5.
75. Farge, *Subversive Words*; Graham, *If only the King Knew*.
76. Rogers, "Crowds and Political Festival."
77. Chartier, "Chimera of the Origin"; Baker, "Foucauldian French Revolution?"
78. Foucault, *Discipline and Punishment*, 23–28.
79. Kim, "Archeology"; Adas, *Machines*.
80. Shapin and Schaffer, *Leviathan*; Kim, "Archeology"; Kim, *Affinity*.
81. Arasse, *Guillotine*; Outram, *Body and the French Revolution*, 106–23; Gerould, *Guillotine*.
82. Orgel, "Poetics of Spectacle"; Orgel, *Illusion of Power*; Bradburne, "Local Heroes."
83. Tresch, *Romantic Machine*.
84. Wise, "Mediating Machines"; Wise and Wise, "Staging an Empire"; Tresch, *Romantic Machine*; Brain, *Pulse of Modernism*.
85. McMahon, *Divine Fury*; Jefferson, *Genius in France*.
86. Campbell and Milner, *Artistic Exchange*, 1–13.
87. Jean-Pierre Blanchard mentions the existence of London Hotel in almost every city he visited. On Grand Tour, see Chard, *Pleasure and Guilt*; Black, *British abroad*.
88. Laugero, "Infrastructure."
89. Harland-Jacobs, *Builders of Empire*.
90. Turner, *Dramas*, 13–14.
91. "Fête aérostatique (1790)."
92. Arasse, *Guillotine*; Gerould, *Guillotine*.

PART I: INVENTION IN THEATRICAL POLITY

1. Marshall, *Figure of Theater*; Cowart, "Carnival"; Cowart, *Triumph of Pleasure*.
2. Turner, *Dramas*, 13–14; Turner, *From Ritual to Theater*, 28; Turner, *Dramas*, 16.
3. Marshall, "Rousseau," 86.
4. Bryson, *Chastised Stage*; Kessler, "'New System' of Government," 110. On the debate over theatrical space, see Camp, *First Frame*; Lawrenson, "Ideal Theatre"; Thomas, *Aesthetics of Opera*.

5. Baker, *Inventing the French Revolution*, 112. See also Dziembowski, *Un nouveau patriotisme*; Echeverria, *Maupeou Revolution*.
6. Ravel, *Contested Parterre*, 7. See also Russo, *Styles of Enlightenment*, 221–51.
7. Johnson, "Musical Experience"; Johnson, *Listening in Paris*.
8. Mercier, *Du Théâtre* (1773), ix–x; Gelbart, "'Frondeur' Journalism," 498.
9. Mercier, *Du Théâtre* (1773), v–vi (added emphasis). On Mercier, see Rufi, *Le rêve laïque*.
10. Isherwood, *Farce and Fantasy*, 49–55, 81–97; Root-Bernstein, *Boulevard Theater*.
11. Isherwood characterizes material spectacles as the "entertainments focused almost entirely on the human body, the symbol of material existence, the universal denominator of humanity." I would like to follow Rétif de la Bretonne's characterization of them as curiosities such as machines, exotic animals, optical illusions, and bodily performances "proportioned" to amuse the "people." Isherwood, *Farce and Fantasy*, 40; Rétif de la Bretonne, *La Mimographe* (1770), 448.
12. Hoffmann, *Society of Pleasures*.
13. Lynn, *Popular Science*; Sutton, *Polite Society*; Isherwood, *Farce and Fantasy*, 44–50.
14. Diderot, *Rameau's Nephew*, 8; Schaffer, "Enlightened Automata"; Standage, *The Turk*.
15. Ferrone, "Man of Science"; Terrall, *Man Who Flattened the Earth*; Spary, *Utopia's Garden*.
16. Takats, *Expert Cook*, 1.
17. Mokyr, *Enlightenment Economy*; Mokyr, *Gifts of Athena*; Jones, *Industrial Enlightenment*.
18. On the three genres of technologies, see Shapin, "Pump and Circumstance"; Shapin and Schaffer, *Leviathan*.

1: A RUPTURE OF THE EQUILIBRIUM

Epigraph: "Impromptu de Monsieur," *Grimm's Corr.* 14 (Dec. 1783): 415.
1. Rostaing, *La famille*, 187–228; Digonnet, *L'invention de l'Aérostation*.
2. Russell, *Gibraltar*; McGuffie, *Siege of Gibraltar*; Chartrand, *Gibraltar*; Murphy, *Charles Gravier*, 261–79, 358–67.
3. D'Alembert to Frederick the Great, Oct. 11, 1782, in Frederick the Great, *Correspondance*, 10:237.
4. De Gérando, "Notice sur M. Joseph Montgolfier." For other stories, see FM I.8; Maravelas, "Historiography."
5. *Lettre à M. de**** (1783), 4.
6. Tombs and Tombs, *That Sweet Enemy*; Black, *Natural and Necessary Enemies*; Hancock, *Citizens of the World*, 25–39.
7. Rétif de la Bretonne, *La Mimographe* (1770), 55; Dziembowski, *Un nouveau patriotisme*, 472–86.
8. Harris, *Industrial Espionage*; Alder, *Engineering the Revolution*, 23–55; Bertucci, "Enlightened Secrets."
9. *Réflexions amusantes* (1782), 18–19.
10. *Lettre à un ami* (1783), 3; *Annales politiques* 10 (1783): 368.
11. Diderot, *Salons*, 3:228; Weinshenker, "Diderot's Use."

12. Written in 1787, the poem remained unpublished; Cherpack, "Structure of Chénier's *L'Invention*"; Scarfe, *André Chénier*; Smernoff, *André Chénier*. Also see Bingham, *Marie-Joseph Chénier*.
13. Faujas de Saint-Fond, *Description* (1783), 1–4; Erickson, "Methodical Invention."
14. Biagioli, *Galileo*; Terrall, *Man Who Flattened the Earth*; Spary, *Utopia's Garden*.
15. Mercier, "Titres de noblesse," *Tableau*, 1273–77, quote on p. 1274.
16. *Grand Registre du secretariat de la société d'émulation* (AN T* 160/5, no. 199, Jan. 1779). The term "artiste" referred to both artisans and painters; Hilaire-Pérez, "Diderot's Views."
17. *Journal de Monsieur* 4 (1783): 65–72.
18. Reynard, "Manufacturing Quality."
19. Duret, "Notices historiques"; Boissy d'Anglas, "Étienne Montgolfier"; "Règles à observer," in Rostaing, *La famille*, 508–11.
20. Reynaud, *Les moulins*, 64–88; Rosenband, *Papermaking*, 22–29.
21. Montgolfier, "La Corbeille ailée," 29.
22. Quotation is from FM I.2–4. See also Rostaing, *La famille*, 71–84; Reynaud, *Les frères*, 7–12.
23. Besançon Academy, Dec. 10, 1760 (AN F 12 1477, dossier V).
24. Rosenband, *Papermaking*, 37; Reynard, "Manufacturing Quality."
25. Reynaud, *Les moulins*, 94; Ballot, *L'Introduction du machinisme*, 554–63.
26. Guignet, "Royal Manufactures"; Minard, *La fortune du colbertisme*; Crouzet, *Britain Ascendant*; Rosenband, "Competitive Cosmopolitanism"; Rosenband, "Becoming Competitive."
27. Rostaing, *La famille*, 89–102; Reynaud, *Les moulins*, 94–121; Rosenband, *Papermaking*, 39–46.
28. Dakin, *Turgot*; Fox-Genovese, *Origins of Physiocracy*; Rothschild, "Commerce and the State."
29. FM XXVII.1; Rostaing, *La famille*, 469–507.
30. Rostaing, *La famille*, 86–88; Sauvaire-Jourdan, *Isaac de Bacalan*.
31. Desmarest, "Premier mémoire" (1771); Desmarest, "Second mémoire" (1774).
32. Horn, *Path Not Taken*.
33. Étienne Montgolfier to Beaumarchais, Sept. 20, 1781, Dec. 12, 1782 (TC 7.5); Brown, *Literary Sociability*.
34. Adams, *Taste for Comfort*; Barber, *Bourgeoisie*, 99–140; Forster, *Merchants*; Maza, *Myth of the French Bourgeoisie*.
35. Pierre attended a seminary in Lyon. His brothers Augustin (1711–1793) and Étienne (1712–1791) entered religious orders and settled down in Toulouse and Montreal, Canada, respectively; Rostaing, *La famille*, 41–186; *Dictionary of Canadian Biography*, 4:1771–800.
36. Rothschild, *Inner Life of Empires*.
37. Saint-Léon, *Le compagnonage*; Truant, "Independent and Insolent"; Sonenscher, "Journeymen's Migrations"; Roche, *Journal of My Life*.
38. On January 31, 1762, Jean-Pierre married Pierrette Charlotte Girault (1736–1814), daughter of André Girault who is identified as Chancellor Maupeou's secretary in Rostaing, *La famille*, 162.

39. Reynaud, *Les frères*, 31–33.
40. Restif de la Bretonne, *Monsieur Nicolas* 2:158; Coward, *Philosophy of Restif*, 63–89.
41. Duret, FM I.1; Rostaing, *La famille*, 187–228, 327–40; Reynaud, *Les moulins*, 261.
42. Rostaing, *La famille*, 76–78; Reynaud, *Les moulins*, 29–34.
43. Palmer, *School of the French Revolution*.
44. Spary, *Eating the Enlightenment*; Level, *Le Caveau*.
45. Rosenband, *Papermaking*, 130.
46. Étienne was in Paris by January 5, 1760 (FM XXXIII.1). His mother died on March 11.
47. Their cousin Mathieu-Louis-Pierre Duret in his "Notices historiques" claimed that Étienne was schooled under Jacques-Germain Soufflot (1713–1780), the famous architect of the Sainte-Geneviève (now Panthéon). I found no documentary evidence for this oft-recycled claim. Soufflot was appointed in 1755 as contrôleur des Bâtiments du roi à Marly and then in 1756 to the département de Paris, which might help explain the family's allusion.
48. Descat, *Le Voyage d'Italie*; Howarth, *French Theater*, 476–78; Thomas, *Aesthetics of Opera*; Lawrenson, "Ideal Theatre"; Gruber, *Les grandes fêtes*, 115–25; Braham, *Architecture*, 83–108.
49. Schmidt, "Expose Ignorance," 7.
50. Blondel, "Discours sur la manière d'étudier l'architecture (1747)"; Benhamou, "Cours public"; Benhamou, "Education of the Architect."
51. Laugier, *Essai sur l'architecture* (1753); Vidler, *Writing of the Walls*; Vidler, *Claude-Nicolas Ledoux*; Picon, *French Architects*.
52. Monneron to Étienne, Feb. 12, 1767 (FM XXIV.5).
53. Torlais, *Un physicien*; Payen, *Capital et machine*; Gillispie, *Science and Polity*; Belhoste, *Paris savant*; Kim, *Affinity*; Spary, *Eating the Enlightenment*.
54. Brown, "Scripting the Patriotic Playwright"; Bonnet, *Louis Sébastien Mercier*.
55. Poirier, *Lavoisier*; Conner, *Jean Paul Marat*, 33; Gillispie, *Montgolfier Brothers*, 11.
56. Darnton, *Great Cat Massacre*, 215–56.
57. Barny, *Comte d'Antraigues*, 9; Duckworth, *D'Antraigues Phenomenon*.
58. Levy, *Ideas and Careers*, 15–22.
59. Monneron to Étienne, Jan. 17, 1767 (FM XXIV.3).
60. There is a Monneron in the founding document of the Masonic lodge dated April 1, 1766; Rostaing, *Les anciennes loges*, 6–8.
61. Monneron to Étienne, Jan. 22, 1767 (FM XXIV.4).
62. Gordon, *Citizens without Sovereignty*, 196–99; Shovlin, *Political Economy*, 123–25; Margerison, "Commercial Liberty."
63. Mahuet and Louis le fils letters, between 1767 and 1773 (FM XXXI, XXXII).
64. Mahuet to Étienne, Nov. 12, 1767 (FM XXXII.1). On the development of man-as-automaton, see Kang, *Sublime Dreams*, 146–84.
65. Details in FM XXV.1–19.
66. Details in FM XII.3–5; Rostaing, *La famille*, 148–56; Reynaud, *Les moulins*, 31–35; Reynaud, *Les frères*, 29.
67. Sauvé to Étienne, n.d. (FM XXV.26); Reynaud, *Les frères*, 22–28, 56–57; Rostaing, *La famille*, 130.
68. Diderot, *La Religieuse* (1796).

69. Details in FM XVIII.1; Kwass, *Politics of Taxation*.
70. Loiselle, *Brotherly Love*.
71. Details in FM XXVI.1–2; Rostaing, *La famille*, 129–34. There were two Montgolfiers among the founding members of La Vraie Vertu in 1766, one was secretary (probably Antoine François, who died in 1785) and the other was Chanoine (probably the abbé); Rostaing, *Les anciennes loges*, 6–8.
72. I have not found the certificate of induction for Étienne, but after the balloon ascent he was invited to attend several Parisian lodges: Loge Écossoise, Loge des neuf soeurs, Adoption de la Candeur (FM XXVI). There is one Montgolfier, "négotiant," listed in the list of the Annonay lodge, n.d. (BNF Richelieu Fonds maçonique, FM2 276).
73. Smith and Wise, *Energy and Empire*; Coen, *Vienna*; Terrall, *Catching Nature*.
74. Rosenband, *Papermaking*, 26.
75. Details in FM I.1–2, 8; La Vaulx and Tissandier, *Joseph et Étienne*, 8.
76. Bret, "Power, Sociability and Dissemination of Science"; Perkins, "Creating Chemistry."
77. Cavallo, *Treatise* (1781); Faujas, *Description* (1783), xxviii–xxix.
78. Details in FM I.1, art. 4.
79. Kim, "Invention as a Social Drama"; Graber, "Inventing Needs."
80. Reynolds, *Stronger than a Hundred Men*; Sawday, *Engines of the Imagination*, 4–69.
81. Brandstetter, "Most Wonderful Piece"; AN O[1] 1293, ff. 226–45. Joseph eventually designed a hydraulic ram for the Machine de Marly. See Gillispie, *Montgolfier Brothers*, 145–59.
82. Deparcieux, "Mémoire sur la possibilité," (1762), 342; Beaumont-Maillet, *L'Eau à Paris*, 96–115; Bouchary, *L'eau à Paris*; Mercier, *Memoirs* (1772), 1:45-46.
83. See BM Bordeaux, Ms. 828.XXI, XXV, XXXV, LXXXIX.
84. Mercier, *Tableau*, 1:615; *Journal de Paris*, Nov. 17, 1779; AN O[1] 1293; also Bouchary, *L'eau à Paris*.
85. Reynaud, *Les moulins*, 94–121; Rosenband, *Papermaking*, 39–46; Darnton, *Business of Enlightenment*, 185–96.
86. Creveaux and Alibaux, *Un grand ingénieur papétier*.
87. Rostaing, *La famille*, 90; Desmarest, "Mémoire sur l'état actuel (1779)"; FM I.1.
88. These included the archbishop of Toulouse who served as president of the Languedoc Estates and the syndic of the Vivarais Estates. Letters from January 3 to April 8, 1780 (FM XXVII.5, 16–19), and McCormmick Collection 8.1, Princeton University Library.
89. Rosenband, *Papermaking*, 4. Also see Rostaing, *La famille*, 71–124; FM XXVII.2–4, 22.
90. Letters between Desmarest and the Montgolfiers, Jan.–July 1780 (FM XXVII.16–21, 31; McCormick Collection 8.1); Rome to Étienne, Jan. 19, 1780 (FM XXVII.2).
91. Beik, *Absolutism*; Johnson, *Life and Death*; Miller, *State and Society*; Mukerji, *Impossible Engineering*.
92. Étienne reported the beginning of the work on May 9, 1780 (McCormmick Collection 8.1, letter 1); Desmarest's letter, July 17, 1780 (FM XXVII.21).
93. Rosenband, *Papermaking*, 4–7, 49–68; Rostaing, *La famille*, 229–52, 331–34.

94. Joseph apparently obtained degrees in law and theology (a Bachelier en droit and a Licence de théologie) on July 31 and August 2, 1781, respectively; Reynaud, *Les frères*, 33.

95. This information on Blanchard's youthful inventions was published in an attempt to defend his reputation against Hamburg libels and is not verified by any other source; *Wer ist Blanchard*. His recent biographer assumes that he had no formal education; Debove, *Blanchard*, 18.

96. Coutil, *Jean-Pierre Blanchard*, 3–4; AN O^1 1293, f. 75.

97. Regular performances were scheduled at 5, 6, 7 PM every Monday, Wednesday, Saturday, and Sunday (charging twelve and thirty sous); *Journal de Paris*, Aug. 17, 1779. On the Colisée, see Isherwood, *Farce and Fantasy*, 131–60.

98. Beales, *Prosperity and Plunder*.

99. *Journal de Paris*, Aug. 28, 1781; Fontaine, *La Manche*, 7–11.

100. "Facsimile d'une lettre de Jacque Étienne de Montgolfier," Jan. 26, 1782 (Gallica); *Mémoires secrets*, March 26, 1782.

101. *Mémoires secrets*, May 5–6, 1782.

102. Blanchard's lecture, n.d. (TC 1.21); Coutil, *Jean-Pierre Blanchard*, 4, 31–38.

103. Rétif de la Bretonne, *Découverte australe* (1781); Poster, *Utopian Thought*; Wagstaff, *Memory and Desire*.

104. Darnton, *Literary Underground*, 1–40; Mason, *French Writers*; Mason, *Darnton Debate*; McMahon, "Counter-Enlightenment"; Gordon, *Citizens without Sovereignty*.

105. *Journal de Paris*, May 23, July 8, 1782; *Mémoires secrets*, March 26, Oct. 2, 1782.

106. Joseph Montgolfier's friendship with Pierre-Louis Johannot made him a target in the legal proceeding by Jean-Baptiste Johannot. He received a sentence of *conservation*, but the family hired Vittet to run a public campaign; *Précis pour le sieur Montgolfier* (1782); *Mémoires de Mathieu et Pierre-Louis Johannot* (1782); Rostaing, *La famille*, 192–96; FM I.1 art. 5; Notice of release to Jean-Pierre on June 8, 1782 (TC 6.16).

107. Tomory, "Let It Burn"; Langins, "Hydrogen Production."

108. Lana de Terzi, *Prodromo* (1670); Duhem, *Musée aéronautique*, 124–29.

109. Faujas, *Description* (1783), xiii–xxii.

110. Rostaing, *La famille*, 192–96; Moulinas, *L'imprimerie*; Gagnière, *Histoire d'Avignon*.

111. Reynaud, *Les frères*, 35–51; Calculations in FM II.8–9.

112. The abbé to the chevalier du Bourg, n.d. (FM XXI.1).

113. Montgolfier, "Mémoire lu à l'académie de Lyon (1784)."

114. Long, *Openness, Secrecy, Authorship*; Hesse, "Enlightenment Epistemology"; Woodmanse, "Genius."

115. Morand's letter, April 17, 1778 (AN O^1 1293, ff. 66–68).

116. For other routes of invention, see Hilaire-Pérez, "Invention and the State"; Hilaire-Pérez, *L'Invention technique*.

117. Parker, *Bureau of Commerce*; Parker, *Administrative Bureau*.

118. Étienne to Desmarest, Dec. 16, 1782; reproduced in La Vaulx and Tissandier, *Joseph et Étienne*, Planche VI and 26–27; discussed in Gillispie, *Montgolfier Brothers*, 21–23.

119. Desmarest, March 6, April 12, 1783; Auisson, May 23, 1783 (FM XXVII.23–

24, 51). For Réveillon, see Procès-verbaux de l'Académie royal des sciences, Jan. 22, 1783 (Les Archives de l'Académie des sciences, Paris); Rosenband, "Jean-Baptiste Réveillon."

120. Faujas, *Kurze Nachricht* (1784), 39.

121. La Vaulx and Tissandier, *Joseph et Étienne*, 28–29; FM XXI.1.

122. Tournon de la Chapelle, *La vie*; Buron Pilâtre, *Pilâtre de Rozier*; Lipp, *Noble Strategies*.

123. *Mémoires secrets*, Dec. 10, 1781, Jan. 3, Feb. 17, Nov. 17, 1782; Bégin, *Pilatre de Rozier*, 14–23; Lynn, "Musée de Monsieur"; Lilti, *Le monde des salons*; Cabanes, "Histoire du premier Musée"; Smeaton, "Early Years."

124. Tournon de la Chapelle, *La Vie*; Duval, "Pilâtre de Rozier"; Buron Pilâtre, *Pilâtre*.

125. Letters of the duc de Chaulnes and Pilâtre, Jan. 17, 1783 (TC 9.13–15).

126. Abbé d'Arnal, *Mémoire sur les moulins à feu établis a Nîmes* (1783).

127. Payen, *Capital et machine*, 47; Joseph to Étienne, May 16, 1783 (FM VIII.1).

128. FM I.2.

129. Shapin and Schaffer, *Leviathan*; Shapin, "House of Experiment"; Shapin, *Social History*.

130. Schaffer, "The Charter'd Thames"; Kim, *Affinity*.

131. The procès-verbal and "Observations des inventeurs de la Machine," *Mercure de France*, Aug. 9, 1783.

132. Wise, *Values of Precision*.

133. D'Ormesson to Condorcet, June 28, 1783 (TC 9.8).

134. *Almanach des ballons* (1784), 17.

135. Étienne to Desmarest, n.d. (FM III.9); added emphasis.

136. Johnson, "Musical Experience," 224.

137. Blagden to Banks, July 4, 1783 (*SCBanks*, 2:102); Desmarest to Étienne, n.d. (FM XXVII.25).

138. The abbé to Étienne, July 21, 31, 1783 (FM VIII.6, 11); Payen, *Capital et machine*, 59–97; Ray, *L'expérience de Jouffroy d'Abbans*.

139. Condorcet to d'Ormesson, dossier July 19, 1783 (Les Archives de l'Académie des sciences, Paris). On technical issues, see Deparcieux, *Dissertation* (1783).

140. Le Roy to Étienne, n.d. (FM XIII.10).

141. Kim, "Public Science."

142. An extract in (Louis-Théodore) Chomel's letter, June 26, 1783 (FM XXI.36).

143. *Affiches de France*, July 10, 1783. On *Mercure de France*, see Acomb, *Mallet du Pan*, 172.

144. *Journal de Paris*, July 27, 1783; Berman, "Cadet Circle"; Faujas, *Description* (1783), 110–28.

145. Boissy d'Anglas, "Étienne Montgolfier" (FM I.4); Jean-Pierre to Étienne, July 12, 1783 (FM VIII.2).

146. Joseph wondered if they could find a "curious rich man" who would pay "fair profit." To secure privileges and profit, it was essential to ward off "counterfeiters" (FM VIII.2, 23).

147. Richard, *Histoire naturelle* (1770), 5:76; Fontenelle to Étienne, n.d. (FM XX.19).

148. The abbé and Jean-Pierre to Étienne, July 31, 1783 (FM VIII.11–12).

149. The abbé to Étienne, August 21, 1783 (FM VIII.26).

150. Joseph to Étienne, July 25, 1783 (FM VIII.7).
151. Chomel to editors, July 30, 1783 (FM XX.1).
152. Roberts, Schaffer, and Dear, *Mindful Hand*, xiii–xxvii.
153. *Le Triomphe* (1784), 8. On the "society of orders," see Jones, *Great Nation*, 1–18.

2: BALLOON TRANSCRIPTS

1. On artificial pleasures, see "Voyage qui n'est point sentimental (1784)." On brokers, see Kettering, *Patrons*; Schaffer, Roberts, Raj, and Delbourgo, *Brokered World*.
2. Shennan, *Philippe*; *Journal de Monsieur* 4 (1783): 65.
3. Kavanagh, *Esthetics of the Moment*, 1.
4. Darnton, "News in Paris"; McMahon, "Birthplace of the Revolution."
5. Crow, *Painters and Public Life*; Ravel, *Contested Parterre*; Pucci, *Sites of the Spectator*; Brennan, *Public Drinking*; Spary, *Eating the Enlightenment*; Goodman, *Republic of Letters*; Belhoste, *Paris savant*.
6. Baker, *Condorcet*; Poirier, *Lavoisier*; Gillispie, *Science and Polity*; Adkins, *Idea of the Sciences*, 89–112; Alder, *Engineering the Revolution*.
7. Walton, *Policing Public Opinion*; Chisick, "Pubic Opinion."
8. Burke, "Freemasonry"; Burke and Jacob, "French Freemasonry"; Halévi, *Les loges maçonniques*; Jacob, *Radical Enlightenment*; Amiable, *Une loge*.
9. Auricchio, "Pahin de la Blancherie's Commercial Cabinet"; Lynn, *Popular Science*, 72–93; Kirsop, "Cultural Networks."
10. Chisick, *Limits of Reform*.
11. Anderson, *Imagined Communities*; Jones, "Great Chain of Buying."
12. Shapin, "Pump and Circumstance"; Shapin and Schaffer, *Leviathan*.
13. Censer, *French Press*; Popkin, *News and Politics*; Censer and Popkin, *Press and Politics*; Feyel, *L'Annonce et la nouvelle*; Darnton, *Literary Underground*, 3–7; Francalanza, *Jean-Baptiste-Antoine Suard*; Berman, "Cadet Circle."
14. Trouillot, *Silencing the Past*.
15. Darnton, *Business of Enlightenment*; Darnton, *Mesmerism*; Roger, *Buffon*.
16. *Mémoires secrets*, March 26, 1782.
17. *Journal de Paris*, Aug. 28, 1783.
18. The relationship between Faujas and Buffon began with Faujas's gift of volcanic substances for the cabinet du roi; Buffon to Faujas, March 28, 1777 (Lettre CCLXVIII, in Buffon, *Correspondance*); Freycinet, *Essai sur la vie*.
19. Faujas, *Description* (1783), 5–15 (added emphasis); Lilti, *Le monde des salons*.
20. Rushton, "Royal Agamemnon."
21. Mercier, *Tableau*, 1:1328; Darnton, "News in Paris."
22. Gordon, *Citizens without Sovereignty*.
23. Mercier, *Tableau*, 2:930–43.
24. *Grimm's Corr.* 13 (June 1784), 552–56; Isherwood, *Farce and Fantasy*, 217–49.
25. Darnton, "News in Paris"; McMahon, "Birthplace of the Revolution"; Lever, *Philippe Égalité*, 85–108; Rétif de la Bretonne, *Le Palais royal*.
26. The Roberts are identified as Charles's students in *Mémoires secrets*, Aug. 23, 1783, or as his friends in *Journal encyclopédique* 56 (1783): 125. Turgot recommended them as "quite good mechanics [*méchaniciens*]" to the comte d'Angiviller, Feb. 24, 1782 (AN O^1 1916[1], f. 61).

27. Payen, *Capital et machine.*
28. Kim, *Affinity*; Kim, "Public Science."
29. Rémond, *John Holker.* Sulfuric acid cost 10 sols per livre at Javel's, which would have come to 900 livres for Charles's balloon, in addition to 60 livres for the iron filings, assuming that the envelope was tightly sealed. A leaky balloon would require at least 40 percent more material, as an estimate for Blanchard's balloon (26 foot in diameter) in the following spring indicates (Cuthbert Collection, Ib-3).
30. *Journal de Paris*, Sept. 28, 1783; Schiebinger, *Plants and Empire.*
31. Coates, *Commerce in Rubber*, 3–19; Reisz, "Curiosity and Rubber."
32. Bélanger to Charles, Dec. 11, 1783 (TC 3.8).
33. Faujas, *Description* (1783), 6, 128–31.
34. Palmer, "Posterity."
35. *Journal encyclopédique* 56 (Oct. 1783): 124–33.
36. Franklin to Banks, Aug. 30, 1783 (*SCBanks*, 2:127). He identifies French *taffeta gommé* with English oiled silk, the silk impregnated with a solution of elastic gum in linseed oil.
37. *Mémoires secrets*, Aug. 25, 1783; Gillispie, *Montgolfier Brothers*, 27–33.
38. *Journal de Paris*, Aug. 28, 1783.
39. Lenoir, *Détail sur quelques établissemens* (1780); Garrioch, "The Police of Paris"; Milliot, "Qu'est-ce qu'une police éclairée?"
40. There are conflicting reports, but a reasonable estimate would be that the 500 subscribers each paid one louis (24 livres) for 8 admission tickets, which would have raised 12,000 livres and turned up 4,000 ticket holders; *Mémoires secrets*, Aug. 25, 27, 1783; *Corr. secrète*, Sept. 3, 1783.
41. Guillaume Le Gentil de la Galaisière (1725–1792) and Edme-Sébastien Jeaurat (1724–1803). They estimated 339 and 327 toises, respectively; *Affiches France*, Sept. 3, 1783; Meusnier to Faujas, Oct. 31, 1783, in Faujas, *Description* (1783), 33–110.
42. Franklin to Banks, Aug. 30, 1783 (*SCBanks*, 2:127).
43. *Journal de Paris*, Aug. 28, 1783.
44. Franklin to Banks, Aug. 30, 1783 (*SCBanks*, 2:127).
45. *Courier de l'Europe*, Sept. 9, 1783.
46. Kaplan, *Provisioning Paris*, 458–63; Kaplan, *Bakers of Paris*; Gillispie, *Science and Polity*, 24–33.
47. *Journal encyclopédique* 6 (Oct. 1783): 127–33.
48. A short performance, *Guillot physicien ou la chute du globe volant à Gonnesse* is mentioned in the letter of Jean Étienne Montucla to Charles Hutton, Dec. 26, 1783 (TC 9.13).
49. *Mercure de France*, Sept. 13, 1783.
50. Williams, *Police of Paris*; Darnton, *Poetry.*
51. *Journal de Paris*, Aug. 27; *Mercure de France*, Sept. 6, 1783. On public fear, see Stewart, "Science and Superstition."
52. *Journal de Paris*, Sept. 13, 1783; Baker, *Inventing the French Revolution*, 167–99.
53. Scott, *Domination*, 15. Although the *nouvelles* were distributed mostly in Paris, which precluded their direct circulation outside city limits, they were a common source of information for the Parisian public.
54. *Journal de Paris*, Aug. 27, 28, 1783; *Mercure de France* and *Courier de l'Europe*, Sept. 9, 1783; *Journal de Monsieur* 6 (1783): 57–72.

55. Blagden to Banks, Sept. 20, 25, 1783 (*SCBanks*, 2:140, 146).
56. *Journal de Paris*, Aug. 27, 29, Sept. 13, 1/83; "Réponse aux plaisantes de MM. Robert contre les souscripteurs de l'éxperience faite au champ des mars" (FM V.7); *Considérations sur le globe aérostatique* (1783), 1.
57. *Journal de Paris*, Aug. 29, 1783. See also Sept. 14, 18, 1783.
58. *Journal de Paris*, Sept. 28, 1783.
59. *Corr. secrète*, Sept. 3, 1783.
60. *Mémoires secrets*, Aug. 28, 1783. See also "Machine aërostatique," *Journal de Monsieur* 4 (1783): 69; Fort and Popkin, *Mémoires secrets*; Merrick, "Sexual Politics."
61. Charles, "Mémoire sur l'aérostatique" (TC 3.6); Orléans, *Lettres*.
62. *Corr. secrète*, Sept. 24, 1783; *Mémoires secrets*, Aug. 23, Sept. 24, Nov. 24, 1783.
63. *Journal encyclopédique* 56 (Oct. 1783): 124–33.
64. Walton, *Policing Public Opinion*.
65. Charles, "Mémoire sur l'aérostatique" (TC 3.6).
66. Lever, *Philippe Égalité*, 220–22; Darnton, *Devil in the Holy Water*, 344–59.
67. *Grimm's Corr.* 13 (1783): 369, states that this open letter was commissioned by Charles; Cointat, *Rivarol*.
68. *Lettre à Monsieur le Président de ****, dated Sept. 20, 1783; *Affiches* (or *Journal*) *de France*, November 12, 1783.
69. Faujas, *Description* (1783), 132–75.
70. Walton, *Policing Public Opinon*.
71. Kelly, "Lèse-Majesté to Lèse-Nation"; Graham, *If only the King Knew*.
72. Smith, *Nobility Reimagined*.
73. Gordon, *Citizens without Sovereignty*; Walton, *Policing Public Opinion*.
74. Hardy, "Mes Loisirs," Sept. 16, 19, 1783.
75. *Journal de Paris*, Sept. 3, 1783.
76. *Mémoires secrets*, Sept. 9, 1783.
77. *Grimm's Corr.* 13 (Aug. 1783): 344.
78. Goldbeater's skin (*peau de beaudruche*) was the interior lining of the large intestine of ox, cleaned of fat and other uneven parts, dried, and afterward softened to produce a thin, flexible, and impermeable envelope. It was used for separating the leaves of metal in goldbeating. Humidity increased its weight. It also did not hold inflammable air for long due to invisible pores; Faujas de Saint-Fond, *Description*, 15–19. On the baron de Beaumanoir's trials, see *Journal de Paris*, Sept. 11, 12, 1783.
79. *Almanach des ballons* (1784), 26–28.
80. Franklin to Banks, Aug. 30, 1783 (*SCBanks*, 2:127–29).
81. Réveillon to Franklin, April 4, 1783 (found at http://founders.archives.gov/documents/Franklin/01-39-02-0265, Founders Online).
82. Jean-Pierre and the abbé to Étienne, Aug. 39–Sept. 6, 1783 (FM VIII.30–34) and n.d. (FM XXXIII.4). On Réveillon's maneuver, see *Papers of Benjamin Franklin*, 39:433–35.
83. Caradonna, "Monarchy of Virtue."
84. Rosenband, "Jean-Baptiste Réveillon." On the issue of access, see Shapin and Schaffer, *Leviathan*, 112–15.
85. Luc-Vincent Thiery, *Almanach du voyageur à Paris* (1785), 318.
86. On the faubourg St. Antoine, see Kaplan, "Paris Bread Riot of 1725." On guild organization, see Fitzsimmons, *From Artisan to Worker*.
87. Rosenband, "Jean-Baptiste Réveillon"; Velut, "L'industrie dans la ville."

88. AN F 12 1477, dossier 6.
89. Malesherbe, Sept. 10, 1783 (FM XIV.1); Boissy d'Anglas to Étienne, 1786–1791 (FM XVI. 1–45).
90. Boyer to Étienne, Sept. 12, 1783 (FM XX.41).
91. Monneron, presumably a brother of Étienne's old friend, letter dated Sept. 18, 1783 (FM XXIV.23).
92. *Mémoires secrets*, Sept. 9, 1783; Fajuas to Banks, July 28, 1783 (*SCBanks*, 2:113).
93. Joseph to Étienne, n.d., responding to the letter of Aug. 18, 1783 (FM VIII.25).
94. Franklin, *Writings*, 9:82.
95. *Gazette de Leyde*, Sept. 23, 1783.
96. Faujas, *Description* (1783), 21.
97. Letters between Étienne and Joseph, Sept. 6–13, 1783 (FM VII.1–2, VIII.36–37).
98. The procès-verbal is in FM V.14 and Franklin's ticket, FM XIII.37.
99. Étienne to Joseph, Sept. 13, 1783 (FM VII.2); Faujas, *Description* (1783), 34.
100. Joseph and the abbé to Étienne, Sept. 13, 1783 (FM VIII.38–39).
101. Étienne to Joseph, Sept. 13, 1783 (FM VII.2); *Gazette de Leyde*, Sept. 23, 1783; Faujas, *Description* (1783), 36–37.
102. *London Magazine* 2 (1784):12–16; *Mercure de France*, Aug. 14, 1784. On textile printers, see Chapman and Chassagne, *European Textile Printers*; Chassagne, *Oberkampf*; Jenkins, *History of Western Textiles*.
103. The abbé to Étienne, Sept. 13, 15, 1783 (FM VIII.39–40).
104. Details in FM IV; Étienne to Joseph, Sept. 13, 1783 (FM VII.2).
105. Étienne to his wife, Sept. 19, 1783 (FM VII.3); translated in Gillispie, *Montgolfier Brothers*, 40–43.
106. Gruber, *Les grandes fêtes*, 7–9; Mercier, *Tableau*, 2:813–16.
107. Pingeron, *L'Art de faire soi-même* (1783), 2–4. Pingeron exaggerates the dimension as 100 feet wide and 10 feet high, which is countered in Bombelles, *Journal*, 1:261.
108. Faujas, *Description* (1783), 25–27.
109. Étienne to his wife, Sept. 19, 1783 (FM VII.3).
110. Pingeron, *L'Art de faire soi-même* (1783), 15.
111. *Gazette de Leyde*, Sept. 26, 1783; *Journal encyclopédique* 56 (1783): 326.
112. Pingeron, *L'Art de faire soi-même* (1783), 12.
113. Bombelles, *Journal*, 1:261–62.
114. Étienne to his wife, Sept. 19, 1783 (FM VII.3).
115. Hardy, "Mes Loisirs," Sept. 16, 1783; *Le mouton, le canard, et le coq* (1783).
116. The abbé and Jean-Pierre to Étienne, Sept. 22–29, 1783 (FM VIII.46–50).
117. *Mercure de France*, Sept. 20; *Journal politique de Bruxelles*, Sept. 16, 1783.
118. The Redoute opened in 1781 in the faubourg St. Martin to replace the failed Colisée (1771–1778); Mercier, *Tableau*, 1:536, 1658.
119. Ruggieri to Étienne, n.d. (FM XXI.12); Lavoisier, "Procédés d'artifice proposés par M. Ruggieri," *Oeuvres de Lavoisier*, 4:417–18; Lavoisier to Lenoir, Aug. 29, 1784 in *Correspondance*, 4:31; Faujas to Étienne, n.d. (FM XV.18); Werrett, *Fireworks*, 220–23.
120. Pleinchesne to Étienne, Sept. 23, 1783 (FM XXI.19); Madame Réveillon to Étienne, Sept. 24, 1783 (FM XII.6).

121. *Corr. secrète*, Oct. 1, 1783; Hardy, "Mes Loisirs," Sept. 25, 1783.
122. Franklin to Banks, Oct. 8, 1783 (*SCBanks*, 2:162–63).
123. *Courier de l'Europe* 14 (1783): 258–59.
124. *Annales politiques* 10 (1783): 358–59; Gagnière, "Lettre sur le Globe aérostatique."
125. Fabry, *Réflexions* (1784), 4–5.
126. Klein and La Vopa, *Enthusiasm and Enlightenment*.

3: TRUE COLUMBUS

Epigraph: "Vers de M. le Vicomte de Ségur à MM. Charles et Robert," *Grimm's Corr.* 14 (Dec. 1783): 415.
1. *Grimm's Corr.* 13 (Aug. 1783): 345.
2. *Journal de Paris*, Sept. 19, 1783.
3. Gudin de la Brénellerie, "Sur le Globe Ascendant," *Journal de Paris*, Aug. 28, 1783.
4. Campbell, "Our Eagles."
5. Raynal, *Histoire* (1770), 1; Riley, *Seven Years War*.
6. Bideaux and Faessel, "Introduction."
7. Diderot, *Political Writings*, 166–214; Muthu, *Enlightenment against Empire*, 72–121.
8. Edelstein and Lerner, *Myth and Modernity*, 1–4.
9. Appelbaum, *Literature*, 24–49; Rétif de la Bretonne, *La Découverte australe* (1781).
10. Schiff, *Great Improvisation*; Schoenbrun, *Triumph in Paris*; Kramer, *Lafayette*.
11. Echeverria, *Mirage in the West*, 42. See also St. John de Crèvecoeur, *Letters from an American Farmer* (1782).
12. Upon his return from exile, Raynal established a prize competition on this subject at the Lyon Academy in 1787; Lüsebrink and Mussard, *Avantages et désavantages*; Echeverria, "Condorcet's *The Influence of the American Revolution*."
13. *Annales politiques* 11 (1784): 4. See also Adams, *Diary* 1:192 (Sept. 20, 1783).
14. *Morning Herald*, Sept. 17, 1783.
15. Linguet, *Mémoire sur la Bastille* (1783); Evans, *Réfutation* (1783); Dusaulx, *Observations* (1783); Servan, *Apologie de la Bastille* (1784); Vyverberg, "Limits of Nonconformity"; Venturi, *End of the Old Regime* 1:379–410.
16. Mercier, *Memoirs* (1772), 1:43.
17. Joseph's letter, Feb. 10, 1784 (Lyon Academy Ms. 268, ff. 176–78).
18. Kaplan, "Paris Bread Riot"; Tilly, "Food Riot"; Réveillon, "Exposé justificatif" (1821).
19. *Mémoires secrets*, Oct. 12, 1783.
20. *Gazette d'Amsterdam*, Nov. 7, 1783; Price, *Preserving the Monarchy*, 94–104; Bossenga, "Financial Origins."
21. Hardman and Price, *Louis XVI*, 324–29.
22. Franklin to Banks, Oct. 8, 1783 (*SCBanks*, 2:163). See also Pingeron, *L'Art de faire soi-même* (1783), 18.
23. Étienne to Joseph, Oct. 2, 1783 (FM VII.4). See also Kim, "Invention as a Social Drama."
24. *Journal de Paris*, Oct. 11, 1783 (added emphasis).
25. FM XX.34–35, XXI.5–10.

26. *Journal de Paris,* Oct. 3, 1783.
27. Gillispie, *Montgolfier Brothers,* 46.
28. Étienne to Joseph, Oct. 8, 1783 (FM VII. 6).
29. Cassini de Thung to Étienne, Oct. 6, 1783 (FM XIII.18); *Mercure de France,* Oct. 25, 1783.
30. Subscriptions were priced at six livres to fund a new balloon; *Journal de Paris,* Sept. 3, 7, 1783.
31. *Journal de Paris,* Nov. 3, 1783; *Mémoires secrets,* Oct. 15, 1783; *Grimm's Corr.* 13 (Sept. 1783): 365.
32. *Mémoires secrets,* Oct. 18, 1783; Hunn, "Balloon Craze," 120–24.
33. Pilâtre's invitation in FM XV.1; *Mémoires secrets,* Oct. 21, 1783.
34. Fleur, "Les grands pharmaciens"; Perkins, "Creating Chemistry," 48–59.
35. Bégin, *Pilâtre de Rozier,* 1–13; Dorveaux "Pilâtre de Rozier."
36. "Extrait d'un memoire contenant une suite d'expériences sur les gaz," in Tournon de la Chapelle, *La Vie,* 110–45; *Procès-verbaux de l'Académie,* Dec. 13, 1780, Feb. 24, 1781, June 18, 1783, May 5, 1784.
37. *Mercure de France,* Oct. 25, 1783. This reports that Étienne rose to about 30–40 feet, which caused considerable anxiety back at home.
38. *Journal de Paris,* Oct. 21–26, 1783; FM I.36, XII.9; Faujas, *Description* (1783), 180–88.
39. *Journal de Paris,* Nov. 3, 1783; *Journal encyclopédique* 56 (1783): 325; *Grimm's Corr.* 13 (Sept. 1783): 364.
40. Pilâtre to Étienne, Oct. 30, 1783 (FM XV.2).
41. Fort, "Voice of the Public," 388; Crow, *Painters and Public Life,* 95.
42. The abbé to Étienne, Oct. 27, 1783 (FM IX.17).
43. Joseph and the abbé to Étienne, Oct. 2–11, 1783 (FM IX.1–7); *Supplément à l'Art de voyager* (1784), 3.
44. She sent her secretary Mr. de la Grëze; Faujas to Étienne, Oct. 21, 1783 (FM XV.14).
45. *Journal de Paris,* Dec. 9, 1783; *Journal encyclopédique* 57 (1784): 115.
46. Franklin, *Writings,* 9:114–15; Faujas, *Première suite* (1784), 11–22; *Journal de Paris,* Nov. 22, 1783.
47. *Grimm's Corr.* 13 (Nov. 1783): 393–94.
48. *Mémoires secrets,* November 22, 1783; Franklin, *Writings,* 9:115.
49. "Lettre à M. l'Abbé de Fontenai," *Affiches de France,* Dec. 10, 1783 (added emphasis).
50. *Journal de Monsieur* 6 (1783): 57–72 (quotation on p. 57).
51. Blumenberg, *Shipwreck with Spectator.*
52. *Journal de Paris,* Nov. 29, 1783. See also Gillispie, *Montgolfier Brothers,* 52–56.
53. Picon, *French Architects,* 211–55; *Mémoires secrets,* Nov. 26, 1783.
54. *Journal de Paris,* Nov. 27–28, Dec. 10, 1783.
55. *Journal encyclopédique* 57 (1784): 189; *Gazette de Leyde,* Dec. 12, 1783; Bombelles, *Journal,* Nov. 30, 1783.
56. Biagioli, *Galileo,* 323–29.
57. Bell, *Lawyers and Citizens*; Maza, *Private Lives*; Levy, *Ideas and Careers.*
58. *Annales politiques* 10 (1783): 355. See also Greene, "Ursule's Road"; Rétif de la Bretonne, *Le Paysan perverti* (1775).
59. Kessler, "'New System' of Government"; Levy, *Ideas and Careers.*

60. Charles, "Mémoire sur l'aérostatique" (TC 3.6), ff. 15–17; *Mémoires secrets*, Nov. 30, 1783.
61. Tissandier, *Histoire*, 1:31–32; Gillispie, *Montgolfier Brothers*, 60.
62. Brussonet to Banks, Dec. 4, 1783 (*SCBanks*, 2:226).
63. *Journal de Paris*, Nov. 19, 1783 (added emphasis).
64. "Souscription pour l'exécution d'un Globe Aérostatique, construit par MM. Robert." It was limited to one hundred subscriptions at four louis each, which provided thirty tickets and a privileged seat near the balloon. They intended to charge six livres for two tickets for the amateurs who could not afford the subscription; Hunn, "Balloon Craze," 131.
65. Blagden to Banks, Oct. 9, 1783 (*SCBanks*, 2:164–66; added emphasis). On Maupeou Revolution, see Echeverria, *Maupeou Revolution*.
66. AN O^1 1917(1), ff. 365–66; O^1 1214, f. 238 (Nov. 18, 1783); O^1 1682, f. 220; Hunn, "Balloon Craze,"132.
67. Lever, *Philippe Égalité*, esp. 220–22; Ambrose, *Godfather of the Revolution*; Darnton, *Devil in the Holy Water*, 344–59.
68. *Journal de Paris*, Nov. 26, *Mercure de France*, Nov. 29, 1783; *Mémoires secrets*, Nov. 30, 1783.
69. TC 3.8. Unless otherwise noted, this and next paragraph follow this manuscript.
70. *Journal de Paris*, Nov. 26, 28, 1783; *Mémoires secrets*, Nov. 30, 1783.
71. Quoted in Walton, *Policing Public Opinion*, 39.
72. Hardy, "Mes Loisirs," Dec. 1, 1783.
73. Cradock, *La vie française*, 1; *Journal de Paris*, Jan. 3, *Journal politique de Bruxelles*, Jan. 3, 10, 1784.
74. Franklin to Banks, Dec. 1, 1783. See also Franklin, *Writings*, 9:119.
75. The summer of 1783 was marked by high temperatures, frequent thunderstorms, and dry fog; Hochadel, "In nebula nebulorum."
76. *Journal de Paris*, December 13, 1783.
77. Charles, "Mémoire sur l'aérostatique" (TC 3.6), f. 26.
78. Charles, Mémoire manuscrit inédit" (TC 3.8).
79. Charles, "Second Mémoire" (BIF 2104).
80. The elder Robert could not go up due to his pregnant wife's plea to Lenoir; Hardy, "Mes Loisirs," Dec. 1, 1783.
81. Faujas, *Première suite*, 31–55; Gillispie, *Montgolfier Brothers*, 56–62; Hunn, "Balloon Craze," 130–45; Thébaud-Sorger, *L'Aérostation*, 54–57.
82. "Le premier décembre," Mercier, *Tableau*, 2:886–87, translated in Popkin, *Panorama of Paris*, 196.
83. Charles, "Mémoire sur l'aérostatique" (TC 3.6), ff. 11–25; *Journal de Paris*, Dec 13, 1783.
84. *Questions et conjectures* (1786).
85. Charles, "Mémoire sur l'aérostatique" (TC 3.6), ff. 11–25.
86. The barometer changed from 28 inches (*pouces*) 4 lignes to 18 inches 10 lignes (1 pouce = 12 lignes), and the thermometer, from 7.5 degrees above to 5 degrees below the freezing point.
87. Charles, "Mémoire sur l'aérostatique" (TC 3.6), ff. 50–55; *Journal de Paris*, Dec. 14, 1783.
88. *Almanach des ballons* (1784), 59; Charles, "Mémoire manuscrit inédit" (TC 3.8).

89. Bombelles, *Journal*, Dec. 1, 1783; *Mémoires secrets*, Dec. 3, 1783.
90. *Journal de Paris*, Dec. 2, 25, 1783; Faujas, *Première suite* (1784), 56–62.
91. Lever, *Philippe Égalité*, 216–20; *Journal de Paris*, Dec. 3, 13, 14, 1783.
92. *Annales politiques* 11 (1784): 1–19.
93. Charles, "Mémoire sur l'aérostatique" (TC 3.6), ff. 56; Meusnier, *Mémoire sur l'équilibre* (ca. 1783).
94. *Journal de Paris*, Dec. 9, 1783. See also *Mémoires secrets*, Dec. 9, 1783.
95. In 1783, Buffon received about 3,587 livres, while payments to other Academicians ranged from 300 to 1,000 livres, totaling about 12,457 livres (Dépense, année 1783, Les Archives de l'Académie des sciences).
96. Letter of Condorcet to Étienne, n.d., in La Vaulx and Tissandier, *Joseph et Étienne*, 44, and Planche XXIX.
97. Procès-Verbaux de l'Académie, Dec. 20, 1783.
98. *Rapport fait à l'Académie des sciences* (1784); *Journal de Paris*, Jan. 28, 1784.
99. *Journal de Paris*, Dec. 13, 1783.
100. *Mémoires secrets*, Dec. 9, 1783; *Mercure de France*, Dec. 20, 1783.
101. Joseph to Étienne, Dec. 18, 1783, in La Vaulx and Tissandier, *Joseph et Étienne*, 44–45, 53–55; *Mémoires secrets*, Dec. 22, 1783.
102. *Mercure de France*, Dec. 20, 27, 1783.
103. Hardy, "Mes Loisirs," Dec. 16, 1783; Remington, "A Monument."
104. AN O¹ 1214, f. 255, entry for Dec. 24, 1783.
105. *Mémoires secrets*, Dec. 20, 21, 1783; *Mercure de France*, Dec. 27, 1783; *Journal encyclopédique* 57 (1784): 40; *Journal de Paris*, Jan. 3, 1784.
106. Diderot, *Rameau's Nephew*, 181 (written in the 1760s and 1770s, but first published in 1805).
107. Diderot, *Political Writings*, 170.
108. *Journal de Paris*, Dec. 4, 1783; *Mémoires secrets*, Dec. 8, 1783.
109. *Mercure de France*, Dec. 13, 27, 1783; *Journal de Paris*, Dec. 29, 1783.
110. Bélanger to Charles, Dec. 11, 1783 (TC 3.8).
111. Letters dated Dec. 8, 1783, Jan. 11, 15, 1784 (TC 3.8).
112. Letters in BIF 2014.
113. Jean Étienne Montucla at Versailles to Charles Hutton at Woolwich, Dec. 26, 1783 (TC 9.13).
114. *Précis des expériences faites par MM. Alban & Vallet* (1785).
115. Luce de Lancival, *Poëme sur le globe* (1784). Also see Arnaud de Saint-Maurice, *L'Observatoire volant*.
116. Dubroca de Rouges to Étienne, Feb. 3, 1784 (FM XXII.13).
117. Lodges of Neuf Soeurs, La Candeur, Ecossoisse, and Contrat social, where Étienne was often invited along with Pilâtre and in the presence of the duc de Chartres (FM XXVI.3–16, XXII.46).
118. *Lettre à Monsieur le President* (1783), 19.

PART II: PHILOSOPHICAL NATION

1. *Procès-verbaux et détails* (1783), 3; *Mémoire sur les expériences aérostatiques faites par MM. Robert frères* (1784), 1.
2. *Lettre à Mr. M. Saint-Just* (1784), 3–4, 9–12.
3. Arnaud de Saint-Maurice, *L'Observatoire Volant* (1784); Bertholon, *Des avantages* (1784), 7.

4. Mercier, "Ballon-Montgolfier" (1784).
5. Diderot, *Mémoires pour Catherine II*, 149.
6. Montesquieu, *Reflections* (1734); *Persian Letters* (1722); *Spirit of Laws* (1750).
7. Rousseau, *Social Contract*, book 1, chapter 8. See also Baker, "Transformations of Classical Republicanism."
8. Linton, *Politics of Virtue*.
9. Bell, *Cult of the Nation*.
10. Fontenelle, *République des philosophes* (1768), 35–52, 129–52. See also Rendall, "Fontenelle and His Public."
11. Mercier, *Memoirs* (1772), 37. See also Wilkie Jr., "Mercier's *L'An 2440*."
12. Mably, *Des Droits et des devoirs du citoyen*, 29 (Written ca. 1758, it was published only in 1789.)
13. La Harpe, "Réflexions sur le drame," in *Oeuvres*, 1:175-77; quoted and translated in Gelbart, "'Frondeur' Journalism," 507.
14. Marshall, "Rousseau," 96. Also see Fontana, *Invention of the Modern Republic*; Maloy, "Rousseau's Tutorial Republicanism."
15. Mercier, *Memoirs* (1772), 1:166-70.
16. [Saunier], *Le Triomphe* (1784), 4.
17. Forrest, *Paris;* Kumar, "Nation-States as Empires."
18. Popkin, "Pamphlet Journalism"; Maza, *Private Lives*, 167–211.
19. McMahon, "Narratives of Dystopia"; McMahon, *Enemies of the Enlightenment*.
20. Edelstein, *Terror of Natural Right*.
21. Israel, *Revolutionary Ideas*; Bell, "Different French Revolution"; Hunt, "Louis XVI."

4: BALLOON SPECTATORS

1. *Lettre à Mr. M. de Saint-Just* (1784), 23.
2. Gordon, *Citizens without Sovereignty;* Pucci, *Sites of the Spectator*.
3. This definition of scientific hegemony is informed by Scott, *Domination*, although he is interested in a system of domination without hegemony, or domination without ideological articulation.
4. For the articulation of critical history, as opposed to monumental or antiquarian varieties, see Nietzsche, *Uses and Abuses of History*; Emden, *Friedrich Nietzsche*.
5. Davis, *Fiction in the Archive*.
6. Popkin, "Pamphlet Journalism"; Burrows, *Blackmail*.
7. Maza, *Private Lives*, 2; Mercier, *Le Charlatan* (1780), 29.
8. Darnton, "News in Paris."
9. Muralt, *Lettres sur les Anglois*. For Montesquieu's *Persian Letters*, see Hundert and Nelles, "Liberty and Theatrical Space"; Romani, *National Character*.
10. La Bruyère, "Of the City," in *Characters*; Marivaux, *Journaux*, 5–39. On the link between ancient Theophrastus and modern English flaneur, see Brand, *Spectator*, 14–40.
11. Marivaux, *Le Spectateur français*, April 10, 1722, in *Journaux*, 132–37; Pucci, *Sites of the Spectator*, 36–37.
12. Pucci, *Sites of the Spectator*.
13. Chisick, "Public Opinion"; Popkin, "Pamphlet Journalism"; Margerison, *Pamphlets*; Hanson, "Monarchist Clubs"' Cowans, *To Speak for the People*; Walton, *Policing Public Opinion*.

14. Eley, "What Is Cultural History?"
15. Etkind, *Internal Colonization.*
16. Pratt, *Imperial Eyes,* 6–7.
17. Novick, *That Noble Dream.*
18. Haraway, "Situated Knowledges."
19. Miller and Reill, *Visions of Empire.*
20. Rétif de la Bretonne, *La Découverte australe* (1781), 71. See also Lo Tufo, "Images"; Porter, *Restif's Novels*; Poster, *Utopian Thought.*
21. On the rhetorical strategy of contraries, see Rex, *Attraction of the Contrary*; Rex, *Diderot's Counterpoints.*
22. Bell, "Unbearable Lightness." On Morellet, see Gordon, *Citizens without Sovereignty,* 177–241.
23. Echeverria, *Maupeou Revolution*; Conlin, "Wilkes"; Maza, *Private Lives.*
24. *Réflexions amusantes* (1782), 12–13. On hostage princesses, see Thomas, *Wicked Queen,* 29–42.
25. Laclos, "Education of Women" (1783).
26. Darnton, *Forbidden Best-Sellers*; Maza, *Private Lives,* 178–83; Campbell, *Power and Faction.*
27. Kelly, "Machine of the duc d'Orléans"; McMahon, "Birthplace of the Revolution"; Gruder, "Question of Marie-Antoinette."
28. Hunt, "Many Bodies of Marie Antoinette"; Price, "Vie privées et scandaleuses"; Thomas, *La Reine scélérate*; Colwill, "Just Another Citoyenne?"; Burrows, *Blackmail,* 147–70.
29. Grieder, "Kingdoms of Women"; Pohl and Tooley, *Gender and Utopia*; Winn and Kuizenga, *Women Writers*; Thomas, *Wicked Queen,* 184–90; Revel, "Marie-Antoinette."
30. Baker, *Inventing the French Revolution,* 188.
31. Schwartz, "F. M. Grimm"; Schroder, "Going Public."
32. Thomas, *Wicked Queen,* 81–103; Weber, *Queen of Fashion*; Feydeau, *Scented Palace.*
33. Jones, *Sexing la mode.*
34. Hyde, "'Makeup' of the Marquise"; Hellman, "Furniture, Sociability"; Martin, *Selling Beauty.*
35. Weber, *Queen of Fashion.*
36. Maza, "Diamond Necklace Affair," 76. See also Brown, "Diamond Necklace Affair Revisited"; Merrick, "Sexual Politics."
37. Grieder, "Kingdoms of Women"; Hunt, "Many Bodies of Marie Antoinette."
38. It seems possible that the duc de Chartres sponsored this, having observed Rivarol's performance at the Café du Caveau in 1778. Ironically, Rivarol's sharp pen would earn him a royal pension to discourage him from following "his inclination towards those who are dangerous." LeBreton, *Rivarol,* 12. See also Tourneux, "Un projet," 283.
39. Schaffer, "Natural Philosophy."
40. Faujas de Saint-Fond, *Description* (1783), 132–75.
41. *Lettre à Mr. M. de Saint-Just* (1784), 12.
42. This character trait is also evident in the play, *Cassandre mécanicien* (1783).
43. Diderot, *Supplément au Voyage de Bougainville,* 37.
44. Rosenfeld, "Tom Paine's Common Sense."

45. [Cornélie de Vasse,] *Le Char volant* (1783), 83.
46. [Cornélie de Vasse,] *Le Char volant* (1783), 13, 26.
47. Berlanstein, *Daughters of Eve*.
48. Kavanagh, *Shadows of Chance*; Kaiser, "Money, Despotism"; Lande, *Rise and Fall of John Law*; Shennan, *Philippe*.
49. [Cornélie de Vasse,] *Le Char volant* (1783), 39.
50. [Cornélie de Vasse,] *Le Char volant* (1783), 76.
51. Elliott, *Shape of Utopia*.
52. Davenant, *Essay* (1756); Pincus, *Prostestantism*, 256–68; Dwyer, "Napoleon."
53. *Histoire intéressante* (1784), 71–74.
54. Shackleton, "Evolution of Montesquieu's Theory of Climate"; Wolloch, *History and Nature*, 108–11.
55. *Lettre à Mr. M. de Saint-Just* (1784), 20.
56. *Lettre à Mr. M. de Saint-Just* (1784), 24.
57. Rousseau, *Social Contract*, especially I.9.
58. The republican ideals and polity had diverse advocates who interpreted them loosely; see Kwass, "Consumption."
59. It was approved for publication on December 12, 1783.
60. [Saunier], *Le Triomphe* (1784), 8, 16.
61. [Saunier], *Le Triomphe* (1784), 5–6, 24.
62. [Saunier], *Le Triomphe* (1784), 12, 16.
63. This machination became the task of Jacques-Pierre Brissot when he joined the staff of Ducrest after the duc de Chartres inherited the family fortune. See Brissot, *Memoires*, 2:64–65; Kelly, "Machine of the duc d'Orléans."
64. Lever, *Philippe Égalité*, esp. 220–22; Darnton, *Devil in the Holy Water*, 344–59.
65. "Le premier voyage aérien," in *Histoire intéressante* (1784).
66. *Le mouton* (1783), 8–9.
67. Sahlins, "Royal menageries."
68. *Réclamation du mouton* (1784), 13.
69. Mercier, *Memoirs* (1772), 1:170-72.
70. Brissot de Warville, *De la verité* (1782), 165–66. See also Darnton, *Mesmerism*, 91; Darnton, "Grub Street Style"; Luna, "Dean Street Style"; Burrows, "Innocence of Jacques-Pierre Brissot."
71. The literature is too large to cite, but one can easily gauge its scope from the aeronautica collection of the Huntington Library.
72. Trouillot, *Silencing the Past*.
73. Mercier, *Memoirs* (1772), 1:166-72.

5: FERMENTATION AND DISCIPLINE

1. *Mercure de France*, Oct. 4, 1783.
2. *Journal de Paris*, Dec. 29, 1783; *Mémoires secrets*, Dec. 28, 1783.
3. The subscription by Jan Ingenhousz (1730–1799) failed, as did one by François Gattey (1753–1819) and Pierre Megnié (1751–1807) supposedly of the Dijon Academy. The two latter sought to raise ten thousand livres at twenty-four livres each by the end of March for an ascent in May (NASM Archive, xxxx-0823 and FM V.27).
4. See their application in AN F 17 1021B, dossier 1; *Journal de Paris*, Jan. 9, Feb.

4, 1784; Hunn, "Balloon Craze," 385-94. *Précis des expériences faites par MM. Alban & Vallet* (1785) includes an image of their balloon (Gimbel collection XL-5-1172).

5. La Vaulx and Tissandier, *Joseph et Étienne*, planches XLVII–XLVIII; *Journal encyclopédique* 57 (1784): 523–24.

6. Brisson, *Observations* (1784); Carra, *Essai sur la nautique aérienne* (1784); Guyot, *Essai sur la construction* (1784); Robert, *Mémoire présenté* (1784); Salle, *Moyen de diriger l'Aérostat* (1784); Seconds, *Navigateur aérien* (1784); *Dissertation sur les Aérostates des anciens et des modernes* (1784).

7. Pilâtre's letter, Feb. 6, 1784 (TC 9.13); Lyon Academy Ms. 233; Robert, *Mémoire présenté* (1784), 3.

8. Crouch, *Taking Flight*, 61–62.

9. Peltonen, "Clues, Margins, and Monads."

10. *Grimm's Corr.* 14 (Aug. 1784): 20.

11. *Grimm's Corr.* 13 (April 1784): 517.

12. Weinstein, *Subversive Tradition*, 1:66–72 (quote, 66). For a more sophisticated analysis, while acknowling the play's transgressive character, see Jones, *Great Nation*, 322–35.

13. Howarth, *Beaumarchais*; Morton and Spinelli, *Beaumarchais*; Proschwitz and Proschwitz, *Beaumarchais*.

14. Darnton, *Devil in the Holy Water*, 177; Graham, *If only the King Knew*.

15. Maza, *Private Lives*, 130–40, 290–95.

16. Schaffer, "Natural Philosophy."

17. Ravel, *Contested Parterre*.

18. Fairchilds, "Populuxe Goods"; Coquery, *Tenir boutique à Paris*.

19. Bailey, *Patriotic Taste*; Smith, *Nobility Reimagined*.

20. Sewell, "Empire of Fashion"; Sonenscher, "Fashion's Empire"; Sama, "Liberty, Equality, Frivolity!"; Kwass, "Ordering the World"; Cottom, "Taste and the Civilized Imagination."

21. Jean Étienne Montucla to Charles Hutton, Dec. 26, 1783 (TC 9.13). See also Cornu, *Galerie des modes*.

22. Gril-Mariotte, "Topical Themes"; Lynn, "Consumerism."

23. *Journal de Paris, Affiches de France*, Nov. 5, 1783; Pingeron, *L'art de faire soi-même* (1783).

24. Lavoisier, *Oeuvres de Lavoisier*, 3:740.

25. Kim, "Public Science."

26. *Rapport fait à l'Académie des sciences* (1784).

27. "Exposition d'un moyen de diriger les ballons aérostatiques," Dossier, December 10, 1783 (Les Archives de l'Académie des sciences, Paris).

28. Lavoisier, "Réflexions sur les points principaux qui doivent occuper les commissaires nommés pour les machine aérostatiques," *Oeuvres de Lavoisier*, 2:741–44.

29. For contrasting views on provincial academic culture, see Roche, *Le siècle des lumières*; Brockliss, *Calvet's Web*.

30. Gillispie, *Montgolfier Brothers*, 260; Lyon Academy Ms. 233; Robert, *Mémoire présenté* (1784), 3.

31. Terrall, "Gendered Spaces"; Conner, *Jean Paul Marat*; Gillispie, *Science and Polity*.

32. Xavier de Maistre, *Les premiers essais*, 24–25. See also *Annales politiques* 11

(1784): 28. Maistre raised the funds for the subscription; see Carncastle, *More Moderate Side*, 176–77.
33. Clay, *Stagestruck*.
34. *Journal de Paris*, Nov. 8, 1783.
35. Jones, "Great Chain of Buying."
36. Mason, *Singing the French Revolution*; Šmidchens, *Power of Song*.
37. *Affiches de France*, Nov. 12, 1783.
38. Bryant, *The King and the City*; Wintroub, "Taking Stock."
39. Latour, *Pasteurization of France*.
40. Boyé, *Les premières expériences*.
41. *Annonces Normandie*, Sept. 19, Oct. 17, Nov. 7, 1783.
42. *Extrait de la Gazette de France, du Mardi 2 Septembre, 1783*.
43. Debièvre, *Notes*, 2–3, 22–24; AM Lille, AG 703.10.
44. The date is given in *Supplément à l'Art de Voyager* (1784), 3. They sold 283 subscriptions to 174 people and collected 3,396 livres; "Lettre de Mathon de la Cour," in Faujas, *Première suite* (1784), 84–85; "État des souscriptions pour la grande expérience du ballon aérostatique," in Cazenove, *Premiers voyages aériens*, 60–63.
45. Joseph to Etienne, Nov. 15, 1783 (FM IX.27). A local jewelry designer composed a song to "immortalize" the Montgolfiers (FM XXI.25).
46. *Journal de Paris*, Nov. 26, 1783.
47. Gillispie, *Montgolfier Brothers*, 69–79; *Rapport Lyon* (1784); *Dissertation sur le fluide* (1784); *Supplément à l'Art de voyager* (1784); *Histoire du Ballon de Lyon* (1784).
48. Minkelers, *Mémoire* (1784).
49. "Procès-verbal de l'expérience aérostatique faite à Nantes (1784)"; "Précis du voyage de l'Aérostat *Le Suffren* (1784)." Although the chief experimenters were identified in public newspapers as M. Lévêque, correspondent of the Royal Academy of Sciences, assisted by M. Coustard de Massi, Chevalier de l'Ordre Royal & Militaire de Saint-Louis, and M. Mouchet, professor of natural philosophy at the Oratorian college and many others, the Oratorian professor of philosophy at the University of Nantes Budan claimed Mouchet's leadership in the entire affair; *Lettre au Rédacteur; Réponse du Rédacteur; Réponse de M. Louvrier; Seconde et derniere Lettre* (1784).
50. *Description de la second Expérience Aérostatique faite à Nantes* (ca. 1784).
51. *Journal de Paris*, Feb. 5, 21, 1784; *Mémoires secrets*, Feb. 4, 1784; *Mercure de France*, Feb. 14, 1784.
52. Arnaud de Saint-Maurice, *L'Observatoire volant* (1784), 9.
53. The balloon was made at Tourillon's taffeta factory at the rue Pavée, St. André des Arts; *Journal de Paris*, Feb. 5, 1784; *Mémoires secrets*, Feb. 4, 1784; *Mercure de France*, Feb. 14, 1784.
54. *Journal de Paris*, Feb. 21, 1784.
55. Dupont's association with Napoleon at Brienne has not been confirmed definitively.
56. *Mercure de France*, Feb. 14, 1784. See also *Mémoires secrets*, Feb. 4, March 1, 2, 3, 14, 1784.
57. Beales, *Prosperity and Plunder*.
58. Fort, "Voice of the Public," 388; Crow, *Painters and Public Life*, 95.
59. Duchosal, *Blanchard, poëme* (1786); Johann Moritz Graf von Brühl to Banks, March 8, 1784 (Dawson, *Banks Letters*, 180).

60. *Journal de Paris*, March 3, 4, 1784; *Mémoires secrets*, March 3, 9, 1784; *Mercure de France*, March 13, 1784.
61. *Grimm's Corr.* 13 (March 1784): 504.
62. Gillispie, *Montgolfier Brothers*, 89–91.
63. Gillispie, *Montgolfier Brothers*, 91–94; *Mercure de France* 33 (Aug 14. 1784), 87–89; Castelli, *Relation* (1784); Letter of Charles Castelli, Feb. 25, 1784, in Faujas, *Première suite* (1784), 128–60; Arecco, *Montgolfiere*.
64. *Grimm's Corr.* 14 (July 1784): 8–17.
65. Xavier de Maistre, *Les premiers essais*, 49–53.
66. *Journal de Paris*, March 11, 20, *Feuille de Flandres*, March 16, 1784.
67. Magistrates of Lille and Desjobert, March 17, 25, May 4, 1784 (AM Lille, 703.13).
68. *Journal de Paris*, April 30, 1784; *Mémoires secrets*, April 25, 1784.
69. *Gazette de Leyde*, supplementary no. 38, May 11, 1784.
70. Emgart to the Magistrates, May 4, 1784 (AM Lille, 703.13).
71. Kaplan, *Bread*; Kaplan, *Provisioning Paris*; Bouton, *Flour War*.
72. The abbé to Étienne, Dec.1–18, 1783 (FM IX.33–38).
73. The factory would recover the pre-balloon level only in 1788; Rosenband, *Papermaking*, 123–24.
74. *Journal de Paris*, May 29, 1784; FM XXII.56.
75. Amadou, *Franz-Anton Mesmer*, 15–28; Buranelli, *Wizard from Vienna*; Steptoe, "Mozart."
76. Venturi, *End of the Old Regime*, 1:354–57.
77. Amadou, *Franz Anton Mesmer*, 59–79; Schaffer, "Astrological Roots."
78. Letter from Servan to M.A. Julien, August 17, 1781, in Darnton, *Mesmerism*, 60.
79. Delson, *Observations* (1780).
80. Mesmer, *Mémoire sur la découverte du magnétisme animal* (1779); Darnton, *Mesmerism*, 40–125.
81. Mesmer's letter to the queen is translated in Buranelli, *Wizard from Vienna*, 139–41; Gillispie, *Science and Polity*, 261–89.
82. Edelstein, *Terror of Natural Right*; Wright, *Classical Republican*.
83. Maza, *Private Lives*; Darnton, *Mesmerism*, 83–100.
84. Bergasse, *Lettre d'un médecin* (1781), 66–67.
85. Mesmer's letter to Philip, dean of the Faculté (1782) in Amadou, *Franz Anton Mesmer*, 229–32.
86. Sutton, "Electric Medicine."
87. On their meetings, see Darnton, *Mesmerism*, 180–82.
88. The two would later form a splinter group. Bergasse defended Kornmann in the latter's divorce proceedings against Beaumarchais who represented Mrs. Kornmann; Maza, *Private Lives*, 297–311.
89. Gillispie, *Montgolfier Brothers*, 277; *Grimm's Corr.* 13 (April 1784): 510.
90. Court de Gébelin, *Lettre de l'auteur du monde primitif* (1784).
91. *Mémoires secrets*, April 9, 1784.
92. *Journal de Paris*, Jan. 10, Feb. 13–17, 1784; Paulet, *L'Antimagnétisme* (1784).
93. Initially delivered as a lecture to the Musée de Paris in November 1783. Also see Hervier, *Lettre aux habitants de Bordeaux* (1784).
94. Darnton, *Mesmerism*, 75; Buranelli, *Wizard from Vienna*, 143–56, 169–79; Pattie, *Mesmer*, 117–41, 197–215.

95. Girard, *Mesmer blessé* (1784); *Mesmer guéri* (1784); Pattie, *Mesmer*, 185-90.
96. *Grimm's Corr.* 13 (Jan. 1784): 456–64.
97. *Grimm's Corr.* 13 (Jan. 1784): 448-52.
98. Russo, *Styles of Enlightenment*, 194–220; Wade, *The "Philosophe" in the French Drama*.
99. On acceptable notions of civility and exemplary careers, see Gordon, *Citizens without Sovereignty*.
100. *Grimm's Corr.* 13 (Jan. 1784): 467 (added emphasis). Ducis had previously adapted *Hamlet* (1769), *Romeo and Juliet* (1772), and *King Lear* (1783); see Golder, *Shakespeare*, 163–230; Monaco, *Shakespeare and the French Stage*.
101. Argenson, *Journal*, 6:464; quoted in Hammersley, *English Republican Tradition*, 1.
102. Edelstein, *Terror of Natural Right*, 61–62.
103. *Grimm's Corr.* 13 (Feb. 1784): 483. See also Vendrix, "La notion de revolution"; Rushton, "Theory and Practice of Piccinnisme."
104. *Grimm's Corr.* 13 (March 1784): 505–7.
105. La Harpe, *Correspondance littéraire*, 4:266.
106. Hardy, "Mes Loisirs"; *Journal politique de Bruxelles*, May 1, 1784.
107. *Mémoires secrets*, May 26, 1784; *Journal politique de Bruxelles*, May 22, 1784.
108. *Grimm's Corr.* 13 (April 1784): 519.
109. MacArthur, "Embodying the Public Sphere"; *Grimm's Corr.* 13 (April 1784): 517–25.
110. Weinstein, *Subversive Tradition*, 1:71–72.
111. The next in staging popularity were La Mott's *Inès de Castro* (staged thirty-two times), and Voltaire's *Zaïre* (staged thirty times) and *Mérope* (staged twenty-eight times).
112. Maza, *Private Lives*, 290–95.
113. *Grimm's Corr.* 13 (April 1784): 519.
114. Parisian police spent substantial resources to provide wet nurses for the babies of working mothers. This kept the babies away from their own mothers on both ends. Parisian working families often incurred considerable debt to pay for wet nurses, which often put the fathers in jail. Beaumarchais proposed providing a small subsidy for Parisian working mothers so they might nurse their own babies, which would alleviate problems on all sides. The Institut de bienfaisance maternelle, as it developed in Lyon, became a huge success; *Journal de Paris*, Aug. 15, 1784; *Journal de Lyon*, Nov. 24, 1784; *Almanach astronomique et historique de la ville de Lyon et des provinces de lyonnois* (1788), 72–75; Kite, *Beaumarchais*, 2:212–28.
115. Darnton, *Mesmerism*.
116. Schaffer, "Astrological Roots"; Fulford, "Conducting the Vital Fluid."
117. Beik, *Urban Protest*.

6: PROVINCIAL CITIZENS

1. *Lettre à Mr. M. de Saint-Just* (1784), 3–4, 9–12.
2. *Histoire du Ballon de Lyon* (1784), 15.
3. Chartier, "Culture as Appropriation."
4. Wood, "State and Popular Sovereignty," 85.
5. Mercier, "Grandeur démesurée de la capitale," in *Tableau*, 1:32–33, translated in Popkin, *Panorama of Paris*, 32–33.

6. Shovlin, *Political Economy of Virtue*.
7. Prakash, *Another Reason*, 49–50.
8. Kumar, "Nation-States as Empires"; Gerson, "Parisian Litterateurs."
9. Schama, *Citizens*; Roche, *Le siècle des lumières*; Clay, *Stagestruck*.
10. Swann, *Provincial Power*; Reynard, *Ambitions Tamed*.
11. By the 1780s the city (population 130,000) possessed sixteen lodges with total membership nearing a thousand; Garden, *Lyon et les Lyonnais*; Beaurepaire, "Universal Republic"; Joly, *Un mystique lyonnais*.
12. Guyonnet, *Jacques de Flesselles*; Fricke, "Le Prince Charles-Joseph de Ligne."
13. Adëlaide to Jean-Pierre, Jan. 3, 1784 (FM X.3); Joseph to the comte d'Artois, Feb. 5, 1784, in Freycinet, *Essai sur la vie*, 22–24.
14. Reynard, *Ambitions Tamed*.
15. *Supplément à l'Art de voyager*, 8; AM Lyon, Ms. 14 II 19 [4], Mss. 10882, 10889.
16. *Rapport fait à l'Académie des sciences* (1784), 2.
17. *Description de deux machines*, 2. The author of this description experimented with four different materials—taffetas soaked in elastic gum, *papier brouillard* normally used for curling women's hair; extremely fine *toile de Rouen*, and well-tanned leather.
18. Linen cost ten sols per aune; AM Lyon, Ms. 1 II 466 [1]; Miller, "Paris-Lyon-Paris."
19. *Supplément à l'Art de voyager* (ca. 1784), 4; Gillispie, *Montgolfier Brothers*, 71–72.
20. Joseph to Étienne, Dec. 6, 18, 1783 (FM IX.35, 39); La Vaulx and Tissandier, *Joseph et Étienne*, 53–55 and planches XL–XLI; Gillispie, *Montgolfier Brothers*, 69–79.
21. Augustin to Étienne, Dec. 27, 1783 (FM IX.45); *Histoire du Ballon de Lyon* (1784), 5.
22. The abbé to Étienne, Jan. 24, 1784 (FM X.26).
23. Joseph and the abbé to Jean-Pierre, Dec. 28, 1783 (FM IX.46); the abbé to Jean-Pierre, n.d. (FM X.7).
24. Jean-Pierre to the abbé, Jan. 5, 1784 (FM X.6).
25. Joseph to the Lyon Academy, February 10, 1784 (Lyon Academy Ms. 268, ff. 176–78).
26. *Journal de Paris*, Jan. 9, 15, 1784; *Journal encyclopédique* 57 (Jan. 1784): 317–18.
27. *Mémoires secrets*, Jan. 22, 1784; *Gazette de Leyde*, Jan. 27, 1784; Middleton, "Brief History"; Bennett and Talas, *Cabinets of Experimental Philosophy*.
28. Jean-Pierre, Jan. 3, the abbé and Joseph to Pierre, Jan. 4, and to Jean-Pierre, Jan. 9, 1784 (FM X.2, 4, 11).
29. One pamphleteer exaggerated that the actionnaires du Pont Morand would have made almost 24,000 francs if they had charged normal fees for crossing; *Histoire du Ballon de Lyon* (1784), 3–6; *Supplément à l'Art de voyager* (ca. 1784), 8–9.
30. Latour and Weibel, *Making Things Public*.
31. Ruplinger, *Charles Bordes*.
32. Edmonds, *Jacobinism*, 9–37; Garden, *Lyon et les Lyonnais*.
33. "Avertissement," *Histoire du Ballon de Lyon* (1784).
34. Adelaïde to Jean-Pierre, Jan. 3, 1784 (FM X.3); *Rapport fait à l'Académie des sciences* (ca. 1784), 4–5.
35. The abbé to Jean-Pierre, Jan. 9, 1784 (FM X.11); Jean-Pierre to Joseph, Jan. 12, 1784 (FM X.14).

36. *Histoire du Ballon de Lyon* (1784), 6–8.
37. AM Lyon, 1 II 466 1
38. *Supplément à l'Art de voyager* (ca. 1784), 15; *Observations sur le Rapport* (1784), 3; *Journal de Paris*, Jan. 31, 1784.
39. AM Lyon, 1 II 112, marked No. 63.
40. *Histoire du Ballon de Lyon* (1784), 18.
41. Southern, *Treatise* (1785), 9–24.
42. Joseph Montgolfier to the comte d'Artois, Feb. 5, 1784 in Freycinet, *Essai sur la vie*, 22–24.
43. *Rapport fait à l'Académie des sciences* (ca. 1784), 6; AM Lyon, 1 II 466 1.
44. *Mercure de France*, Jan. 31, 1784; *Histoire du Ballon de Lyon* (1784), 20; *Supplément à l'Art de voyager* (ca. 1784), 21–24.
45. AM Lyon, 1 II 466 1.
46. *Histoire du Ballon de Lyon* (1784), 23–24.
47. The abbé to Étienne, Jan. 20, 1784 (FM X.20).
48. Lyon Academy Mss. 268, 272.
49. *Registre des arrêtes du consulats*, January 27, 1784 (AM Lyon, BB362).
50. *Registre des séances de la Société royale des sciences*, March 4, 1784 (AD Hérault, D123). See also Legrand, "Chemistry in a Provincial Context."
51. The abbé to Étienne, March 6, 23, 1784 (FM XI.2, 10).
52. Joseph's letters, Feb. 12, 19, 1784 (FM X.35, 40).
53. Digonnet, *L'Invention de l'Aérostation*; Gillispie, *Montgolfier Brothers*, 83–86; Brantes to Joseph, April 7, 1784 (FM XV.33). Pingeron's *L'art de faire soi-même*, dated Sept. 22, 1783, was addressed to the marquise de Brantes.
54. Joseph was planning to visit Avignon and cut the *fuseaux* at Vidalon; Joseph to the comte d'Artois, Feb. 5, 1784, in Freycinet, *Essai sur la vie*, 22–24; Joseph to Étienne, March 24, 1784 (FM III.5).
55. Thébaud-Sorger, *L'Aérostation*, 236.
56. *Hollandsche Historische Courant*, July 13, 1784; Lensink, "Science & Spectacle," 52.
57. *Histoire du Ballon de Lyon* (1784), 21–22; Guyton knew Pilâtre as a "reputable colleague" from the year he learned pneumatic chemistry in Paris; Guyton de Morveau, Maret, and Durande, *Elemens de Chymie* (1777), iv.
58. *Mémoires secrets*, Jan. 25, Feb. 1, 9, 1784.
59. *Journal de Paris*, Jan. 23, 25, *Mercure de France*, Jan. 24, 1784; *Mémoires secrets*, Jan. 22, 24, 1784; *Journal encyclopédique* 57 (March 1784): 304.
60. Pilâtre to Faujas, Jan. 28, 1784, in Faujas, *Première Suite* (1784), 77–79.
61. *Grimm's Corr.* 13 (Jan. 1784): 454, (Feb. 1784): 480.
62. *Journal de Paris*, Jan. 31, *Courier de l'Europe*, Feb. 17, 20, 1784; *Observations sur le Rapport* (1784); *Journal encyclopédique* 57 (March 1784): 305–7; Faujas, *Première Suite* (1784), 80–97.
63. Jean-Pierre and the abbé to Étienne, March 7–8, 1784 (FM XI.2-3).
64. A notable exception is Thébaud-Sorger, *L'Aérostation*.
65. Combe, *Histoire de la Franc-maçonnerie*, 103–4.
66. Laurencin, *Lettre* (1784).
67. AM Lyon, Ms. 1 II 19 [4].
68. Du Coudray, *Voyage du comte de Haga* (1784).

69. Roberts, "Great Britain and the Swedish Revolution"; Schuchard, *Emanuel Swedenborg*, 726–48.
70. Murphy, *Charles Gravier*, 172–207, 333–44; Murphy, *Diplomatic Retreat*, 41–45, 126–33; Geffroy, *Gustave III*, 2:43–44.
71. Skuncke, "Press and Political Culture"; Barton, "Gustav III."
72. AM Lyon, 14 II 19 [4].
73. Soufflot spent about fifteen years in Lyon, first for twelve years (1738–1749) when he came back from Rome, then intermittently afterwards. He designed many buildings that defined the Lyon cityscape in the neoclassical style; [Actes du colloque], *Soufflot et l'architecture des lumières*; Pérouse de Montclos, *Jacques-Germain Soufflot*.
74. Laurencin, *Lettre* (1784), 23–24.
75. Laurencin, *Lettre* (1784), 25.
76. *Journal de Lyon*, June 9, 23, July 7, 21, 1784.
77. Swann, *Provincial Power*.
78. Kim, "De l'érudition à la science."
79. Note the coverage in the *Affiches, Annonces et Avis Divers de Bourgogne, Bresse, Bugey et Pays de Gex*, edited by Louis-Nicolas Frantin between 1776 and 1779.
80. André Villot (1743–1800) resumed *Affiches, Annonces et Avis Divers de Dijon, ou Journal de la Bourgogne* [hereafter cited as *Affiches de Dijon*] on November 4, 1783.
81. RAD, Dec. 4, 1783; Delmasse, *Histoire comique*, ff. 11–20.
82. *Affiches de Dijon*, Dec. 9, 30, 1783; *Mercure Dijonnais*, 306.
83. Delmasse, *Histoire comique*, f. 33bis.
84. Barquuncourt, a conseiller d'État, paid five louis (RAD Jan. 3, 1784).
85. *Journal de Paris*, Jan. 7, *Affiches de Dijon*, Jan. 13, 1784.
86. *Journal encyclopédique* 57 (1784): 124–25.
87. RAD, Jan. 3, 15, 1784.
88. RAD, Jan. 3, 1784.
89. *Journal de Paris*, Jan. 28, *Affiches de Dijon*, Feb. 17, 1784.
90. RAD, Jan. 29, Feb. 5, 19, 1784.
91. Minkelers, *Mémoire* (1784); Guyton de Morveau, *Description* (1784).
92. *Mercure de France*, Feb. 7, *Journal de Paris*, Feb. 10, 1784.
93. *Gazette de Leyde*, Supplément à l'*Art de voyager dans les Airs* No. 10, Feb. 3, 1784.
94. RAD, Jan. 22, 1784.
95. RAD, Jan. 30, Feb. 12, 1784.
96. *Affiches de Dijon*, Feb. 17, 1784.
97. RAD Dec. 27, 1783; *Affiches de Dijon*, Jan. 20, 27, Feb. 17, 24, 1784.
98. AM Dijon, I 40; Deliberations, Feb. 18, 1784 (AM Dijon, B 418).
99. Sonnet, "Le palais abbatial de Saint-Bénigne."
100. *Mercure dijonnais*, 309–10; *Affiches de Dijon*, March 2, *Mercure de France*, March 13, 1784.
101. RAD March 18, 1784; *Affiches de Dijon*, March 23, 1784.
102. Bryant, *The King and the City*; Wintroub, "Taking Stock"; Bryant, "Royal Ceremony."
103. *Affiches de Dijon*, May 4, 1784. See also *Journal de Paris*, May 2, 1784; Bryant, *King and the City*.
104. RAD, April 29, May 6, 13, 1784.
105. *Mémoires secrets*, May 2, 1784; *Gazette de Leyde*, supp. No. 37, May 17, 1784.

106. Guyton, *Description* (1784), 189–96.
107. *Journal de Paris*, June 13, 1784.
108. *Mercure dijonnais*, 312.
109. *Affiches de Dijon*, June 22, 1784; Procès-verbal in Guyton de Morveau, *Description* (1784), 197–217.
110. *Journal de Paris*, June 23, 1784; *Mémoires secrets*, June 29, 1784. The landing location is given in *Mercure dijonnais*, 314.
111. AM Nantes, FF 276, f. 1.
112. Edmonds, *Jacobinism*, 1–37; Reynard, *Ambitions Tamed*.

7: THE FALL OF A NATIONAL ARTIFACT

1. Anderson, *Imagined Communities*, 4. In focusing on the balloon's material agency, I am disputing in part Anderson's emphasis on the role of newspapers in the consolidation of a nation-state. For the religious contribution to modern nationalism, see Bell, *Cult of the Nation*.
2. *Annales politiques* 11 (1784): 296–310.
3. *Journal de Paris*, May 17, 1784, May 23, 1785.
4. La Bruyère, *Characters*; Elias, *Court Society*; Biagioli, *Galileo*.
5. "Le premiere décembre," in Mercier, *Tableau*, 2:886–89.
6. Franta, *Romanticism*; Bentham, *Essay on Political Tactics* (1791).
7. *Annales politiques* 10 (1783): 355–68, 11 (1784): 1–19.
8. Forrest, *Society and Politics*, 1–29; Poussou, *Bordeaux*; Kingston, *Montesquieu*.
9. Duck, "Plantation/Empire"; Mignolo, "Geopolitics of Knowledge."
10. Rothschild, "Isolation."
11. Bordes, *L'administration provinciale*; Doyle, *Parlement of Bordeaux*; Lhéritier, *La révolution à Bordeaux*, 50–67.
12. The population doubled to one hundred thousand between 1715 and 1790; Forrest, "Condition of the Poor."
13. Clay, *Stagestruck*, 50–52.
14. Marionneau, *Victor Louis*.
15. "Ordonnance . . . concernant la Police des Spectacles, du 31 Mars, 1780" (AM Bordeaux, FF70).
16. Mouchy to Mayor, May 3, 1782 (AM Bordeaux FF70); Ravel, *Contested Parterre*, 178–83.
17. This section is built on Hunn, "Balloon Craze," 253–88.
18. BM Bordeaux, Ms. 712 II, f. 11; AM Bordeaux, BB188 (dossier 1783), FF70.
19. BM Bordeaux, Ms. 713 XLVII, ff. 220–24. On retribution, see Beik, *Urban Protest*.
20. Jurats, Dec. 6, 20, 1783 (AM Bordeaux, BB179); Mouchy, Dec. 11, 15, 1783 (AM Bordeaux, FF70, BB194).
21. AN O^1 573, f. 367.
22. BM Bordeaux, Ms. 713.2III.no.42, Ms. 713.1.47, f. 243.
23. Darbelet's letter, Feb. 12, 1784 (BM Bordeaux, Ms. 829X, f. 320).
24. Delisle Thibaut and Étienne Montgolfier, Dec. 23, 1783, Jan. 17, 1784 (FM VI. 20–22).
25. *Extrait des registres de l'Académie royale des belles-lettres, sciences et arts de Bordeaux du 14 Décembre, 1783* (BM Bordeaux, H5139); Devaux, "Jean-André Cazalet."

26. "Première liste de messieurs les souscripteurs, pour le globe aerostatique proposé par M. Cazalet," Feb. 23, 1784 (TC 3.3).
27. *Comptes rendus des séances du Musée*, April 10, 1783—March 25, 1789 (BM Bordeaux, Ms. 829XIII); Céleste, "La société philomatique de Bordeaux"; Bouyssy, *Le Musée de Bordeaux*.
28. Forrest, *Society and Politics*, 26–27; Loiselle, "Living the Enlightenment"; McLeod, "A Bookseller."
29. BM Bordeaux, Ms. 829XIII, Dec. 23, 1783; Chauvet's letter in Ms. 829X, f. 326.
30. St.-Maur to François de Lamontaigne, Jan. 6, 1784 (AD Gironde, C3604).
31. "Projet d'une machine aérostatique"; Philippe to Étienne, Jan. 1, 1784 (FM VI.27).
32. Jean-Michel Le Bris, "Biographie de De Grassi, médecin installé à Bordeaux," (thesis, Bordeaux, 1979).
33. AM Bordeaux, BB180, f. 4, BB140, f. 39.
34. AM Bordeaux, XL-A/54.
35. "Arrêt du conseil d'etat du roi," Oct. 4, 1779 (AM Bordeaux, F82b).
36. Jurats to Vergennes, May 4, 1784 (AM Bordeaux, BB180).
37. Lamontaigne, *Chronique bordelaise*, 149; BM Bordeaux, Ms. 713.47, f. 244.
38. Jurats, report to Vergennes, n.d. (AM Bordeaux, BB180).
39. Vergennes, May 10, 1784 (AM Bordeaux, FF70). See also Lamontaigne, *Chronique bordelaise*, 151.
40. *Courier de l'Europe*, May 28, 1784; Hunn, "Balloon Craze," 267–78.
41. BM Bordeaux, Ms. 713.47, f. 246, Ms. 829 X, f. 320.
42. Darbelet and Chalifour, "Relation de deux voyages aériens (1787)," 112–29; Lamontaigne, *Chronique bordelaise*, 154; BM Bordeaux, Ms. 713 XLVII, ff. 248–50; AM Bordeaux, Ms. 440, f. 337.
43. Sept. 30, 1784 (FM V.23).
44. Darbelet, Desgranges, and Chalifour, "Relation de deux voyages aériens," 129–47; AM Bordeaux, BB140, ff. 50–51.
45. BM Bordeaux, Ms. 713, no. 47, f. 256.
46. This was probably the *Temple of the Muses* now at the Metropolitan Museum of Art, New York.
47. *Journal de Paris*, May 30, 1784.
48. Geffroy, *Gustave III*.
49. *Grimm's Corr.* 13 (June 1784): 537–48, 556.
50. Du Coudray, *Voyage du comte de Haga* (1784).
51. Étienne to the duc de Castries, n.d. (FM VII.16–18). The duc de Castries (Armand Charles Augustin de la Croix, 1756–1842) was a new creation, different from the marquis de Castries (Charles Eugène Gabriel de la Croix, 1727–1801), the Navy minister during the American War.
52. Calonne to Étienne Montgolfier, June 11, 1784 (FM XIVbis.7).
53. Réveillon to Étienne, June 17, 1784 (FM XII.15).
54. The comte d'Antraigues to Étienne, June 21, 1784 (FM XVII.30).
55. Pilâtre de Rozier, *Première expérience* (1784), 4.
56. *Journal de Paris*, June 24, 1784; procès-verbal of the prince de Condé, June 23, 1784 (TC 3.3).
57. *Grimm's Corr.* 14 (July 1784): 9.

58. Pilâtre to Étienne, July 30, 1784 (FM XV.10); *Annales politiques* 11 (1784): 298.
59. Pilâtre, *Première expérience* (1784).
60. Réveillon to Étienne, June 24, 1784 (FM XII.16).
61. Réveillon to Étienne, June 26, Aug. 12, 1784 (FM XII.17–19).
62. Étienne to Calonne, July 1 (FM IV.1), to De Castries, July 1, Aug. 23, 1784 (FM IV. 2–3, 8); Gillispie, *Montgolfier Brothers*, 93–94.
63. De la Roche to Étienne, July 27, Oct. 6, 1784 (FM XX.38–39).
64. The Montgolfiers' appeal to the duchess de Polignac also failed (FM XVIbis.12-13). Only the death of Pilâtre would change this cold reception. In March 1786, Étienne submitted a new proposal for sixty thousand livres and was granted forty thousand livres; letter corrected by the comte d'Antraigues (FM VII.9); Calonne's response dated March 23, 1786 (FM XVIbis.8); dossier, "Expériences aérostatiques" (AN F 17 1021B).
65. *Grimm's Corr.* 14 (July 1784): 9–11; *Annales politiques* 11 (1784): 299.
66. *Journal de Paris*, Feb. 26, March 21, 29, 1784.
67. *Gazette de Leyde*, July 20, 1784.
68. *Mémoires secrets*, Sept. 28, 1778.
69. *Journal de Paris*, April 1, 21, 1784.
70. *Journal de Paris*, June 20, 25, 1784; *Grimm's Corr.* 14 (July 1784): 9–10.
71. *Journal de Paris*, July 2, 1784; *Annales politiques* 11 (1784): 299.
72. "Voyage qui n'est point sentimental," July 11, 1784; Cradock, *La vie française*, July 11, 1784.
73. Réveillon to Étienne, July 17, 1784 (FM XII.18).
74. Mercier, *Tableau*, 2:1196–98; *Grimm's Corr.* 14 (July 1784): 9–11.
75. Beik, *Urban Protest*.
76. *Gazette de Leyde*, July 20, 1784; Miolan, July 21, 1784 (TC 7.4).
77. Voyage qui n'est point sentimental," July 12, 1784; *Grimm's Corr.* 14 (July 1784): 9–11.
78. Miolan and Étienne Montgolfier, Dec. 6, 1786 (FM XV.30–31).
79. Meusnier to Étienne Montgolfier, July 27 (FM XIII.56); *Journal de Paris*, July 12, 1784.
80. Kelly, "Machine of duc d'Orléans."
81. *Journal encyclopédique* 57 (March 1784): 307.
82. Kim, "Public Science."
83. *Annales politiques* 11 (1784): 301–2; Gillispie, *Montgolfier Brothers*, 100–103.
84. *Description de deux machines*.
85. *Mémoires secrets*, July 4, 1784; *Gazette de Leyde*, July 20, 1784.
86. Mercier, *Tableau*, 2:741–46.
87. *Gazette de Leyde*, July 27, 1784; *Grimm's Corr.* 14 (July 1784): 11–13.
88. *Mémoires secrets*, July 16, 1784; *Annales politiques* 11 (1784): 302–3.
89. Suspicions for the authorship of this libel fell on Théveneau de Morande, and for its patronage on the Monsieur and the king himself; Lever, *Philippe Égalité*, 220–22; Darnton, *Devil in the Holy Water*, 344–59.
90. *Mémoires secrets*, July 16, 1784; *Annales politiques* 11 (1784): 303–6; *Gazette de Leyde*, July 27–Aug. 3, 1784.
91. Gillispie, *Science and Polity*, 103–6.
92. *Grimm's Corr.* 14 (July 1784): 8–17.

93. *Annales politiques* 11 (1784): 296–310.
94. Franklin, *Rapport des commissaires* (1784).
95. Lavoisier, *Oeuvres de Lavoisier*, 3:508–10.
96. *Grimm's Corr.* 14 (Aug. 1784): 20–26; Poissonnier, *Rapport des commissaires de la société Royale de Médecine* (1784).
97. Darnton, *Mesmerism*, 64, 85; Gillispie, *Science and Polity*, 261–89.
98. *Courier de l'Europe*, Oct. 5, 1784.
99. Jefferson, Feb. 5, 1785 in *Papers of Thomas Jefferson*, 7:635.
100. *Grimm's Corr.* 14 (Dec. 1784): 76–78; *Mémoires secrets*, Jan. 17, 1785.
101. *Journal politique de Bruxelles*, Dec. 11, 1784. [Brissot de Warville], *Un mot à l'oreille des académicians de Paris*, 8–9; Fournel, *Remonstrances des malades* (1785); Bonnefoy, *Analyse raisonnée* (1784); Servan, *Doutes* (1784), 101–2.
102. Gillispie, *Science and Polity*, 278; Amadou, *Le magnétisme animal*, 361–99.
103. Weiner, *Citizen-Patient*; Thouret, *Extrait de la correspondance de la Société royale de Médecine*.
104. Bentham, *Essay on Political Tactics* (1791). See also Franta, *Romanticism*, 1–3.

PART III: MATERIAL EMPIRE

1. *Ordonnance de l'empereur* (1786); Alexander, "Aeromania" (see chapter 10 for the German case).
2. Spary, *Utopia's Garden*; Brockliss, *Calvet's Web*.
3. Andréani, *Relation* (1784); Arecco, *Montgolfiere*.
4. Faujas, *Raccolta* (1784); Faujas, *Descrizione* (1784).
5. Noorden, *Korte Verhandeling* (1784); Faujas, *Beschryving* (1784); Damen, *Naturen Wiskundige Beschouwing* (1784); van Lier, *Verhandeling* (1784); Lensink, "Science and Spectacle," 24–28.
6. *Annual Register* (1783): 65–75; *New Annual Register* (1784): 154–66; Copeland, "Burke and Dodsley's Annual Register."
7. Southern, *Treatise* (1785); Cavallo, *History* (1785).
8. Faujas, *Beschreibung* (1784). Compare Faujas's *Kurze Nachricht* (1784) to Southern's *Treatise* (1785) for different native capacities.
9. Colley, *Britons*.
10. Berlanstein, *Daughters of Eve*.
11. Eagles, *Francophilia*; Lottum, "Labour Migration"; Nelson, *Royalist Revolution*.
12. Blanchard, *Relation de la vingtième ascension* (1786), 12–13; Lindemann, *Patriots and Paupers*.
13. Levinger, *Enlightened Nationalism*.
14. Daston, "Nationalism"; Sama, "Liberty, Equality, Frivolity!"; Zimmer, *Contested Nation*; Alter, "Playing with the Nation."
15. MacAloon, "Olympic Games"; Barnes, "Soccer Nation"; Onwemechili and Akindes, *Identity and Nation*.
16. Clifford, *Routes*, 2–7.

8: MODERN ATLANTIS

1. Eagles, *Francophilia*; Brewer and Porter, *Consumption*; Bermingham and Brewer, *Consumption of Culture*; McKendrick, Brewer, and Plumb, *Birth of a Consumer Society*; Berg, "From Imitation to Invention."

2. Rogers, "Crowd and People"; Haywood and Seed, *Gordon Riots*; Rudé, *Paris and London*, 268–92; Rudé, *Ideology and Popular Protest*; Sainsbury, *Disaffected Patriots*; Palmer, *Police and Protest*.
3. Hoock, *Empires of the Imagination*, 23–36; Wilson, *Sense of the People*, 237–84.
4. British Museum online Images 1868,0808.5067 and 1868,0808.5070, respectively.
5. Bullion, "George III on Empire," 306–7.
6. Colley, "Apotheosis of George III," 99; Cannon, *Fox-North Coalition*; Langford, *Polite and Commercial People*, 518–64.
7. Bayly, *Imperial Meridian*, 80.
8. Keen, "Balloonomania," 508–9. See also Keen, *Literature*; Brewer, "Most Polite Age"; Campbell, *Romantic Ethic*. For the strategies of commodification, see Lynn, "Consumerism."
9. Colley, *Britons*, 242–52; Robinson, *Edmund Burke*, 53–78.
10. Hoock, *Empires of the Imagination*, 117–29; Weber, "1784 Handel Commemoration."
11. Hodgson, *History of Aeronautics*, 197–204.
12. Green, "Balloon and Seraglio"; O'Quinn, *Staging Governance*; Christie, *Stress and Stability*.
13. Johnson, *John Nelson*; Bredin, *Pale Abyssinian*.
14. *The Modern Atlantis* (1784), 82.
15. Harkness, *Jewel House*; Ogborn, *Spaces of Modernity*.
16. Wilson, "The Good, the Bad"; Klein, "Politeness"; Brewer, *Pleasures of the Imagination*.
17. *Gentleman's Magazine* 54 (Sept. 1784): 711.
18. Henry Smeathman to Banks, Feb. 11, 1784 (*SCBanks*, 2:258–61).
19. Gillespie, "Ballooning."
20. Gascoigne, *Joseph Banks*; Gascoigne, *Science in the Service of Empire*; Fara, "Benjamin West's Portrait."
21. Chambers, *Joseph Banks*; Blagden to Banks, Oct. 14, 1783 (*SCBanks*, 2:172–73).
22. Blagden to Banks, March 1, Banks to Franklin, March 23, 1784 (*SCBanks*, 2:263, 270).
23. Banks to Franklin, Sept. 13, 1783 (*SCBanks*, 2:132).
24. Blagden and Banks, June 11–July 23, 1783 (*SCBanks*, 2:86–110); Crosland, "Relationships"; Fauque, "Englishman Abroad."
25. Franklin to Banks, July 27, 1783 (*SCBanks*, 2:111).
26. Franklin to Banks, Aug. 30, 1783 (*SCBanks*, 2:127–29).
27. Blagden to Banks, Sept. 11, 1783 (*SCBanks*, 2:131). See also Geikie, *Annals of the Royal Society Club*.
28. Banks to Franklin, Sept. 13, 1783 (*SCBanks*, 2:132). See also *Morning Chronicle*, Sept. 8, *London Chronicle*, Sept. 9, 23, *Whitehall Evening Post*, Sept. 9, *Morning Herald*, Sept. 17, 23, *Public Advertiser*, Sept. 24, 1783. They often carried identical reports.
29. *Morning Herald*, Sept. 17, 1783.
30. *St. James's Chronicle*, Sept. 30, 1783.
31. *European Magazine* 4 (Oct. 1783): 272.
32. Franklin to Banks, Oct. 8, 1783 (*SCBanks*, 2:162–63).
33. Blagden to Banks, Sept. 20, 1783 (*SCBanks*, 2:138–40).
34. Banks to George Cumberland, March 6, 1784, in Dawson, *Banks Letters*, 245.

35. Dryander to Banks, Nov. 1, 1783 (*SCBanks*, 2:204–6).
36. Blagden to Banks, Oct. 16–23, 1783 (*SCBanks*, 2:179, 190–92).
37. Broussonet to Banks, Oct. 23, Nov. 10, 1783, in Dawson, *Banks Letters*, 157.
38. Romani, *National Character*; Colley, *Britons*, 242–52.
39. Hodgson, *History of Aeronautics*, 99; *London Chronicle*, *St. James's Chronicle*, August 19–21, 1783.
40. *General Evening Post*, Sept. 25, *London Chronicle*, *Morning Herald*, *Whitehall Evening Post*, *Morning Chronicle*, *Public Advertiser*, all Sept. 27–29, 1783.
41. *European Magazine* 4 (Sept. 1783): 233; *Gentleman's Magazine* 53 (Sept. 1783): 795.
42. *Morning Herald*, Sept. 17, 1783.
43. Banks to Franklin, Nov. 7, 1783 (*SCBanks*, 2:209).
44. Inglis, *Georgian London*, 130.
45. *General Evening Post*, *London Chronicle*, *Morning Herald*, *Whitehall Evening Post*, *Morning Chronicle*, *Public Advertiser*, Nov. 25–27, 1783; *Gazetteer*, Nov. 26, 1783.
46. The author is revealed in *Monthly Review* 70 (1784): 78–79; Brown, *Fables of Modernity*, 121.
47. Probably adapted from the earlier one without the air balloon; *Morning Herald*, Nov. 7, *Gazetteer*, *Public Advertiser*, Nov. 8–10, 1783.
48. *General Evening Post*, *London Chronicle*, *Whitehall Evening Post*, Nov. 25–27, *Gazetteer*, Nov. 28, 1783. They report Argand as Argene and a Prussian.
49. Blagden to Banks, Oct. 18, 30 (*SCBanks*, 2:184, 201); Faujas and Broussonet to Banks, Nov. 1–2, 1783, in Dawson, *Banks Letters*, 321, 157; Argand to Étienne, Nov. 21, 1783 (FM XIII.39); Schrøder, *Argand Burner*; Wolfe, *Brandy, Balloons, & Lamps*.
50. Levere and Turner, *Discussing Chemistry*, 79–81.
51. Morton and Wess, *Public and Private Science*; Morton, *Science in the 18th Century*; Watkin, *Architect King*; Hoskin, "Herschel's 40 Ft Reflector."
52. [Cooke], *Air Balloon* (1783), 24.
53. Altick, *Shows of London*, 85; Argand's letter, n.d., in Faujas, *Première suite* (1784), 191–92.
54. Barbauld, *The Works*, 2:22–23; Thaddeus, *Frances Burney*.
55. *Gentleman's Magazine*, 53 (Nov. 1783): 977. See also *European Magazine* 4 (Nov. 1783): 395.
56. *Morning Chronicle*, Jan. 10, *Morning Herald*, Jan. 13, 1784.
57. [Cooke], *Air Balloon*, 2nd ed. (1783); *London Chronicle*, Dec. 13–16, 1784.
58. *European Magazine* 4 (Dec. 1783): 406, 5 (Jan. 1784): 11.
59. For various balloon hats, see Cornu, *Galerie des modes*.
60. Samuel Johnson to Hester Thrale, Jan. 12, 1784, in Johnson, *Letters*, 4:272.
61. Fabbroni and Strange to Banks, Dec. 30, 31, 1783 (*SCBanks*, 2:247, 252).
62. Banks to Franklin, Nov. 28, Dec. 9, 1783 (*SCBanks*, 2:217, 230).
63. *European Magazine* 5 (Jan. 1784): 25–27.
64. A long extract of Faujas's book appeared in the 1783 issue; *Annual Register* (1783): 65–75. See also *New Annual Register* (1784): 154–66; Copeland, "Burke and Dodsley's Annual Register."
65. *Gentleman's Magazine* 53 (Dec. 1783): 1059, 987–89.
66. Walpole to Mann, Dec. 2, 1783, in *Horace Walpole's Correspondance*, 449–51.
67. William Cowper to John Newton, Dec. 15, 1783 in *Private Correspondence of William Cowper*, 1:294–96 (added emphasis).

68. Greig, *Beau Monde*; Berg, *Luxury and Pleasure*; Fox and Turner, *Luxury Trades*.

69. The subscription was one guinea for four persons (on foot or in a carriage) and a half-guinea for two persons (on foot) to be collected by Coghlan (bookseller at Duke St. Grosvenor Square), Booker (stationer at 56 New Bond St.), Samuel Hayes (bookseller at 332 Oxford St.), Debrett (successor to Almon, opposite Burlington House, Piccadilly), Thomas Payne (bookseller, Mew's Gate), Adams (mathematician to his Majesty, Fleet St.), William Nicoll (bookseller, 51 St. Paul's Church Yard), Nairne and Blunt (mathematical and philosophical instrument makers, 10 Cornhill, opposite the Royal Exchange), and John Hamilton Moore (mathematician, 104 Minorites); *Morning Herald*, Jan. 3, *Morning Post*, Jan. 16, *Gazetteer*, Jan. 20, 1784.

70. *Morning Chronicle*, March 1, *Morning Herald*, March 11, *Morning Post*, June 8, 1784.

71. *Gazetteer*, Feb. 17, *Morning Herald*, Feb. 23, 1784.

72. *Morning Herald*, Jan. 3, Feb. 9, May 3, 1784. Even the ailing Samuel Johnson subscribed to one of these balloons; John to Hester Maria Thrale, Jan. 31, Johnson to William Bowles, Feb. 3, Johnson to Richard Brocklesby, Aug. 21, 1784 (*Lettres*, 4:279, 281, 377).

73. [Cooke], *Air Balloon*, 4th ed. (1784), 33–35; Proudfoot, *Biographical Memoir*.

74. *Gentleman's Magazine* 54 (March 1784): 228, 245–46, 329.

75. *European Magazine* (May 1784): 395; *Gentleman's Magazine* 54 (June 1784): 433.

76. They are easily accessible now through the *British Newspaper Archive*. For provincial scientific culture, see Porter, "Science."

77. Hodgson, *History of Aeronautics*, 102; *Morning Post*, *Daily Advertiser*, Dec. 17, 1783; *St. James's Chronicle*, Jan. 31–Feb. 3, March 13–16, 1784; Lunney, "Dinwiddie"; Proudfoot, *Biographical Memoir*.

78. James Watt to Joseph Black, Sept. 25, 1783 (City Library of Birmingham Ms. 3219/4/127); Robinson and McKie, *Partners in Science*; Uglow, *Lunar Men*; Jones, *Industrial Enlightenment*.

79. Priestley to Wedgwood, Jan. 16, 1784, in Bolton, *Scientific Correspondence*, 66.

80. Elliott, *Derby Philosophers*, 69–73. There are some factual errors in this source.

81. Darwin's letters to Wedgwood and Boulton, Jan. 9–Feb. 25, 1784, in King-Hele, *Collected Letters of Erasmus Darwin*, 222–26.

82. Pradeaux to Boulton, July 30, 1784 (City Library of Birmingham Ms. 3782/12/29).

83. James Watt to James Lind, Dec. 26, 1784 (City Library of Birmingham Ms. 3219/4/56).

84. Southern, *Treatise* (1785).

85. Powell, "Scottophobia versus Jacobitism."

86. Suibhne, "Whiskey"; Mirala, "A Large Mob"; Stewart, *Deeper Silence*.

87. *Volunteers Journal*, Oct. 17, 1783.

88. *Finn's Leinster Journal*, Sept. 26, *Freemans Journal*, Nov. 10, 1783. Even the *Volunteers Journal* had begun to carry balloon news by February 4, 1784.

89. Brown, "French Scientific Innovation," 111–13.

90. MacMahon, *Ascend or Die*, 60–62; McDowell, *Trinity College*.

91. McDowell, *Ireland*, 239–92; Whelan, *Tree of Liberty*; Bartlett, *Fall and Rise of the Irish Nation*.

92. MacMahon, *Ascend or Die*, 62.
93. Hodgson, *History of Aeronautics*, 187–89.
94. *Faulkner's Dublin Journal*, Sept. 21–25, *Dublin Evening Post*, Sept. 28, 1784; Lunney, "Dinwiddie," 76–78.
95. *Freemans Journal*, Sept. 16–18, 1784; Hodgson, *History of Aeronautics*, 187–89.
96. McBride, *Scripture Politics*, 134–61.
97. Wilkins quoted from McMahon, *Ascend or Die*, 68–70.
98. British Museum online image 1868,0808.5377.
99. On the different styles for the integration of Scotland and Ireland, see Bayly, *Imperial Meridian*, 77–89; Hechter, *Internal Colonialism*.
100. Lenman, *Integration*; Colley, *Britons*. On the limitations of Colley's thesis, see Devine and Young, *Eighteenth-Century Scotland*; Phillipson, "Edinburgh."
101. Doig, Kafker, Loveland, and Trinkle, "James Tytler's Edition." On Scottish science, see Wood, "Science"; Morell, "University of Edinburgh."
102. In addition to the Lunar Society members, John Grieve wrote frequently during his Parisian sojourn, in part to claim the credit of invention for Black; Ramsay, *Life and Letters*, 75–82.
103. Tytler's own account in Lunardi, *Five Aerial Voyages* (1786), 105–14; Fergusson, *Balloon Tytler*.
104. Meek, *Biographical Sketch*; Stern, "A Salem Author."
105. Imbruglia, *Naples*; Deldonna, *Opera*; Astarita, *Salt Water and Holy Water*; Robertson, *Case for the Enlightenment*.
106. Lunardi, *Account of the First Aërial Voyage* (1784), 1–5. The ambassador moved to Paris in 1784, which left Lunardi to his own devices.
107. Green, "Balloon and Seraglio."
108. Lunardi, *Account of the First Aërial Voyage* (1784), 1–5.
109. Lunardi to Banks, June 21, 1784, in Dawson, *Banks Letters*, 558; *Gentleman's Magazine* 54 (Oct. 1784): 770–73.
110. *Morning Herald*, *Morning Post*, July 26, *Public Advertiser*, Aug. 5, *Morning Chronicle*, Aug. 6, 12, *Morning Herald*, *Public Advertiser*, Aug. 13, *Morning Chronicle*, Aug. 14, 1784. Subscriptions were taken by bookseller Deret, stationer Booker, engraver Barnes, mathematician (to his Majesty) Adams, and instrument makers Nairne and Blunt.
111. *Gentleman's Magazine* 54 (Oct. 1784): 770–73; *Universal Magazine* 75 (Sept. 1784): 161; Gardiner, *Man in the Clouds*, 29–33.
112. *Morning Post*, *Morning Chronicle*, *Morning Herald*, *Public Advertiser*, Aug. 4–7, *Gazetteer*, Aug. 9, *Morning Chronicle*, Aug. 10, 1784.
113. *St. James's Chronicle*, Aug. 11, *Morning Post*, *Public Advertiser*, Aug. 13, *Morning Herald*, Aug. 16, *Morning Post*, Aug. 31, 1784.
114. *Morning Herald*, *Public Advertiser*, *Morning Post*, Aug. 16–23, *Gazetteer*, Aug. 24, 1784.
115. *Gentleman's Magazine* 54 (Oct. 1784): 770.
116. *Public Ledger*, Sept. 2, *Morning Herald*, *Public Advertiser*, Aug. 28, 1784.
117. Lunardi, *Account of the First Aërial Voyage* (1784), 16–19.
118. *Gentleman's Magazine* 54 (Oct. 1784): 771; *Universal Magazine* (Sept. 1784): 162.
119. Banks to Blagden, Sept. 22, 1784 (*SCBanks*, 2:308).
120. *Gentleman's Magazine* 54 (Sept. 1784): 711.

121. Lunardi, *Account of the First Aërial Voyage* (1784), 26; *Universal Magazine* (Sept. 1784): 360.

122. *Gentleman's Magazine* 54 (Sept. 1784). 711.

123. *London Chronicle, Morning Chronicle, Morning Herald, Morning Post, Public Advertiser, General Evening Post, Whitehall Evening Post, St. James's Chronicle, Gazetteer,* etc. Sept. 16–18, 1784.

124. *Oxford Journal*, Sept. 18, 1784.

125. *Gentleman's Magazine* 54 (Sept. 1784): 711.

126. Tissandier, *Histoire*, 105–9; *Hibernian Magazine* (Oct. 1784): 558–60.

127. One estimate stood at 150 pounds. Although the surrounding crowd was estimated at from thirty to forty thousand, the Artillery Ground could hold only about a thousand, but very few sat in the half-guinea seats and still fewer in the guinea seats. *Oxford Journal*, Sept. 18, 1784.

128. Lunardi, *Account of the First Aërial Voyage* (1784).

129. Extracted in *Universal Magazine* 75 (Oct. 1784): 207–9.

130. *Gentleman's Magazine* 54 (Sept. 1784): 650, 711; Johnson to Reynolds, Sept. 18, 1784, in *Letters*, 4:407.

131. *Oxford Journal*, Sept. 18, *Caledonian Mercury*, Sept. 20, 1784.

132. Lunardi, *Grand Aerostatic Voyage* (1784).

133. Blagden to Banks, Oct. 14, 1784 (*SCBanks*, 2:314); *Morning Herald*, Sept. 25, *Morning Post*, Oct. 1, 1784.

134. Banks to Blagden, Sept. 22, 1784 (*SCBanks*, 2:308).

135. Vidal-Naquet and Lloyd, "Atlantis and the Nations."

136. *The Modern Atlantis* (1784), 4, 87.

137. Brewer, *Party Ideology*, 163–200; Cash, *John Wilkes*.

138. *The Modern Atlantis* (1784), 109–11.

139. *The Modern Atlantis* (1784), 11–15.

140. *The Modern Atlantis* (1784), 74–75, 91–93.

141. *The Modern Atlantis* (1784), 113–16.

142. Romani, *National Character*, 19–62, 159–200.

143. Mercier, *Tableau*, 2:704.

144. Mercier, *Tableau*, 1:61, 537.

145. Rudé, *Paris and London*; Thompson, *English Working Class*; Thompson, "Moral Economy"; Christie, *Stress and Stability*; Goodwin, *Friends of Liberty*.

146. Blagden to Banks, Oct. 14, 1783 (*SCBanks*, 2:173).

147. Brewer, *Sinews of Power*.

9: CROSSING THE CHANNEL

1. *Air-Balloon*, 24.

2. The project came to a screeching halt in 1788 due to the cost and was completed only in the 1850s; Gaudillot, *Le Voyage de Louis XVI*.

3. Deslile to Étienne Montgolfier, Dec. 6, 1783 (FM XX.21).

4. Many more can be found in Gallica.

5. *Journal de Paris*, May 29, June 2, 1784; *Mémoires secrets*, May 31, July 29, 1784.

6. *Mémoires secrets*, May 19, 1784. Blanchard had apparently attended the Collège de St. Nicaise in Rouen. See *Wer ist Blanchard* (n.d.).

7. *Annonces Normandie*, Sept. 19, Oct. 17, Nov. 7, 1783. This newspaper fulfilled the more commercial function of advertising goods, and the space for *Avis Divers*

was limited until it was supplemented by the *Journal de Normandie*, which began publication in January 1785.

8. Vergennes and Rouen authorities, April 25, 29, 1784 (AD Rouen, C910). The price of entry was three livres to visit the works at the Célestins, rue Eau-de-Robec (9–12 AM and 3–7 PM from May 18) and six livres to be in the enclosure on the day of ascension. By May 21, the tickets were sold out and the city swelled with people; *Annonces Normandie*, April 30, 1784; *Mémoires secrets*, May 19, 1784.

9. *Annonces Normandie*, May 21, 28, July 2, 1784; Coutil, *Jean-Pierre Blanchard*, 5.

10. *Annonces Normandie*, June 18, 25, 1784.

11. On Rouen theater, see Dewald, *Formation*; Élart, "Les origines."

12. *Mercure de France* complained about the excessive number of signatures (eighty-five) in four procès-verbaux. See also *Journal de Paris*, May 29, June 2, 1784; *Mémoires secrets*, May 31, July 29, 1784.

13. Blanchard's letter, Aug. 16, 1784 (TC 1.20).

14. Harkness, *Jewel House*; Ogborn, *Spaces of Mondernity*.

15. Werrett, *Fireworks*.

16. AD Seine, C910.

17. Dias, "World of Pictures."

18. *Universal Magazine* 75 (Sept. 1784): 156–62; *Gentleman's Magazine* 54 (Oct. 1784): 769–73. Earlier reports of Blanchard's ascents had appeared in provincial newspapers; *Caledonian Mercury*, March 29, *Oxford Journal*, June 26, 1784.

19. *Annonces Normandie*, April 30, June 18, 25, 1784; *Mémoires secrets*, May 19, 1784.

20. Blanchard, *Exact and Authentic Narrative of Mr. Blanchard's Third Aerial Voyage* (1784). The quote is supposedly from Bacon's *Novum Organum*.

21. Blagden to Banks, Oct. 17, 1784 (*SCBanks*, 2:315–16); McConnell, *Jesse Ramsden*.

22. Sheldon, *Journal and Certificates* (1784).

23. Hoock, *Empires of the Imagination*, 117–29; Burney, *An Account*.

24. Keen, "Balloonomania," 529–30; Cowley, *More Ways* (1784); *Balloon Jester* (1784).

25. [Alcock], *Air Balloon; or Flying Mortal*.

26. Pilon, *Aerostation* (1785).

27. *Air-Balloon* (n.d.).

28. Blake, *Island in the Moon* (written 1782–1785), 28–31.

29. *Gentleman's Magazine* 54 (Oct. 1784): 769–73, (Nov. 1784): 873–74.

30. *European Magazine* 6 (1784): 339; *Critical Review* 58 (Dec. 1784): 417.

31. Blagden to Banks, Oct. 14, 24, 1784 (*SCBanks*, 2:315, 320–21).

32. *Morning Herald*, *Whitehall Evening Post*, Jan. 3, *General Evening Post*, Jan. 4, *London Chronicle*, Jan. 6, 1784. The list would increase to twenty-one the following year; *London Chronicle*, *Whitehall Evening Post*, Jan. 4, *Public Advertiser*, Jan. 7, 1785. On Sadler's ballooning career, see Davies, *King of All Balloons*.

33. Wall, *Dissertations* (1783); Williams, Chapman, and Rowlinson, *Chemistry at Oxford*, 65–66; Ewing, *Lost World*, 57–64.

34. Wall, *Syllabus* (1782), 51–52.

35. Smith, *Man with His Head in the Clouds*; Davies, *King of All Balloons*.

36. *Gazetteer*, Feb. 24, *Morning Post*, Feb. 25, *Felix Farley's Bristol Journal*, Feb. 28, *Oxford Journal*, May 15, June 12, Sept. 11, 1784.

37. *St. James's Chronicle, Whitehall Evening Post*, Oct. 7, *Gazetteer*, Oct. 9, *Morning Herald, Gloucester Journal*, Oct. 11, 1784.
38. *Oxford Journal*, Nov. 13, 20, 1784; *Northampton Mercury*, Nov. 15; *Hibernian Magazine*, Nov. 24, *London Chronicle, Whitehall Evening Post*, Nov. 13, *Gazetteer*, Nov. 15, *Felix Farley's*, Nov. 20, 1784.
39. Pye, *Aerophorion* (1784).
40. *Oxford Journal*, Nov. 27, 1784; *Gazetteer*, Dec. 7, 1784, Jan. 3 1785; *London Chronicle*, Jan. 1, *Morning Chronicle*, Jan. 3, *Morning Post*, Jan. 4, 1785.
41. *Morning Post*, Jan. 7, 11, 1785.
42. Beddoes had been educated at Oxford (1776–1781), then in London (1781–1784) and Edinburgh (1784–1788), but it is not clear whether he was involved with Sadler's early ballooning; Jay, *Atmosphere of Heaven*; Stansfield, *Thomas Beddoes*.
43. Jeffries, *Narrative* (1786); Sorrenson, *Perfect Mechanics*.
44. Jeffries, *Narrative* (1786).
45. Blanchard, *Relation de la vingtième ascension*, 33–34.
46. *Critical Review*, 58 (Dec. 1784): 424–25.
47. *Mémoires secrets*, Oct. 22, 1784.
48. *Times*, Jan. 11, *Morning Chronicle*, Jan. 17, 1785.
49. Coutil, *Jean-Pierre Blanchard*, 9; Fontaine, *La Manche*, 129–30.
50. *London Chronicle*, Jan. 1–4, *Morning Herald, Morning Chronicle, Gazetteer*, Jan. 3, 1785.
51. *General Evening Post, London Chronicle, St. James's Chronicle, Whitehall Evening Post*, Jan. 8, *Morning Herald, Morning Post*, Jan. 10, 1785.
52. "Extract of a Letter from Dr. Jeffries," *London Chronicle*, Jan. 11, 1785.
53. Duchosal, *Blanchard, poëme* (1986).
54. Breteuil's letter, Jan. 27, 1785 (TC 2.4–5).
55. Jeffries, *Narrative* (1786).
56. Jeffries to Banks, Jan. 13, 1785 (*SCBanks*, 3:6–9).
57. Jeffries, *Narrative* (1786), 76.
58. Hunn, "Balloon Craze," 401–2.
59. Blanchard, *Relation de la vingtième ascension*, 33–34.
60. *Air-Balloon* (n.d.), 11, 14.
61. *European Magazine* 7 (1785): 149; *Ballooniad* (1785).
62. George Ogle, William Caldbeck, William Downes, Oliver Carlton, Revd Dr Ussher, Richard Cuthbert, Revd Mr Ledwich, John Macauley, William B. Dunn, Mr. Crosthwaite, Lord Charlement, and Duke of Leinster. Dublin police materialized in September 1786; Palmer, *Police and Protest*, 92–162.
63. MacMahon, *Ascend or Die*, 92–93.
64. *Times*, Jan. 26, 1785.
65. *Aerial Voyage* (1785), vi.
66. *Hibernian Magazine* (Jan. 1785): 1–3; "Advertisement," dated Jan. 24, 1785, in *Aerial Voyage* (1785); MacMahon, *Ascend or Die*, 95.
67. Potain, *Relation aérostatique* (1824).
68. *Freemans Journal*, May 18, 23, 1785.
69. Potain, *Relation aérostatique* (1824). Newenham spent a number of years on the continent and returned in summer 1783 for the new parliamentary election, missing the Parisian ascents.

70. Wecter, "Benjamin Franklin," 229; Coyle, "Sir Edward Newenham"; Kelly, *Sir Edward Newenham*.
71. *Public Advertiser*, July 21, 1785.
72. *Times*, May 9, 1785.
73. Hudson, "Samuel Johnson"; Ogborn, *Spaces of Modernity*.
74. Jeffries, *Narrative*, 29–30.
75. *Thoughts on the Farther Improvement of Aerostation* (1785); Southern, *Treatise* (1785); *Critical Review* 59 (1785): 340.
76. *Monthly Review* 73 (1785): 262, in Keen, "Balloonomania," 515.
77. *Times*, May 4, 1785.
78. *Morning Chronicle*, Nov. 11, *Morning Herald*, Dec. 24, 1784.
79. Wilson, "Empire, Trade and Popular Politics."
80. Blagden to Banks, Nov. 6, 1784 (*SCBanks*, 2:328); *Morning Herald*, Jan. 20, *General Advertiser*, Jan. 20, 1785.
81. *Gazetteer*, Dec. 11, 1784; *Morning Post*, Feb. 9, 1785.
82. *Morning Post*, March 26, *Public Advertiser*, April 1, 1785.
83. *British Balloon* (1785); *Town and Country Magazine* 17 (March 1785): 116–17; *Morning Post*, March 19, 30, 1785.
84. *General Evening Post*, *Whitehall Evening Post*, and *Morning Post*, March 24, 1785.
85. *Morning Chronicle*, *Morning Herald*, *Morning Post*, March 25, *General Evening Post*, *Whitehall Evening Post*, *London Chronicle*, March 26, 1785.
86. *Times*, *Morning Post*, *Public Advertiser*, May 4, 1785.
87. *Gentleman's Magazine* 55 (May 1785): 481.
88. *Times*, May 14, 1785.
89. Sage, *Letter*, 4.
90. *Gentleman's Magazine* 55 (May 1785): 481.
91. *Mr. Lunardi's Account* (1785); Lunardi, *Five Aerial Voyages* (1786). His dedications to the Earl of Orford and the Duke and Duchess of Buccleugh, respectively, indicate aristocratic patronage. For a defense of Lunardi over Blanchard, see *Morning Chronicle*, Jan. 6, *Morning Post*, Jan. 7, 1786.
92. Brockliss, *French Higher Education*.
93. Conlin, "Afterlife," 722; Ogborn, *Spaces of Modernity*, 116–57; Edelstein, *Vauxhall Gardens*; Coke and Borg, *Vauxhall Gardens*.
94. Brewer, "Most Polite Age"; Bermingham, "Elegant Females."
95. George Walpole to Lord Townsend, April 4, 1785 (NSAM archive, xxxx-0995).
96. *Times*, May 9, 1785.
97. *Times*, May 4, 23, 1785.
98. *Morning Post*, June 3, *General Evening Post*, June 4, 1785.
99. *Public Advertiser*, June 4, *Morning Chronicle*, June 10, 1785.
100. *Morning Herald*, *Whitehall Evening Post*, June 6, 1785.
101. *Morning Post*, June 4, 7, 1785.
102. *Morning Post*, June 9, 10, 13, *Morning Herald*, June 10, 14, 15, *General Advertiser*, June 14, *Gazetteer*, *Morning Chronicle*, *Public Advertiser*, June 15, 1785.
103. *General Evening Post*, *General Advertiser*, *Morning Post*, *Gazetteer*, June 16–17, 1785.
104. *Morning Post*, June 20–21, 1785.

105. *European Magazine* 8 (1785): 234, 476. Also see *General Advertiser*, June 25, *Morning Post*, June 27, *Gazetteer*, June 27, 29, *General Evening Post*, Aug. 13–16, 1785.

106. Fontaine, *La Manche*, 119–50; *General Evening Post*, June 20, 1785.

107. *Morning Post*, June 1, *Morning Chronicle*, June 13, *General Evening Post*, June 21, 1785.

108. *Journal de France*, June 18, *General Evening Post*, June 20, *Whitehall Evening Post*, June 18, *Morning Chronicle*, June 20, *Morning Post*, June 21, 1785.

109. [Marat], *Lettres de l'Observateur Bon-sens* (1785), 5. This collection contains six letters dated between June 22 and July 1, 1785. See also *Journal de Normandie*, July 2, 1785.

110. [Marat], *Lettres de l'Observateur Bon-sens* (1785), 22-23.

111. [Marat], *Lettres de l'Observateur Bon-sens* (1785), 18.

112. For an early design (dated Sept. 27, 1783), see *Description de deux machines* (n.d.).

113. "Procès-verbal sur la cause de l'accident arrivé a M. Pilâtre du Rosier," in Blanchard, *Relation de la vingtième ascension*.

114. [Marat], *Lettres de l'Observateur Bon-sens* (1785).

115. Marat, *Les Charlatans modernes* (1791).

116. Hodgson, *History*, 145–48. These ascents were broadly reported in *General Evening Post*, *Whitehall Evening Post*, *Morning Chronicle*, and *Morning Herald*.

117. *Morning Post*, *Public Advertiser*, June 4, *Whitehall Evening Post*, *London Recorder or Sunday Gazette*, June 6, 1785.

118. Rigby, *Account* (1785).

119. *St. James's Chronicle*, *Whitehall Evening Post*, July 26, *Morning Chronicle*, *Whitehall Evening Post*, July 28, *Morning Herald*, *Public Advertiser*, July 29–30, *General Evening Post*, *Whitehall Evening Post*, Oct. 1, *Morning Herald*, Oct. 21, 1785. The Zambeccari balloon was on sale by October. See the *Morning Post*, Oct. 26, 1785.

120. Walpole to Mann, May 13, 1785, in *Horace Walpole's Correspondance*, 578–79.

121. Cavallo, *Theory* (1785); Southern, *Treatise* (1785); *Thoughts on the Farther Improvement of Aerostation* (1785).

122. Clifford, *Routes*, 7, 9.

10: A LIMINAL GEOGRAPHY

1. Harris, *Diaries*, 2:68–73.

2. German reports in 1783–1784 include Kramp, *Geschichte*; Lichtenberg, "Ueber die neuerlich"; Lichtenberg, "Vermischte Gedanken"; Lichtenberg, "Kurze Geschichte"; Wieland, "Die Äropetomanie"; Watermeyer, "Erläuterung"; Faujas, "Beschreibung"; Ehrmann, *Montgolfier'sche Luftkörper*; Müller, *Etwas zur Erklärung*.

3. Schmidt, *What Is Enlightenment?*

4. *Gazette des deux-ponts*, Sept. 17, 1788 (TC 1.21).

5. Levinger, *Enlightened Nationalism*.

6. Harris, *Diaries*, 2:102–6.

7. *Hollandsche Historische Courant*, *Rotterdamse Courant*, June 23, *Zuid-Hollandsche Courant*, June 24, *Leeuwarder Courant*, June 25, *Zuid-Hollandsche Courant*, June 27, *Middelburgsche Courant*, June 28, 1785.

8. *Amsterdamse Courant*, July 7, *Zuid-Hollandsche Courant*, July 8, 1785.

9. Schama, *Patriots*, 100–5; Murphy, *Charles Gravier*, 405–16; Prak, "Citizen Radicalism."

10. Harris, *Diaries*, 2:66, 78.
11. Whitney, *Jean Ternant*, 115–18.
12. Murphy, *Charles Gravier*, 459–72; Hardman and Price, *Louis XVI*, 332–73.
13. Te Brake, "Popular Politics"; Te Brake, "Violence."
14. The author is indebted to Sjoukje Atema at The Hague city archive for the information on digitized Dutch newspapers at http://kranten.kb.nl/. The timely online appearance of Rachel Lensink's thesis has also significantly enhanced the author's understanding of the Dutch scene.
15. Jean Nicolas Sebastien Allamand (Leiden) to Banks, Nov. 8, 1783, in Dawson, *Banks Letters*, 15.
16. Liender to Étienne Montgolfier, Nov. 3, 1783 (FM XX.47).
17. Noorden, *Korte Verhandeling* (1784).
18. *Nederlandsche Courant*, Dec. 1, *Hollandsche Historische Courant*, Dec. 13, *Oprechte Haarlemsche Courant*, Dec. 13, *Leeuwarder Courant*, Dec. 13, *Amsterdamse Courant*, Dec. 20, 1783. On Diller, see Tomory, *Progressive Enlightenment*, 30–33; *General Evening Post*, Dec. 23, 1783.
19. According to Lensink, "Science & Spectacle," 50–53. See also *Hollandsche Historische Courant*, Oct. 30, Dec. 5, 11, 1783, Jan. 31, Feb. 5, 1784; Berkhey, *Op de roekeloze proefnemingen*.
20. *Hollandsche Courant*, Dec. 2, 1783; *Natuurlyke Historie* (1761–1785); Sprunger, "Frans Houttuyn."
21. Roberts, "Arcadian Apparatus"; Jacob and Mijnhardt, *Dutch Republic*; Mitchell, *Landscape and Power*.
22. Schama, *Patriots*, 86.
23. Lensink, "Science & Spectacle," 38, 46, 53–56; *Nederlandsche Courant*, August 13, 1784.
24. *Groninger Courant*, July 19, *Whitefall Evening Post*, July 16, *St. James's Chronicle*, *Public Advertiser*, July 21, 1785.
25. *General Evening Post, Morning Chronicle, Morning Herald, Public Advertiser*, Aug. 10, *Whitehall Evening Post*, Aug. 6–9, *Journal de Paris*, Aug. 18, 1785.
26. Lensink, "Science & Spectacle," 48; *Oprechte Haerlemsche Courant*, July 14, *Middelburgsche Courant*, July 16, 19, *Hollandsche Historische Courant*, July 14, 19, *Nederlandsche Courant*, July 15, *Zuid-Hollandsche Courant*, July 15, *Leeuwarder Courant*, July 16, 20, *Whitehall Evening Post, Public Advertiser, Morning Post, Morning Herald and Daily Advertiser, Morning Chronicle*, July 16–22, 1785.
27. *Whitehall Evening Post, General Advertiser, Morning Chronicle, Public Advertiser*, July 2, 1785.
28. *New London Magazine* (July 1785): 47–48.
29. *General Evening Post*, July 16, *Morning Herald*, July 21, 22, 1785.
30. *Groninger Courant*, July 19, 22, *Hollandsche Historische Courant*, July 21, *Rotterdamsche Courant*, July 19, 28, 1785; Lensink, "Science & Spectacle," 71.
31. Blanchard, *Prospectus de la 14e experience* (1785).
32. He apparently lost sixty-five hundred livres in The Hague, while breaking even at Rotterdam; Blanchard, *Relation de la vingtième ascension* (1786), 7.
33. *General Evening Post*, Aug.13–16, 1785; Breteuil to Blanchard, Jan. 27, 1785 (TC 2.4–5). Calais notables would organize a dedication ceremony on the anniversary of the Channel crossing; Debove, *Blanchard*, 211–15.

34. Bossenga, *Politics of Privilege*, 17–21; Bosher, *Single Duty Project*, 1–7; Reddy, *Rise of Market Culture*.

35. Debièvre, *Notes*, 3–4.

36. Lottin, *Lille*; Trénard, *Histoire d'une métropole*; Duplessis, *Lille*; Braure, *Lille*.

37. Esmangart wrote on May 2, and Desjobert of Paris police on May 4, 1784 (AM Lille 703.13). For the general impact of the ordinance, see Thébaud-Sorger, *L'Aérostation*, 164–66.

38. *Revue savoisienne* 29 (1888): 304; *Feuilles de Flandres*, Dec. 26, 1783, March 16, 1784. For a longer biography, see Pâris de l'Épinard, *Mon retour* (n.d.).

39. "Lettre de M. Blanchard . . . Aug. 2, 1785," Supplement au No. 4 des *Feuilles de Flandres* (1785): 25–27.

40. Burke, *Philosophical Enquiry* (1757); Stormer, "Addressing the Sublime."

41. Blanchard to Lille authorities, Aug. 11, 1785 (AM Lille 703.20).

42. *Registres aux Résolutions des MM. du magistrat de la ville de Lille*, Aug. 12, 1785 (AM Lille, 703.20).

43. Blanchard, *Prospectus de la 14e expérience* (1785).

44. Debièvre, *Notes*, 13.

45. "Prospectus de l'expérience de M. Blanchard," approved by the mayor, Aug. 12, 1785; "List du villes auxquelles on a écrit le 13 aout 1785" (AM Lille 703.20).

46. "Réglement de Police" (AM Lille, 703.20); Bossenga, *Politics of Privilege*, 91.

47. Maës, *Les Watteau*, 36, 239–41.

48. *General Evening Post, London Chronicle, St. James's Chronicle, Whitehall Evening Post, Morning Post*, all Sept. 6–8, 1785.

49. "A Lille. Suite des Notes" (AM Lille, 703.13).

50. Letters of Blanchard and l'Épinard, Aug. 30, 1785; *Extrait des Régistres aux Résolutions des M.M. des magistrate de la ville de Lille*, Aug. 30, 1785; Blanchard to Lille Magistrates, Sept. 17, 1785 (AM Lille, 703.20).

51. *General Evening Post*, Sept. 9, 1785.

52. *Lettre à Messieurs Blanchard* (1785).

53. *Seconde Lettre* à Messieurs Blanchard (1785), 8–9.

54. François-Joseph Plancq (1736–1795) submitted a project to the city authorities on November 16, 1785, which was examined by Nicolas Joseph Saladin (1733–1829); Debièvre, *Notes*, 22–24.

55. Bossenga, *Politics of Privilege*, 17; Thibaut, "Les Voies navigables."

56. Lamblin, "L'apothicaire lillois Carette"; Carette, "Dissertation" (1785); Saladin, "Mémoire" (1784).

57. Blanchard, *Relation de la vingtième ascension* (1786), 29.

58. Soliday, *Community in Conflict*.

59. *Journal de Paris*, Oct. 1, 7, 1785; *Mémoires secrets*, Oct. 8, 1785.

60. Blanchard, *Relation du quinzième voyage* (1785), 11–12; *Morning Herald, Morning Post, Public Advertiser*, Oct. 17, 1785.

61. *General Evening Post*, Oct. 25, *Whitehall Evening Post*, Oct. 22–25, *General Advertiser*, Oct. 24, *Morning Herald and Daily Advertiser*, Oct. 24, *Journal de Paris*, Oct. 31, 1785.

62. Blanchard, *Relation du quinzième voyage* (1785).

63. Blanchard, *Relation du seizième voyage* (1786); *Journal de Paris*, Dec. 5, 1785; *Seconde Lettre* à Messieurs Blanchard (1785), 6.

64. *Ordonnance de l'empereur* (1786).

65. *Journal de Paris*, June 5, 1786; *Mémoires secrets*, June 6, 19, 1786.
66. Blanchard, *Relation de la vingtième ascension* (1786), 2–6.
67. 111-1_CI VII_Lit Fl_No. 11_Vol. 2, Staatsarchiv Hamburg.
68. Quoted in Lindemann, *Patriots and Paupers*, 3. See also Lindemann, *Merchant Republics*.
69. Blanchard, *Relation de la vingtième ascension* (1786).
70. Although he stated August 10 as the date, the abbé Kentzinger's letter on July 17 states that Blanchard had already inspected and approved the site of the ascent, which was later changed to the Sternschanze; Blanchard, *Relation de la vingtième ascension* (1786); 111-1_CI VII_Lit Fl_No 11_ Vol 2, Staatsarchiv Hamburg.
71. Lindemann, *Liaisons dangereuses*, esp. 72–76.
72. Hess was the author of *Hamburg, topograpisch, politisch, und historisch beschrieben* (1787); Jenkins, *Provincial Modernity*, 15–16.
73. *Journal de Paris*, Sept. 18, 1786. This must have been either Johann Gottfried Misler or Johann Hartmann Misler.
74. 111-1_CI VII_Lit Fl_No 11_Vol 2, Staatsarchiv Hamburg.
75. Blanchard, *Relation de la vingtième ascension* (1786). On the comtesse de Bentinck, misspelled as Beinteinkc, see Le Blond, *Charlotte Sophie*.
76. Hohendahl, *Patriotism*; Whaley, *Religious Toleration*, 148–51; Jenkins, *Provincial Modernity*, 26–30.
77. One anonymous defense in German was entitled *Wer ist Blanchard* (n.d.).
78. Leckey, *Patrons of Enlightenment*, 124–32; Billington, *Icon and the Axe*.
79. Blanchard to the comte d'Escherny, Sept. 18, 1786 (TC 1.21).
80. D'Escherny and Joseph II to Blanchard, Oct. 4, Nov. 2, 1786 (TC 1.21).
81. Standage, *The Turk*; Schaffer, "Enlightened Automata."
82. *Monthly Review* (May 1784): 408; *Essex Journal*, Sept. 3, 1784.
83. *Gazette littéraire de Berlin*, March 18, 1784.
84. Schmidt, *What Is Enlightenment?* 49–52, 420–44.
85. Mirabeau, *Histoire secréte* (1789); Johnston, "Mirabeau's Secret Mission"; Welschinger, *La mission*.
86. William's undated refusal, which cites Blanchard's letter of Oct. 23, was published in *European Magazine* (Dec. 1786).
87. Bruhns, *Life of Alexander von Humboldt*, 1:60.
88. Über die bewundernswürdige Luftfahrt (n.d.).
89. *Journal de Paris*, Dec. 29, 1786.
90. *Mémoires secrets*, March 27, Oct. 27, 1787; *Journal de Paris*, July 15, 1787; circular of the ascent from Valenciennes using four balloons, dated March 6, 1787 (TC 1.21); Debove, *Blanchard*, 228–35.
91. Blanchard, *Prospectus de la vingt-sixième Expérience* (1787).
92. Thébaud-Sorger, *L'Aérostation*, 158–64 (AM Strasbourg, Serie AA No. 2053).
93. *Avis concernant la 26e éxperience aérostatique de Mr. Blanchard*, Aug. 13, 1787; Blanchard's letter, Oct. 8, 1787 (AM Strasbourg, Serie AA No. 2053, dossier "Aérostat").
94. Ford, *Strasbourg*, 235–64.
95. Blanchard, "Description detaillée au 28ème voyage (1787)," 4.
96. Blanchard, "Description detaillée au 28ème voyage (1787)." For the 1788 ascents, see *Journal de Paris*, May 6 (Basel), May 25 (Mulhouse), Aug. 29 (Brunswick), 1788; Debove, *Blanchard*, 235–72.

97. Crouch, *Eagle Aloft*, 102–18; Leary, "Phaeton in Philadelphia"; Pethers, "Balloon Madness."

EPILOGUE: Revolutionary Metamorphoses

1. Marat, *Chains of Slavery* (1774). Although this work was written in the English political context, the rhetoric against slavery had developed in French political discourse ranging from Montesquieu to Rousseau; Hammerseley, "Jean-Paul Marat's *The Chains of Slavery*." On Marat, see Germani, *Jean-Paul Marat*; Brian, *La Mesure de l'État*. On the crosscurrent of ministerial despotism between Britain and France, see Conlin, "Wilkes."
2. [Saunier], *Le Triomphe* (1784), 8.
3. For a thoughtfully constructed sequence of revolutionary events that synthesizes many historiographical strands, see Jones, *Great Nation*, 395–580.
4. Sieyès, *Political Writings*, 95, 156. The essay's impact is divergently assessed in Sewell, *Rhetoric of Bourgeois Revolution*; Margerison, *Pamphlets*, 92–107. On exclusion as a powerful motive for revolutionary action, see Sonenscher, *Sans-Culottes*.
5. Condorcet, *Pluralité des voix* (1785); Condorcet, *Essai sur la constitution* (1788); Baker, *Condorcet*; Urbinati, "Condorcet's Democratic Theory"; McLean and Hewitt, *Condorcet*; Williams, *Condorcet and Modernity*; Thompson, *Popular Sovereignty*.
6. Rudé, *Crowd*, 34. For a comprehensive critique of the revisionist depoliticization of the *menu peuple*, see Comninel, "Political Context."
7. Mercier, *Memoirs* (1772), 1:43; Linguet, *Mémoires sur Bastille* (1783).
8. Schama, *Citizens*, 326–31; Godechot, *Taking of the Bastille*, 204–46.
9. The incident took place on August 10, 1792; Hunn, "Balloon Craze," 148-49.
10. Margerison, *Pamphlets*.
11. Lever, *Philippe Égalité*.
12. Ozouf, *Festivals*, 42–49; Margerison, "History"; Nora, *Lieux de mémoire*.
13. Schama, *Citizens*, 502.
14. Tackett, *Becoming a Revolutionary*, 297–301.
15. Hanson, "Monarchist Clubs"; Higonnet, *Goodness beyond Virtue*; Hanson, *Jacobin Republic*; Hanson, *Contesting the French Revolution*.
16. "Fête aérostatique (1790)."
17. Baecque, *Body Politic*, 251–52. It was prepared by André-Jacques Garnerin (1770–1823).
18. Andress, *Massacre*.
19. L'Allemand de Saint-Crox, "Procès-verbal très intéressant du voyage aérien."
20. Hunn, "Balloon Craze," 421–49.
21. Hunt, *Politics*, 1–15. I am labeling Hunt's own approach as performative for paying attention to the reflux of ideology and practice, along with Tackett's *Coming of the Terror* for its focus on the process of the revolution.
22. Friedland, *Seeing Justice Done*, 218–65; Outram, *Body and the French Revolution*, 106–23.
23. Arasse, *Guillotine*; Gerould, *Guillotine*.
24. Darnton, *Devil in the Holy Water*, 344–59.
25. Joseph Pâris de l'Épinard, "Humanity Trampled Upon," in *Reign of Terror*, 1:218–60.
26. Turner, *Dramas*, 13.
27. For the transitional process, see Tackett, *Coming of the Terror*, 121–41.

28. For the use of Foucauldian genealogy, see Kim, *Affinity*; Kim, "Archeology." For a full repertoire of the social technologies that provided continuity with the Old Regime, see Tocqueville, *Old Regime*. Literary technologies are shown in pamphlet and journal literature.

29. Ozouf, *Festivals*. See also Farge, *Fragile Lives*; Farge, *Subversive Words*; Roche, *People of Paris*.

30. Crosland, *Society of Arcueil*; Fox, *The Savant and the State*; Gillispie, *Science and Polity*.

31. On the pros and cons of this "revolution of the mind" and the need to combine Enlightenment intellectual legacy with revolutionary contingencies, see Tackett, *Becoming a Revolutionary*, 48–76. For a recent example of the ideological approach, see Edelstein, *Terror of Natural Right*.

32. Schnapp, "Fascist Mass Spectacle."

33. Jones and Wahrman, *Age of Cultural Revolutions*.

34. Kim, *Affinity*; Kim, "Archeology."

35. Tocqueville, *Old Regime*; Furet, *Interpreting*.

36. Hunt, *Politics*; Sutherland, *French Revolution and Empire*.

37. Walton, *Policing Public Opinion*.

38. For the difficulties of writing a Foucauldian French Revolution, see Chartier, "Chimera of the Origin"; Baker, "Foucauldian French Revolution?"

39. Alder, *Engineering the Revolution*; Alder, *Measure of All Things*.

40. Ozouf, *Festivals*; Mosse, *Nationalization*; Hazareesingh, "Common Sentiment"; Levinger, *Enlightened Nationalism*.

41. Greenhalgh, *Ephemeral Vistas*, 3–6; Rydell, *World's a Fair*.

42. Jainchill, *Reimagining Politics after the Terror*; Harrison, "Planting Gardens"; Pyenson, *Civilizing Mission*; Conklin, *Mission to Civilize*.

GLOSSARY

bourgeois: an urban resident with accompanying legal rights.

Charlière: a hydrogen balloon, made after Charles's method.

écu: a silver coin equal to six livres.

the fashionable society: a translation of *le monde*, Parisian high society formed around the court society and nobility, yet plastic enough to include other notable figures in their social circles. In England, a similar formation was called *beau monde*.

gallery: in ballooning, this was the most often used name (both in French and English) for the gondola that carried the aeronauts.

league: translation of *lieue*, the unit of distance equal to 2,000 toises.

livre: a unit of French money and weight. For the latter, it is translated as pound.

louis (d'or): a gold coin, equal to 24 livres.

mécanicien: a machinist in the context of ballooning, seen as inferior to a physicien.

Montgolfière: a hot-air balloon, made after the Montgolfiers' method.

natural philosophy: translation of *la physique*, which initially meant a systematic, causal study of nature (e.g., Aristotelian or Cartesian). In the eighteenth-century, it came to encompass other branches of knowledge such as chemistry and medicine in so far as they related to the truth of nature.

parlement: the judicial court that had the privilege of *remonstrance*, or the right to refuse registering royal edicts, which was used as the primary means of opposition to the Crown throughout the eighteenth century.

parterre: the ground part of the theater where people stood to watch the show.

patrie: fatherland or homeland, the region to which one had natural allegiance. Since the concept derived from classical republican literature, the regional unit was normally a province rather than the nation, although the boundary often became blurred. *L'amour de la patrie*, often translated as patriotism, carried an ambiguous reference to either a provincial region or to the nation.

populace: translation of *peuple*, which mostly referred in eighteenth-century usage to the illiterate population with no traditional privileges, often derogatively called *menu peuple*.

philosophe: philosopher, a term that in eighteenth-century usage referred initially to those who were associated with Diderot's *Encyclopédie* or advocated similar ideas. By the 1780s, according to the *Mémoires secrets*, contemporaries identified three types of philosophes — encyclopédists, patriots, and economists (Physiocrats).

physicien: natural philosopher, or an expert in *la physique*. In the context of ballooning, it often designated those who could command the new gases.

politique: a person inclined to engage in political debates or maneuvers.

président: was a venal magistrate at the parlement, which was inherited. In most other cases (e.g., academies) where it referred to the person in charge of proceedings, it is translated as president.

procès-verbal (or verbaux in the plural): a brief summary of an event or a meeting.

the public good: translation of *le bien public*.

salon: referred in the eighteenth century to the annual exhibition of paintings at the Palais des Tuileries. Historians have used the word to designate the soirées where witty conversation was the main purpose.

savant: scholar or scientist (avant la lettre).

sous (or **sol**, in copper coin): a unit of French money equal to 1/20th of a livre or 12 deniers.

spectacle: in eighteenth-century French usage referred most often to theatrical performances, although it also referred to other kinds of performance.

toise: a unit of distance, equal to six feet (about 1.9 meters).

BIBLIOGRAPHY

MAJOR AERONAUTICAL COLLECTIONS

Bibliothèque nationale de France, Paris

In addition to the numerous pamphlets and published literature housed at the Tolbiac location, the library holds a large collection of printed illustrations in the Estamps collection at the Richelieu location. Many are now digitized and easily available from Gallica.bnf.fr.

British Museum

This institution has many objects and images that are available online.

Deutsches Museum, München

This outstanding science museum holds an extensive collection, judging from Elske Neidhardt-Jensen and Ernst H. Berninger, *Katalog der Ballonhistorischen Sammlung*.

Gimbel Collection, McDermott Library, U.S. Air Force Academy, Colorado Springs

This is the most coherent collection on early ballooning conveniently housed in one room, which provides an excellent overview. The library also has an extensive range of printed images that are scanned (with low resolution), but they are not yet freely available online. An excellent introduction with many images, *The Genesis of Flight*, was published by a team of specialists working with Tom D. Crouch.

Huntington Library, Pasadena

This beautiful library holds an extensive collection of "aeronautica" (about nine hundred items) that can be searched through its online catalogue, which includes a large collection of images. The pamphlet collection includes French, Italian, and German publications.

Musée de l'Air et de l'Espace, Bourget, France

In addition to the Fonds Montgolfier (FM), the main bulk of the family's balloon correspondence, this facility near the Charles de Gaulle Airport contains records on the entire history of aviation that matches its museum collection. One can attain a glimpse of its eighteenth-century image collection through their publication, *Le temps des ballons: Art et Histoire*.

National Aerospace Library, Farnborough, U.K.

This is a comprehensive library that traces the entire history of flight. The Cuthbert-Hodgson Collection on ballooning contains most of the newspaper cuttings used by John Edmund Hodgson in his book *History of Aeronautics in Great Britain*, although their utility has been eclipsed by digitized databases such as the British Newspaper Archive and the Burney Newspaper Collection. Many images are available from their website.

National Air and Space Museum, Smithsonian Institution, Washington, D.C.

In addition to a large collection of pamphlets, publications, and images that are housed at the Smithsonian Library in downtown, this museum maintains an extensive archive of rare documents and objects at their airport facility. Many of their balloon images are available online through the Smithsonian website with detailed commentaries prepared by their aeronautic specialist Tom D. Crouch. Some of the items are listed in the exhibition catalog, *Ballooning 1782–1972*, prepared by Roger Pineau.

Tissandier Collection, Library of Congress, Washington, DC

This extensive collection of documents and printed material by Gaston Tissandier, purchased by the Library of Congress in the 1930s, has remained mostly forgotten and unused even by balloon historians. A large number of precious original documents still remain in boxes without an itemized catalogue. Many of the images have been digitized and are available through the online catalogue.

PRIMARY SOURCES.

Periodicals and Unattributed Sources

The Aerial Voyage, A Poem inscribed to Richard Crosbie, Esq. Dublin: Marchbank, 1785.
Affiches, Annonces et Avis divers de Dijon, ou Journal de la Bourgogne.
Affiches, Annonces, et Avis divers, ou Journal général de France.
Air-Balloon, or Blanchard's Triumphal Entry into the Etherial World: A Poem. N.d. [ECCO]
Almanach des ballons, ou globes aérostatiques. Annonay and Paris: Langlois, 1784. [BNF Tolbiac Vz 2147]
Annales politiques, civiles et littéraires.
Annonces, Affiches et Avis Divers de la Haute et Basse Normandie.
The Annual Register, or a View of the History, Politics, and Literature for the Year 1783. London: J. Dodsley, 1785.

Le Ballon, ou la Physicomanie, comédie en 1 acte et en vers. Paris: Cailleau, 1783. [BNF Tolbiac 8 YTH 1664]

The Ballooniad. In Two Cantos. Birmingham, 1785.

British Balloon, which ascended from Tottenham court road the 23rd of March last, with its constructor, Count Zambeccari. London, 1785.

Clovis au premier champ de mars, ou origine, établissement et révolutions des loix fondamentales de la monarchie françoise, par M. L. J., avocat. 1789. [Maclure Collection, University of Pennsylvania]

Considérations sur le globe aérostatique, par M. D. Paris: Le Jay, 1783. [BNF Tolbiac Vz 2115]

Correspondance littéraire, philosophique et critique par Grimm, Diderot, Raynal, Meister, etc. Edited by Maurice Tourneux. 16 vols. Paris: Garnier Frères, 1877–1882.

Correspondance secrète, politique et littéraire, ou Mémoires pour servir à l'Histoire des Cours, des Sociétés, & de la Littérature en France, depuis la mort de Louis XV. Attributed to François Métra. 18 vols. London: J. Adamson, 1787–1790.

Le courrier historique, politique, litteraire, gallant et moral d'Avignon.

Description de deux machines propres à la navigation aérienne, avec figures; par M. B. . . . . [Paris, 1783]. [BNF Tolbiac Vz 2141]

Description de la seconde expérience aérostatique faite à Nantes, le 6 Septembre 1784, sous la direction de M. Lévêque, correspondant de l'Académie Royale des Sciences, Professeur Royal d'Hydrographie & de Mathematiques. Nantes: Brun, n.d. [BNF Tolbiac Vz 2118]

Dissertation sur les aérostats des anciens et des modernes, par A.G. RO. Geneva: Libraires des nouveautés, 1784. [BNF Tolbiac Vz 2154]

"Fête aérostatique, qui sera célébrée aujourd'hui au champ de Mars." [BNF Tolbiac Mfiche LB39–3763]

Gazetteer and New Daily Advertiser.

The Hibernian Magazine, or Compendium of Entertaining Knowledge.

Histoire du Ballon de Lyon, suivie d'une autre pièce non moins piquant. 1784. [BNF Tolbiac Vz 2117bis]

Histoire intéressante d'un nouveau voyage à la lune, et de la descente à Paris d'une jolie dame de cette terre etrangère. Whiteland and Paris: F. G. Deschamps, 1784. [BNF Tolbiac Vz 1850]

Journal de Lyon, ou Annonces et Variétés littéraires, pour servir de suite aux Petites Affiches de Lyon.

Lettre à M. de ***. Sur son Projet de voyager avec la Sphère Aërostatique de M. de Montgolfier. Aeropolis et Paris : marchands de Feuilles volantes, 1783. [BNF Tolbiac Vz 2136]

Lettre à Messieurs Blanchard & le Chevalier l'Épinard, sur leur Voyage Aérien, commencé à Lille en Flandre le 26 Août 1785, sur les onze heures du matin, & terminé à six heures précises du même jour, au Village de Servan dans le Clermontois en Champagne. N.d. [BNF Tolbiac Vz 2111]

Lettre à Mr. M. de Saint-Just, sur le Globe aërostatique de MM. Montgolfier, et sur la révolution que cette découverte peut produire dans les Sciences & dans les Arts. Amsterdam and Paris: Mérigot l'aîné and Royez, 1784. [BNF Tolbiac Vz 2108]

Lettre à un ami, sur l'utilité des globes volans de M. de Montgolfier, & sur la possibilité de la prise de Gibraltar. Amsterdam and Paris: Gueffier, 1783. [BNF Tolbiac Vp 5353]

Lettre au Rédacteur du Procès verbal de l'expérience Aérostatique faite à Nantes le 14 Juin 1784. [BNF Tolbiac Vz 649]

Lunardi's Grand Aerostatic Voyage through the Air, containing a complete and circumstantial Account of the Grand Aerial Flight made by that enterprising Foreigner, in his Air Balloon, on September 15, 1784, from the Time of his being launched in it, in the Artillery-Ground, to his descending with it, on the same Day, near Ware, in Hertfordshire; together with a Variety of Particularism during his Abode in that truly wonderful and magnificent Machine, that have never yet transpired. London: J. Bew, 1784. [ECCO]

Mémoires secrets pour servir à l'histoire de la République des lettres en France, depuis MDCCLXII jusqu'à nos jours; ou Journal d'un observateur, contenant les Analyses des Pièces de Théâtre qui ont paru durant cet intervalle; les Relations des Assemblées Littéraires; les notices des Livres nouveaux, clandestins, prohibés; les Pieces fugitives, rares ou manuscrites, en prose ou en vers; les Vaudevilles sur la Cour; les Anecdotes & Bons Mots; les Eloges des Savans, des Artistes, des Hommes de Lettres morts, &c.&c.&c. Attributed to Louis Petit de Bachaumont. 36 vols. London: J. Adamson, 1780–1789.

Mesmer Guéri, ou Lettre d'un provincial au R.P.N***, en réponse à sa lettre intitulé, Mesmer blessé. London and Paris: Marchands des nouveautés, 1784.

The Modern Atlantis: or, the Devil in an Air Balloon. London: G. Kearsley, 1784.

Morning Chronicle and London Advertiser.

Morning Herald and Daily Advertiser.

Morning Post and Daily Advertiser.

Le mouton, le canard, et le coq. Fable dialoguée par M. C***. Bruxelles and Paris: Hardouin, 1783. [BNF Tolbiac Vz 2139]

The New Annual Register or General Repository of History, Politics, and Literature, for the Year 1785. London: G. G. J. and J. Robinson, 1786.

Observations sur le Rapport fait à l'Académie des Sciences, Belles-Lettres & Arts de Lyon, le 19 Mars, à l'occasion de l'Expérience de l'Aérostat de M. Joseph de Montgolfier, faite aux Broteaux, près de cette Ville, le 19 Janvier, 1784.

Ordonnance de l'empereur, concernant les Ballons, ou Machines aérostatiques à feu, nommées Montgolfières du 26 May 1786. Bruxelles: Pauwels, 1786. [Huntington Library rare book 3554]

Précis des expériences faites par MM. Alban & Vallet; & Souscription proposé pour un Cours de Direction Aérostatique. Paris: la Veuve Valade, 1785. [BNF Vz 2121]

"Précis du voyage de l'Aérostat *Le Suffren*, lancé à Nantes le 14 Juin 1784, dans le jardin des Enfants-Trouvés." *Supplément aux Affiches de la Province de Bretagne* 228 (June 14, 1784). [BNF Tolbiac Vz 647]

Précis pour le sieur Montgolfier, négociant à Annonay, détenu dans les prisons de Lyon, à la Requête du Sieur Jean-Baptiste Johannot, son débiteur, et demandeur en Révocation. Contre ledit sieur Jean-Baptiste Johannot, ci-devant négociant audit Annonay, Failli & Défendeur en Révocation, signé Vitet, avocat, suivi de quatre pièces justificatives, January 1, 1782. Lyon: J. M. Barret, 1782.

"Procès-verbal de l'expérience Aérostatique faite à Nantes le 14 Juin 1784." *Supplément aux Affiches de la Province de Bretagne* 227 (June 14, 1784). [BNF Tolbiac Vz 647]

Procès-verbaux et détails des deux voyages aériens faits d'après la découverte de MM. Montgolfier. Bruxelles and Paris: Bailly, 1783.

Questions et conjectures sur l'application de l'électricité à l'aérostatique aux aérostats et à l'aérostation. Rodez: Marin Devic, 1786. [BNF Tolbiac Vz 2130]

Rapport fait à l'Académie des sciences, belles-lettres et arts de Lyon, sur l'Expérience de l'Aérostat, faite le 19 Janvier 1784; auquel on a joint une Dissertation de quelques Académiciens, sur le Fluide, principe de l'ascension des Machines aérostatiques developpées par l'action du feu. [BNF Tolbiac Vz 652 and 653]

Rapport fait à l'Académie des sciences, sur la machine aérostatique, inventée par MM. de Montgolfier. Signed by Le Roy, Tillet, Brisson, Cadet, Lavoisier, Bossut, the marquis de Condorcet and Desmarest on December 23, 1783. Paris: Moutard, 1784. [BNF Tolbiac Vz 645]

Réclamation du mouton, premier navigateur aërien. London and Paris: Cailleau, 1784. [BNF Tolbiac Vz 2117]

Réflexions amusantes et intéressantes sur le vaisseau volant, suivies des hommes volants. Fiction. Bruxelles and Paris: Marchands des nouveautés, 1782. [BNF Tolbiac Zp 2543]

The Reign of Terror: A Collection of Authentic Narratives. 2 vols. London: W. Simpkin and R. Marshall, 1826.

Réponse de M. Louvrier, Apothicaire, Rédacteur du Procès-verbal de l'Expérience Aérostatique faite à Nantes, le 14 Juin 1784, à la deuxieme & derniere Lettre du P. Budan, Professeur de Philosophie du Collège de Nantes. [BNF Tolbiac Vz 651]

Réponse du rédacteur du procès-verbal de l'Expérience Aérostatique du 14 Juin 1784, au P. Budan, professeur de Philosophie au Collège de l'Oratoire de Nantes. [BNF Tolbiac Vz 646]

Seconde et dernière Lettre au Rédacteur du Procès-Verbal de l'Expérience Aérostatique, faite à Nantes, le 14 Juin 1784, en réplique à sa Réponse. [BNF Tolbiac Vz 650]

Seconde Lettre à Messieurs Blanchard & le Chevalier l'Épinard, sur leur Voyage Aérien de Lille & autres semblables. N.d. [BNF Tolbiac Vz 2011]

"Souscription pour l'exécution d'un Globe Aérostatique, construit par MM. Robert, sous la direction de M. Charles, Professeur de Physique." [Gallica]

St. James' Chronicle or the British Evening Post

Supplément à l'Art de voyager dans les Airs, contenant le Précis historique de la grande Expérience faite à Lyon le 19 Janvier 1784 et l'Exposé d'un moyen ingénieux pour diriger à volonté les Ballons aérostatiques. N.d. [BNF Tolbiac Vz 2104]

Thoughts on the Farther Improvement of Aerostation, or the Art of Travelling in the Atmosphere, with a Description of a Machine. London: Printed for the Author, 1785. [ECCO]

The Town and Country Magazine, or, Universal Repository of Knowledge, Instruction and Entertainment.

Vie privée ou Apologie de très-sérénissime Prince Monseigneur le duc de Chartres, contre un Libel diffamatoire écrit en mil sept cent quatre-vingt un, mais qui n'a point parut à cause des menaces que nous avons faites à l'Auteur de le déceler, par une Société d'Amis du Prince. A Cent lieues de la Bastille, 1784.

"Voyage qui n'est point sentimental comme ceux de Mr. Stern anglois, Journal d'un provincial à Paris, 25 juin–1er âout, 1784." [BNF Richelieu Ms. Naf. 18903]

Wer ist Blanchard? Etwas über die ärostatische Kunst. N.d. [Staatsbibliothek zu Berlin 46 MA 1703]

ATTRIBUTED SOURCES

Adams, John Quincy. *Diary of John Quincy Adams*. Edited by David Grayson Allen, Robert J. Taylor, Marc Friedlaender, and Celeste Walker. 2 vols. Cambridge, MA: Belknap Press of Harvard University Press, 1981.

[Alcock, Mary]. *The Air Balloon; or Flying Mortal. A Poem*. London: E. Macklew, 1784.

Amadou, Robert, ed. *Franz-Anton Mesmer: le magnétisme animal*. Paris: Payot, 1971.

Argenson, René-Louis de Voyer de Paulmy, marquis d'. *Journal et mémoires du marquis d'Argenson*. Edited by J.-B. Rathéry. 9 vols. Paris: J. Renouard, 1859–1867.

Argenson, René-Louis de Voyer de Paulmy, marquis d'. "Invention des Ballons." In *Mémoires et Journal inédit du marquis d'Argenson*, edited by marquis d'Argenson, 5:390–91. 5 vols. Paris: P. Jannet, 1857–1858.

Arnaud de Saint-Maurice. *L'Observatoire volant et le triomphe héroïque de la navigation aérienne, et des vésicatoires amusants et célestes, poëme en quatre chants*. Paris: Cussac, 1784. [BNF Tolbiac YE 10155]

Barbauld, Anna Laetitia. *The Works of Anna Laetitia Barbauld*. Edited by Luch Aikin. 2 vols. London: Longman, 1825.

Bélidor, Bernard Forest de. *Architecture hydraulique, ou L'art de conduire, d'élever et de ménager les eaux pour les différens besoins de la vie*. 4 vols. Paris, 1737–1739.

Bentham, Jeremy. *Essay on Political Tactics*. London: T. Payne, 1791.

Bergasse, Nicolas. *Lettre d'un médecin de la Faculté de Paris à l'un médecin de collège de Londre*. La Haye, 1781.

Bergasse, Nicolas. *Considérations sur le magnétisme animal, ou sur la théorie du monde et des êtres organizés.* La Haye, 1784.

Bernardin de Saint-Pierre, Henri. *Voyage à l'Isle de France, à l'Isle de Bourbon, au Cap de Bonne-Espérance, etc. avec des observations sur la nature et sur les hommes, par un officier du roi.* Neufchâtel: Imprimerie de la Société typographique, 1773.

Bernardin de Saint-Pierre, Henri. *A Voyage to the Island of Mauritius, (or, Isle of France) the Isle of Bourbon, the Cape of Good Hope, &c.* Translated by John Parish. London: Printed for W. Griffin, 1775. [ECCO]

Bertholon, Pierre. *Des avantages que la physique, et les arts qui en dépendent, peuvent retirer des globes aérostatiques.* Montpellier: Jean Martel, 1784. [BNF Tolbiac Rz 3480]

Blake, William. *An Island in the Moon.* A facsimile of the manuscript (w. 1782–1785), annotated by Michael Phillips. Cambridge: Cambridge University Press, 1987.

Blanchard, Jean-Pierre. *An Exact and Authentic Narrative of M. Blanchard's Third Aerial Voyage, from Rouen in Normandy, on the 18th of July, 1784. Accompanied by M. Boby; in which they traversed a space of forty-five miles in two hours and a quarter, inclusive of the time employed in raising and depressing the Machine in the Air to which are added four certificates, testifying the truth of the relation and signed by several respectable characters.* London: C. Heydinger, 1784. [ECCO]

Blanchard, Jean-Pierre. *Journal and Certificates on the Fourth Voyage of Mr. Blanchard, who ascended from the Royal Military Academy, at Chelsea, the 16th of October, 1784, at 9 Minutes past Twelve o'clock, and was accompanied, as far as Sunbury, by John Sheldon.* London: Baker and Galabin, 1784. [ECCO]

Blanchard, Jean-Pierre. *Journal et procès verbaux du quatrième voyage aérien de M. Blanchard.* London: Galabin et Baker, 1784. [ECCO]

Blanchard, Jean-Pierre. *Procès-Verbaux lors du troisième voyage aérien de M. Blanchard, accompagné de M. Boby, Gressier au Parlement, fait le 18 Juillet 1784.* Rouen: Vve Machuel, 1784. [BNF Tolbiac Vz 1246]

Blanchard, Jean-Pierre. *Prospectus de la 14e expérience de M. Blanchard, pensionnaire du Roi & Citoyen de Calais, qui aura lieu à Lille du 20 au 25 de ce present mois d'Août 1785.* Supplement au no. 4 des *Feuilles de Flandres*, 1785.

Blanchard, Jean-Pierre. *Relation du quinzième voyage aërien de Mr. Blanchard, fait à Francfort sur le Meyn, le 3 Octobre 1785, dédié à son altesse sérénissme Monseigneur Charles, prince Palatin, duc de deux-ponts.* Frankfurt am Main: J. G. Eslinger, 1785. [Gallica]

Blanchard, Jean-Pierre. *Relation du seizième voyage aërien de Mr. Blanchard fait à Gand, le 20 Novembre 1785, dédié à son altesse sérénissme Monseigneur le Prince de Ligne.* Gand, 1786. [Gallica]

Blanchard, Jean-Pierre. *Relation de la vingtième ascension de M. Blanchard, qui a eu lieu à Hambourg le 23 Août 1786. Dédié à son excellence M. le chevalier de Viviers, ministre plénipotentiaire de France.* Aix-la-Chapelle, aux dépens de l'auteur, 1786.

Blanchard, Jean-Pierre. *Procès-Verbal du vingt-unieme Voyage Aërien de M. Blanchard, Citoyen de Calais, &c.&c.&c. fait à Aix-la-Chapelle, le 9 Octobre 1786.* [AM Strasbourg Serie AA No. 2053]

Blanchard, Jean-Pierre. *Prospectus de la vingt-sixième Expérience aérostatique de M. Blanchard* (dated Aug. 1, 1787). [AM Strasbourg Serie AA no. 2053]

Blanchard, Jean-Pierre. "Description detaillée au 28ème voyage aerien que mr. Blanchard entreprit et executa sans accident à Nuremberg, le 12 nov. 1787." [TC 2.3]

Blanchard, Jean-Pierre. *Journal of my forth-fifth ascension, being the first performed in America, on the Ninth of January, 1793.* Philadelphia: Charles cist, 1793. [ECCO]

Blanchard, Jean-Pierre. *The Principles, History & Use of Air-Balloons: Also, a Prospectus of Blanchard & Baker's intended aerial voyage from the City of New York.* New York: C. C. Van Alen, 1796.

Blanchard, Jean-Pierre. *Troisème voyage aérien de Blanchard (18 Juillet 1784), Introduction posthume de Charles Lefebvre.* Rouen: Albert Lainé, 1921.

Blanchard, Jean-Pierre. *The Forty-fifth aerial flight of the universally celebrated Mr. Blanchard, at Philadelphia.* N.d.

Blondel, Jacques-François. "Discours sur la manière d'étudier l'architecture, et les arts qui sont relatifs à celui de bastir, prononcé par M. Blondel, Architecte à Paris, à l'ouverture de son deuxième cours public sur l'Architecture, le 16 Juin 1747." *Mercure de France* (Aug. 1747): 57–74.

Bodard de Tezay, Nicolas-Marie-Felix. *Le Ballon, ou la Physicomanie.* Paris: Cailleau, 1783. [BNF Tolbiac 8-YTH-1664]

Boissy d'Anglas, François Antoine, comte de. "Étienne Montgolfier." [FM I.4]

Bombelles, Marc Marie, marquis de. *Journal de Marquis de Bombelles.* Edited by Jean Grassion and Frans Durif. 4 vols. Geneva: Droz, 1977.

Bonnefoy, Jean-Baptiste. *Analyse raisonnée des rapports des commissaires chargées par le Roi de l'examen du magnétisme animal.* Lyon and Paris: Prault, 1784.

Borde, Charles. *Discours sur les avantages des sciences et des arts.* Geneva: Barillot et fils, 1752.

Bourgeois, David. *Recherches sur l'art de voler, depuis la plus haute antiquité jusqu'à ce jour, pour servir de supplément à la Description des expériences aérostatiques de M. Faujas de Saint-Fond.* Paris: Cuchet, 1784. [BNF Tolbiac Vz 2158]

Brisson, Mathurin Jacques. *Observations sur les nouvelles découvertes aërostatiques, et sur la probabilité de pouvoir diriger les Ballons.* Paris: Le Boucher, 1784. [BNF Tolbiac Vz 2113]

Brissot de Warville, Jacques-Pierre. *De la verité, ou méditations sur les moyens de parvenir à la vérité dans toutes les connoissances humaines.* Neufchâtel: Imprimerie de la Société typographique, 1782.

[Brissot de Warville, Jacques-Pierre]. *Le Philadelphien à Genève, ou Lettres d'un Américain sur la dernière révolution de Genève, sa Constitution nouvelle, l'émigration en Irlande, &c. pouvant servir de tableau politique de Genève jusqu'en 1784.* Dublin, 1783.

Brissot de Warville, Jacques-Pierre. *Correspondance et papiers.* Edited by Cl. Perroud. Paris: Picard, n.d.

Brissot de Warville, Jacques-Pierre. *Mémoires (1754–1793).* Edited by Cl. Perroud. 2 vols. Paris: Alphonse Picard & fils, n.d.

Brunt, Samuel. *A Voyage to Cacklogallinia: with a description of the religion, policy, customs and manners, of that country.* London: J. Watson, 1727.

Buffon, Georges-Louis Leclerc, comte de. *Correspondance générale.* Edited by Henri Nadalt de Buffon. 2 vols. Geneva: Slatkine Reprints, 1971.

Burke, Edmund. *A Philosophical enquiry into the origin of our ideas of the sublime and beautiful.* London: R. and J. Dodsley, 1757.

Burke, Edmund. *Recherches philosophiques sur l'origine des idées que nous avons du beau et du sublime.* 2 vols. Paris, 1765.

Burney, Charles. *An Account of the Musical Performances in Westminster Abbey and the Pantheon, May 26th, 27th, 29th; and June the 3d, and 5th, 1784 in Commemoration of Handel.* Dublin, 1785.

Burton, Robert. *The Anatomy of Melancholy.* Edited by Holbrook Jackson. New York: Nyrb, 2001.

Carette, Pierre Louis Joseph. "Dissertation sur l'inflammation spontanée des matieres tirées du règne végétal & animal." *Journal de physique, de chimie, de l'histoire naturelle et des arts* 27 (1785): 92–94.

Carnus, Charles. *Lettre de M. l'Abbé Carnus, professeur de Philosophie à Rodez, contenant la rélation du voyage aérien fait, le 6 Aout 1784, sur la Montgolfière La Ville de Rodez.* Rodez: Marin Devic, 1784. [BNF Tolbiac Vz 2142]

Carra, Jean-Louis. *Essai sur la nautique aérienne, contenant l'art de diriger les Ballons aérostatiques à volonté, & d'accélérer leur course dans les plaines de l'air; avec le Précis de deux expériences particulieres de météorologie à faire. Lu à l'Académie royale des sciences de Paris, le 14 janvier, 1784.* Paris: E. Onfroy, 1784. [BNF Tolbiac Vz 2144]

Castelli, Charles. *Relation des Expériences de la Machine Aérostatique de Don Paul Andréani; exposée dans une Lettre de Ch. Castelli, chanoine, adrésée à M. Faujas de Saint-Fond, traduit de l'Italien.* 1784. [BNF Tolbiac Vz 2111]

Cavallo, Tiberius. *A Treatise on the Nature and Properties of Air, and other permanently elastic fluids.* London: Printed for the author, 1781.

Cavallo, Tiberius. *The History and Practice of Aerostation.* London: Printed for the author, 1785. [ECCO]

Charles, Jacques-Alexandre-César. "Seconde Mémoire." [BIF Ms. 2104]

Charles, Jacques-Alexandre-César. "Mémoire de Mr Charles sur l'aérostatique, comprenant la relation de son voyage du 1er Décembre, 1783." Dated 1784. [TC 3.6]

Charles, Jacques-Alexandre-César. "Mémoire manuscrit inédit du plus haut intérêt sur les préparatifs de la première ascension de Charles, et Robert dans le jardin des Tuileries le 1er Décembre, 1783." [TC 3.8]

Charles, Jacques-Alexandre-César. "Notices communiquées aux auteurs de ce Journal, sur l'expérience du globe ascendant, faite à Paris, au champ de Mars, le 27 Août dernier, par MM. Charles & Robert." *Journal encyclopédique* 17 (Oct. 1783): 124–33.

Charles, Jacques-Alexandre-César. "Cours de Physique en 60 Leçons." 1802. [BIF Ms. 2104]

Chénier, André. *Oeuvres complètes*. Edited by Gérard Walter. Paris: Gallimard, 1950.

Chénier, André. *L'Invention*. 1784. Edited by Paul Dimoff. Paris: Nizet, 1966.

Choderlos de Laclos, Pierre-Ambroise-François. *Les liaisons dangereuses*. 2 vols. Paris: Durand neveu, 1782.

Condorcet, Jean-Antoine-Nicolas de Caritat, marquis de. *Correspondance inédite de Condorcet et Madame Suard, 1771–1791*. Edited by Elisabeth Badinter. Paris: Fayard, 1988.

[Cooke, William]. *The Air Balloon, or A Treatise on the Aerostatic Globe, lately invented by the celebrated Mons. Montgolfier of Paris*. London: Printed for G. Kearsley, 1783. [ECCO]

[Cooke, William]. *The Air Balloon, or A Treatise on the Aerostatic Globe, lately invented by the celebrated Mons. Montgolfier of Paris*. 4th ed. London: Printed for G. Kearsley, 1784. [ECCO]

[Cornélie de Vasse, la baronne de]. *Le Char volant ou voyage dans la lune*. Paris, 1783. [Arsenal Library, Paris, 8 BL 19278]

[Cornélie de Vasse, la baronne de]. *Le Char volant, avec la relation d'un voyage dans la lune*. Paris, l'An V (1796–1797). [BNF Tolbiac 16 Y2 42461]

Cornu, Paul, ed. *Galerie des modes et costumes français dessinés d'après nature, 1778–1787*. 4 vols. Paris: Émile Lévy, 1912.

Cowper, William. *Private Correspondence of William Cowper, esq., with several of his most intimate friends*. Edited by John Johnson. 2 vols. London: H. Colburn, 1824.

Cradock, Anna Francesca. *La vie française à la veille de la révolution (1783–1786): Journal inédit de Madame Cradock*. Paris: Perrin, 1911.

Cradock, Joseph. *Literary and Miscellaneous Memoirs*. 4 vols. London: J. Nichols and Sons, 1826.

Croy, Emmanuel duc de. *Journal inédit du duc de Croÿ, 1718–1784*. Edited by Vicomte de Grouchy and Paul Cottin. 4 vols. Paris: Ernest Flammarion, 1906.

Court de Gébelin, Antoine. *Lettre de l'auteur du monde primitif a messieurs les souscripteurs sur le magnétisme animal*. 2nd ed. Paris: Valleryre, l'aine, 1784.

Cyrano de Bergerac, Savinien. *Histoire comique: contenant les états et empires de la lune*. Paris: Charles de Sercy, 1657.

Cyrano de Bergerac, Savinien. *The Comical History of the States and Empires of the worlds of the Moon and Sun*. Translated by A. Lovell. London: Printed for Henry Rhodes, 1687.

Damen, Christiaan Hendrik. *Natur- en Wiskundige Beschouwing van den Lugtbol, Tot eene betere Kennis en Beoordeling dier berugte Ontdekking.* Utrecht, 1784.

Darbelet, Desgranges, and Chalifour. "Relation de deux voyages aériens, fait à Bordeaux, les 16 Juin et 26 Juillet 1784." In *Recueil des ouvrages du Musée de Bordeaux, dédié à la Reine: Annee 1787*, 112–53. Bordeaux: Michel Racle, [1787]. [Huntington Library rare book 47547]

Davenant, Charles. *An Essay upon Universal Monarchy.* London: Baldwin, 1756.

De Boer, C., ed. *Ovide moralisé en prose.* Amsterdam: North-Holland, 1954. For a full bibliography on this anonymous text, see http://www.arlima.net/mp/ovide_moralise.html.

De Gérando, Joseph, "Notice sur M. Joseph Montgolfier." *Bulletin de la société d'encouragement pour l'industrie nationale* 13 (1814): 91–108.

Deparcieux, Antoine. "Mémoire sur la possibilité d'amener à Paris, à la même hauteur à laquelle y arrivent les eaux d'Arcueil, mille à douze cents pouces d'eau, belle & de bonne qualité, par un chemin facile & par un seul canal ou aqueduc." *Mémoires de l'Académie royale des sciences* (1762): 337–401.

De Parcieux, Antoine. *Dissertation sur les globes aérostatiques.* Paris: Chez l'auteur, 1783. [BNF Tolbiac Vz 2138]

Desmarest, Nicolas. "Premier mémoire sur les principales manipulations qui sont en usage dans les papeteries de Hollande, avec l'explication physique des résultats de ces manipulations." *Mémoires de l'Académie royale des sciences* (1771): 335–64. Published 1774.

Desmarest, Nicolas. "Second mémoire sur la papeterie, dans lequel on traite de la nature & des qualités des Pâtes hollandoises & françoises: de la manière dont ells se comportent." *Mémoires de l'Académie royale des sciences* (1774): 599–687. Published 1778.

Diderot, Denis. *La Religieuse.* Paris: Buisson, 1796.

Diderot, Denis. *Salons (1759–1781).* Edited by Jean Seznec and Jean Adhémar. 4 vols. Oxford: Clarendon Press, 1957–1963.

Diderot, Denis. *Rameau's Nephew and Other Works.* Translated by Jacques Barzun and Ralph H. Bowen. Indianapolis: Bobbs-Merrill, 1964.

Diderot, Denis. *Mémoires pour Catherine II.* Edited by Paul Vernière. Paris: Garnier, 1966.

Diderot, Denis. *Diderot: Political Writings.* Edited by John Hope Mason and Robert Wokler. Cambridge: Cambridge University Press, 1992.

Diderot, Denis. *The Supplément au Voyage de Bougainville.* In *Diderot: Political Writings*, 35–75.

Dinwiddie, James. *Syllabus of a Course of Lectures on Experimental Philosophy.* London: A. Grant, 1789. [ECCO]

Dinwiddie, James. *Biographical Memoir of James Dinwiddie.* Edited by William Proudfoot. Liverpool: E. Howell, 1868.

Duchosal, Marie-Émilie-Guillaume. *Blanchard, poëme en quatre chants, dédié à Messieurs les maire et citoyens de Calais*. Bruxelles, 1786. [BNF Tolbiac YE 20629]

Du Coudray, Alexandre-Jacques. *Voyage du comte de Haga en France, recueilli et mis en ordre par M. le chevalier du Coudray*. Paris: Belin, 1784.

Duret, Mathieu-Louis-Pierre. "Notices historiques sur Étienne Montgolfier lue à la rentrée de l'école centrale de Tournon de 3 Brumaire an huit." [FM 1.2 and 1.9]

Dusaulx, Jean. *Observations sur l'histoire de la Bastille*. London: Aux dépens de l'Auteur, 1783.

[Ebertin, Johann Samuel]. *Über die bewundernswürdige Luftfahrt welche Herr Blanchard, aufgenommener Bürger zu Calais, pensioner des Königs von Fankreich*. [Staatsbibliothek Berlin, Rare Books Ym 5086]

Ehrmann, Friedrich Ludwig. *Montgolfier'sche Luftkörper oder aërostatische Maschinen: Eine Abhandlung, worinn die Kunst sie zu verfertigen und die Geschichte der bisher damit angestellten Versuche beschrieben warden*. Strasbourg: Johann Georg Treuttel, 1784.

Evans, Thomas. *Réfutation des Mémoires de la Bastille*. London: Edward Cox, 1783.

Fabry, [Jacques Joseph] de. *Réflexions sur la relation du voyage aérien de MM. Charles & Robert, et la Brochure intitulée: Méthode aisée de faire la Machine aérostatique*. Paris: Libraires des Nouveautés, 1784. [BNF Tolbiac Vz 2140]

Faujas de Saint-Fond, Barthélemy. *Mémoire sur des bois de cerfs fossiles*. Grenoble: J. Cuchet, 1776. [Gallica]

Faujas de Saint-Fond, Barthélemy. *Recherches sur les volcans éteints du Vivarais et du Velay*. Paris: J. Cuchet, 1778. [Gallica]

Faujas de Saint-Fond, Barthélemy. *Description des expériences de la machine aérostatique de MM de Montgolfier, et de celles auxquelles cette découverte a donné lieu*. Paris: Cuchet, 1783. [Gallica]

Faujas de Saint-Fond, Barthélemy. *Beschreibung der Versuch mit den aerostatischen Maschinen der Herren von Montgolfier nebst verschiedenen zu dieser Materie gehörigen Abhandlungen*. Leipzig: Weidmanns Erben und Reich, 1784. [Huntington Library rare books #3591]

Faujas de Saint-Fond, Barthélemy. "Beschreibung der Versuche mit den ärostatischen Maschinen der Herren von Montgolfier, nebst verschiedenen zu dieser Materie gehörigen Abhandlungen." *Hannoverisches Magazin* (1784): 337–400 (March 15) and 405–32 (March 29).

Faujas de Saint-Fond, Barthélemy. *Beschryving der Proefneemingen met konstige Lugtbollen*. Translated by Martinus Houttuyn. Amsterdam: Jacobus van der Burgh en Zoon, 1784. [Huntington Library rare books #491286]

Faujas de Saint-Fond, Barthélemy. *Descrizione delle esperienze della macchina aerostatica del signori di Montgolfier*. Venice: Graziosi si vende, 1784. [Huntington Library rare books #3001]

Faujas de Saint-Fond, Barthélemy. *Kurze Nachricht von den aerostatischen Maschinen, ihrem Baue, und den bisher damit angestellten Versuchen.* Kehl: Bei der Expedition der gelehrten Zeitung; Strasbourg: J. F. Stein, 1784. [BNF Tolbiac R 158347]

Faujas de Saint-Fond, Barthélemy. *Première suite de la description des expériences de la machine aérostatique de MM de Montgolfier, et de celles auxquelles cette découverte a donné lieu.* Paris: Cuchet, 1784. [Gallica]

Faujas de Saint-Fond, Barthélemy. *Raccolta universale di tutte le esperienze, osservazioni, riflessioni ecc.: all'occasione delle machine o palloni aerostatici.* Genova: Stamperia Gesiniana, 1784. [Huntington Library rare books #496557]

Farge, Arlette, ed. *Flagrants délits sur les Champs-Élysées: Les dossiers de police du Gardien Federici (1777–1791).* Paris: Mercure de France, 2008.

[Fontenelle]. *La République des philosophes, ou Histoire des Ajaoiens. Ouvrage posthume de Mr. de Fontenelle.* Geneva, 1768.

Franklin, Benjamin. *The Writings of Benjamin Franklin.* Edited by Albert Henry Smyth. 10 vols. New York: Macmillan, 1907.

Franklin, Benjamin, et al. *Rapport des commissaires chargés par le roi de l'examen du Magnétisme animal.* Paris: L'Imprimerie royale, 1784.

Frederick the Great. *Correspondance de Frédéric II, roi de Prusse.* 12 vols. Berlin: Rodolphe Decker, 1850–1856.

Gagnière, Abbé des Granges. "Lettre sur le Globe aérostatique." *Journal de Monsieur* 6 (1783): 57–72.

Galien, R. P. Joseph. *L'Art de naviguer dans les airs: Amusement physique et géometrique.* Avignon, 1757.

Girard, J.-F. *Mesmer Blessé, ou Réponse à la lettre du R. P. Hervier sur le magnétisme animal, par M***.* London and Paris: Couturier, 1784. [BNF Tolbiac 8-TB64-14]

Godwin, Francis. *The Man in the Moone: or A Discourse of a Voyage thither by Domingo Gonsales the speedy Messenger.* London: John Norton, 1638.

Godwin, Francis. *L'Homme dans la Lune, ou le Voyage chimérique fait au monde de la Lune, nouvellement découvert par Dominique Gonzalès, aventurier espagnol, autrement dit le Courrier Volant.* Paris: Anthoine de Sommaville, 1648.

Goulard, Jean-François-Thomas. *Cassandre mécanicien, ou le Bateau volant, comédie parade, en un acte et en vaudevilles.* Paris: Brunet, 1783. [BNF Tolbiac FB 19363]

Guyot, Edme-Gilles. *Essai sur la construction des ballons aérostatiques et sur la maniere de les diriger.* Paris: Gueffier, 1784. [BNF Tolbiac Vz 2122]

Guyton de Morveau, Louis-Bernard, Hugues Maret, and Jean-François Durande. *Elémens de chymie, théorique et pratique, rédigés dans un nouvel ordre, d'après les découvertes modernes.* 3 vols. Dijon: L. N. Frantin, 1777–1778.

Guyton de Morveau, Louis-Bernard. *Description de l'aérostate, L'Académie de Dijon, contenant le détail des procédés, la théorie des opérations, les dessins des machines & les procès-verbaux d'expériences.* Dijon: Causse, 1784.

Happel, Eberhard Werner. *Grösste Denkwürdigkeiten der Welt*. Hamburg: Thomas von Viering, 1689.

Hardy, Siméon-Prosper. "Mes Loisirs, ou Journal d'événemens tells qu'ils parviennent à ma connoissance (1753–1789)." 5 vols. [BNF Richelieu Ms. Fr. 6681-85]

Harris, James. *Diaries and Correspondence of James Harris, First Earl of Malmesbury*. Edited by his grandson, the third earl. 4 vols. London: Richard Bentley, 1844.

Hervier, Charles. *Lettre du P. Hervier aux habitants de Bordeaux sur le magnétisme animal*. Bordeaux, 1784. [BNF Tolbiac 8-LN27-9768]

Hervier, Charles. *Lettre sur la découverte du magnétisme animal à M. Court de Gebelin*. Paris: Couturier, 1784. [BNF Tolbiac 8-TB64-13]

Hess, Jonas Ludwig von. *Hamburg, topograpisch, politisch, und historisch beschrieben*, 3 vols. Hamburg, 1787.

Holbach, Paul Henri Thiry baron d'. *Le bon-sens ou les idées naturelles opposées aux idées surnaturelles*. London, 1772.

Imbert de la Platière, Sulpice, comte d'. *L'Invention des globes aérostatiques: Hommage à MM. de Montgolfier*. London and Paris: Cailleau, 1784. [BNF Tolbiac Vz 2126]

Inchbald, Elizabeth. *The Mogul Tale: or the Descent of the Balloon. A Farce*. [Dublin], 1788. [ECCO]

Jeffries, John. *A Narrative of the Two Aerial Voyages of Doctor Jeffries with Mons. Blanchard; with meteorological observations and remarks*. London: Printed for the author and sold by J. Robon, 1786.

Johnson, Samuel. *The Letters of Samuel Johnson*. Edited by Bruce Redford. 5 vols. Princeton, NJ: Princeton University Press, 1992–1994.

Jupp, Peter, and Eoin Magennis, eds. *Crowds in Ireland, c. 1720–1920*. New York: Macmillan, 2000.

Kindermann, Eberhard Christian. *Die Geschwinde Reise auf dem Lufft-Schiff nach der obern Welt*. Berlin, 1744.

Kramp, Christian. *Geschichte der Aerostatik, historisch, physisch and mathematisch ausgeführt*. Strasbourg: Verlag der akademischen Buchhandlung, 1784.

Kramp, Christian. *Anhang zu der Geschichte der Aërostatik*. 1786.

La Bruyère, Jean de. *Les Caractères ou les moeurs de ce siècle*. Paris, 1688.

La Bruyère, Jean de. *Characters, or the Manners of the Age*. London, 1702. [ECCO]

Laclos, Pierre Choderlos de. "On the Education of Women." 1783. Translated in *The Libertine Reader: Eroticism and Enlightenment in Eighteenth-Century France*, edited by Michel Feher, 129–66. New York: Zone Books, 1997.

La Condamine, Charles-Marie de. *Relation abrégée d'un voyage fait dans l'intérieur de l'Amérique méridionale, depuis la côte de la mer du Sud, jusques aux côtes du Brésil & de la Guiane, en descendant la rivière des Amazones*. Paris: Imprimerie royale, 1749.

Lafayette, Gilbert du Motier, marquis de. *Mémoires, correspondance et manuscripts de Général Lafayette*. 2 vols. Bruxelles, 1837.

[La Folie, Louis Guillaume de]. *Le Philosophe sans prétention ou l'homme rare. Ouvrage physique, chymique, politique et moral, dédié aux savans.* Paris: Clousier, 1775.

La Harpe, Jean François de. *Œuvres de M. de La Harpe*, 6 vols. Paris: Pissot, 1778.

Lalande, Joseph-Jérôme Lefrançois de. *Art de faire le papier.* [Paris, 1761]. Translated by Richard Atkinson as *The Art of Papermaking*. Kilmurry, Ireland: Ashling Press, 1976.

[L'Allemand de Saint-Croix, B.] *Procès-verbal très-intéressant du voyage aérien qui a eu lieu aux Champs-Elysées le 18 Septembre 1791, jour de la proclamation de la Constitution.* Paris: Imprimerie du Patriote François, 1791. [BNF Tolbiac Vz 2134]

Lana de Terzi, Francesco. *Prodromo ouero saggio di alcune inventioni nuove premesso dell'arte maestro.* Brescia: Li Rizzardi, 1670.

[Lancival, Luce de]. *Poëme sur le globe.* Paris, 1784. [BNF Tolbiac YE 26879]

Latude, Henri Masers de. *Le Déspotisme dévoilé ou Mémoires de Henri Masers de Latude.* Paris: Beuret, 1790.

[Launoy and Bienvenu]. *Instruction sur la nouvelle machine inventée par MM. Launoy, naturaliste, & Bienvenu, machinist-physicien, qui a été annoncée dans le Journal de Paris, le 19 Avril 1784.*

Laurencin, Jean-Baptiste de. *Lettre de M. le comte de Laurencin à M. Joseph de Montgolfier, sur l'Expérience Aérostatique faite à Lyon le 4 Juin 1784, en présence du Roi de Suede.* [BNF Tolbiac Vz 2107]

La Vaulx, André Foulon, comte de, and Paul Tissandier. *Joseph et Étienne de Montgolfier.* Annonay, France, 1926.

Lavoisier, Antoine-Laurent. *Oeuvres de Lavoisier.* 6 vols. Paris: Imprimerie impériale, 1862–1893.

Lavoisier, Antoine-Laurent. *Correspondance.* 7 vols. Paris: A. Michel, 1955–2012.

Lenoir, Jean-Charles-Pierre. *Détail sur quelques établissemens de la ville de Paris, demandé par sa majesté imperiale la reine de Hongrie à M. Le Noir.* Paris, 1780. [Gallica]

Lenoir, Jean-Charles-Pierre. *Ordonnance de Police, qui fait défenses de fabriquer & faire enlever des Ballons & autres Machines Aérostatiques auxquelles seroient adaptés des Réchauds à l'esprit-le-vin, de l'Artifice, & autres matieres dangereuses pour le feu; & ordonne que tous autres Ballons Aérostatiques ne pourront être enlevés sans en avoir prealablement obtenu la permission.* Paris: P. G. Simon, 1784.

Le Roy, Jean-Baptiste. *Rapport fait à l'Académie des sciences, sur la machine aérostatique, inventée par MM. de Montgolfier.* Paris: Moutard, 1784. Reproduced in Faujas, *Description* (1784), 200–231.

Levere, Trevor H., and Gerard L'Estrange Turner, eds. *Discussing Chemistry and Steam: The Minutes of a Coffee-House Philosophical Society, 1780–1787.* Oxford: Oxford University Press, 2002.

Lichtenberg, Georg Christoph. "Kurze Geschichte einiger der merkwuerdigsten Luftarten." *Hannoverisches Magazin* (Dec. 15, 19, 1783): 1585–610.

Lichtenberg, Georg Christoph. "Ueber die neuerlich in Frankreich angestellten Versuche, groBe hohle Körper in der Luft aufsteigen zu machen, und damit Lasten auf eine groBe Höhe zu heben." *Göttingisches Magazin der Wissenschaften und Literatur* (1783): 783–93.

Lichtenberg, Georg Christoph. "Vermischte Gedanken ueber die aërostatischen Maschinen." *Göttingisches Magazin der Wissenschaften und Literatur* (1783): 930–53.

Linguet, Simon-Nicolas-Henri. *Mémoires sur la Bastille.* London: Thomas Siplsbury, 1783.

Lottin, Antoine-Prosper. *Discours contre le luxe: Il corrompt les moeurs et détruit les Empires.* Paris, 1783.

Lucian of Samosata. "Icaromenippus, an Aerial Expedition." In *Marcus Aurelius and His Times: The Transition from Paganism to Christianity,* edited by Irwin Edman, 220–38. Roslyn, NY: Walter J. Black, 1945.

Lucian of Samosata. *Certaine select dialogues of Lucian together with his true historie, translated from the Greeke into English by Mr Francis Hickes. Whereunto is added the life of Lucian gathered out of his owne writings, with briefe notes and illustrations upon each dialogue and booke, by T.H. Mr of Arts of Christ-Church in Oxford.* Oxford: Printed by William Turner, 1634.

Lucian of Samosata. "My Dream [Somnium]" and "A True Story [Verae Historiae]." In *Selected Satires of Lucian,* edited by Likonel Casson, 3–10. Chicago: Aldine, 1962.

Lunardi, Vincent. *An Account of the First Aërial Voyage in England, in a series of letters to his guardian, Chevalier Gherardo Compagni.* London, 1784. [ECCO]

Lunardi, Vincent. *Lunardi's Grand Aerostatic Voyage through the Air.* London: printed for J. Bew, 1784.

Lunardi, Vincent. *An Account of Five Aerial Voyages in Scottland, in a series of letters to his guardian, Chevalier Gherardo Compagni.* London: printed for the author and sold by J. Bell, 1786. [ECCO]

Lunardi, Vincent. *Mr. Lunardi's Account of his Ascension and Aerial Voyage, from the New Fort, Liverpool, on Wednesday the 20th of July, 1785, in Two letters to George Biggin, esq.* N.d. [ECCO]

Lunardi, Vincent. *Mr. Lunardi's Account of his Second Aerial Voyage from Liverpool, on Tuesday the 9th of August, 1785.* N.d. [ECCO]

Marat, Jean-Paul. *The Chains of Slavery.* London: T. Becket, 1774. [ECCO]

[Marat, Jean-Paul]. *Lettres de l'Observateur Bon-sens, à M. de ***, sur la fatale catastrophe des infortunés Pilatre de Rosier & Romain, les Aéronautes & L'Aérostation.* London and Paris: Méquignon, 1785. [ECCO]

Marat, Jean-Paul. *Les Charlatans modernes, ou lettres sur le charlatanisme académique.* Paris: L'imprimerie de Marat, 1791.

Marat, Jean-Paul. *La correspondance de Marat.* Edited by Charlres Vellay. Paris: Charpentier et Fasquelle, 1908.

Marivaux, Pierre de. *Journaux et oeuvres diverses de Marivaux.* Paris: Garnier, 1988.

Marmontel, Jean-François. *Mémoires de Marmontel.* Edited by Maurice Tourneux. 3 vols. Paris: librairie de bibliophiles, 1891.

Martyn, Thomas. *Hints of Important Uses, to be derived from Aerostatic Globes.* London: Printed for the Author, 1784.

Mably, Gabriel Bonnot de. *Principes de Morale.* Paris: Alexandre Jombert jeune, 1784.

Mably, Gabriel Bonnot de. *Des droits et des devoirs du citoyen.* Kell, 1789 [Gallica]. Edited by Jean-Louis Lecercle. Paris: Marcel Didier, 1972.

Ménétra, Jacques-Louis. *Journal of My Life.* Translated by Arthur Goldhammer. New York: Columbia University Press, 1986.

Mercier, Louis-Sébastien. *L'An deux mille quatre cent quarante, Rêve s'il en fuit jamais; suivi de l'homme de fer, songe.* London, 1771.

Mercier, Louis-Sébastien. *Memoirs of the Year Two Thousand Five Hundred.* Translated by W. Hooper. 2 vols. London: G. Robinson, 1772. [ECCO]

Mercier, Louis-Sébastien. *Du Théâtre, ou nouvel essai sur l'art dramatique.* Amsterdam: Harevelt, 1773.

Mercier, Louis-Sébastien. *Le Charlatan ou le docteur Sacroton.* The Hague and Paris: Ballard, 1780.

Mercier, Louis-Sébastien. *Le Tableau de Paris.* Amsterdam, 1781–1783. Edited by Jean-Claude Bonnet. 2 vols. Paris: Mercure de France, 1994. Select translations in *Panorama of Paris,* edited by J. D. Popkin. University Park: Pennsylvania State University Press, 1999.

Mercier, Louis-Sébastien. "Le Ballon-Montgolfier." 1784. In *Mon Bonnet de Nuit, suivi de Du Théâtre,* edited by Jean-Claude Bonnet and Pierre Frantz, 555–71. Paris: Mercure de France, 1999.

Mirabeau, Honoré Gabriel de Riquetti. *Histoire secrete de la cour de Berlin.* 2 vols. London: S. Bladon, 1789.

Mesmer, Anton. *Le magnétisme animal.* Edited by Robert Amdou. Paris: Payot, 1971.

Mesmer, Anton. *Mémoire sur la découverte du magnétisme animal.* Paris: Fr. Didot le jeune, 1779. [Gallica]

Mesmer, Anton. *Mesmerism.* Edited by George Bloch. Los Altos, CA: William Kaufmann, 1980.

Meusnier, Jean Baptiste. *Mémoire sur l'équilibre des Machines aérostatiques, sur les différens moyens de les faire monter & descendre, & spécialement sur celui d'exécuter ces manoeuvres, sans jeter de lest, & sans perdre d'air inflammable, en ménageant dans le ballon une capacité particulière, destiné à renfermer de l'air atmosphérique, présenté à l'Académie, le 3 Décembre.* With "Addition au mémoire précédent, contenant une application de la théorie qui y est exposée, à un exemple particulier." [BNF Tolbiac Vz 654]

Meusnier, Jean-Baptiste. "Lettre à M. Faujas de Saint-Fond." Dated Oct. 31, 1783. In Faujas, *Description* (1783), 33–110.

Minkelers, Jean-Pierre [Johann Peter]. *Mémoire sur l'Air inflammable tiré de différentes substances, rédigé par M. Minkelers, Professeur de Philosophie au Collège du Faucon, Université de Louvain*. Louvain, 1784.

Montesquieu, Charles-Louis de Secondat, baron de. *Lettres persanes*. Amsterdam: P. Brunel, 1721.

Montesquieu, Charles-Louis de Secondat, baron de. *Persian Letters*, 2 vols. London: Printed for J. Jonson, 1722. [ECCO]

Montesquieu, Charles-Louis de Secondat, baron de. *Reflections on the Causes of the grandeur and declension of the Romans*. London: W. Innys and R. Manby, 1734. [ECCO]

Montesquieu, Charles-Louis de Secondat, baron de. *The Spirit of Laws*, 2 vols. London: printed for J. Nourse and P. Vaillant, 1750.

Montgolfier, Adélaïde. *Contes devenus histoires*. Paris: Au Bureau de la Ruche, 1835.

Montgolfier, Joseph-Michel. "Mémoire lu à l'académie de Lyon." In Faujas de Saint-Fond, *Première suite* (1784), 98–111.

Müller, Friedrich Christopher. *Etwas zur Erklärung der Luftmaschinen des Herrn Montgolfier*. Frankfurt, 1784.

Muralt, Béat-Louis de. *Lettres sur les Anglois et les François et sur les voiages*. [London,] 1725.

Muralt, Béat-Louis de. *Letters describing the Character and Customs of the English and French Nations. With a curious Essay on Travelling; and a criticism on Boileau's Description of Paris*. London: Tho. Edlin, 1726.

Murie, Yves. *La Digue qui a fait Cherbourg*. Cherbourg-Octeville: Isoète, 2006.

Nietzsche, Friedrich. "Vom Nutzen und Nachteil der Historie für das Leben." In *Unzeitgemäss Betrachtungen*. Vol. 2. Leipzig, 1874.

Nietzsche, Friedrich. "On the Uses and Disadvantages of History for Life." In *Untimely Meditations*, 59-123, translated by R. J. Hollingdale. Cambridge: Cambridge University Press, 1983.

Noorden, Johannes van. *Korte Verhandeling over de Lugtweegkundige Bol*. Rotterdam, 1784.

Nougaret, Pierre Jean Baptiste. *Tableau mouvant de Paris, ou variétés amusantes; Ouvrages enrichi de Notes historiques & critiques*. 3 vols. London: Thomas Hookham, 1787.

Oberkirch, Henriette Louise de Waldner de Freundstein, baronne d'. *Mémoires de la Baronne d'Oberkirch sur la cour de Louis XVI et la société française avant 1789*. Paris: Mercure de France, 1970.

Orléans, Louis-Philippe-Joseph, duc d'. *Lettres de L.- P.-J. d'Orléans, duc de Chartres à Nathaniel Parker Forth (1778–1785)*. Edited by Amédée Britsch. Paris: Société d'histoire diplomatique, 1926.

Ovid. *The Art of Love*. Translated by Charles D. Young. New York: Liveright Publishing, 1931.

Ovid. *Metamorphoses*. New York: Penguin, 2004.

Paltock, Robert. *The Life and Adventures of Peter Wilkins*. London: Printed for J. Robinson, 1751.

Pâris de l'Épinard, Joseph. *Mon retour à la vie après quinze mois d'agonie, anecdote qui peut servir à la connaissance de l'homme*. Paris, n.d.

Paulet, Jean-Jacques. *L'Antimagnétisme, ou Origine, progrès, décadence, renouvellement et réfutation du magnétism animal*. London, 1784.

Pérouse de Montclos, Jean-Marie. *Jacques-Germain Soufflot*. Paris: Monum, 2004.

Pilâtre de Rozier, Jean-François. *Première expérience de la Montgolfière construite par ordre du roi, lancée en présence de Leurs Majestés, de la Famille Royale, et de Monsieur le Comte d'Haga*. Paris: Imprimerie de Monsieur, 1784. [BNF Tolbiac Vz 644]

Pingeron, Jean-Claude. *L'Art de faire soi-même les ballons aérostatiques, conformes à ceux de M. de Montgolfier*. Amsterdam and Paris: Hardouin, 1783. [BNF Tolbiac Vz 2129]

Poissonnier, Pierre-Isaac, Claude-Antoine Caille, Pierre-Jean-Claude Mauduyt, and Charles-Louis François Andry. *Rapport des Commissaires de la Société Royale de Médecine, nommés par le Roi, pour faire l'examen du Magnétisme animal*. Paris : L'Imprimerie Royale, 1784.

Potain. *Relation aérostatique dédiée à la Nation irlandaise*. Paris: Delaunay, 1824.

Priestley, Joseph. *Experiments and Observations on Different Kinds of Air*. London: Printed for J. Johnson, 1774.

Priestley, Joseph. *Expériences et Observations sur différentes espèces d'air*. 5 vols. Paris, 1777–1780.

Pye, Henry James. *Aerophorion: A Poem*. Oxford: Printed for D. Prince and J. Cooke, 1784. [ECCO]

Raynal, Guillaume-Thomas. *Histoire philosophique et politique des établissemens et du commerce des européens dans les deux Indes*. Amsterdam, 1770.

Rétif de la Bretonne, Nicolas-Edme. *La Mimographe, ou Idées d'une honnête-femme pour la réformation du théâtre national*. Amsterdam, 1770.

Rétif de la Bretonne, Nicolas-Edme. *La Découverte australe par un Homme-volant ou le Dédale français*. Leipzig, 1781.

Rétif de la Bretonne, Nicolas-Edme. *Le Palais-Royal*. Paris, 1790. [ECCO]

Restif de la Bretonne, Nicolas-Edme. *Monsieur Nicolas or the Human Heart Laid Bare*. New York: Clarkson N. Potter, 1966.

Restif de la Bretonne, Nicolas-Edme. *Monsieur Nicolas ou le Coeur humain dévoilé*. 2 vols. Paris: Gallimard, 1989.

Réveillon, Jean-Baptiste. "Exposé justificatif pour le sieur Réveillon, entrepreneur de la Manufacture royale de papiers peints, faubourg St.-Antoine." In *Mé-*

moires du Marquis de Ferrières, edited by Albin de Berville and François Barrière, 1:427–38. 3 vols. Paris: Baudouin Frères, 1821.

Reynard, Pierre Claude. "Manufacturing Quality in the Pre-industrial Age: Finding Value in Diversity." *Economic History Review*, n.s. 53 (2000): 493–516.

Richard, Jérôme. *Histoire naturelle de l'Air et des météores*. 6 vols. Paris: Saillant & Nyon, 1770.

Rigby, Edward. *An Account of Mr. James Deeker's Two Aerial Expeditions from the City of Norwich*. Norwich: John Crouse, 1785. [Cambridge University Library]

[Rivarol, Antoine de]. *Lettre à Monsieur le Président de ***, sur le Globe aërostatique, sur les Têtes parlantes, & sur l'état present de l'opinion publique à Paris*. London and Paris: Cailleau, 1783. [BNF Tolbiac Vz 2125]

Robert. *Mémoire présenté a l'Académie des sciences, arts et belles-lettres de Lyon, sur la maniere le plus sûre, la moins dispendieuse & la plus efficace de diriger à volonté les Machine Aérostatiques*. Dijon and Paris: Belin, 1784. [BNF Tolbiac Vz 2128]

[Robert]. *Mémoire sur les expériences aérostatiques faites par MM. Robert frères, Ingénieurs-Pensionnaires du roi*. Paris: Imprimerie de Philippe-Denys Pierres, Imprimeur ordinaire du roi, 1784. [BNF Tolbiac Vz 648]

Rousseau, Jean-Jacques. *Les institutions chimiques de Jean-Jacques Rousseau*. Edited by Théophile Dufour. Geneva: Journal de Genève, 1905.

Rousseau, Jean-Jacques. *Le Nouveau Dédale*. Geneva: Jullien, 1910.

Rousseau, Jean-Jacques. *Julie ou La nouvelle Hëloise*. Paris: Garnier, 1960.

Rousseau, Jean-Jacques. *Lettre à d'Alembert sur les spectacles*. Paris, 1758. Reprint, Geneva: Droz, 1948.

Rousseau, Jean-Jacques. *Politics and the Arts: Letter to M. d'Alembert on the Theatre*. Ithaca: Cornell University Press, 1960. [translation of *Lettre à d'Alembert*]

Rousseau, Jean-Jacques. *The Discourses and Other Early Political Writings*. Edited by Victor Gourevitch. 2 vols. Cambridge: Cambridge University Press, 1997.

Sage. *A Letter addressed to a female friend by Mrs. Sage, The first English Female Aerial Traveller; describing the General Appearance and Effects of her Expedition with Mr. Lunardi's Balloon; which ascended from St. George's Fields on Wednesday, 29th June, 1785, accompanied by George Biggin, Esq.* London: Printed for the writer, [1785]. [ECCO]

Saladin, N. J. "Mémoire sur l'inflammation spontanée des herbes cuites dans des corps gras." *Observations et mémoires sur la physique* 25 (1784): 370–72.

Salle. *Moyen de diriger l'Aérostat*. Peking and Paris: Couturier, 1784. [Gallica]

[Saunier, Pierre Maurice]. *Le Triomphe de la machine aérostatique ou l'Anti-Balloniste, converti par l'expérience. Dialogues entre un Envieux & des Amateurs de Physique*. Athens and Paris: Cailleau, 1783. [BNF Tolbiac Vz 2112]

Seconds, Jean-Louis. *Le Navigateur aérien ou nouveaux globes aérostatiques, avec les moyens de les gouverner a volonté*. Rodez: Marin Devic, 1784. [BNF Vz 2127]

Ségur, Louis Philippe comte de. *Mémoires, souvenirs et anecdotes par M. le comte de Ségur*. 3 vols. Paris: Alexis Eymery, 1824.

Servan, Joseph-Michel-Antoine. *Apologie de la Bastille, pour servir de réponse aux Mémoires de M. Linguet sur la Bastille*. Philadelphia, 1784.

Servan, Joseph-Michel-Antoine. *Doutes d'un provincial, proposés à MM. les médecins-commissaires, chargées par le roi de l'examen du magnétisme animal*. Lyon and Paris: Prault, 1784.

Sheldon, John. *Journal and Certificates on the Fourth Voyage of Mr. Blanchard*. London: Printed for the author, 1784. [ECCO]

Sheridan, Charles Francis. *A History of the Late Revolution in Sweden*. Dublin: M. Mills, 1778.

Sheridan, Charles Francis. *Histoire de la dernière révolution de Suède, contenant le récit de ce qui s'est passé dans les trois dernières diètes, et un précis de l'histoire de Suède*. London, 1783.

Southern, John. *A Treatise upon aerostatic machines*. Birmingham: Pearson and Rollason, 1785.

St. John de Crèvecoeur, J. Hector. *Letters from an American Farmer*. London, 1782.

Thiery, Luc-Vincent. *Almanach du voyageur à Paris, contenant une description sommaire mais exacte de tous les monumens, chefs d'oeuvres des arts*. Paris: Hardouin, Gattey, 1785.

Thouret, Michel-Augustin. *Recherches et doutes sur le magnétism animal*. Paris, 1784.

Tournon de la Chapelle, Antoine. *La Vie et les Mémoires de Pilâtre de Rozier, écrits par lui-même & publiés par M.T.* Paris, 1786.

Voltaire. *Essai sur les moeurs et l'esprit des nations*. 2 vols. Paris: Garnier, 1963.

Wall, Martin. *A Syllabus of a Course of Lectures in Chemistry*. Oxford: D. Prince and J. Cooke, 1782. [ECCO]

Walpole, Horace. *Private Correspondence of Horace Walpole, Earl of Orford*. 4 vols. Vol. 4, *1775–1797*. London: Printed for Rodwell and Martin, 1820.

Walpole, Horace. *Horace Walpole's Correspondence with Sir Horace Mann*. Edited by W. S. Lewis, Warren Hunting Smith, and George L. Lam. 11 vols. New Haven: Yale University Press, 1954–1971.

Watermeyer, A. A. "Erläuterung zu der ärostatischen Maschine und der damit gemachten Luftfahrt." *Hannoverisches Magazin* (1784): 49–96.

Watson, Elkanah. *Men and Times of the Revolution: Memoirs of Elkanah Watson, Journals of Travels in Europe and America, from the year 1777 to 1842*. Edited by Winslow Watson. New York: Dana, 1857.

Wieland, Christoph Martin. "Die Äropetomanie, oder Die neuesten Schritte der Franzosen zur Kunst zu fliegen." *Der Teutsche Merkur* (Sept. 1783): 69–96.

Wilkins, John. *The Discovery of a World in the Moone; or, A discourse tending to prove, that 'tis probable there may be another habitable world in that planet*. London: E. Griffin, 1638.

Wille, Jean Georges [Johann Georg]. *Mémoires et Journal de J.-G. Wille, graveur du roi*. Edited by Georges Duplessis. 2 vols. Paris: Jules Renouard, 1857.

Xavier de Maistre. *Lettre . . . contenant une relation de l'expérience aérostatique de Chambéry*. Chambery: M. F. Gorrin, 1784. In Xavier de Maistre, *Les premiers essais*, 49–66.

Xavier de Maistre. "Lettre de l'Hermite de Nivolet sur l'expérience aérostatique faite a Chambéry le 22 Avril 1784." In Xavier de Maistre, *Les premiers essais*, 35–41.

Xavier de Maistre. *Les premiers essais*. Edited by Jules Philippe, Chambéry: A. Perrin, 1874.

Xavier de Maistre. *Prospectus de l'expérience aérostatique de Chambéry*. Chambéry: F. Puthod, 1784. In Xavier de Maistre, *Les premiers essais*, 17–29.

SECONDARY SOURCES

Acomb, Frances. *Mallet du Pan (1749–1800): A Career in Political Journalism*. Durham, NC: Duke University Press, 1973.

[Actes du colloque]. *Soufflot et l'architecture des lumières*. Paris: CNRS, 1980.

Adams, Christine, Jack R. Censer, and Lisa Jane Graham, eds. *Visions and Revisions of Eighteenth- Century France*. University Park: Pennsylvania State University Press, 1997.

Adams, Christine. *A Taste for Comfort and Status: A Bourgeois Family in Eighteenth-Century France*. University Park: Pennsylvania State University Press, 2000.

Adas, Michael. *Machines as the Measure of Men: Science, Technology, and Ideologies of Western Dominance*. Ithaca: Cornell University Press, 1989.

Adkins, Gregory Matthew. *The Idea of the Sciences in the French Enlightenment: A Reinterpretation*. Newark: University of Delaware Press, 2014.

Alcover, Madeleine. *La pensée philosophique et scientifique de Cyrano de Bergerac*. Geneva: Droz, 1970.

Alder, Ken. *Engineering the Revolution: Arms and Enlightenment in France, 1763–1815*. Princeton, NJ: Princeton University Press, 1997.

Alder, Ken. *The Measure of All Things: The Seven-Year Odyssey and Hidden Error that Transformed the World*. New York: Free Press, 2002.

Al-Douri, Taha. "The Constitution of Pleasure: François-Joseph Belanger and the Château de Bagatelle." *Anthropology and Aesthetics* 48 (2005): 155–62.

Alexander, John T. "Aeromania, 'Fire-Balloons,' and Catherine the Great's Ban of 1784." *The Historian* 58 (1996): 497–516.

Allwod, John. *The Great Exhibitions*. London: Studio Vista, 1977.

Alpaugh, Micah. *Non-violence and the French Revolution: Political Demonstrations in Paris, 1787–1795*. Cambridge: Cambridge University Press, 2015.

Alter, Peter. "Playing with the Nation: Napoleon and the Culture of Nationalism."

In *Unity and Diversity in European Culture*, edited by Tim Blanning and Hagen Schultz, 61–75. Oxford: Oxford University Press, 2006.

Altick, Richard. *The Shows of London*. Cambridge: Cambridge University Press, 1978.

Ambrose, Tom. *Godfather of the Revolution: The Life of Philippe Egalité, duc d'Orléans*. London: Peter Owen, 2008.

Amiable, Louis. *Une loge maçonnique d'avant 1789*. Paris: Germer Baillière & C. Félix Alcan, 1897.

Anderson, Benedict Richard O'Gorman. *Imagined Communities: Reflections on the Origin and Spread of Nationalism*. London: Verso, 1983.

Andress, David. *Massacre at the Champ de Mars: Popular Dissent and Political Culture in the French Revolution*. Woodbridge, Suffolk: Boydell Press, 2000.

Apostolidès, Jean-Marie. *Le roi-machine: Spectacle et politique au temps de Louis XIV*. Paris: Minuit, 1981.

Appelbaum, Robert. *Literature and Utopian Politics in Seventeenth-Century England*. Cambridge: Cambridge University Press, 2002.

Arasse, Daniel. *The Guillotine and the Terror*. London: Penguin, 1989.

Aravamudan, Srinivas. *Enlightenment Orientalism: Resisting the Rise of the Novel*. Chicago: University of Chicago Press, 2012.

Arecco, Davide. *Montgolfiere, scienze e lumi nel tardo Settecento: Cultura accademica e conoscenze tecniche dalla vigilia della Rivoluzione francese all'età napoleonica*. Bari: Cacucci, 2003.

Astarita, Tommaso. *Between Salt Water and Holy Water: A History of Southern Italy*. New York: Norton, 2005.

Auricchio, Laura. "Pahin de la Blancherie's Commercial Cabinet of Curiosity (1779–1787)." *Eighteenth-Century Studies* 36 (2002): 47–61.

Baecque, Antoine de. *The Body Politic: Corporeal Metaphor in Revolutionary France, 1770–1800*. Stanford, CA: Stanford University Press, 1997.

Bailey, Colin B. *Patriotic Taste: Collecting Modern Art in Pre-revolutionary Paris*. New Haven: Yale University Press, 2002.

Bain, Robert Nisbet. *Gustavus III and His Contemporaries*. London: Kegan Paul, 1894.

Baker, Keith Michael. *Condorcet: From Natural Philosophy to Social Mathematics*. Chicago: University of Chicago Press, 1975.

Baker, Keith Michael, ed. *The Political Culture of the Old Regime*. Oxford: Pergamon Press, 1987.

Baker, Keith Michael. *Inventing the French Revolution: Essays on French Political Culture in the Eighteenth Century*. Cambridge: Cambridge University Press, 1990.

Baker, Keith Michael. "A Foucauldian French Revolution?" In Goldstein, *Foucault and the Writing of History*, 187–205.

Baker, Keith Michael. "Transformations of Classical Republicanism in Eighteenth-Century France." *Journal of Modern History* 73 (2001): 32–53.

Bakhtin, Mikhail Mikhaïlovich. *Rubelais and His World*. Cambridge, MA: MIT Press, 1968.

Ballot, Charles. *L'Introduction du machinisme dans l'industrie française*. Lille: O. Marquant, 1923.

Barber, Benjamin R. "Rousseau and the Paradoxes of the Dramatic Imagination." *Daedalus* 107 (1978): 79–92.

Barber, Elinor G. *The Bourgeoisie in 18th Century France*. Princeton, NJ: Princeton University Press, 1955.

Barkan, Leonard. *The Gods Made Flesh: Metamorphosis and the Pursuit of Paganism*. New Haven: Yale University Press, 1986.

Barker, Nancy N. "Philippe d'Orléans, frère unique du roi: Founder of the Family Fortune." *French Historical Studies* 13 (1983): 145–71.

Barnes, Teresa. "Soccer Nation/Corporation." *Journal of Sports and Social Issues* 35 (2011): 101–6.

Barny, Roger. *Le comte d'Antraigues: Un disciple aristocrate de J.-J. Rousseau. De la fascination au reniement 1782–1797*. Oxford: Voltaire Foundation, 1991.

Barolsky, Paul. "As in Ovid, So in Renaissance Art." *Renaissance Quarterly* 51 (1988): 451–74.

Barrière, Pierre. *L'Académie de Bordeaux: Centre de culture internationale au XVIIIe siècle (1712–1792)*. Bordeaux: Editions Bière, 1951.

Bartlett, Robert. *The Natural and the Supernatural in the Middle Ages*. Cambridge: Cambridge University Press, 2008.

Bartlett, Thomas. *The Fall and Rise of the Irish Nation: The Catholic Question, 1690–1830*. Dublin: Gill and Macmillan, 1992.

Barton, H. Arnold. "Gustav III of Sweden and the Enlightenment." *Eighteenth-Century Studies* 6 (1972): 1–34.

Bayly, Christopher Alan. *Imperial Meridian: The British Empire and the World, 1780–1830*. London: Longman, 1989.

Beachy, Robert. *The Soul of Commerce: Credit, Property, and Politics in Leipzig, 1750–1840*. Leiden: Brill, 2005.

Beales, Derek. *Prosperity and Plunder: European Catholic Monasteries in the Age of Revolution, 1650–1815*. Cambridge: Cambridge University Press, 2003.

Beaumont-Maillet, *L'Eau à Paris*. Paris: Hazan, 1991.

Beaurepaire, Pierre-Yves. "The Universal Republic of the Freemasons and the Culture of Mobility in the Enlightenment." *French Historical Studies* 29 (2006): 407–31.

Bégin, Émile-Auguste. *Pilâtre de Rozier et les aérostats*. N.d. [Gallica]

Beik, William. *Absolutism and Society in Seventeenth-Century France: State Power and*

Provincial Aristocracy in Languedoc. Cambridge: Cambridge University Press, 1985.

Beik, William. *Urban Protest in Seventeenth-Century France: The Culture of Retribution.* Cambridge: Cambridge University Press, 1997.

Belhost, Bruno. *Paris savant: Parcours et rencontres au temps des Lumières.* Paris: A. Colin, 2011.

Bell, David A. *Lawyers and Citizens: The Making of a Political Elite in Old Regime France.* New York: Oxford University Press, 1994.

Bell, David A. *The Cult of the Nation in France: Inventing Nationalism, 1680–1800.* Cambridge, MA: Harvard University Press, 2001.

Bell, David A. "The Unbearable Lightness of Being French: Law, Republicanism and National Identity at the End of the Old Regime." *American Historical Review* 106 (2001): 1215–35.

Bell, David A. "A Very Different French Revolution." *New York Review of Books*, July 10, 2014.

Benhamou, Reed. "Cours public: Elective Education in the Eighteenth Century." *Studies on Voltaire and the Eighteenth Century* 241 (1986): 365–76.

Benhamou, Reed. "From Curiosité to Utilité: The Automaton in Eighteenth-Century France." *Studies in Eighteenth-Century Culture* 17 (1987): 91–105.

Benhamou, Reed. "The Education of the Architect in Eighteenth-Century France." *British Journal for Eighteenth-Century Studies* 12 (1989): 187–99.

Bennett, Jim A., and Sofia Talas, eds. *Cabinets of Experimental Philosophy in Eighteenth-Century Europe.* Leiden: Brill, 2013.

Bensaude-Vincent, Bernadette, and Christine Blondel, eds. *Science and Spectacle in the European Enlightenment.* Aldershot: Ashgate, 2008.

Beretta, Marco, ed. *Lavoisier in Perspective.* Munich: Deutsches Museum, 2005.

Berg, Maxine. "From Imitation to Invention: Creating Commodities in Eighteenth-Century Britain." *Economic History Review* 55 (2002): 1–30.

Berg, Maxine. *Luxury and Pleasure in Eighteenth-Century Britain.* Oxford: Oxford University Press, 2005.

Berlanstein, Lenard R. *Daughters of Eve: A Cultural History of French Theater Women from the Old Regime to the Fin-de-siècle.* Cambridge, MA: Harvard University Press, 2001.

Berman, Alex. "The Cadet Circle: Representatives of an Era in French Pharmacy." *Bulletin of the History of Medicine* 40 (1966): 101–11.

Bermingham, Ann, and John Brewer, eds. *The Consumption of Culture, 1600–1800: Image, Object, Text.* London: Routledge, 1995.

Bermingham, Ann. "Elegant Females and Gentlemen Connoisseurs: The Commerce in Culture and Self-Image in Eighteenth-Century England." In Bermingham and Brewer, *The Consumption of Culture,* 489–513.

Bertucci, Paola. "Enlightened Secrets: Silk, Intelligent Travel, and Industrial Espionage in Eighteenth-Century France." *Technology and Culture* 54 (2013): 820–52.

Biagioli, Mario. *Galileo, Courtier: The Practice of Science in the Culture of Absolutism.* Chicago: University of Chicago Press, 1993.

Bideaux, Michel, and Sonia Faessel. "Introduction." In *Voyage autour du monde*, by Bougainville, 1–46. Paris: Sorbonne, 2001.

Billington, James H. *The Icon and the Axe: An Interpretive History of Russian Culture.* New York: Knopf, 1966.

Bingham, Alfred Jepson. *Marie-Joseph Chénier, Early Political Life and Ideas, 1789–1794.* New York: Privately printed, 1939.

Birn, Raymond. *Royal Censorship of Books in Eighteenth-Century France.* Stanford, CA: Stanford University Press, 2012.

Birn, Richard. "Reinventing le Peuple in 1789." *History Teacher* 23 (1990): 363–73.

Bishop, Karen. "Myth Turned Monument: Documenting the Historical Imaginary in Buenos Aires and Beyond." *Journal of Modern Literature* 30 (2007): 151–62.

Black, Jeremy. *Natural and Necessary Enemies: Anglo-French Relations in the Eighteenth Century.* Athens: University of Georgia Press, 1987.

Black, Jeremy. *The British Abroad: The Grand Tour in the Eighteenth Century.* New York: St. Martin's Press, 1992.

Black, Robert. *Humanism and Education in Medieval and Renaissance Italy.* Cambridge: Cambridge University Press, 2001.

Blumenberg, Hans. *Shipwreck with Spectator: Paradigm of a Metaphor for Existence.* Cambridge, MA: MIT Press, 1997.

Bonnet, Jean-Claude. *Louis Sébastien Mercier 1740–1814: Un hérétique en litérature.* Paris: Mercure de France, 1995.

Bordes, Maurice. *L'administration provinciale et municipale en France au XVIIIe siècle.* Paris: Société d'édition d'enseignement supérieur, 1972.

Bosher, J. F. *The Single Duty Project: A Study of the Movement for a French Customs Union in the Eighteenth Century.* London: Athlone Press, 1964.

Bossenga, Gail. *The Politics of Privilege: Old Regime and Revolution in Lille.* Cambridge: Cambridge University Press, 1991.

Bossenga, Gail. "Financial Origins of the French Revolution." In Kaiser and Van Kley, *From Deficit to Deluge*, 37–66.

Bouchary, J. *L'eau à Paris, à la fin du XVIIIe siècle: La Compagnie des eaux de Paris et l'entreprise de l'Yvette.* Paris: M. Rivière, 1946.

Bourdieu, Pierre. *Outline of a Theory of Practice.* Cambridge: Cambridge University Press, 1977.

Bourdieu, Pierre. *Distinction: A Social Critique of the Judgement of Taste.* Cambridge, MA: Harvard University Press, 1984.

Bourdieu, Pierre. *The Field of Cultural Production: Essays on Literature and Art.* New York: Columbia University Press, 1993.

Bouton, Cynthia. *The Flour War: Gender, Class and Community in Late Ancien Régime French Society.* University Park: Pennsylvania State University Press, 1993.

Bouyssy, Marie-Thèrese. *Le Musée de Bordeaux: 1783–1789, étude psycho-sociologique d'une société des lumières.* Paris: Hachette, 1973.

Boyd, Malcolm, ed. *Music and the French Revolution.* Cambridge: Cambridge University Press, 1992.

Boyé, Pierre. *Les premières experiences aérostatiques faites en Lorraine (1783–1788).* Paris: Berger-Levrault & Cie, 1909.

Braham, Allan. *The Architecture of the French Enlightenment.* Berkeley: University of California Press, 1980.

Brand, Dana. *The Spectator and the City in Nineteenth-Century American Literature.* Cambridge: Cambridge University Press, 1991.

Brandstetter, T. "The Most Wonderful Piece of Machinery the World Can Boast of: The Water-Works at Marly, 1680–1830." *History and Technology* 21 (2005): 205–20.

Braure, Maurice. *Lille et la Flandre Wallonne au XVIIIe siècle.* 2 vols. Lille: Émile Raoust, 1932.

Bredin, Miles. *The Pale Abyssinian: A Life of James Bruce, African Explorer and Adventurer.* London: HarperCollins, 2000.

Brennan, Thomas. *Public Drinking and Popular Culture in Eighteenth-Century Paris.* Princeton, NJ: Princeton University Press, 1988.

Bret, Patrice. *L'état, l'armée, la science, l'invention de la recherche publique en France, 1763–1830.* Rennes: Presses universitaires de Rennes, 2002.

Bret, Patrice. "Power, Sociability and Dissemination of Science: Lavoisier and the Learned Societies." In Beretta, *Lavoisier in Perspective*, 129–52.

Brewer, John. *Party Ideology and Popular Politics at the Accession of George III.* Cambridge: Cambridge University Press, 1976.

Brewer, John. *Sinews of Power: War, Money and the English State, 1688–1783.* New York: Knopf, 1988.

Brewer, John, and Roy Porter, eds. *Consumption and the World of Goods.* Princeton, NJ: Princeton University Press, 1993.

Brewer, John. "'The Most Polite Age and the Most Vicious': Attitudes towards Culture as a Commodity, 1600–1800." In Bermingham and Brewer, *Consumption of Culture*, 341–61.

Brewer, John. *The Pleasures of the Imagination: English Culture in the Eighteenth Century.* Chicago: University of Chicago Press, 1997.

Brewer, David A. *The Afterlife of Character, 1726–1825.* Philadelphia: University of Pennsylvania Press, 2005.

Brey, Philip. "Artifacts as Social Agents." In *Inside the Politics of Technology: Agen-*

cy and Normativity in the Co-production of Technology and Society, edited by Hans Harbers, 61–84. Amsterdam: Amsterdam University Press, 2005.

Brian, Éric. *La Mesure de l'État: Administrateurs et géomètres au XVIIIe siècle*. Paris: Albin Michel, 1994.

Bristol, Michael D. *Carnival and Theater: Plebian Culture and the Structure of Authority in Renaissance England*. New York: Methuen, 1985.

Britsch, Amédée. *La maison d'Orléans à la fin de l'ancien régime: La Jeunesse de Philippe-Égalité (1747–1785) d'après des documents inédits*. Paris: Payot, 1926.

Brockliss, L. W. B. *French Higher Education in the Seventeenth and Eighteenth Centuries: A Cultural History*. Oxford: Clarendon Press, 1987.

Brockliss, L. W. B. "The Scientific Revolution in France." In *The Scientific Revolution in National Context*, edited by Roy Porter and Mikuláš Teich, 11–54. Cambridge: Cambridge University Press, 1992.

Brockliss, L. W. B. *Calvet's Web: Enlightenment and the Republic of Letters in Eighteenth-Century France*. Oxford: Oxford University Press, 2002.

Brooke, John L. "Reason and Passion in the Public Sphere: Habermas and the Cultural Historians." *Journal of Interdisciplinary History* 29 (1998): 43–67.

Brown, Barbara Traxler. "French Scientific Innovation in Late Eighteenth-Century Dublin: The Hydrogen Balloon Experiments of Richard Crosbie (1783–1785)." In *Ireland and the French Enlightenment, 1700–1800*, edited by Graham Gargett and Geraldine Sheridan, 107–26. London: Macmillan, 1999.

Brown, Gregory S. "Scripting the Patriotic Playwright in Enlightenment-Era France: Louis-Sébastien Mercier's Self-Fashionings between Court and Public." *Historical Reflections/Réflexions historiques* 26 (2000): 31–57.

Brown, Gregory S. *Literary Sociability and Literary Property in France, 1775–1793: Beaumarchais, the Société des auteurs dramatiques and the Comédie Française*. Aldershot: Ashgate, 2006.

Brown, Laura. *Fables of Modernity: Literature and Culture in the English Eighteenth Century*. Ithaca: Cornell University Press, 2001.

Brown, Rory. "The Diamond Necklace Affair Revisited." *Renaissance and Modern Studies* 33 (1989): 21–40.

Brownlee, Kevin. "Phaeton's Fall and Dante's Ascent." *Dante Studies* 102 (1984): 135–44.

Bruhns, Karl, ed. *Life of Alexander von Humboldt*. 2 vols. London: Longmans Green, 1873.

Bryant, Lawrence M. *The King and the City in the Parisian Royal Entry Ceremony*. Geneva: Droz, 1986.

Bryant, Lawrence M. "Royal Ceremony and the Revolutionary Strategies of the Third Estate." *Eighteenth-Century Studies* 22 (1989): 413–50.

Bryson, Scott S. *The Chastised Stage: Bourgeois Drama and the Exercise of Power*. Saratoga, CA: Anma Libri, 1991.

Bullion, John. "George III on Empire, 1783." *William and Mary Quarterly* 51 (1994): 305–10.

Buranelli, Vincent. *The Wizard from Vienna*. New York: Coward, McCann and Geoghegan, 1975.

Burke, Janet M. "Freemasonry, Friendship and Noblewomen: The Role of the Secret Society in Bringing Enlightenment Thought to Pre-revolutionary Women Elites." *History of European Ideas* 10 (1989): 283–93.

Burke, Janet M., and Margaret Jacob. "French Freemasonry, Women and Feminist Scholarship." *Journal of Modern History* 68 (1996): 513–49.

Burke, Peter. *Popular Culture in Early Modern Europe*. New York: NYU Press, 1978.

Burke, Peter. *The Fabrication of Louis XIV*. New Haven: Yale University Press, 1992.

Buron Pilâtre, Philippe. *Pilâtre de Rozier, un Lorrain d'exception, 1754–1785*. Metz: Serpenoise, 2006.

Burrows, Simon. "The Innocence of Jacques-Pierre Brissot." *Historical Journal* 46 (2003): 843–71.

Burrows, Simon. *Blackmail, Scandal, and Revolution: London's French Libellistes, 1758–92*. Manchester: Manchester University Press, 2006.

Bynum, Caroline Walker. *Metamorphosis and Identity*. New York: Zone Books, 2001.

Cabanes, Charles. "Histoire du premier Musée autorisé par le Gouvernement." *La Nature* 3015 (1937): 577–83.

Calhoun, Craig, ed. *Habermas and the Public Sphere*. Cambridge, MA: MIT Press, 1994.

Camcastle, Cara. *The More Moderate Side of Joseph de Maistre: Views on Political Liberty and Political Economy*. Montreal: McGill-Queen's University Press, 2005.

Camp, Pannill. *The First Frame: Theater Space in Enlightenment France*. Cambridge: Cambridge University Press, 2015.

Campbell, Colin. *The Romantic Ethic and the Spirit of Modern Consumerism*. Oxford: Blackwell, 1987.

Campbell, Peter Robert. *Power and Faction in Louis XV's France*. Princeton, NJ: Princeton University Press, 1994.

Campbell, Stephen J., and Stephen J. Milner, eds. *Artistic Exchange and Cultural Translation in the Italian Renaissance City*. Cambridge: Cambridge University Press, 2004.

Campbell, Stephen J. "'Our Eagles always Held Fast to Your Lilies': The Este, the Medici, and the Negotiation of Cultural Identity." In Campbell and Milner, *Artistic Exchange*, 138–61.

Cannon, John. *The Fox-North Coalition: Crisis of the Constitution, 1782–4*. Cambridge: Cambridge University Press, 1969.

Caradonna, Jeremy L. "The Monarchy of Virtue: The 'Prix de Vertu' and the Economy of Emulation in France, 1777–91." *Eighteenth-Century Studies* 41 (2008): 443–58.

Carlson, Marvin. *Places of Performance: The Semiotics of Theatre Architecture.* Ithaca: Cornell University Press, 1989.

Carlson, Marvin. "Theater Audiences and the Reading of Performance." In *Interpreting the Theatrical Past: Essays in the Historiography of Performance,* edited by Thomas Postlewait and Bruce A. McConachie, 82–98. Iowa City: University of Iowa Press, 1989.

Carroll, Patrick. *Science, Culture and Modern State Formation.* Berkeley: University of California Press, 2006.

Case, Sue-Ellen, and Janelle Reinelt. *The Performance of Power: Theatrical Discourse and Politics.* Iowa City: University of Iowa Press, 1991.

Cash, Arthur H. *John Wilkes: The Scandalous Father of Civil Liberty.* New Haven: Yale University Press, 2006.

Casid, Jill H. "Queer(y)ing Georgic: Utility, Pleasure, and Marie-Antoinette's Ornamented Farm." *Eighteenth-Century Studies* 30 (1997): 304–18.

Cazenove, Raoul de. *Premiers voyages aériens à Lyon en 1784.* Lyon: Pitrat aîné, 1887.

Céleste, Raymond. "La société philomatique de Bordeaux de 1783 à 1808." *Revue philomatique de Bordeaux et du Sud-Ouest* 1 (1897–1898): 65–83.

Censer, Jack R., and Jeremy D. Popkin, eds. *Press and Politics in Pre-revolutionary France.* Berkeley: University of California Press, 1987.

Censer, Jack R. *The French Press in the Age of Enlightenment.* Princeton, NJ: Princeton University Press, 1994.

Chambers, Neil. *Joseph Banks and the British Museum: The World of Collecting, 1770–1830.* London: Pickering and Chatto, 2007.

Chapin, Seymour L. "A Legendary Bon Mot? Franklin's 'What Is the Good of a Newborn Baby?'" *Proceedings of the American Philosophical Society* 129 (1985): 278–90.

Chapman, S. D., and S. Chassagne. *European Textile Printers in the Eighteenth Century: A Study of Peel and Oberkampf.* London: Heinemann, 1981.

Chard, Chloe. *Pleasure and Guilt on the Grand Tour: Travel Writing and Imaginative Geography, 1600–1830.* New York: Manchester University Press, 1999.

Chartier, Roger. "Culture as Appropriation." In *Understanding Popular Culture: Europe from the Middle Ages to the Nineteenth Century,* edited by S. L. Kaplan, 229–53. The Hague: Mouton, 1984.

Chartier, Roger. *The Cultural Origins of the French Revolution.* Durham, NC: Duke University Press, 1991.

Chartier, Roger. "The Chimera of the Origin: Archeology, Cultural History and the French Revolution." In Goldstein, *Foucault and the Writing of History,* 167–86.

Chartrand, Rene. *Gibraltar, 1779–1783: The Great Siege.* Oxford: Osprey, 2006.

Chassagne, Serge. *Oberkampf: Un entrepreneur capitaliste au siècle des lumières.* Paris: Aubier Montaigne, 1980.

Christie, Ian R. *Stress and Stability in Late Eighteenth-Century Britain: Reflections on the British Avoidance of Revolution.* Oxford: Clarendon Press, 1984.

Cherpack, Clifton. "The Structure of Chénier's *L'Invention.*" *PMLA* 72 (1957): 74–83.

Chisick, Harvey. *The Limits of Reform in the Enlightenment.* Princeton, NJ: Princeton University Press, 1981.

Chisick, Harvey. "Public Opinion and Political Culture in France during the Second Half of the Eighteenth Century." *English Historical Review* 470 (2002): 48–77.

Clark, William, Jan Golinski, and Simon Schaffer, eds. *The Sciences in Enlightened Europe.* Chicago: University of Chicago Press, 1999.

Clay, Lauren. "Patronage, Profits, and Public Theaters: Rethinking Cultural Unification in Ancien Régime France." *Journal of Modern History* 79 (2007): 729–71.

Clay, Lauren. *Stagestruck: The Business of Theater in Eighteenth-Century France and Its Colonies.* Ithaca: Cornell University Press, 2013.

Clegg, Brian. *The First Scientist: A Life of Roger Bacon.* New York: Caroll and Graf, 2003.

Clifford, James. *Routes: Travel and Translation in the Late Twentieth Century.* Cambridge, MA: Harvard University Press, 1997.

Coates, Austin. *The Commerce in Rubber: The First 250 Years.* New York: Oxford University Press, 1987.

Cobb, R. C. *The Police and the People: French Popular Protest, 1789–1820.* Oxford: Clarendon Press, 1970.

Cohen, Claude, Lauren Pelpel, and Marie-Pierre Perdrizet, eds. *La formation architecturale au dix-huitième siècle en France.* Paris, 1980.

Cointat, Michel. *Rivarol (1753–1801): Un écrivain controversé.* Paris: L'Harmattan, 2001.

Coke, David, and Alan Borg. *Vauxhall Gardens.* New Haven: Yale University Press, 2011.

Colley, Linda. "The Apotheosis of George III: Loyalty, Royalty and the British Nation, 1760–1820." *Past & Present* 102 (1984): 94–129.

Colley, Linda. *Britons: Forging the Nation, 1707–1837.* New Haven: Yale University Press, 1992.

Colwill, Elizabeth. "Just Another Citoyenne? Marie-Antoinette on Trial, 1790–1793." *History Workshop* 28 (1989): 63–87.

Combes, André. *Histoire de la Franc-maçonnerie à Lyon des origines à nos jours.* Brignais, France: Traboules, 2006.

Comninel, George C. "The Political Context of the Popular Movement in the French Revolution." In Krantz, *History from Below*, 115–40.

Conklin, Alice. *A Mission to Civilize: The Republican Idea of Empire in France and West Africa, 1895–1930.* Stanford, CA: Stanford University Press, 1997.

Conlin, Jonathan. "Wilkes, the Chevalier d'Éon, and 'the Dregs of Liberty': An Anglo-French Perspective on Ministerial Despotism, 1762–1771." *English Historical Review* 120 (2005): 1251–88.

Conlin, Jonathan. "The Afterlife of a London Pleasure Garden, 1770–1859." *Journal of British Studies* 45 (2006): 718–43.

Conner, Clifford D. *Jean Paul Marat: Scientist and Revolutionary*. Atlantic Highlands, NJ: Humanities Press, 1998.

Cooper, James. *Colonialism in Question*. Berkeley: University of California Press, 2005.

Cooter, R., and S. Pumfrey. "Separate Spheres and Public Places." *History of Science* 32 (1994): 237–67.

Copeland, Thomas W. "Burke and Dodsley's Annual Register." *PMLA* 54 (1939): 223–45.

Coquard, Olivier. *Jean-Paul Marat*. Paris: Fayard, 1993.

Coquery, Natacha. *Tenir boutique à Paris au XVIIIe siècle: Luxe et demi-luxe*. Paris: Comité des travaux historiques et scientifiques, 2011.

Cornu, Paul, ed. *Galerie des modes et costumes français dessinés d'après nature, 1778–1787*. 4 vols. Paris: Émile Lévy, 1912.

Cottom, Daniel. "Taste and the Civilized Imagination." *Journal of Aesthetics and Art Criticism* 39 (1981): 367–80.

Coutil, Léon. *Jean-Pierre Blanchard, physicien-aéronaute*. Évreux, France: Charles Hérissey, 1911.

Cowans, John. *To Speak for the People: Public Opinion and the Problem of Legitimacy in the French Revolution*. Princeton, NJ: Princeton University Press, 2001.

Coward, David. *The Philosophy of Restif de la Bretonne*. Oxford: Voltaire Foundation, 1991.

Cowart, Georgia J. "Carnival in Venice or Protest in Paris? Louis XIV and the Politics of Subversion at the Paris Opéra." *Journal of the American Musicological Society* 54 (2001): 265–302.

Cowart, Georgia J. *The Triumph of Pleasure: Louis XIV and the Politics of Spectacle*. Chicago: University of Chicago Press, 2008.

Coyle, Eugene A. "Sir Edward Newenham: The 18th Century Dublin Radical." *Dublin Historical Record* 46 (1993): 15–30.

Creveaux, Eugène, and Henri Alibaux. *Un grand ingénieur papetier: Jean-Guillaume Écrevisse, collaborateur de Nicolas Desmarest*. Grenoble: Éditions de l'Industrie papetière, 1937.

Crosland, Maurice P. *The Society of Arcueil: A View of French Science at the Time of Napoleon I*. Cambridge, MA: Harvard University Press, 1967.

Crosland, Maurice P. "Relationships between the Royal Society and the Académie des Sciences in the Late Eighteenth Century." *Notes and Records of the Royal Society of London* 59 (2005): 25–34.

Crouch, Tom D. *The Eagle Aloft: Two Centuries of the Balloon in America*. Washington, DC: Smithsonian Institution Press, 1983.

Crouch, Tom D., ed. *The Genesis of Flight: The Aeronautical History Collection of Colonel Richard Gimbel*. Los Angeles: Perpetua Press, 2000.

Crouzet, François. *Britain Ascendant: Comparative Studies in Franco-British Economic History*. Cambridge: Cambridge University Press, 1990.

Crow, Thomas E. *Painters and Public Life in Eighteenth-Century Paris*. New Haven: Yale University Press, 1985.

Crow, Thomas E. *Emulation: Making Artists for Revolutionary France*. New Haven: Yale University Press, 1995.

Dakin, Douglas. *Turgot and the Ancien Régime in France*. London: Methuen, 1939.

Darnton, Robert. "The Grub Street Style of Revolution: J.-P. Brissot, Police Spy." *Journal of Modern History* 40 (1968): 302–27.

Darnton, Robert. *Mesmerism and the End of the Enlightenment in France*. Cambridge, MA: Harvard University Press, 1968.

Darnton, Robert. "The Memoirs of Lenoir, Lieutenant de Police of Paris, 1774–1785." *English Historical Review* 85 (1970): 532–59.

Darnton, Robert. *The Business of Enlightenment: A Publishing History of the Encyclopédie, 1775–1800*. Cambridge, MA: Belknap Press of Harvard University Press, 1979.

Darnton, Robert. *The Literary Underground of the Old Regime*. Cambridge, MA: Harvard University Press, 1982.

Darnton, Robert. *The Great Cat Massacre and Other Episodes in French Cultural History*. New York: Vintage Books, 1984.

Darnton, Robert. "The Brissot Dossier." *French Historical Studies* 17 (1991): 191–205.

Darnton, Robert. *The Forbidden Best-Sellers of Pre-revolutionary France*. New York: Norton, 1995.

Darnton, Robert. "Poetry and the Police in Eighteenth-Century Paris." *Studies on Voltaire and the Eighteenth Century* 371 (1999): 1–22.

Darnton, Robert. "The News in Paris: An Early Information Society." In *George Washington's False Teeth: An Unconventional Guide to the Eighteenth Century*, 25–75. New York: Norton, 2003.

Darnton, Robert. *The Devil in the Holy Water or the Art of Slander from Louis XIV to Napoleon*. Philadelphia: University of Pennsylvania Press. 2010.

Darnton, Robert. *Poetry and Police: Communication Networks in Eighteenth-Century Paris*. Cambridge, MA: Harvard University Press, 2010.

Daston, Lorraine. "Nationalism and Scientific Neutrality under Napoleon." In *Solomon's House Revisited: The Organization and Institutionalization of Science*, edited by Tore Frängsmyr, 95–119. Canton, MA: Science History Publications, 1990.

Davies, Mark. *King of All Balloons: The Adventurous Life of James Sadler, the First English Aeronaut.* Stroud: Amberley, 2015.

Davis, Natalie Zemon. *Fiction in the Archive: Pardon Tales and Their Tellers in Sixteenth Century France.* Stanford, CA: Stanford University Press, 1987.

Dawson, Warren R., ed. *The Banks Letters: A Calendar of the Manuscript Correspondence of Sir Joseph Banks.* London: British Museum, 1958.

De Baecque, Antoine. *Body Politic: Corporeal Metaphor in Revolutionary France, 1777–1800.* Stanford, CA: Stanford University Press, 1997.

Debièvre, Eugène. *Notes sur l'Histoire de l'Aérostation dans la Région du nord de la France (1783 à 1851).* Paris: Revue du Nord, 1895.

Debord, Guy. *The Society of the Spectacle.* New York: Zone Books, 1995.

Debove, Philippe-Jean. *Blanchard ou l'aiguillon de la Gloire.* Paris: Edilivre, 2011.

Dejean, Joan. *Libertine Strategies: Freedom and the Novel in Seventeenth Century France.* Columbus: Ohio State University Press, 1981.

Dejean, Joan. *The Essence of Style: How the French Invented High Fashion, Fine Food, Chic Cafés, Style, Sophistication and Glamour.* New York: Free Press, 2005.

Deldonna, Anthony R. *Opera, Theatrical Culture and Society in Late Eighteenth-Century Naples.* Aldershot: Ashgate, 2012.

Desan, Suzanne. *Reclaiming the Sacred: Lay Religion and Popular Politics in Revolutionary France.* Ithaca: Cornell University Press, 1990.

Descat, Sophie. *Le Voyage d'Italie de Pierre-Louis Moreau: Journal intime d'un architecte des lumieres (1754–1757).* Bordeaux: Presses universitaires de Bordeaux, 2004.

Devaux, Guy. "Jean-André Cazalet, pittoresque apothicaire bordelaise du XVIIIe siècle." *Bulletin de la société de la pharmacie de Bordeaux* 144 (2005): 333–48.

Devine, T. M., and J. R. Young, ed. *Eighteenth Century Scotland: New Perspectives.* East Lothian, Scotland: Tuckwell Press, 1999.

Dewald, Jonathan. *The Formation of a Provincial Nobility: The Magistrates of the Parlement of Rouen, 1499–1610.* Princeton, NJ: Princeton University Press, 1980.

Dias, Rosie. "A World of Pictures: Pall Mall and the Topography of Display, 1780–99." In *Georgian Geographies: Essays on Space, Place and Landscape in the Eighteenth Century,* edited by Miles Ogborn and Charles W. J. Withers, 92–113. Manchester: Manchester University Press, 2004.

Digonnet, Félix. *L'invention de l'aérostation à Avignon en 1782 et les premières ascensions dans cette ville.* Avignon: François Séguin, 1906.

Doig, K. H., F. A. Kafker, J. Loveland, and D. A. Trinkle. "James Tytler's Edition (1777–1784): A Vast Expansion and Improvement." In *The Early Britannica (1768–1803): The Growth of an Outstanding Encyclopedia,* 69–155. Oxford: Voltaire Foundation, 2009.

Dorveaux, Paul. "Les grands pharmaciens: IX. Pilatre de Rozier." *Bulletin de la Société d'histoire de la pharmacie* 27 (1920): 209–20.

Dorveaux, Paul. "Apothicaires membres de l'Académie des Sciences: X. Louis-Claude Cadet, dit Cadet de Gassicourt, alias Cadet-Gassicourt." *Revue d'histoire de la pharmacie* 22 (1934): 385–97.

Doyle, William. *The Parlement of Bordeaux and the End of the Old Regime, 1771–1790.* New York: St. Martin's Press, 1974.

Duck, Leigh Anne. "Plantation/Empire." *New Centennial Review* 10 (2010): 77–87.

Duckworth, Colin. *The D'Antraigues Phenomenon: The Making and Breaking of a Revolutionary Royalist Espionage Agent.* Newcastle-upon-Tyne: Avero Publications, 1986.

Duhem, Jules. *Histoire des idées aéronautiques avant Montgolfier.* Paris: Fernand Sorlot, 1943.

Duhem, Jules. *Musée aéronautique avant Montgolfier: Recueil de figures et de documents pour servir a l'histoire des idées aéronautiques avant l'invention des aerostats.* Paris: Fernand Sorlot, 1944.

Dumas, Antoine, ed. *Le Temps des ballons: Art et histoire.* Paris: Martinière, 1994.

Duplessis, Robert S. *Lille and the Dutch Revolt: Urban Stability in an Era of Revolution, 1500–1582.* Cambridge: Cambridge University Press, 1991.

Duval, Clément. "Pilâtre de Rozier (1754–1785), Chemist and First Aeronaut." *Chymia* 12 (1967): 99–117.

Dwyer, Philip. "Napoleon and the Universal Monarchy." *History* 95 (2010): 293–307.

Dziembowski, Edmond. *Un nouveau patriotisme français, 1750–1770: La France face à la puissance anglaise à l'époque de la guerre de Sept Ans.* Oxford: Voltaire Foundation, 1998.

Eagles, Robin. *Francophilia in English Society, 1748–1815.* London: Macmillan, 2000.

Echeverria, Durand. *Mirage in the West: A History of the French Image of American Society to 1815.* Princeton, NJ: Princeton University Press, 1957.

Echeverria, Durand. "Condorcet's *The Influence of the American Revolution on Europe*." *William and Mary Quarterly* 25 (1968): 85–108.

Echeverria, Durand. *The Maupeou Revoluton: A Study in the History of Libertarianism in France, 1770–1774.* Baton Rouge: Louisiana State University Press, 1985.

Edelstein, Dan, and Bettina R. Lerner, ed. *Myth and Modernity.* New Haven: Yale University Press, 2007.

Edelstein, Dan. *The Terror of Natural Right: Republicanism, the Cult of Nature, and the French Revolution.* Chicago: University of Chicago Press, 2009.

Edelstein, T. J. *Vauxhall Gardens.* New Haven: Yale University Press, 1983.

Edmonds, W. D. *Jacobinism and the Revolt of Lyon, 1789–1793.* Oxford: Clarendon Press, 1990.

Élart, Joann. "Les origines du concert public à Rouen à la fin de l'Ancien Régime." *Revue de Musicologie* 93 (2007): 53–73.

Eley, Geoffrey. "What Is Cultural History?" *New German Critique* 65 (1995): 19–36.

Elias, Norbert. *The Court Society.* Oxford: Blackwell, 1983.

Ellergy, Eloise. *Brissot de Warville: A Study in the History of the French Revolution.* Boston: Houghton Mifflin, 1915.

Elliott, John H. *Empires of the Atlantic World: Britain and Spain in America, 1492–1830.* New Haven: Yale University Press, 2006.

Elliott, Paul A. *The Derby Philosophers: Science and Culture in British Urban Society, 1700–1850.* Manchester: Manchester University Press, 2009.

Elliott, Robert C. *The Shape of Utopia: Studies in a Literary Genre.* Chicago: University of Chicago Press, 1970.

Ellis, Harold A. *Boulainvilliers and the French Monarchy: Aristocratic Politics in Early Eighteenth-Century France.* Ithaca: Cornell University Press, 1988.

Emden, Christian. *Friedrich Nietzsche and the Politics of History.* Cambridge: Cambridge University Press, 2008.

Erickson, Lars O. "Methodical Invention: Scientific Imagination in the French Enlightenment." *Symposium* 58 (2004): 3–14.

Etkind, Alexander. *Internal Colonization: Russia's Imperial Experience.* Cambridge: Polity, 2011.

Ewing, Heather. *The Lost World of James Smithson: Science, Revolution, and the Birth of the Smithsonian.* New York: Bloomsbury, 2007.

Fairchilds, Cissie. "The Production and Marketing of Populuxe Goods in Eighteenth-Century Paris." In Brewer and Porter, *Consumption and the World of Goods*, 228–48.

Fara, Patricia. "Benjamin West's Portrait of Banks." *Endeavour* 24 (2000): 1–3.

Farge, Arlette. *Fragile Lives: Violence, Power and Solidarity in Eighteenth-Century Paris.* Cambridge, MA: Harvard University Press, 1993.

Farge, Arlette. *Subversive Words: Public Opinion in Eighteenth-Century France.* University Park: Pennsylvania State University Press, 1995.

Fauque, Danielle M. E. "An Englishman Abroad: Charles Blagden's Visit to Paris in 1783." *Notes and Records of the Royal Society of London* 62 (2008): 373–90.

Fergusson, James B. *Balloon Tytler.* London: Faber and Faber, 1972.

Ferrone, Vincent. "Man of Science." In *Enlightenment Portraits*, edited by Michel Vovelle, 190–225. Chicago: University of Chicago Press, 1997.

Feydeau, Elisabeth de. *A Scented Palace: The Secret History of Marie Antoinette's Perfumer.* New York: I. B. Tauris, 2006.

Feyel, Gilles. *L'Annonce et la nouvelle: La presse d'information en France sous l'ancien régime (1630–1788).* Oxford: Voltaire Foundation, 2000.

Fitzsimmons, Michael P. *From Artisan to Worker: Guilds, the French State, and the Organization of Labor, 1776–1821.* Cambridge: Cambridge University Press, 2010.

Fleur, Elie. "Les grands pharmaciens: XVIII. Jean-Baptiste Thyrion, apothicaire à Metz au XVIIIe siècle." *Bulletin de la Société d'histoire de la pharmacie* 13 (1925): 81–88, 129–43.

Fogu, Claudio. *The Historical Imaginary: Politics of History in Fascist Italy.* Buffalo, NY : University of Toronto Press, 2003.
Fontaine, Raymond. *La Manche en Ballon: Blanchard contre Pilâtre de Rozier.* Dunkirk: Westhoek, 1982.
Fontana, Biancamaria, ed. *The Invention of the Modern Republic.* Cambridge: Cambridge University Press, 1994.
Ford, Franklin Lewis. *Strasbourg in Transition, 1648–1789.* Cambridge, MA: Harvard University Press, 1958.
Forrest, Alan. "The Condition of the Poor in Revolutionary Bordeaux." *Past & Present* 59 (1973): 147–77.
Forrest, Alan. *Society and Politics in Revolutionary Bordeaux.* Oxford: Oxford University Press, 1975.
Forrest, Alan. *Paris, the Provinces and the French Revolution.* London: Arnold, 2004.
Forster, Robert. *Merchants, Landlords, Magistrates: The Depont Family in Eighteenth-Century France.* Baltimore: Johns Hopkins University Press, 1980.
Fort, Bernadette. "Voice of the Public: The Carnivalization of Salon Art in Prerevolutionary Pamphlets." *Eighteenth-Century Studies* 22 (1989): 368–94.
Fort, Bernadette, and Jeremy Popkin, eds. *The* Mémoires secrets *and the Culture of Publicity in Eighteenth-Century France.* Oxford: Voltaire Foundation, 1998.
Foucault, Michel. *Discipline and Punish: The Birth of the Prison.* New York: Vintage Books, 1979.
Foucault, Michel. *Power/Knowledge: Selected Interviews & Other Writings, 1972–1977.* New York: Pantheon Books, 1980.
Foucault, Michel. *Madness and Civilization: A History of Insanity in the Age of Reason.* New York: Vintage Books, 1988.
Fox, Robert, and Anthony Turner, eds. *Luxury Trades and Consumerism in Ancien Regime Paris: Studies in the History of the Skilled Workforce.* Aldershot: Ashgate, 1998.
Fox, Robert. *The Savant and the State: Science and Cultural Politics in Nineteenth-Century France.* Baltimore: Johns Hopkins University Press, 2012.
Fox-Genovese, Elizabeth. *The Origins of Physiocracy: Economic Revolution and Social Order in Eighteenth-Century France.* Ithaca: Cornell University Press, 1976.
Francalanza, Eric. *Jean-Baptiste-Antoine Suard, journaliste des lumières.* Paris: Champion, 2002.
Franta, Andrew. *Romanticism and the Rise of the Mass Public.* Cambridge: Cambridge University Press, 2007.
Freeman, E., H. Mason, M. O'Regan, and S. W. Taylor, eds. *Myth and Its Making in the French Theatre: Studies Presented to W. D. Howarth.* Cambridge: Cambridge University Press, 1988.
Freycinet, Louis de. *Essai sur la vie, les opinions et les ouvrages de Barthélemy Faujas de St. Fond.* Valence: Jacques Montal, 1820.

Fricke, Ernest-Jean. "Le Prince Charles-Joseph de Ligne, franc-maçon." *Nouvelles Annales Prince de Ligne* 16 (2005): 7–123.

Fried, Michael. *Absorption and Theatricality: Painting and Beholder in the Age of Diderot*. Berkeley: University of California Press, 1980.

Friedland, Paul. *Political Bodies and Theatricality in the Age of the French Revolution*. Ithaca: Cornell University Press, 2002.

Friedland, Paul. *Seeing Justice Done: The Age of Spectacular Capital Punishment in France*. Oxford: Oxford University Press, 2012.

Fulford, Tim. "Conducting the Vital Fluid: The Politics and Poetics of Mesmerism in the 1790s." *Studies in Romanticism* 43 (2004): 57–78.

Furet, François. *Interpreting the French Revolution*. Cambridge: Cambridge University Press, 1981.

Furet, François. *In the Workshop of History*. Translated by Jonathan Mandelbaum. Chicago: University of Chicago Press, 1984.

Furet, François. *The French Revolution, 1770–1814*. London: Blackwell, 1988.

Gagnière, Sylvain, et al. *Histoire d'Avignon*. Aix-en-Provence: Édisud, 1979.

Garden, Maurice. *Lyon et les lyonnais au XVIIIe siècle*. Paris: Flammarion, 1975.

Gardiner, Leslie. *Man in the Clouds: The Story of Vincenzo Lunardi*. Edinburgh: W. and R. Chambers, 1963.

Garrioch, David. "The Police of Paris as Enlightened Social Reformers." *Eighteenth-Century Life* 16 (1992): 43–59.

Garrioch, David. *The Making of Revolutionary Paris*. Berkeley: University of California Press, 2002.

Gascoigne, John. *Joseph Banks and the English Enlightenment: Useful Knowledge and Polite Culture*. Cambridge: Cambridge University Press, 1994.

Gascoigne, John. *Science in the Service of Empire: Joseph Banks, the British State and the Uses of Science in the Age of Revolution*. Cambridge: Cambridge University Press, 1998.

Gaudillot, Jean Marie. *Le Voyage de Louis XVI en Normandie, 21–29 juin 1786: Texts et Documents réunis*. Cherbourg: Société nationale académique de Cherbourg, 1967.

Geertz, Clifford. *Negara: The Theater State in Nineteenth-Century Bali*. Princeton, NJ: Princeton University Press, 1980.

Geertz, Clifford. "Centers, Kings, and Charisma: Reflections on the Symbolics of Power." In *Rites of Power: Symbolism, Ritual, and Politics since the Middle Ages*, edited by Sean Wilentz, 13–38. Philadelphia: University of Pennsylvania Press, 1985.

Geffroy, Auguste. *Gustave III et la cour de France*. Paris: Didier, 1867.

Geikie, Sir Archibald. *Annals of the Royal Society Club: The Record of a London Dining Club in the Eighteenth and Nineteenth Centuries*. London: Macmillan, 1917.

Gelbart, Nina R. "'Frondeur' Journalism in the 1770s: Theater Criticism and

Radical Politics in the Prerevolutionary French Press." *Eighteenth-Century Studies* 17 (1984): 493–514.

Gelbart, Nina R. *Feminine and Opposition Journalism in Old Regime France: Le Journal des Dames*. Berkeley: University of California Press, 1987.

Gérando, Joseph de. "Notice sur M. Joseph Montgolfier." *Bulletin de la société d'encouragement pour l'industrie nationale* 13 (1814): 91–108.

Germani, Ian. *Jean-Paul Marat: Hero and Anti-hero of the French Revolution*. New York: Edwin Mellen, 1992.

Gerould, Daniel. *Guillotine: Its Legend and Lore*. New York: Blast Books, 1992.

Gerson, Stephane. "Parisian Litterateurs, Provincial Journeys and the Construction of National Unity in Post-revolutionary France." *Past & Present* 151 (1996): 141–73.

Gillespie, Richard. "Ballooning in France and Britain, 1783–1786: Aerostation and Adventurism." *Isis* 75 (1984): 248–68.

Gillispie, Charles Coulston. *Science and Polity in France: The End of the Old Regime*. Princeton, NJ: Princeton University Press, 1980.

Gillispie, Charles Coulston. *The Montgolfier Brothers and the Invention of Aviation, 1783–1784*. Princeton, NJ: Princeton University Press, 1983.

Gillispie, Charles Coulston. *Science and Polity in France: The Revolutionary and Napoleonic Years*. Princeton, NJ: Princeton University Press, 2004.

Godechot, Jacques. *The Taking of the Bastille, July 14th, 1789*. New York: Charles Scribner's Sons, 1970.

Golder, John. *Shakespeare for the Age of Reason: The Earliest Stage Adaptations of Jean-François Ducis, 1769–1792*. Oxford: Voltaire Foundation, 1992.

Goldstein, Jan. *Foucault and the Writing of History*. Cambridge: Blackwell, 1994.

Goldstein, Laurence. *The Flying Machine and Modern Literature*. Bloomington: Indiana University Press, 1986.

Golinski, Jan. *Science as Public Culture: Chemistry and Enlightenment in Britain, 1760–1820*. Cambridge: Cambridge University Press, 1992.

Golinski, Jan. *British Weather and the Climate of Enlightenment*. Chicago: University of Chicago Press, 2007.

Goodman, Dena. *The Republic of Letters: A Cultural Hisory of the French Enlightenment*. Ithaca: Cornell University Press, 1994.

Gordon, Daniel. *Citizens without Sovereignty: Equality and Sociability in French Thought, 1670–1789*. Princeton, NJ: Princeton University Press, 1995.

Gove, Philip Babcock. *The Imaginary Voyage in Prose Fiction: A History of Its Criticism and a Guide for Its Study, with an Annotated Check List of 215 Imaginary Voyages from 1700 to 1800*. London: Holland Press, 1961.

Gowland, Angus. "The Problem of Early Modern Melancholy." *Past & Present* 191 (2006): 77–120.

Graber, Frédéric. "Inventing Needs: Expertise and Water Supply in Late Eighteenth-

and Early Nineteenth-Century Paris." *British Journal for the History of Science* 40 (2007): 315–32.

Graham, Lisa Jane. *If only the King Knew: Seditious Speech in the Reign of Louis XV.* Charlottesville: University Press of Virginia, 2000.

Greig, Hannah. *The Beau Monde: Fashionable Society in Georgian London.* Oxford: Oxford University Press, 2013.

Green, Katherine. "Balloon and Seraglio: Burkean Anti-imperialism in Elizabeth Inchbald's *The Mogul Tale.*" *Restoration and 18th Century Theater Research* 25 (2010): 5–24.

Greenblatt, Stephen Jay. *Renaissance Self-Fashioning: From More to Shakespeare.* Chicago: University of Chicago Press, 1980.

Greenblatt, Stephen Jay. *Marvelous Possessions: The Wonder of the New World.* Chicago: University of Chicago Press, 1991.

Greene, J. P. "Ursule's Road to Ruin: Carriages in Rétif de la Bretonne's *La Paysanne pervertie.*" *Studies on Voltaire and the Eighteenth Century* 371 (1999): 175–88.

Greenhalgh, Paul. *Ephemeral Vistas: The Expositions universelles, Great Exhibitions and World's Fairs, 1851–1939.* Manchester: Manchester University Press, 1988.

Grieder, Josephine. *Anglomania in France, 1740–1789: Fact, Fiction and Political Discourse.* Geneva: Droz, 1985.

Grieder, Josephine. "Kingdoms of Women in French Fiction of the 1780s." *Eighteenth-Century Studies* 23 (1989–1990): 140–56.

Gril-Mariotte, Aziza. "Topical Themes from the Oberkampf Textile Manufactory, Jouy-en-Josas, France, 1760–1821." *Studies in the Decorative Arts* 17 (2009–2010): 162–97.

Gruber, Alain-Charles. *Les grandes fêtes et leurs décors à l'époque de Louis XVI, 1763–1790.* Geneva: Droz, 1972.

Gruder, Vivian R. "The Question of Marie-Antoinette: The Queen and Public Opinion before the Revolution." *French History* 16 (2002): 269–98.

Guha, Ranajit. "The Prose of Counter-Insurgency." In *Subaltern Studies II: Writings on South Asian History and Society,* 1–40. Delhi: Oxford University Press, 1983.

Guha, Ranajit. *Dominance without Hegemony: History and Power in Colonial India.* Cambridge, MA: Harvard University Press, 1997.

Guignet, Pierre Deyon-Philippe. "The Royal Manufactures and Economic and Technological Progress in France before the Industrial Revolution." *Journal of Economic History* 9 (1980): 611–32.

Guyonnet, Marie-Claire. *Jacques de Flesselles, intendant de Lyon, 1768–1784.* Lyon: Guillotière, 1956.

Habermas, Jürgen. *The Structural Transformation of the Public Sphere: An Inquiry into a Category of Bourgeois Society.* Cambridge, MA: MIT Press, 1991.

Hackett, Jeremiah. *Roger Bacon and the Sciences: Commemorative Essays.* Leiden: Brill, 1997.

Hafid-Martin, Nicole. *Voyage et connaissance au tournant des Lumières (1780–1820)*. Oxford: Voltaire Foundation, 1995.

Halévi, Ran. *Les loges maçonniques dans la France d'Ancien Régime: Aux origines de la sociabilité démocratique*. Paris: A. Colin, 1984.

Hallion, Richard P. *Taking Flight: Inventing the Aerial Age from Antiquity through the First World War*. Oxford: Oxford University Press, 2003.

Hammersley, Rachel. "Jean-Paul Marat's *The Chains of Slavery* in Britain and France, 1774–1833." *Historical Journal* 48 (2005): 641–60.

Hammersley, Rachel. *The English Republican Tradition and Eighteenth-Century France: Between the Ancients and the Moderns*. Manchester: Manchester University Press, 2010.

Hancock, David. *Citizens of the World: London Merchants and the Integration of the British Atlantic Community, 1735–1785*. Cambridge: Cambridge University Press, 1995.

Hanson, Paul R. "Monarchist Clubs and the Pamphlet Debate over Political Legitimacy in the Early Years of the French Revolution." *French Historical Studies* 21 (1998): 299–324.

Hanson, Paul R. *The Jacobin Republic under Fire: The Federalist Revolt*. University Park: Pennsylvania State University Press, 2003.

Hanson, Paul R. *Contesting the French Revolution*. Oxford: Wiley-Blackwell, 2009.

Haraway, Donna. "Situated Knowledges: The Science Question in Feminism and the Privilege of Partial Perspective." *Feminist Studies* 14 (1988): 575–99.

Hardman, John, and Munro Price. *Louis XVI and the Comte de Vergennes: Correspondence 1774–1787*. Oxford: Voltaire Foundation, 1998.

Harkness, Deborah. *The Jewel House: Elizabethan London and the Scientific Revolution*. New Haven: Yale University Press, 2007.

Harland-Jacobs, Jessica L. *Builders of Empire: Freemasons and British Imperialism, 1717–1927*. Chapel Hill: University of North Carolina Press, 2007.

Harris, John R. *Industrial Espionage and Technology Transfer: England and France in the Eighteenth Century*. Aldershot: Ashgate, 1998.

Harrison, Carol E. "Planting Gardens, Planting Flags: Revolutionary France in the South Pacific." *French Historical Studies* 34 (2011): 243–77.

Hart, Clive. *The Prehistory of Flight*. Berkeley: University of California Press, 1985.

Hart, Clive. "Printed Books, 1489–1850." In *The Genesis of Flight*, edited by Tom D. Crouch, 37–90. Los Angeles: Perpetua Press, 2000.

Harth, Erica. *Cyrano de Bergerac and the Polemics of Modernity*. New York: Columbia University Press, 1970.

Haskell, Yasmin Annabel. *Prescribing Ovid: The Latin Works and Networks of the Enlightened Dr Heerkens*. London: Bloomsbury Academic, 2013.

Hawkins, Peter S. "The Metamorphosis of Ovid." In *Dante and Ovid: Essays in Intertextuality*, edited by Madison U. Sowell, 17–34. Binghamton, NY: Medieval & Renaissance Texts and Studies, 1991.

Hays, David. "Carmontelle's Design for the Jardin de Monceau: A Freemasonic Garden in Late-Eighteenth-Century France." *Eighteenth-Century Studies* 32 (1999): 447–62.

Haywood, Ian, and John Seed, eds. *The Gordon Riots: Politics, Culture and Insurrection in Late Eighteenth-Century Britain*. Cambridge: Cambridge University Press, 2012.

Hazareesingh, Sudhir. "'A Common Sentiment of National Glory': Civic Festivities and French Collective Sentiment under the Second Empire." *Journal of Modern History* 76 (2004): 280–311.

Hechter, Michael. *Internal Colonialism: The Celtic Fringe in British National Development, 1536–1966*. Princeton, NJ: Princeton University Press, 1975.

Hellman, Mimi. "Furniture, Sociability, and the Work of Leisure in Eighteenth Century France." *Eighteenth-Century Studies* 32 (1999): 415–45.

Higonnet, Patrice. *Goodness beyond Virtue: Jacobins during the French Revolution*. Cambridge, MA: Harvard University Press, 1998.

Hilaire-Pérez, Liliane. "Invention and the State in 18th-Century France." *Technology and Culture* 32 (1991): 911–31.

Hilaire-Pérez, Liliane. *L'Invention technique au siècle des lumières*. Paris: Albin Michel, 2000.

Hilaire-Pérez, Liliane. "Diderot's Views on Artists' and Inventors' Rights: Invention, Imitation and Reputation." *British Journal for the History of Science* 35 (2002): 129–50.

Hirschauer, Charles. *Notes sur l'histoire de l'aérostation dans le Pas-de-Calais*. Arras: Répessé, Cassel & Cie, 1910.

Hochadel, Oliver. *Öffentliche Wissenschaft: Elektrizität in der deutschen Aufklärung*. Göttingen: Wallstein Verlag, 2003.

Hochadel, Oliver. "In Nebula Nebulorum: The Dry Fog of the Summer of 1783 and the Introduction of Lightning Rods in the German Empire." In *Playing with Fire: Histories of the Lightning Rod*, edited by Peter Heering, Oliver Hochadel, and David J. Rhees, 45–70. Philadelphia: American Philosophical Society, 2009.

Hodgson, John Edmund. *The History of Aeronautics in Great Britain from the Earliest Times to the Latter Half of the Nineteenth Century with 150 Illustrations from Contemporary Sources Chronology, Bibliography, etc*. Oxford: Oxford University Press, 1924.

Hodgson, John Edmund. *Doctor Johnson on Ballooning and Flight*. London: Elkin Mathews, 1925.

Hoffmann, Kathryn A. *Society of Pleasures*. New York: St. Martin's Press, 1997.

Hohendahl, Peter Uwe, ed. *Patriotism, Cosmopolitanism, and National Culture: Public Culture in Hamburg, 1700–1933*. Amsterdam: Rodopi, 2003.

Hoock, Holger. *Empires of the Imagination: Politics, War, and the Arts in the British World, 1750–1850*. London: Profile Books, 2010.

Horn, Jeff. *The Path Not Taken: French Industrialization in the Age of Revolution, 1750–1830*. Cambridge, MA: MIT Press, 2008.

Hoskin, Michael. "Herschel's 40 Ft Reflector." *Journal for the History of Astronomy* 34 (2003): 1–32.

Howarth, William D. *Beaumarchais and the Theatre*. Princeton, NJ: Princeton University Press, 1995.

Howarth, William D., ed. *French Theater in the Neo-classical Era, 1550–1789*. Cambridge: Cambridge University Press, 1997.

Hudson, Nicholas. "Samuel Johnson, Urban Culture, and the Geography of Postfire London." *Studies in English Literature, 1500–1900* 42 (2002): 577–600.

Hufton, Olwen H. *The Poor of Eighteenth-Century France, 1750–1789*. Oxford: Oxford University Press, 1974.

Hundert, E. J., and Paul Nelles. "Liberty and Theatrical Space in Montesquieu's Political Theory: The Poetics of Public Life in the Persian Letters." *Political Theory* 17 (1989): 223–46.

Hunn, James Martin. "The Balloon Craze in France, 1783–1799: A Study in Popular Science." Unpublished PhD dissertation, Vanderbilt University, 1982.

Hunt, Lynn. *Politics, Culture, and Class in the French Revolution*. Berkeley: University of California Press, 1984.

Hunt, Lynn, ed. *The New Cultural History*. Berkeley: University of California Press, 1989.

Hunt, Lynn. "The Many Bodies of Marie-Antoinette: Political Pornography and the Problem of the Feminine in the French Revolution." In *Eroticism and the Body Politic*, 108–30. Baltimore: Johns Hopkins University Press, 1991.

Hunt, Lynn. "Louis XVI Wasn't Killed by Ideas." *New Republic*, June 27, 2014.

Hyde, Melissa. "The 'Makeup' of the Marquise: Boucher's Portrait of Pompadour at Her Toilette." *Art Bulletin* 82 (2000): 453–75.

Imbruglia, Gerolamo. "From Utopia to Republicanism: The Case of Diderot." In *The Invention of the Modern Republic*, edited by Biancamaria Fontana, 63–85. Cambridge: Cambridge University Press, 1994.

Imbruglia, Gerolamo, ed. *Naples in the Eighteenth Century: The Birth and Death of a Nation State*. Cambridge: Cambridge University Press, 2000.

Inglis, Lucy. *Georgian London: Into the Streets*. London: Viking, 2013.

Isherwood, Robert M. *Farce and Fantasy: Popular Entertainment in Eighteenth-Century Paris*. Oxford: Oxford University Press, 1986.

Israel, Jonathan I. *Radical Enlightenment: Philosophy and the Making of Modernity, 1650–1750*. Oxford: Oxford University Press, 2001.

Israel, Jonathan I. *Revolutionary Ideas: An Intellectual History of the French Revolution from the Rights of Man to Robespierre*. Princeton, NJ: Princeton University Press, 2014.

Jacob, Margaret C. *The Radical Enlightenment: Pantheists, Freemasons and Republicans.* London: George Allen and Unwin, 1981.

Jacob, Margaret C., and Wijnand Mijnhardt, eds. *The Dutch Republic in the Eighteenth Century: Decline, Enlightenment and Revolution.* Ithaca: Cornell University Press, 1992.

Jacob, Margaret C. "The Mental Landscape of the Public Sphere: A European Perspective." *Eighteenth-Century Studies* 28 (1994): 95–113.

Jainchill, Andrew J. S. *Reimagining Politics after the Terror: The Republican Origins of French Liberalism.* Ithaca: Cornell University Press, 2008.

Jardine, Nicolas, James A. Secord, and Emma C. Spary, eds. *Cultures of Natural History.* Cambridge: Cambridge University Press, 1996.

Jay, Mike. *The Atmosphere of Heaven: The Unnatural Experiments of Dr. Beddoes and His Sons of Genius.* New Haven: Yale University Press, 2009.

Jefferson, Ann. *Genius in France: An Idea and Its Uses.* Princeton, NJ: Princeton University Press, 2014.

Jenkins, David, ed. *The History of Western Textiles.* 2 vols. Cambridge: Cambridge University Press, 2003.

Jenkins, Jennifer. *Provincial Modernity: Local Culture and Liberal Politics in Fin-de-Siècle Hamburg.* Ithaca: Cornell University Press, 2003.

Jobé, J. *The Romance of Ballooning: The Story of the Early Aeronauts.* New York: Viking Press, 1971.

Johnson, Christopher H. *Life and Death in Industrial Languedoc: Politics of Deindustrialization, 1700–1920.* Oxford: Oxford University Press, 1995.

Johnson, James H. "Musical Experience and the Formation of a French Musical Public." *Journal of Modern History* 64 (1992): 191–226.

Johnson, James H. *Listening in Paris: A Cultural History.* Berkeley: University of California Press, 1995.

Johnston, R. M. "Mirabeau's Secret Mission to Berlin." *American Historical Review* 6 (1901): 235–53.

Johnson, Richard R. *John Nelson: Merchant Adventurer: A Life between Empires.* Oxford: Oxford University Press, 1991.

Joly, Alice. *Un mystique lyonnais et les secrets de la franc-maçonnerie, 1730–1824.* Mâcon, France: Protat frères, 1938.

Jones, Christopher M. S. *Antonio Canova and the Politics of Patronage in Revolutionary and Napoleonic Europe.* Berkeley: University of California Press, 1998.

Jones, Colin. "The Great Chain of Buying: Medical Advertisement, the Bourgeois Public Sphere, and the Origins of the French Revolution." *American Historical Review* 101 (1996): 13–40.

Jones, Colin, and Dror Wahrman, eds. *The Age of Cultural Revolutions: Britain and France, 1750–1820.* Berkeley: University of California Press, 2002.

Jones, Colin. *The Great Nation: France from Louis XV to Napoleon.* New York: Penguin, 2002.
Jones, Jennifer M. *Sexing la Mode: Gender, Fashion and Commercial Culture in Old Regime France.* New York: Berg, 2004.
Jones, Peter M. "Living the Enlightenment and the French Revolution: James Watt, Matthew Boulton, and Their Sons." *Historical Journal* 42 (1999): 157–82.
Jones, Peter M. *Industrial Enlightenment: Science, Technology and Culture in Birmingham and the West Midlands, 1760–1820.* Manchester: Manchester University Press, 2008.
Jones, Philip. *The Italian City-State: From Commune to Signoria.* Oxford: Clarendon Press, 1997.
Kaiser, Thomas E. "This Strange Offspring of Philosophie: Recent Historiographical Problems in Relating the Enlightenment to the French Revolution." *French Historical Studies* 15 (1988): 549–62.
Kaiser, Thomas E. "Money, Despotism, and Public Opinion in in Early Eighteenth-Century France: John Law and the Debate on Royal Credit." *Journal of Modern History* 63 (1991): 1–28.
Kaiser, Thomas E., and Dale K. Van Kley, eds. *From Deficit to Deluge: The Origins of the French Revolution.* Stanford, CA: Stanford University Press, 2011.
Kang, Minsoo. *Sublime Dreams of Living Machines: The Automaton in the European Imagination.* Cambridge, MA: Harvard University Press, 2011.
Kantorowicz, Ernst Hartwig. *The King's Two Bodies: A Study in Mediaeval Political Theology.* Princeton, NJ: Princeton University Press, 1957.
Kaplan, Steven Laurence. *Bread, Politics and Political Economy in the Reign of Louis XV.* 2 vols. The Hague: Martinus Nijhoff, 1976.
Kaplan, Steven Laurence. *Provisioning Paris: Merchants and Millers in the Grain and Flour Trade during the Eighteenth Century.* Ithaca: Cornell University Press, 1984.
Kaplan, Steven Laurence. "The Paris Bread Riot of 1725." *French Historical Studies* 14 (1985): 23–56.
Kaplan, Steven Laurence, and Cynthia J. Koepp, eds. *Work in France: Representations, Meaning, Organization, and Practice.* Ithaca: Cornell University Press, 1986.
Kaplan, Steven Laurence. *The Bakers of Paris and the Bread Question, 1700–1775.* Durham, NC: Duke University Press, 1996.
Kavanagh, Thomas M. *Enlightenment and the Shadows of Chance: The Novel and the Culture of Gambling in Eighteenth-Century France.* Baltimore: Johns Hopkins University Press, 1993.
Kavanagh, Thomas M. *Esthetics of the Moment: Literature and Art in the French Enlightenment.* Philadelphia: University of Pennsylvania Press, 1996.
Keen, Paul. "The 'Balloonomania': Science and Spectacle in 1780s England." *Eighteenth-Century Studies* 39 (2006): 507–35.

Keen, Paul. *Literature, Commerce, and the Spectacle of Modernity, 1750–1800*. Cambridge: Cambridge University Press, 2011.

Kelly, George Armstrong. "The Machine of the Duc d'Orléans and the New Politics." *Journal of Modern History* 51 (1979): 667–84.

Kelly, George Armstrong. "From Lèse-Majesté to Lèse-Nation: Treason in Eighteenth-Century France." *Journal of the History of Ideas* 42 (1981): 269–86.

Kelly, James. *Sir Edward Newenham MP, 1734–1814: Defender of the Protestant Constitution*. Dublin: Four Courts Press, 2003.

Kessler, Amalia D. "Searching for a 'New System' of Government: Linguet and the Rise of the Centralized, Administrative State." *Historical Reflections/Réflexions historique* 28 (2002): 93–117.

Kettering, Sharon. *Patrons, Brokers and Clients in Seventeenth-Century France*. Oxford: Oxford University Press, 1986.

Kim, Mi Gyung. *Affinity, That Elusive Dream: A Genealogy of the Chemical Revolution*. Cambridge, MA: MIT Press, 2003.

Kim, Mi Gyung. "Balloon Mania: News in the Air." *Endeavour* 28 (2004): 149–55.

Kim, Mi Gyung. "Lavoisier, the Father of Modern Chemistry?" In Beretta, *Lavoisier in Perspective*, 167–91.

Kim, Mi Gyung. "Public Science: Hydrogen Balloons and Lavoisier's Decomposition of Water." *Annals of Science* 63 (2006): 291–318.

Kim, Mi Gyung. "De l'érudition à la science: La mode des lumières à Dijon." *Annales de Bourgogne* 85 (2013): 241–55.

Kim, Mi Gyung. "Invention as a Social Drama: From an Ascending Machine to the Aerostatic Globe." *Technology and Culture* 54 (2013): 853–87.

Kim, Mi Gyung. "Archeology, Genealogy, and Geography of Experimental Philosophy." *Social Studies of Science* 44 (2014): 151–63.

Kim, Mi Gyung. "Material Enlightenments." *Historical Studies in the Natural Sciences* 44 (2014): 424–33.

Kingston, Rebecca. *Montesquieu and the Parlement of Bordeaux*. Geneva: Droz, 1996.

Kirsop, Wallace. "Cultural Networks in Pre-revolutionary France: Some Reflexions on the Case of Antoine Court de Gébelin." *Australian Journal of French Studies* 18 (1981): 231–47.

Kite, Elizabeth Sara. *Beaumarchais and the War of American Independence*. 2 vols. Boston: Gorham, 1918.

Klein, Lawrence E. "Politeness for Plebes: Consumption and Social Identity in Early Eighteenth-Century England." In Bermingham and Brewer, *Consumption of Culture*, 362–82.

Knapp, Jeffrey. *An Empire Nowhere: England, America, and Literature from Utopia to the Tempest*. Berkeley: University of California Press, 1992.

Koppisch, Michael S. *The Dissolution of Character: Changing Perspectives in La Bruyère's Caractères*. Lexington, KY: French Forum, 1981.

Kramer, Lloyd S. *Lafayette in Two Worlds: Public Cultures and Personal Identities in an Age of Revolutions*. Chapel Hill: University of North Carolina Press, 1996.

Krantz, Frederick, ed. *History from Below: Studies in Popular Protest and Popular Ideology*. Oxford: Blackwell, 1988.

Kumar, Krishan. "Nation-States as Empires, Empires as Nation-States: Two Principles, One Practice?" *Theory and Society* 39 (2010): 119–43.

Kümin, Beat. "Popular Culture and Sociability." In *A Companion to Eighteenth-Century Europe*, edited by Peter H. Wilson, 192–207. Oxford: Blackwell, 2008.

Kwass, Michael. *Privilege and the Politics of Taxation in Eighteenth-Century France*. Cambridge: Cambridge University Press, 2000.

Kwass, Michael. "Ordering the World of Goods: Consumer Revolution and the Classification of Objects in Eighteenth-Century France." *Representations* 82 (2003): 87–116.

Kwass, Michael. "Consumption and the World of Ideas: Consumer Revolution and the Moral Economy of the Marquis de Mirabeau." *Eighteenth-Century Studies* 37 (2004): 187–213.

Lamblin, Edmond. "L'apothicaire lillois Carette et les débuts de l'aérostation." *Revue d'histoire de la pharmacie* 215 (1972): 268–69.

Lande, Lawrence M., ed. *The Rise and Fall of John Law, 1716–1720*. Montreal: McGill University Press, 1982.

Landes, Joan B. *Visualizing the Nation: Gender, Representation, and Revolution in Eighteenth-Century France*. Ithaca: Cornell University Press, 2001.

Landes, Joan B. "Revolutionary Anatomies." In *Monstrous Bodies/Political Monstrosities in Early Modern Europe*, edited by Laura Lunger Knoppers and Joan B. Landes, 148–76. Ithaca: Cornell University Press, 2004.

Langford, Paul. *A Polite and Commercial People: England, 1727–1783*. Oxford: Clarendon Press, 1989.

Langins, Janis. "Hydrogen Production for Ballooning during the French Revolution: An Early Example of Chemical Process Development." *Annals of Science* 40 (1983): 531–58.

Latour, Bruno. *The Pasteurization of France*. Cambridge, MA: Harvard University Press, 1988.

Latour, Bruno, and Peter Weibel, eds. *Making Things Public: Atmospheres of Democracy*. Cambridge, MA: MIT Press, 2005.

Laugero, Greg. "Infrastructures of Enlightenment: Road-Making, the Public Sphere, and the Emergence of Literature." *Eighteenth-Century Studies* 29 (1995): 45–67.

Laurenza, Domenico. *Leonardo on Flight*. Baltimore: Johns Hopkins University Press, 2007.

La Vaulx, André Foulon, comte de, and Paul Tissandier, *Joseph et Étienne de Montgolfier*. Annonay, France, 1926.

Lawrenson, Tom. "The Ideal Theatre in the Eighteenth Century: Paris and Venice." In *Drama and Mimesis*, edited by James Redmond, 51–64. Cambridge: Cambridge University Press, 1980.

Leary, Lewis. "Phaeton in Philadelphia: Jean Pierre Blanchard and the First Balloon Ascents in America, 1793." *Pennsylvania Magazine of History of Biography* 67 (1943): 49–60.

Le Blond, Aubrey. *Charlotte Sophie, Countess Bentinck, Her Life and Times, 1715–1800*. London: Hutchinson, 1912.

Le Breton, André. *Rivarol; sa vie, ses idées, son talent, d'après des documents nouveaux*. Paris: Hachette, 1895.

Leckey, Colum. *Patrons of Enlightenment: The Free Economic Society in Eighteenth-Century Russia*. Newark: University of Delaware Press, 2011.

Leffler, Phyllis K. "French Historians and the Challenge to Louis XIV's Absolutism." *French Historical Studies* 14 (1985): 1–22.

Legay, Marie-Laure. *Les états provinciaux dans la construction de l'état moderne aux XVIIe et XVIIIe siècles*. Geneva: Droz, 2001.

Legrand, H. E. "Chemistry in a Provincial Context: The Montpellier Société Royal des Sciences in the Eighteenth Century." *Ambix* 29 (1982): 88–105.

Leith, James A. *The Idea of Art as Propaganda in France, 1750–1799*. Toronto: University of Toronto Press, 1965.

Lenman, Bruce. *Integration, Enlightenment and Industrialisation: Scotland, 1746–1832*. London: E. Arnold, 1981.

Lensink, Rachel. "Science and Spectacle: Early Ballooning in the Netherlands (1783–1830)." Thesis, Utrecht University, 2015.

Lepenies, Wolf. *Melancholy and Society*. Cambridge, MA: Harvard University Press, 1992.

Le Roy Ladurie, Emmanuel. *Political Culture of the Old Regime*. Oxford: Pergamon Press, 1987.

Level, Brigitte. *Le Caveau, à travers deux siècles: Société bachique et chantante, 1726–1939*. Paris: Sorbonne, 1988.

Lever, Évelyne. *Philippe Égalité*. Paris: Fayard, 1996.

Levere, Trevor, and G. Turner, ed. *Discussing Chemistry and Steam: The Minutes of a Coffee-House Philosophical Society, 1780–1787*. Oxford: Oxford University Press, 2002.

Levinger, Matthew. *Enlightened Nationalism: The Transformation of Prussian Political Culture, 1806–1848*. Oxford: Oxford University Press, 2000.

Levy, Darlene Gay. *The Ideas and Careers of Simon-Nicolas-Henri Linguet: A Study in Eighteenth-Century French Politics*. Urbana: University of Illinois Press, 1980.

Lewis, John. *Galileo in France: French Reactions to the Theories and Trial of Galileo*. New York: Peter Lang, 2006.

Lhéritier, Michel. *La révolution à Bordeaux dans l'histoire de la révolution française*. Paris: Presses universitaires de France, 1942.

Lilti, Antoine. *Le monde des salons: Sociabilité et mondanité à Paris au XVIIIe siècle.* Paris: Fayard, 2005.

Lilti, Antoine. "The Writing of Paranoia: Jean-Jacques Rousseau and the Paradoxes of Celebrity." *Representations* 103 (2008): 53–83.

Lindberg, David C. "Science as Handmaiden: Roger Bacon and the Patristic Tradition." *Isis* 78 (1987): 518–36.

Lindemann, Mary. *Patriots and Paupers: Hamburg, 1712–1830.* Oxford: Oxford University Press, 1990.

Lindemann, Mary. *Liaisons dangereuses: Sex, Law and Diplomacy in the Age of Frederick the Great.* Baltimore: Johns Hopkins University Press, 2006.

Lindemann, Mary. *The Merchant Republics: Amsterdam, Antwerp, and Hamburg, 1648–1790.* Cambridge: Cambridge University Press, 2014.

Linger, Daniel T. "The Hegemony of Discontent." *American Ethnologist* 20 (1993): 3–24.

Linton, Marisa. *The Politics of Virtue in Enlightenment France.* New York: Palgrave, 2001.

Lipp, Charles T. *Noble Strategies in an Early Modern Small State: The Mahuet of Lorraine.* Rochester: University of Rochester Press, 2011.

Livesey, James. "Agrarian Ideology and Commercial Republicanism in the French Revolution." *Past & Present* 157 (1997): 94–121.

Loiselle, Kenneth. "Living the Enlightenment in an Age of Revolution: Freemasonry in Bordeaux (1788–1794)." *French History* 24 (2009): 60–81.

Loiselle, Kenneth. *Brotherly Love: Freemasonry and Male Friendship in Enlightenment France.* Ithaca: Cornell University Press, 2014.

Long, Pamela O. *Openness, Secrecy, Authorship.* Baltimore: Johns Hopkins University Press, 2001.

Lottin, Alain. *Lille: Citadelle de la Contre-Reforme? (1598–1668).* Dunkirk: Westhoek, 1984.

Lottum, Jelle van. "Labour Migration and Economic Performance: London and the Randsted, 1600–1800." *Economic History Review* 64 (2011): 531–70.

Lo Tufo, Ilaria. "Images of the Natural and Social Universe in Rétif de La Bretonne's *La découverte australe.*" *Studies in History and Philosophy of Biological and Biomedical Sciences* 34 (2003): 1–50.

Lucas, Colin. "The Crowd and Politics between 'Ancien Régime' and Revolution in France." *Journal of Modern History* 60 (1988): 421–57.

Luna, Frederick A. de. "The Dean Street Style of Revolution: J-P Brissot, Jeune Philosophe." *French Historical Studies* 17 (1991): 159–90.

Lunney, Linde. "The Celebrated Mr. Dinwiddie: An Eighteenth-Century Scientist in Ireland." *Eighteenth-Century Ireland* 3 (1988): 69–83.

Lüsebrink, Hans-Jürgen, and Rolf Reichardt. "La 'Bastille' dans l'imaginaire social de la France à la fin du XVIIIe siècle (1774–1799)." *Revue d'histoire moderne et contemporaine* 30 (1983): 196–234.

Lüsebrink, Hans-Jürgen, and Alexandre Mussard, ed. *Avantages et désavantages de la découverte de l'Amérique: Chastellux, Raynal et le concours de l'Académie de Lyon.* Saint-Étienne: Université de Saint-Étienne, 1994.

Lynch, Deidre Shauna. *The Economy of Character: Novels, Market Culture, and the Business of Inner Meaning.* Chicago: University of Chicago Press, 1998.

Lynn, Michael R. "Enlightenment in the Public Sphere: The Musée de Monsieur and Scientific Culture in Late-Eighteenth-Century Paris." *Eighteenth-Century Studies* 32 (1999): 463–76.

Lynn, Michael R. *Popular Science and Public Opinion in Eighteenth-Century France.* Manchester: Manchester University Press, 2006.

Lynn, Michael R. "Consumerism and the Rise of Balloons in Europe at the End of the Eighteenth Century." *Science in Context* 21 (2008): 73–98.

Lynn, Michael R. *The Sublime Invention.* London: Pickering and Chatto, 2010.

MacAloon, John J. "Olympic Games and the Theory of Spectacle in Modern Societies." In *Rite, Drama, Festival, Spectacle: Rehearsals toward a Theory of Cultural Performance*, edited by John J. MacAloon, 241–80. Philadelphia: ISHI, 1984.

MacArthur, Elizabeth J. "Embodying the Public Sphere: Censorship and the Reading Subject in Beaumarchais's *Mariage de Figaro*." *Representations* 81 (1998): 57–77.

MacLeod, Christine. *Heroes of Invention: Technology, Liberalism and British Identity, 1750–1914.* Cambridge: Cambridge University Press, 2007.

MacMahon, Bryan. *Ascend or Die: Richard Crosbie, Pioneer of Balloon Flight.* Dublin: History Press Ireland, 2010.

Maës, Gaëtane. *Les Watteau de Lille.* Paris: Arthena, 1998.

Mah, Harold. "Phantasies of the Public Sphere: Rethinking the Habermas of Historians." *Journal of Modern History* 72 (2000): 153–82.

Major, James Russell. *Representative Institutions in Renaissance France, 1421–1559.* Madison: University of Wisconsin Press, 1960.

Major, James Russell. *Representative Government in Early Modern France.* New Haven: Yale University Press, 1980.

Maloy, J. S. "The Very Order of Things: Rousseau's Tutorial Republicanism." *Polity* 37 (2005): 235–61.

Maravelas, Paul. "Historiography of the Invention of the Balloon." *Annals of Balloon History and Museology* 1 (1991): 15–54.

Margerison, Kenneth. "History, Representative Institutions, and Political Rights in the French Pre-Revolution (1787–1789)." *French Historical Studies* 15 (1987): 68–98.

Margerison, Kenneth. *Pamphlets & Public Opinion: The Campaign for a Union of Orders in the Early French Revolution.* West Lafayette, Indiana: Purdue University Press, 1998.

Margerison, Kenneth. "Commercial Liberty, French National Power, and the Indies Trade after the Seven Years' War." *Historical Reflections* 35 (2009): 52–73.
Marin, Louis. *Portrait of the King.* Minneapolis: University of Minnesota Press, 1988.
Marinis, Marco de, and Paul Dwyer. "Dramaturgy of the Spectator." *Drama Review* 31 (1987): 100–14.
Marionneau, Charles. *Victor Louis, architecte du théâtre de Bordeaux: sa vie, sa travaux, et sa correspondence, 1731–1800.* Bordeaux: G. Gounouilhou, 1881.
Marshall, David. *The Figure of Theater: Shaftesbury, Defoe, Adam Smith, and George Eliot.* New York: Columbia University Press, 1986.
Marshall, David. "Rousseau and the State of Theater." *Representations* 13 (1986): 84–114.
Martin, Morag. *Selling Beauty: Cosmetics, Commerce, and French Society, 1750–1830.* Baltimore: Johns Hopkins University Press, 2009.
Martindale, Charles, ed. *Ovid Renewed: Ovidian Influences on Literature and Art from the Middle Ages to the Twentieth Century.* Cambridge: Cambridge University Press, 1988.
Martines, Lauro. *Power and Imagination: City States in Renaissance Italy.* New York: Knopf, 1979.
Maslan, Susan. *Revolutionary Acts: Theater, Democracy, and the French Revolution.* Baltimore: Johns Hopkins University Press, 2005.
Mason, Haydn. *French Writers and Their Society, 1715–1800.* New York: Macmillan, 1982.
Mason, Haydn. *The Darnton Debate: Books and Revolution in the Eighteenth Century.* Oxford: Voltaire Foundation, 1998.
Mason, Laura. *Singing the French Revolution: Popular Culture and Politics, 1787–1799.* Ithaca: Cornell University Press, 1996.
Masuzawa, Tomoko. *The Invention of World Religions, or How European Universalism Was Preserved in the Language of Pluralism.* Chicago: University of Chicago Press, 2005.
Maza, Sarah. "The Diamond Necklace Affair Revisited, 1785–1786: The Case of the Missing Queen." In *Eroticism and the Body Politic,* edited by Lynn Hunt, 63–89. Baltimore: Johns Hopkins University Press, 1991.
Maza, Sarah. *Private Lives and Public Affairs: The Causes Célèbres of Prerevolutionary France.* Berkeley: University of California Press, 1993.
Maza, Sarah. "Luxury, Morality and Social Change in Pre-revolutionary France." *Journal of Modern History* 69 (1997): 199–229.
Maza, Sarah. *The Myth of the French Bourgeoisie: An Essay in the Social Imaginary, 1750–1850.* Cambridge, MA: Harvard University Press, 2003.
McBride, I. R. *Scripture Politics: Ulster Presbyterians and Irish Radicalism in the Late Eighteenth Century.* Oxford: Clarendon Press, 1998.
McClellan, J. E., and F. Regourd. "The Colonial Machine." *Osiris* 15 (2000): 31–50.

McConnell, Anita. *Jesse Ramsden (1735–1800): London's Leading Scientific Instrument Maker.* Aldershot: Ashgate, 2007.

McDowell, Robert Brendan. *Ireland in the Age of Imperialism and Revolution, 1760–1801.* Oxford: Clarendon Press, 1979.

McDowell, Robert Brendan. *Trinity College, Dublin, 1592–1952: An Academic History.* Cambridge: Cambridge University Press, 1982.

McGuffie, Tom Henderson. *The Siege of Gibraltar, 1779–1783.* Philadelphia: Dufour Editions, 1965.

McKendrick, Neil, John Brewer, and J. H. Plumb, eds. *The Birth of a Consumer Society: The Commercialization of Eighteenth-Century England.* Bloomington: Indiana University Press, 1982.

McLean, Iain, and Fiona Hewitt, eds. *Condorcet: Social Choice and Political Theory.* Aldershot: Edward Elgar, 1994.

McLeod, Jane. "A Bookseller in Revolutionary Bordeaux." *French Historical Studies* 16 (1989): 262–83.

McMahon, Darrin. "The Birthplace of the Revolution: Public Space and Political Community in the Palais-Royal of Louis-Philippe d'Orléans, 1781–1789." *French History* 10 (1997): 1–29.

McMahon, Darrin. "The Counter-Enlightenment and the Low-Life of Literature in Pre-revolutionary France." *Past & Present* 159 (1998): 77–112.

McMahon, Darrin. *Enemies of the Enlightenment: The French Counter-Enlightenment and the Making of Modernity.* Oxford: Oxford University Press, 2001.

McMahon, Darrin. "Narratives of Dystopia in the French Revolution: Enlightenment, Counter-Enlightenment, and the 'Isle des philosophes' of the Abbe Balthazard." *Yale French Studies* 101 (2001): 103–18.

McMahon, Darrin. *Divine Fury: A History of Genius.* New York: Basic Books, 2013.

Meaudre de LaPouyade, Maurice. *Les premiers aéronautes bordelaises, 1783–1799.* Bordeaux: G. Gounouilhou, 1910.

Meek, Robert. *A Biographical Sketch of the Life of James Tytler.* Edinburgh: Denovan, 1805.

Melzer, Sara E., and Kathryn Norberg, eds. *From the Royal to the Republican Body: Incorporating the Political in Seventeenth- and Eighteeenth-Century France.* Berkeley: University of California Press, 1998.

Melzer, Sara E. *Colonizer or Colonized: Hidden Stories of Early Modern France.* Philadelphia: University of Pennsylvia Press, 2012.

Merrick, Jeffrey W. *The Descralization of the French Monarchy in the Eighteenth Century.* Baton Rouge: Louisiana State University Press, 1990.

Merrick, Jeffrey. "Sexual Politics and Public Order in Late Eighteenth-Century France: The *Mémoires Secrets* and the *Correspondance Secrète.*" *Journal of the History of Sexuality* 1 (1990): 68–84.

Merrick, Jeffrey W. "The Cardinal and the Queen: Sexual and Political Disorders in the Mazarinades." *French Historical Studies* 18 (1994): 667–99.

Mettam, Roger. *Power and Faction in Louis XIV's France.* Oxford: Blackwell, 1988.
Meyer, Morgan. "Placing and Tracing Absence: A Material Culture of the Immaterial." *Journal of Material Culture* 17 (2012): 103–10.
Middleton, Knowles W. E. "A Brief History of the Barometer." *Journal of the Royal Astronomical Society of Canada* 38 (1944): 41–64.
Mignolo, Walter. "The Geopolitics of Knowledge and the Colonial Difference." *South Atlantic Quarterly* 101 (2002): 57–96.
Miller, David Philip, and Peter Hanns Reill, eds. *Visions of Empire: Voyages, Botany, and Representations of Nature.* Cambridge: Cambridge University Press, 1996.
Miller, Lesley Ellis. "Paris-Lyon-Paris: Dialogue in the Design and Distribution of Patterned Silks in the 18th Century." In Fox and Turner, *Luxury Trades and Consumerism in Ancien Régime Paris*, 139–67.
Miller, Stephen. *State and Society in Eighteenth-Century France: A Study of Political Power and Social Revolution in Languedoc.* Washington, DC: Catholic University of America Press, 2008.
Milliot, V. "Qu'est-ce qu'une police éclairée? La police 'amélioratrice' selon Jean-Charles-Pierre Lenoir, lieutenant-général à Paris (1775–1785)," *Dix-huitième Siecle* 37 (2005): 117–30.
Minard, Philippe. *La fortune du colbertisme: État et industrie dans la France des Lumières.* Paris: Fayard, 1998.
Mirala, Petri. "'A Large Mob, Calling Themselves Freemasons': Masonic Parades in Ulster." In Jupp and Magennis, *Crowds in Ireland*, 117–38.
Mitchell, W. J. T., ed. *Landscape and Power.* Chicago: University of Chicago Press, 1994.
Mokyr, Joel. *The Gifts of Athena: Historical Origins of the Knowledge Economy.* Princeton, NJ: Princeton University Press, 2002.
Mokyr, Joel. *The Enlightened Economy: An Economic History of Britain, 1700–1850.* New Haven: Yale University Press, 2009.
Monaco, Marion. *Shakespeare and the French Stage in the Eighteenth Century.* Paris: Didier, 1974.
Montgolfier, Adélaïde. "La Corbeille ailée." In *Contes devenus histoires*, 27–71. Paris: La Ruche, ca. 1835.
Morell, J. B. "The University of Edinburgh in the Late Eighteenth Century: Its Scientific Eminence and Academic Structure." *Isis* 62 (1971): 158–71.
Morton, Alan, and Jane Wess. *Public and Private Science: The King George III Collection.* Oxford: Oxford University Press, 1993.
Morton, Alan Q. *Science in the 18th Century: The King George III Collection.* London: Science Museum, 1993.
Morton, Brian N., and Donald C. Spinelli. *Beaumarchais and the American Revolution.* Lanham, MD: Lexington Books, 2003.
Morton, Richard. *The English Enlightenment Reads Ovid: Dryden and Jacob Tonson's 1717 Metamorphosis.* New York: AMS Press, 2013.

Moss, Ann. *Ovid in Renaissance France: A Survey of the Latin Editions of Ovid and Commentaries Printed in France before 1600*. London: Warburg Institute, 1982.

Moss, Ann, ed. *Latin Commentaries on Ovid from the Renaissance*. Signal Mountain, TN: Summertown, 1999.

Mosse, George L. *The Nationalization of the Masses: Political Symbolism and Mass Movements in Germany from the Napoleonic Wars through the Third Reich*. New York: H. Fertig, 1975.

Mukerji, Chandra. *Territorial Ambitions and the Gardens of Versailles*. Cambridge: Cambridge University Press, 1997.

Mukerji, Chandra. *Impossible Engineering: Technology and Territoriality on the Canal du Midi*. Princeton, NJ: Princeton University Press, 2009.

Murphy, Orville T. *Charles Gravier, Comte de Vergennes: French Diplomacy in the Age of Revolution, 1719–1787*. Albany: SUNY Press, 1982.

Murphy, Orville T. *The Diplomatic Retreat of France and Public Opinion on the Eve of the French Revolution, 1783–1789*. Washington, DC: Catholic University of America Press, 1998.

Murray, William J. "A Philosophe in the French Revolution: Dominique-Joseph Garat and the *Journal de Paris*." In Krantz, *History from Below*, 159–85.

Muthu, Sankar. *Enlightenment against Empire*. Princeton, NJ: Princeton University Press, 2003.

Nathans, Benjamin. "Habermas's 'Public Sphere' in the Era of the French Revolution." *French Historical Studies* 16 (1990): 620–44.

Neidhardt-Jensen, Elske, and Ernst H. Berninger, eds. *Katalog der Ballonhistorischen Sammlung: Oberst von Brug in der Bibliothek des Deutschen Museums*. Munich, 1985.

Nelson, Eric. *The Royalist Revolution: Monarchy and the American Founding*. Cambridge, MA: Belknap Press of Harvard University Press, 2014.

Nicholl, Charles. *Leonardo da Vinci: Flights of the Mind*. New York: Viking, 2004.

Nicolson, Marjorie Hope. *Voyages to the Moon*. New York: Macmillan, 1960.

Nielsen, Wendy C. "Staging Rousseau's Republic: French Revolutionary Festivals and Olympe de Gouges." *Eighteenth Century* 43 (2002): 268–85.

Nora, Pierre, ed. *Les lieux de mémoire*. 7 vols. Paris: Gallimard, 1984–1992.

Nora, Pierre, ed. *Rethinking France*. Translated by Mary Trouille. 4 vols. Chicago: University of Chicago Press, 2001–2010.

Novick, Peter. *That Noble Dream: The "Objectivity Question" and the American Historical Profession*. Cambridge: Cambridge University Press, 1988.

O'Brian, Patrick. *Joseph Banks: A Life*. Chicago: University of Chicago Press, 1987.

Ogborn, Miles. *Spaces of Modernity: London's Geography, 1680–1780*. New York: Guilford Press, 1998.

Olivier-Martin, François. *L'administration provinciale à la fin de l'ancien régime*. Paris: Loysel, 1988.

Onwemechili, Chuka, and Gerard Akindes, eds. *Identity and Nation in African Football: Fans, Community and Clubs.* Basingstoke: Palgrave Macmillian, 2014.

O'Quinn, Daniel. *Staging Governance: Theatrical Imperialism in London, 1770–1800.* Baltimore: Johns Hopkins University Press, 2005.

O'Regan, Michael. "The Myth of Place in Seventeenth- and Eighteenth-Century French Theatre." In Freeman, Mason, O'Regan, Taylor, *Myth and Its Making in the French Theatre,* 81–114.

Orgel, Stephen. "The Poetics of Spectacle." *New Literary History* 2 (1971): 367–89.

Orgel, Stephen. *The Illusion of Power: Political Theater in English Renaissance.* Berkeley: University of California Press, 1975.

Osborne, Michael A. *Nature, the Exotic, and the Science of French Colonialism.* Bloomington: Indiana University Press, 1994.

Osborne, Michael A. "Science and the French Empire." *Isis* 96 (2005): 80–87.

Outram, Dorinda. *The Body and the French Revolution: Sex, Class and Political Culture.* New Haven: Yale University Press, 1989.

Ouvard, L. L. *Libertinage et utopies sous le règne de Louis XIV.* Geneva: Droz, 1989.

Ozouf, Mona. *Festivals and the French Revolution.* Cambridge, MA: Harvard University Press, 1988.

Palmer, Robert Roswell. "Posterity and the Hereafter in Eighteenth-Century French Thought." *Journal of Modern History* 9 (1937): 145–68.

Palmer, Robert Roswell, ed. *The School of the French Revolution: A Documentary History of the College of Louis-le-Grand and Its Director Jean François Champagne, 1762–1814.* Princeton, NJ: Princeton University Press, 1975.

Palmer, Stanley H. *Police and Protest in England and Ireland, 1780–1850.* Cambridge: Cambridge University Press, 1988.

Parker, Harold T. *The Bureau of Commerce in 1781 and Its Policies with Respect to French Industry.* Durham, NC: Carolina Academic Press, 1979.

Parker, Harold T. *An Administrative Bureau during the Old Regime: The Bureau of Commerce and Its Relations to French Industry from May 1781 to November 1783.* Newark: University of Delaware Press, 1993.

Pattie, Frank A. *Mesmer and Animal Magnetism.* Hamilton, NY: Edmonston Publishing, 1994.

Payen, Jacques. *Capital et machine à vapeur au XVIIIe siècle: Les frères Périer et l'introduction en France de la machine à vapeur de Watt.* Paris: Mouton, 1969.

Peltonen, Matti. "Clues, Margins, and Monads: The Micro-Macro Link in Historical Research." *History and Theory* 40 (2001): 347–59.

Perkins, John. "Creating Chemistry in Provincial France before the Revolution." *Ambix* 50 (2003): 145–81 and 51 (2004): 43–75.

Pérouse de Montclos, Jean-Marie. *Jacques-Germain Soufflot.* Paris: Monum, 2004.

Pethers, Matthew. "'Balloon Madness': Politics, Public Entertainment, the Transatlantic Science of Flight, and Late Eighteenth-Century America." *History of Science* 48 (2010): 181–226.

Phillipson, N. "Edinburgh and the Scottish Enlightenment." In *The University in Society*, edited by Lawrence Stone, 2:407–48. Princeton, NJ: Princeton University Press, 1974.

Picon, Antoine. *French Architects and Engineers in the Age of Enlightenment*. Cambridge: Cambridge University Press, 1988.

Pincus, Steven C. A. *Protestantism and Patriotism: Ideologies and the Making of English Foreign Policy, 1650–1668*. Cambridge: Cambridge University Press, 1996.

Pineau, Roger, ed. *Ballooning, 1782–1972*. Washington, DC: Smithsonian Institution Press, 1972.

Pohl, Nicole, and Brenda Tooley, eds. *Gender and Utopia in the Eighteenth Century: Essays in English and French Utopian Writing*. Aldershot: Ashgate, 2007.

Poirier, Jean-Pierre. *Lavoisier: Chemist, Biologist, Economist*. Philadelphia: University of Pennsylvania Press, 1996.

Popkin, Jeremy D. *News and Politics in the Age of Revolution: Jean Luzac's Gazette de Leyde*. Ithaca: Cornell University Press, 1989.

Popkin, Jeremy D. "Pamphlet Journalism at the End of the Old Regime." *Eighteenth-Century Studies* 22 (1989): 351–68.

Porter, Charles A. *Restif's Novels, or an Autobiography in Search of an Author*. New Haven: Yale University Press, 1967.

Porter, Roy. "Science, Provincial Culture and Public Opinion in Enlightenment England." *British Journal for Eighteenth-Century Studies* 3 (1980): 20–46.

Poster, Mark. *The Utopian Thought of Restif de la Bretonne*. New York: New York University Press, 1971.

Poussou, Jean-Pierre. *Bordeaux et le Sud-Ouest au XVIIIe siècle: Croissance économique et attraction urbaine*. Paris: J. Touzot, 1983.

Powell, Martyn J. "Scottophobia versus Jacobitism: Political Radicalism and the Press in Late Eighteenth-Century Ireland." In *Cultures of Radicalism in Britain and Ireland*, edited by John Kirk, Michael Brown, and Andrew Noble, 49–62. London: Pickering and Chatto, 2013.

Power, Amanda. "A Mirror for Every Age." *English Historical Review* 121 (2006): 657–92.

Prak, Maarten. "Citizen Radicalism and Democracy in the Dutch Republic: The Patriot Movement of the 1780s." *Theory and Society* 20 (1991): 73–102.

Prakash, Gyan. *Another Reason: Science and the Imagination of Modern India*. Princeton, NJ: Princeton University Press, 1999.

Pratt, Mary Louise. *Imperial Eyes: Travel Writing and Transculturation*. Princeton, NJ: Princeton University Press, 1992.

Price, Leah. "Vies privées et scandaleuses: Marie-Antoinette and the Public Eye." *Eighteenth Century* 33 (1992): 176–92.

Price, Munro. *Preserving the Monarchy: The Comte de Vergennes, 1774–1787.* Cambridge: Cambridge University Press, 1995.

Proschwitz, Gunmar von, and Mavis von Proschwitz, eds. *Beaumarchais et le Courier de l'Europe: Documents inédites ou peu connus.* 2 vols. Oxford: Voltaire Foundation at the Taylor Institute, 1990.

Proudfoot, William Jardine. *Biographical Memoir of James Dinwiddie.* Liverpool: E. Howell, 1868.

Pucci, Suzanne R. *Sites of the Spectator: Emerging Literary and Cultural Practice in Eighteenth-Century France.* Oxford: Voltaire Foundation, 2001.

Pyenson, Lewis. *Civilizing Mission: Exact Sciences and French Overseas Expansion, 1830–1940.* Baltimore: Johns Hopkins University Press, 1993.

Ramsay, William. *The Life and Letters of Joseph Black, M.D.* London: Constable, 1918.

Ranum, Orest. *Artisans of Glory: Writers and Historical Thought in Seventeenth-Century France.* Chapel Hill: University of North Carolina Press, 1980.

Ravel, Jeffrey S. *The Contested Parterre: Public Theater and French Political Culture.* Ithaca: Cornell University Press, 1999.

Ray, Monique. *L'expérience de Jouffroy d'Abbans en 1783 et la navigation à vapeur dans la région lyonnaise.* Lyon: Musée historique de Lyon, 1983.

Reddy, William M. *The Rise of Market Culture: The Textile Trade and French Society, 1750–1900.* Cambridge: Cambridge University Press, 1984.

Reinelt, Janelle G., and Joseph R. Roach, eds. *Critical Theory and Performance.* Ann Arbor: University of Michigan Press, 1992.

Reisz, Emma. "Curiosity and Rubber in the French Atlantic." *Atlantic Studies* 4 (2007): 5–26.

Remington, Preston. "A Monument Honoring the Invention of the Balloon." *Metropolitan Museum of Art Bulletin* 2 (1944): 241–48.

Rémond, André. *John Holker: Manufacturier et grand fonctionnaire en France au XVIIIe siècle, 1719–1786.* Paris: Marcel Rivière, 1948.

Rendall, Steven. "Fontenelle and His Public." *MLN* 86 (1971): 496–508.

Rétat, Pierre. *La Gazette d'Amsterdam, miroir de l'Europe au XVIIIe siècle.* Oxford: Voltaire Foundation, 2001.

Revel, Jacques. "Marie-Antoinette in Her Fictions: The Staging of Hatred." In *Fictions of the French Revolution,* edited by Bernadette Fort, 111–29. Evanston: Northwestern University Press, 1991.

Rex, Walter E. "On the Background of Rousseau's First Discourse." *Studies in Eighteenth-Century Culture* 9 (1979): 131–50.

Rex, Walter E. *The Attraction of the Contrary: Essays on the Literature of the French Enlightenment.* Cambridge: Cambridge University Press, 1987.

Rex, Walter E. *Diderot's Counterpoints: The Dynamics of Contrariety in His Major Works.* Oxford: Voltaire Foundation, 1998.

Rex, Walter E. "Manufacturing Quality in the Pre-industrial Age: Finding Value in Diversity." *Economic History Review* 53 (2000): 493–516.

Reynard, Pierre Claude. *Ambitions Tamed: Urban Expansion in Pre-revolutionary Lyon.* Montreal: McGill-Queen's University Press, 2009.

Reynaud, Marie-Hélène. *Les Moulins à Papier d'Annonay à l'ère pré-industrielle: Les Montgolfier et Vidalon.* Annonay, France: Editions du Vivarais, 1981.

Reynaud, Marie-Hélène. *Les frères Montgolfier et leurs étonnantes machines.* Vals-les-Bains, France: De Plein Vent, 1982.

Reynolds, Terry S. *Stronger than a Hundred Men: A History of the Vertical Water Wheel.* Baltimore: Johns Hopkins University Press, 1983.

Riley, James C. *The Seven Years War and the Old Regime in France: The Economic and Financial Toll.* Princeton, NJ: Princeton University Press, 1986.

Roberts, Lissa. "An Arcadian Apparatus: The Introduction of the Steam Engine into the Dutch Landscape." *Technology and Culture* 45 (2004): 251–76.

Roberts, Lissa, Simon Schaffer, and Peter Dear, eds. *The Mindful Hand: Inquiry and Invention from the Late Renaissance to Early Industrialization.* Amsterdam: Koninklijke Nederlandse Akademie van Wetenschappen, 2007.

Roberts, Michael. "Great Britain and the Swedish Revolution, 1772–73." *Historical Journal* 7 (1964): 1–46.

Robertson, John. *The Case for the Enlightenment: Scotland and Naples, 1680–1760.* Cambridge: Cambridge University Press, 2005.

Robinson, Eric, and Douglas McKie, ed. *Partners in Science: Letters of James Watt and Joseph Black.* London: Constable, 1970.

Robinson, Nicholas K. *Edmund Burke: A Life in Caricature.* New Haven: Yale University Press, 1996.

Roche, Daniel. *Le siècle des lumières en province: Académies et académiciens provinciaux, 1680–1789.* 2 vols. Paris: Mouton, 1978.

Roche, Daniel. *The People of Paris: An Essay in Popular Culture in the 18th Century.* Berkeley: University of California Press, 1987.

Roche, Daniel. *France in the Enlightenment.* Cambridge, MA: Harvard University Press, 1998.

Roche, Daniel. *History of Everyday Things: The Birth of Consumption in France, 1600–1800.* Cambridge: Cambridge University Press, 2000.

Roger, Jacques. *Buffon: A Life in Natural History.* Ithaca: Cornell University Press, 1997.

Rogers, Nicolas. "Crowd and People in the Gordon Riots." In *The Transformation of Political Culture: England and Germany in the Late Eighteenth Century*, edited by Eckhart Hellmuth, 39–55. Oxford: Oxford University Press, 1990.

Rogers, Nicolas. "Crowds and Political Festival in Georgian England." In *The Politics of the Excluded, c. 1500–1850*, edited by Tim Harris, 233–64. New York: Palgrave, 2001.

Romani, Roberto. *National Character and Public Spirit in Britain and France, 1750–1914*. Cambridge: Cambridge University Press, 2002.

Root-Bernstein, Michèle. *Boulevard Theater and Revolution in Eighteenth-Century Paris*. Ann Arbor: UMI Research Press, 1981.

Rosenband, Leonard N. "Hiring and Firing at the Montgolfier Paper Mill." In *The Workplace before the Factory: Artisans and Proletarians, 1500–1800*, edited by Thomas Max Safley and Leonard N. Rosenband, 225–40. Ithaca: Cornell University Press, 1993.

Rosenband, Leonard N. "Jean-Baptiste Réveillon: A Man on the Make in Old Regime France." *French Historical Studies* 20 (1997): 481–510.

Rosenband, Leonard N. "Nicolas Desmarest and the Transfer of Technology in Old Regime France." In *The Modern Worlds of Business and Industry: Cultures, Technology, Labor*, edited by Karen R. Merrill, 103–20. Turnhout, Belgium: Brepols, 1998.

Rosenband, Leonard N. "The Competitive Cosmopolitanism of an Old Regime Craft." *French Historical Studies* 23 (2000): 455–76.

Rosenband, Leonard N. *Papermaking in Eighteenth-Century France: Management, Labor, and Revolution at the Montgolfier Mill, 1761–1805*. Baltimore: Johns Hopkins University Press, 2000.

Rosenband, Leonard N. "Becoming Competitive: England's Papermaking Apprenticeship, 1700–1800." In Roberts, Schaffer, and Dear, *The Mindful Hand*, 379–401.

Rosenfeld, Sophia. "Tom Paine's Common Sense and Ours." *William and Mary Quarterly* 65 (2008): 633–68.

Rostaing, Léon. *Les anciennes loges maçonniques d'Annonay et les clubs, 1766–1815*. Lyon: Louis Brun, 1903.

Rostaing, Léon. *La famille de Montgolfier, ses alliances, ses descendants*. Lyon: A. Rey & Cie, 1910.

Rothschild, Emma. "Commerce and the State: Turgot, Condorcet, and Smith." *Economic Journal* 414 (1992): 1197–210.

Rothschild, Emma. *The Inner Life of Empires: An Eighteenth-Century History*. Princeton, NJ: Princeton University Press, 2011.

Rothschild, Emma. "Isolation and Economic Life in Eighteenth-Century France." *American Historical Review* (2014): 1054–82.

Rudd, Niall. "Daedalus and Icarus." In *Ovid Renewed: Ovidian Influences on Literature and Art from the Middle Ages to the Twentieth Century*, edited by Charles Martindale, 21–53. Cambridge: Cambridge University Press, 1988.

Rudé, George F. E. *The Crowd in the French Revolution*. Oxford: Oxford University Press, 1959.

Rudé, George F. E. *Paris and London in the Eighteenth Century*. New York: Viking Press, 1971.

Rudé, George F. E. *Ideology and Popular Protest.* New York: Pantheon, 1980.

Rufi, Enrico. *Le rêve laïque de Louis-Sébastien Mercier entre littérature et politique.* Oxford: Voltaire Foundation, 1995.

Ruplinger, André. *Charles Bordes: Un représentant provincial de l'esprit philosophique au XVIIIe siècle en France.* Lyon: A. Rey, 1915.

Rushton, Julian. "The Theory and Practice of Piccinnisme." *Proceedings of the Royal Musical Association* 98 (1971–1972): 31–46.

Rushton, Julian. "'Royal Agamemnon': The Two Versions of Gluck's *Iphigénie en Aulide.*" In *Music and the French Revolution*, edited by Malcom Boyd, 15–36. Cambridge: Cambridge University Press, 1992.

Russell, Jack. *Gibraltar Besieged, 1779–1783.* London: Heimann, 1965.

Russo, Elena. *Styles of Enlightenment: Taste, Politics, and Authorship in Eighteenth-Century France.* Baltimore: Johns Hopkins University Press, 2007.

Rydell, Robert W. *All the World's a Fair: Visions of Empire at American International Expositions, 1876–1916.* Chicago: University of Chicago Press, 1984.

Sahlins, Peter. "The Royal Menageries of Louis XIV and the Civilizing Process Revisited." *French Historical Studies* 35 (2012): 237–67.

Said, Edward W. *Orientalism.* New York: Vintage Books, 1979.

Sainsbury, John. *Disaffected Patriots: London Supporters of Revolutionary America, 1769–1782.* Kingston, Ontario: McGill-Queen's University Press, 1987.

Saint-Léon, E. Martin. *Le compagnonage: Son histoire, ses coutumes, ses règlements, et ses rites.* Paris: Armand Colin, 1901.

Saisselin, Rémy G. *The Enlightenment against the Baroque: Economics and Aesthetics in the Eighteenth Century.* Berkeley: University of California Press, 1992.

Salatino, Kevin. *Incendiary Art: The Representation of Fireworks in Early Modern Europe.* Los Angeles: Getty, 1997.

Sama, Catherine M. "Liberty, Equality, Frivolity! An Italian Critique of Fashion Periodicals." *Eighteenth-Century Studies* 37 (2004): 389–414.

Sauvaire-Jourdan, François. *Isaac de Bacalan et les idées libre-échangistes en France vers le milieu du dix-huitième siècle.* Paris: L. Larose, 1903.

Scarfe, Francis. *André Chénier: His Life and Work, 1762–1794.* Oxford: Clarendon Press, 1965.

Schaffer, Simon. "Natural Philosophy and Public Spectacle in the Eighteenth Century." *History of Science* 21 (1983): 1–43.

Schaffer, Simon. "Enlightened Automata." In Clark, Golinski, and Schaffer, *The Sciences in Enlightened Europe*, 126–65.

Schaffer, Simon. "'The Charter'd Thames': Naval Architecture and Experimental Spaces in Georgian Britain." In Roberts, Schaffer, and Dear, *The Mindful Hand*, 279–305.

Schaffer, Simon, Lissa Roberts, Kapil Raj, and James Delbourgo, eds. *The Brokered World: Go-Betweens and Global Intelligence, 1770–1820. Sagamore Beach, MA: Science History Publications, 2009.

Schaffer, Simon. "The Astrological Roots of Mesmerism." *Studies in History and Philosophy of Biological and Biomedical Sciences* 41 (2010): 158–68.

Schama, Simon. *Citizens: A Chronicle of the French Revolution.* New York: Vintage Books, 1990.

Schama, Simon. *Patriots and Liberators: Revolution in the Netherlands, 1780–1813.* New York: Vintage Books, 1992.

Schiebinger, Londa. *Plants and Empire: Colonial Bioprospecting in the Atlantic World.* Cambridge, MA: Harvard University Press, 2004.

Schiff, Stacy. *A Great Improvisation: Franklin, France and the Birth of America.* New York: Henry Holt, 2005.

Schmidt, Freek H. "Expose Ignorance and Revive the 'Bon Goût': Foreign Architects at Jacques-François Blondel's École des Arts." *Journal of the Society of Architectural Historians* 61 (2002): 4–29.

Schmidt, James, ed. *What Is Enlightenment? Eighteenth-Century Answers and Twentieth-Century Questions.* Berkeley: University of California Press, 1996.

Schnapp, Jeffrey T. "18 BL: Fascist Mass Spectacle." *Representations* 43 (1993): 89–125.

Schoenbrun, David. *Triumph in Paris: The Exploits of Benjamin Franklin.* New York: Harper and Row, 1976.

Schroder, Anne L. "Going Public against the Academy in 1784: Mme de Genlis Speaks Out on Gender Bias." *Eighteenth-Century Studies* 32 (1999): 376–82.

Schrøder, Michael. *The Argand Burner, Its Origin and Development in France and England, 1780–1800: An Epoch Illustrated by the Life and Work of the Physicist Ami Argand (1750–1803).* Translated by Hugh Sheperd. Odense, Denmark: Odense University Press, 1969.

Schuchard, Marsha Keith. *Emanuel Swedenborg, Secret Agent on Earth and in Heaven: Jacobites, Jews, and Freemasons in Early Modern Sweden.* Leiden: Brill, 2012.

Schwartz, Leon. "F. M. Grimm and the Eighteenth-Century Debate on Women." *French Review* 58 (1984): 236–43.

Schwartz, Robert M. *Policing the Poor in Eighteenth-Century France.* Chapel Hill: University of North Carolina Press, 1988.

Scott, James C. *Domination and the Arts of Resistance: Hidden Transcripts.* New Haven: Yale University Press, 1990.

Secord, James A. "Knowledge in Transit." *Isis* 95 (2004): 654–72.

Sewell, William H. *A Rhetoric of Bourgeois Revolution: The Abbé Sieyes and What Is the Third Estate?* Durham, NC: Duke University Press, 1994.

Sewell, William H. "The Empire of Fashion and the Rise of Capitalism in Eighteenth-Century France." *Past & Present* 206 (2010): 81–120.

Shackleton, Robert. "The Evolution of Montesquieu's Theory of Climate." *Revue internationale de philosophie* 9 (1955): 317–29.

Shapin, Steven. "Pump and Circumstance: Robert Boyle's Literary Technology." *Social Studies of Science* 14 (1984): 481–520.

Shapin, Steven, and Simon Schaffer. *Leviathan and the Air-Pump*. Princeton, NJ: Princeton University Press, 1985.

Shapin, Steven. "The House of Experiment in Seventeenth-Century England." *Isis* 79 (1988): 373–404.

Shapin, Steven. *A Social History of Truth: Civility and Science in Seventeenth-Century England*. Chicago: University of Chicago Press, 1994.

Shennan, J. H. *Philippe, Duke of Orléans: Regent of France, 1715–1723*. London: Thames and Hudson, 1979.

Shovlin, John. *The Political Economy of Virtue: Luxury, Patriotism, and the Origins of the French Revolution*. Ithaca: Cornell University Press, 2006.

Simoes, Ana, Ana Carneiro, and Maria Paula Diogo, eds. *Travels of Learning: A Geography of Science in Europe*. Dordrecht: Kluwer, 2003.

Skuncke, Marie-Christine. "Press and Political Culture in Sweden at the End of the Age of Liberty." In *Enlightenment, Revolution and the Periodical Press*, edited by Hans-Jürgen Lüsebrink and Jeremy D. Popkin, 81–101. Oxford: Voltaire Foundation, 2004.

Smeaton, W. A. "The Early Years of the Lycée and the Lycée des Arts, I: The Lycée of the Rue de Valois." *Annals of Science* 11 (1955): 257–67.

Smeaton, W. A. "Jean-François Pilâtre de Rozier, the First Aeronaut." *Annals of Science* 11 (1955): 257–67.

Smernoff, Richard A. *André Chénier*. Boston: Twayne Publishers, 1977.

Šmidchens, Guntis. *The Power of Song: Nonviolent National Culture in the Baltic Singing Revolution*. Seattle: University of Washington Press, 2014.

Smith, Crosbie, and M. Norton Wise. *Energy and Empire: A Biographcial Study of Lord Kelvin*. Cambridge: Cambridge University Press, 1989.

Smith, Jay M. "Between Discourse and Experience: Agency and Ideas in the French Pre-revolution." *History and Theory* 40 (2001): 116–42.

Smith, Jay M. *Nobility Reimagined: The Patriotic Nation in Eighteenth-Century France*. Ithaca: Cornell University Press, 2005.

Smith, Richard O. *The Man with His Head in the Clouds: James Sadler, the First Englishman to Fly*. Oxford: Signal Books, 2014.

Soboul, Albert. *The French Revolution, 1787–1799: From the Storming of the Bastille to Napoleon*. London: Unwin Hyman, 1962.

Soliday, Gerland Lyman. *A Community in Conflict: Frankfurt Society in the Seventeenth and Early Eighteenth Centuries*. Hanover, NH: Brandeis University Press, 1974.

Sonenscher, Michael. "Journeymen's Migrations and Workshop Organization in Eighteenth-Century France." In Kaplan and Koepp, *Work in France*, 74–96. Ithaca: Cornell University Press, 1986.

Sonenscher, Michael. *Work and Wages: Natural Law, Politics and the Eighteenth-Century French Trades*. Cambridge: Cambridge University Press, 1989.

Sonenscher, Michael. "Enlightenment and Revolution." *Journal of Modern History* 70 (1998): 371–83.

Sonenscher, Michael. "Fashion's Empire: Trade and Power in Early 18th-century France." In Fox and Turner, *Luxury Trades*, 231–54.

Sonenscher, Michael. *Sans-Culottes: An Eighteenth-Century Emblem in the French Revolution*. Princeton, NJ: Princeton University Press, 2008.

Sonnet, Bernard. "Le palais abbatial de Saint-Bénigne de Dijon." *Annales de Bourgogne* 72 (2000): 237–54.

Sorrenson, Richard. *Perfect Mechanics: Instrument Makers at the Royal Society of London in the Eighteenth Century*. Boston: Docent Press, 2013.

Spary, Emma C. *Utopia's Garden: French Natural History from Old Regime to Revolution*. Chicago: University of Chicago Press, 2000.

Spary, Emma C. *Eating the Enlightenment*. Chicago: University of Chicago Press, 2014.

Sprunger, Keith L. "Frans Houttuyn, Amsterdam Bookseller." Available at https://www.goshen.edu/mqr/pastissues/apr04sprunger.html.

Stalnaker, Joanna. "The New Paris in Guise of the Old: Louis Sébastien Mercier from Old Regime to Revolution." *Studies in Eighteenth-Century Culture* 35 (2006): 223–66.

Standage, Tom. *The Turk: The Life and Times of the Famous Eighteenth-Century Chess-Playing Machine*. New York: Walker, 2002.

Stansfield, Dorothy A. *Thomas Beddoes M.S. 1760–1808*. Dordrecht: D. Reidel, 1984.

Starobinski, Jean. *1789. The Emblems of Reason*. Charlottesville: University Press of Virginia, 1982.

Starobinski, Jean. *Le remède dans le mal: Critique et légitimation de l'artifice à l'âge des Lumières*. Paris: Gallimard, 1989.

Steptoe, Andrew. "Mozart, Mesmer and 'Cosi Fan Tutte.'" *Music & Letters* 67 (1986): 248–55.

Stern, Madeleine B. "A Salem Author and a Boston Publisher: James Tytler and Joseph Nancrede." *New England Quarterly* 47 (1974): 290–301.

Stewart, Anthony T. Q. *A Deeper Silence: The Hidden Origins of the United Irish Movement*. London: Faber and Faber, 1993.

Stewart, Larry R. *The Rise of Public Science: Rhetoric, Technology and Natural Philosophy*. Cambridge: Cambridge University Press, 1992.

Stewart, Philip. "Science and Superstition: Comets and the French Public in the 18th Century." *American Journal of Physics* 54 (1986): 16–24.

Stone, Bailey. *The French Parlements and the Crisis of the Old Regime*. Chapel Hill: University of North Carolina Press, 1986.

Stormer, Nathan. "Addressing the Sublime: Space, Mass Representation, and the Unpresentable." *Critical Studies in Media Communication* 21 (2004): 212–40.

Strathern, Paul. *The Artist, the Philosopher and the Warrior. Leonardo, Machiavelli and Borgia, a Fateful Collusion.* London: Jonathan Cape, 2009.

Suibhne, Brendan Mac. "Whiskey, Potatoes and Paddies: Volunteering and the Construction of the Irish Nation in Northwest Ulster, 1778–1782." In Jupp and Magennis, *Crowds in Ireland*, 45–82.

Sutherland, Donald. *The French Revolution and Empire: The Quest for a Civic Order.* Oxford: Oxford University Press, 2003.

Sutton, Geoffrey V. "Electric Medicine and Mesmerism." *Isis* 72 (1981): 375–92.

Sutton, Geoffrey V. *Science for a Polite Society: Gender, Culture, and the Demonstration of Enlightenment.* Boulder: Westview Press, 1995.

Swann, Julian. *Provincial Power and Absolute Monarchy: The Estates General of Burgundy, 1661–1790.* Cambridge: Cambridge University Press, 2003.

Tackett, Timothy. *Becoming a Revolutionary: The Deputies of the French National Assembly and the Emergence of a Revolutionary Culture (1789–1790).* Princeton, NJ: Princeton University Press, 1996.

Tackett, Timothy. *The Coming of the Terror in the French Revolution.* Cambridge, MA: Belknap Press of Harvard University Press, 2015.

Takats, Sean. *The Expert Cook in Enlightenment France.* Baltimore: Johns Hopkins University Press, 2011.

Taylor, Diana. *The Archive and the Repertoire: Performing Cultural Memory in the Americas.* Durham, NC: Duke University Press, 2003.

Te Brake, Wayne P. "Popular Politics and the Dutch Patriot Revolution." *Theory and Society* 14 (1985): 199–222.

Te Brake, Wayne P. "Violence in the Dutch Patriot Revolution." *Comparative Studies in Society and History* 30 (1988): 143–63.

Terrall, Mary. "Gendered Spaces, Gendered Audiences: Inside and outside the Paris Academy of Sciences." *Configurations* 3 (1995): 207–32.

Terrall, Mary. *The Man Who Flattened the Earth: Maupertuis and the Sciences in the Enlightenment.* Chicago: University of Chicago Press, 2002.

Terrall, Mary. *Catching Nature in the Act: Reaumur and the Practice of Natural History in the Eighteenth Century.* Chicago: University of Chicago Press, 2014.

Thaddeus, Janice Farrar. *Frances Burney: A Literary Life.* New York: St. Martin's Press, 2000.

Thébaud-Sorger, Marie. *L'aérostation au temps des lumières.* Rennes: Presses universitaires de Rennes, 2009.

Thibaut, L. "Les voies navigables et l'industrialisation du Nord de la France avant 1789." *Revue du Nord* 61 (1979): 149–64.

Thomas, Chantal. *La Reine scélérate: Marie Antoinette dans les pamphlets.* Paris: Éditions du Seuil, 1989.

Thomas, Chantal. *The Wicked Queen: The Origins of the Myth of Marie-Antoinette.* New York: Zone Books, 1999.

Thomas, Downing A. *Aesthetics of Opera in the Ancien Régime, 1647–1785.* Cambridge: Cambridge University Press, 2002.

Thompson, Edward Palmer. *The Making of the English Working Class.* London: Gollancz, 1963.

Thompson, Edward Palmer. "The Moral Economy of the English Crowd in the 18 C." *Past & Present* 50 (1971): 76–136.

Thompson, Eric. *Popular Sovereignty and the French Constituent Assembly, 1789–91.* Manchester: Manchester University Press, 1952.

Thorndike, Lynn. "The True Roger Bacon." *American Historical Review* 21 (1916): 237–57, 468–80.

Tilly, Louise. "The Food Riot as a Form of Political Conflict in France." *Journal of Interdisciplinary History* 2 (1971): 23–58.

Tissandier, Gaston. *Bibliographie aéronautique: Catalogue de livres d'histoire, de science, de voyages et de fantaisie, traitant de la Navigation aérienne ou des Aérostats.* Paris: H. Launette, 1887.

Tissandier, Gaston. *Histoire des ballons et des Aéronautes célèbres, 1783–1800.* 2 vols. Paris: H. Launette, 1887.

Tabacco, Giovanni. *The Struggle for Power in Medieval Italy: Structures of Political Rule, 400–1400.* Translated by Rosalind Brown Jensen. Cambridge: Cambridge University Press, 1989.

Tocqueville, Alexis de. *The Old Regime and the Revolution.* 2 vols. Chicago: University of Chicago Press, 1998.

Tombs, Robert, and Isabelle Tombs. *That Sweet Enemy: The French and the British from the Sun King to the Present.* New York: Alfred A. Knopf, 2007.

Tomory, Leslie. "Let It Burn: Distinguishing Inflammable Airs, 1766–1790." *Ambix* 56 (2009): 253–72.

Torlais, Jean. *Un physicien au siècle des lumières: Abbé Nollet (1700–1770).* Paris: Sipuco, 1954.

Tourneux, Maurice. "Un projet d'encouragement aux lettres et aux sciences sous Louis XVI." *Revue d'histoire littéraire de la France* 8 (1901): 281–311.

Trénard, Louis. "Un notable lyonnais pendant la crise révolutionnaire: Pierre-Tousssaint Dechazelle." *Revue d'histoire moderne et contemporaine* 5 (1958): 201–25.

Trénard, Louis, ed. *Histoire d'une métropole: Lille-Roubaix-Tourcoing.* Toulouse, 1977.

Trouillot, Michel-Rolph. *Silencing the Past: Power and the Production of History.* Boston: Beacon Press, 1995.

Truant, Cynthia M. "Independent and Insolent: Journeymen and Their 'Rites' in the Old Regime Workplace." In Kaplan and Koepp, *Work in France*, 131–75.

Turner, Victor. *Dramas, Fields and Metaphors: Symbolic Action in Human Society*. Ithaca: Cornell University Press, 1974.

Turner, Victor W. *From Ritual to Theater: The Human Seriousness of Play*. New York: Performing Arts Journal Publications, 1982.

Turner, Victor. "Are There Universals of Performance in Myth, Ritual and Drama?" In *By Means of Performance: Intercultural Studies of Theater and Ritual*, edited by Richard Schechner and Willa Appel, 8–18. Cambridge: Cambridge University Press, 1990.

Uglow, Jennifer S. *The Lunar Men: Five Friends Whose Curiosity Changed the World*. New York: Farrar, Straus, and Giroux, 2002.

Urbinati, Nadia. "Condorcet's Democratic Theory of Representative Government." *European Journal of Political Theory* 3 (2004): 53–75.

Van Kley, Dale. *The Jansenists and the Expulsion of the Jesuits from France, 1757–1765*. New Haven: Yale University Press, 1975.

Van Kley, Dale. *Religious Origins of the French Revolution*. New Haven: Yale University Press, 1996.

Velut, Christine. "L'industrie dans la ville: Les fabriques de papiers peints du faubourg Saint-Antoine (1750–1820)." *Revue d'histoire moderne et contemporaine*, 49 (2002): 115–37.

Vendrix, Philippe. "La notion de revolution dans les écrits théoriques concernant la musique avant 1789." *International Review of the Aesthetics and Sociology of Music* 21 (1990): 71–78.

Venturi, Franco. *The End of the Old Regime in Europe, 1776–1789*. 2 vols. Princeton, NJ: Princeton University Press, 1991.

Vidal, Mary. "David among the Moderns: Art, Science, and the Lavoisiers." *Journal of the History of Ideas* 56 (1995): 595–623.

Vidal-Naquet, Pierre, and Janet Lloyd. "Atlantis and the Nations." *Critical Inquiry* 18 (1992): 300–326.

Vidler, Anthony. *The Writing of the Walls: Architectural Theory in the Late Enlightenment*. Princeton, NJ: Princeton University Press, 1987.

Vidler, Anthony. *Claude-Nicolas Ledoux: Architecture and Social Reform at the End of the Ancien Régime*. Cambridge: Cambridge University Press, 1990.

Viroli, Maurizio. *For Love of Country: An Essay on Patriotism and Nationalism*. Oxford: Oxford University Press, 1995.

Vyverberg, Henry. "Limits of Nonconformity in the Enlightenment: The Case of Simon-Nicolas-Henri Linguet." *French Historical Studies* 6 (1970): 474–91.

Wade, Ira O. *The "Philosophe" in the French Drama of the Eighteenth Century*. Princeton, NJ: Princeton University Press, 1926.

Wagstaff, Peter. *Memory and Desire: Rétif de la Bretonne, Autobiography and Utopia*. Amsterdam: Rodopi, 1996.

Walton, Charles. *Policing Public Opinion in the French Revolution: The Culture of Calumny and the Problem of Free Speech*. Oxford: Oxford University Press, 2009.

Watkin, David. *The Architect King: George III and the Culture of the Enlightenment*. London: Royal Collections, 2004.

Weber, Caroline. *Queen of Fashion: What Marie Antoinette Wore to the Revolution*. New York: Henry Holt, 2006.

Weber, William. "The 1784 Handel Commemoration as Political Ritual." *Journal of British Studies* 28 (1989): 43–69.

Wecter, Dixon. "Benjamin Franklin and an Irish Enthusiast." *Huntington Library Quarterly* 4 (1941): 205–34.

Wegner, Phillip E. *Imaginary Communities: Utopia, the Nation, and the Spatial Histories of Modernity*. Berkeley: University of California Press, 2002.

Weiner, Dora. *The Citizen-Patient in Revolutionary and Imperial Paris*. Baltimore: Johns Hopkins University Press, 1993.

Weinshenker, Anne Betty. "Diderot's Use of the Ruin-Image." *Diderot Studies* 16 (1973): 309–29.

Weinstein, Leo. *The Subversive Tradition in French Literature*. 2 vols. Boston: Twayne Publishers, 1989.

Welschinger, Henri, ed. *La mission secrète de Mirabeau à Berlin, 1786–1787, d'après des documents originaux des archives des affaires étrangères*. Paris: E. Plon, 1900.

Werrett, Simon. *Fireworks: Pyrotechnic Arts and Sciences in European History*. Chicago: University of Chicago Press, 2010.

Whaley, Joachim. *Religious Toleration and Social Change in Hamburg, 1529–1819*. Cambridge: Cambridge University Press, 1985.

Whelan, Kevin. *The Tree of Liberty: Radicalism, Catholicism and the Construction of Irish Identity, 1760–1830*. Notre Dame, IN: University of Notre Dame Press, 1996.

White, Lynn. "Eilmer of Malmesbury, an Eleventh Century Aviator." *Technology and Culture* 2 (1961): 97–111.

Whitney, Frank. *Jean Ternant and the Age of Revolutions: A Soldier and Diplomat (1751–1833) in the American, French, Dutch and Belgian Uprisings*. Jefferson, NC: MacFarland, 2015.

Wilkie, Everette C., Jr. "Mercier's L'An 2440: Its Publishing History during the Author's Lifetime." *Harvard Library Bulletin* 32 (1984): 5–35, 348–400.

Williams, Alan. *The Police of Paris, 1718–1789*. Baton Rouge: Louisiana State University Press, 1980.

Williams, David. *Condorcet and Modernity*. Cambridge: Cambridge University Press, 2002.

Williams, Robert J. P., Alland Chapman, and John S. Rowlinson. *Chemistry at Oxford: A History from 1600 to 2005*. Cambridge: Cambridge University Press: RSC Publishing, 2009.

Wilson, Kathleen. "Empire, Trade and Popular Politics in Mid-Hanoverian Britain: The Case of Admiral Vernon." *Past & Present* 121 (1988): 74–109.

Wilson, Kathleen. "The Good, the Bad and the Impotent: Imperialism and the Politics of Identity in Georgian England." In Bermingham and Brewer, *Consumption of Culture*, 229–52.

Wilson, Kathleen. *The Sense of the People: Politics, Culture and Imperialism in England, 1715–1785*. Cambridge: Cambridge University Press, 1995.

Winn, Collette H., and Donna Kuizenga. *Women Writers in Pre-revolutionary France: Strategies of Emancipation*. New York: Garland, 1997.

Wintroub, Michael. "Taking Stock at the End of the World: Rites of Distinction and Practices of Collecting in Early Modern Europe." *Studies in History and Philosophy of Science*, Part A, 30 (1999): 395–424.

Wise, M. Norton. "Mediating Machines." *Science in Context* 2 (1988): 77–113.

Wise, M. Norton, ed. *The Values of Precision*. Princeton, NJ: Princeton University Press, 1995.

Wise, M. Norton, and Elaine M. Wise, "Staging an Empire." In *Things That Talk: Object Lessons from Art and Science*, edited by Lorraine Daston, 101–45. New York: Zone Books, 2004.

Wohl, Robert. *The Spectacle of Flight: Aviation and the Western Imagination, 1920–1950*. New Haven: Yale University Press, 2005.

Wokler, Robert. "Rites of Passage and the Grand Tour: Discovering, Imagining and Inventing European Civilization in the Age of Enlightenment." In *Finding Europe: Discourses on Margins, Communities, Images ca. 13th–ca. 18th Centuries*, edited by Anthony Molho, Diogo Ramada Curto, and Niki Koniordos, 205–22. New York: Berghahn Books, 2007.

Wolfe, John J. *Brandy, Balloons, and Lamps: Ami Argand, 1750–1803*. Carbondale: Southern Illinois University Press, 1999.

Wolloch, Nathaniel. *History and Nature in the Enlightenment: Praise of the Mastery of Nature in Eighteenth-Century Historical Literature*. Burlington, VT: Ashgate, 2011.

Wood, Ellen Meiksins. "The State and Popular Sovereignty in French Political Thought: A Genealogy of Rousseau's 'General Will.'" In Krantz, *History from Below*, 117–39.

Wood, Paul. "Science, the Universities, and the Public Sphere in Eighteenth-Century Scotland." *History of Universities* 13 (1994): 99–135.

Woodmanse, Martha. "The Genius and the Copyright: Economic and Legal Conditions of the Emergence of the 'Author.'" *Eighteenth-Century Studies* 17 (1984): 425–48.

Wright, Johnson Kent. *A Classical Republican in Eighteenth-Century France: The Political Thought of Mably*. Stanford, CA: Stanford University Press, 1997.

Yardeni, Myriam. *Utopie et révolte sous Louis XIV*. Paris: Nizet, 1980.

Yates, Frances A. *Astraea: The Imperial Theme in the Sixteenth Century*. London:

Routledge and Kegan Paul, 1973.

Zimmer, Oliver. *A Contested Nation: History, Memory and Nationalism in Switzerland, 1761–1891*. Cambridge: Cambridge University Press, 2003.

Ziolkowski, Theodore. *Ovid and the Moderns*. Ithaca: Cornell University Press, 2005.

INDEX

abbé, 36, 75, 100; Barrière, 61; Dupont de Jumeaux, 177; d'Espagnac, 73; de Fontana, 46; Kentzinger, 273, 339n70; Mably, 105; Miolan, 140, 184; de Raynal, 77–78; Sieyès, 284; de Viennay, 41, 138–39, 236, 243
abbey, 41, 73; Faremoutiers, 37; St. Benigne, 168; Turpenay, 41
academicians, 6, 135, 152, 190; Achard, 167, 278; Dijon, 165–71; Grossart de Virly, 170; Lyon, 153, 156, 161; Thyrion, 82; Saladin, 136; Scanégatty, 136, 160
Academicians (Paris), 35, 70–72, 80, 98, 132, 140–46, 211; Bailly, 146; Bory, 146; Cadet de Gassicourt, 49–50; Cassini, 81, 185; Condorcet, 48, 53; Coulomb, 43, 47; Deparcieux, 213; Desmarest, 45; Franklin, 146; Jeurat, 185; Méchain, 185; Lalande, 43, 161, 239; Lavoisier, 53, 146; Le Roy, 81, 86, 91, 146, 166, 184; Macquer, 58; Périer, 47; Sage, 82, 91
Academy: Aeronautic, 14, 251, 253, 257; Bordeaux, 39, 65, 177–80; Dijon, 65, 164–71, 177; Irish, 245; Lille, 136; lunar, 123; Lyon, 131, 135, 158, 163, 165, 177; Metz, 82; Marseilles, 82, 159; Military (London), 238; Royal (British), 237; Royal (Irish), 245
Academy of Sciences (Paris): balloon ascent and, 59, 70, 86, 98, 211; balloon research and, 80, 97–99, 131–32, 187–88; functions of, 44–45, 48–50; mesmerism and, 132, 190–91; reports to, 47–48, 82, 84, 140–42; status of, 35, 65, 91, 134–37, 160, 166, 212
access, 47, 50, 58, 64, 69, 80, 83, 116, 135, 137
actor: historical, xxiv, 9–10; human, xxiv, 5, 12; political, 4; silent, 9
adventurer, 227, 229, 257; aerial, 245; aerostatic, 266; balloon, 240; English, 240; foreign, 196, 211, 214, 240, 247; French, 267; merchant, 211
aerial, 101; adventure, 245; Argonaut, 77, 86; armies, xxiii; beings, 239; car, 128; carriage, 115; cavalier, 77; chariot, 187, 222, 245; Columbus, 101; competition, 218; conquest, 9, 110; empire, xviii, xxiv, 30, 77–78, 112, 115, 117, 132; expanse, 149; fish, 66; forces, xxiii; invasion, 131; journey, 154, 170, 218, 223, 225; jump, 223; navigation, 44, 67, 117, 240; navigator, 128, 153, 221, 268; ocean, 131; passage, 88; philosopher, 124–25, 221; plane, 96; prince, 261, 265; resistance, 187; rudder, 215; show, 238; spectacle, 88; spectator, 228; travel, 101; vessel, 168; view, 85; vista, 6, 100;

voyage, 68, 85–86, 153–54, 197, 226, 239, 269, 286; voyager, 86, 126–27
aeronaut, 12, 77, 99, 124–26, 218, 239, 255–57, 270, 286; Blanchard, 14, 196–97, 237–38, 241–43, 265–66, 271, 275, 279; Bordeaux, 180–82; Calais, 255; Charles, 88, 91–96; Dijon, 168–70; female, 163, 253; first, 83–88; foreign, 247; heroic, 127; Jeffries, 241–43; Lille, 269–70; London, 228; Lyon, 156–58; Miolan, 185; philosophical, 12; Pilâtre, 82, 154, 160, 183–84, 268; royal, 183, 254; Sadler, 240; Sheldon, 238; St. Cloud, 186
aeronautics, xviii, 29, 139, 237, 257
aerostatic globe (or machine), 67, 81, 157, 213, 219, 231, 254; ban on, 132, 162; impact of, 74–75, 84–85, 117, 127, 135, 149, 283; invention of, 45, 87, 92–97, 196, 283; Lunardi's, 226; as national artifact, 109, 173; research on, 131–34, 169; as scientific invention, 48, 56, 63, 97, 103; as useful invention, 12, 31, 99
agency: of balloon, xviii, xxv, 5, 54, 110, 117, 286; crowd's or people's, 5, 7, 290; human, 8, 14, 289; machine's, 8–9, 14, 290; material, 8, 289; situational, 5; transgressive, 25
air: different kinds of, 38–39; inflammable (*see* inflammable air); mephitic, 46; vital, 57. *See also* gas
Air Balloon, 201, 210, 214, 220–23, 246
Air Balloon, 215–20
Alban, Léonard, 100, 132
amateur, 58, 61, 68, 72, 77, 100, 135–38, 153–54, 184–85, 225, 277–80
America, xix, 5, 34, 78, 175; balloon, 210; Blanchard and, 281; Lafayette and, 144
American War, 5–8, 29–30, 97, 147, 209–10, 221, 231, 235, 283
ancien régime, xxv, 5–10, 25–26, 83, 92, 118, 121, 148–49, 298n40; citizen, 82; economy, 32; elite, 69, 75; institutions, 173; public space, 53, 67; state-body, xxv, 26; technologies, 26, 39, 283, 293
Annonay: ascent, 47–50, 55–56, 64–66; lodge, 36, 38, 153
anti-balloonist, 10, 75, 112, 124–25, 186
Antoinette, Marie, 68, 84, 114–16, 134, 174, 278, 289; *Marie Antoinette,* 181–83, 189, 235
approbation, 47, 134, 148; Parisian, 160, 236; public, 43, 59, 91, 98, 166; universal, 138
appropriate, xviii, 4, 24, 38, 64, 89, 100, 128, 134, 142, 195, 237, 258

415

appropriation, 6, 14, 54, 109, 140, 151, 170, 196, 235, 270
archeology of: calumny, 67; guillotine, 290–91; hidden transcripts, 67, 152; (mass) silence, 8–9, 54, 129; public space, 285; (scientific) artifact, 8, 152, 291
archive, 9–10, 55, 110
Argand, Aimé, 58, 69–70, 213, 216–17, 238
Argonaut, xix, 77, 86, 125
artifact: British, 261; expensive, 158; French, 13, 196; majestic, 55, 288; malleable, 293; marvelous, 13, 63, 117, 197; material, xxiv, 8; monumental, 8, 10, 112; national (*see* national); popular, 228; potent, 54; public, 58–59, 89, 98, 134, 147, 174; royal, 2, 255, 280; scientific, xviii, 10–13, 26, 48, 50, 75, 101, 104, 109, 136, 174, 195, 211, 231, 288, 290–91; spectacular, 69, 279; useful, 237; virtuous, 74; visible, 45
ascent: heroic, 99; human, 134; Montgolfiers,' 12; moral, xxiv; physical, xviii–xix, xxiv; public, 154; spectacular, 12, 86, 101; spiritual, xviii–xix; superhuman, xix
assembly, 5, 70, 84, 98, 105, 160, 182, 188, 263, 280; national, 192, 262, 284–86, 289
Assumption, 12, 103, 117, 188
audience, 11, 14, 65, 80, 85; balloon, 13, 72, 128, 277, 281; composite or mixed, 55, 149, 173, 195; elite, 48, 54, 142, 180, 182, 250, 253, 263, 274; enlightened, 4, 75; for demonstration, 44; German, 72; Irish, 245; journal, 55; mass, 5, 8, 170; middling, 56, 197; Réveillon, 83; royal, 72, 267; theater, 24, 231; transgressive, 25
authority, 32, 48, 117, 143; aristocratic, 262; civil, 171, 176, 180, 268; cultural, 107; government, 147, 191, 218, 274; imperial, 271; inventor's, 48; of Nature, 149, 283; political, 265; princely, 11; public, 147, 197, 266; rational, 4; royal, 55, 69, 92, 106, 111, 126, 133, 261, 283; of science, 6, 60, 67, 106, 125, 132, 134, 152, 212, 283; universal, 115
author, 64, 74, 89, 158, 279
auto-ethnography, 112
automata or automaton, 3, 11–12, 23–26, 36–37, 51; chess-playing, 25, 278
Avignon, 29, 44, 160

balloon, xviii, 2, 13–14, 23, 35, 44, 48, 55; Annonay, 44–47; Calais, 253–54, 271, 280–81; Charles's, 65, 285; commission, 49, 70, 99, 134–35, 184; Dublin, 245; Enlightenment, 10, 14; enthusiasm, xviii; fever, 8, 73, 117, 131, 134, 136, 146, 176, 223, 228, 237, 267, 272, 278; firework, 275; historiography, 5; hot-air, 35; hydrogen, 56–59, 66, 74, 80, 88, 166, 240, 249, 256–57, 263, 265, 281, 286, 292; intelligence, 212; invasion, 262; Jacobin, 287; Lille, 274; madness, 253; mania, 8, 49, 63, 75, 132, 139, 144, 146, 181, 215, 238; miniature, 68, 73, 134, 212–13; news, 44, 48, 55, 63, 239, 278, 281; *Passarola* by Gusmão's, xxii; promoter, 31; public, 4–7, 10, 255, 268, 284; reconnaissance, 286; republican, 286; Réveillon, 69–70; royal, 68–69, 80, 83, 93; Sadler's, 240, 244, 257; spectacle, 7, 13, 59, 262, 278, 280; spectator, 109–12, 121, 129, 268; theater, 13; transcript, 54; travel, 14, 261; tricolor, 286; Versailles, 70–73; visibility, 10; as weapon, 30
balloon ascent, xxv, 2, 4–8, 26, 45, 101, 104, 109, 116, 126, 131–33, 140, 149, 171, 174, 198, 239, 254; in Birmingham, 221; Blanchard's, 197 (*see also* Blanchard); in Bordeaux, 137, 171, 174–81; at Champ de Mars, 59–61; at Champ Elysées, 286; Crillon's, 74; in Dijon, 13, 166–70; in Edinburgh, 223–24, 251; failed, 220; at Festival of Federation, 286; Harper's, 240; heroic, 99; human, 86, 134, 261–62, 270; Italian, 196; at Jardin du Luxembourg, 185; Jeffries,' 241; in Lille, 136; London, 196, 214; Lunardi's (*see* Lunardi); in Lyon, 136–37, 152–64; Milan, 140, 217, 272; Muette, 85, 165, 218, 221, 225, 235; in Nantes, 137; Parisian, 12, 38, 55, 74–75, 154, 188, 192, 267; provincial, 13, 135–37, 151–52, 171; at Tuileries, 91, 99, 110, 221; Tytler's, 223–24
balloon crowd, 6–7, 10, 54–55, 59, 72, 93–97, 149, 155–58, 162, 171, 180, 192, 197, 288; at Aeronauic Academy, 253; at Artillery Ground, London, 226, 228, 251; in Birmingham, 240; at Bordeaux, 176, 178, 180; at Champ de Mars, 1, 59; at Chateau de la Muette, 84; at Chelsea, 238; in Dijon, 169; in Dublin, 245; in Hamburg, 274; at *Les Brotteaux*, 155, 157–58, 162; at Lille, 269; as liminal zone, 198; London, 236; at Palais Royal, 97; at Place des Victoires, 96; at Rotterdam, 266; at Redoute Chinois, 74; at Reveillon, 71, 82; Rouen, 236; at St. Cloud, 187; at Tuileries, 12, 93–94, 218; at Versailles, 185
Banks, Joseph, 59, 74, 210–17, 221–26, 243, 263
baron: de Beaumanoir, 68; de Breteuil, 91–93, 98–99, 126, 132, 146, 244, 267; d'Holbach, 118; de Montesquieu, 143–44; d'Ogny, 63
baroque, xxiii, 2, 11
Bastille, 6, 69, 78–79, 83, 285
Beaumarchais, 33, 132–33, 147–48
Bélanger, François-Joseph, 58, 100
Benedictine, 41, 82, 139, 168, 180, 236
Bergasse, Nicolas, 142–43, 191
Blagden, Charles, 63, 89, 212–13, 231, 238–39, 241
Blanchard, Jean-Pierre, 41–45; Aeronautic Academy, 14, 251–54; ascents from Berlin, 278–79, Champ de Mars, 137–39, Dover, 242–43, Frankfurt, 271–73, Hamburg, 273–77, Holland, 261–66, Lille, 267–71, London, 237–39, Rouen, 235–36, Vienna, 277–78; Channel crossing, 14, 240–46; flying carriage, 12, 41, 55, 66, 113–15
body politic, 7, 10, 24, 48–49
Bollioud de Brogieux, 37, 45, 73
Bordeaux, 13, 137, 171–81, 191, 268
Bougainville, 78, 99, 218
bourgeois, 3, 6, 24, 35, 158–59, 180, 243, 288, 298n43

Boyle, Robert, 75, 137, 214
broker, 35, 46, 53, 56, 190, 212
Broussonet, Pierre Marie Auguste, 212–16
Buffon, 31, 35, 55, 80, 122, 141
Bureau de Commerce, 32, 45
Burke, Edmund, 210, 218, 223–29
Burton, Robert, xix–xx

Cadet: de Gassicourt, 49–50; de Vaux, 49, 54
café, 3, 33–34, 53–56, 67, 74, 131, 176–78, 185; politics, 5; sociability, 56
Café: de la Régence, 25; du Caveau, 55–56, 186, 243, 315n38
Caisse d'Escompte, 79, 145
calembour, 116, 124
Calonne, Charles Alexandre de, 80, 82, 182–83, 280
calumny, 88, 92, 95, 122; archeology of, 67; culture of, 54, 65, 67, 192, 291
Caoutchouc, 57, 167
caprice, 115–16, 128
carnival, 3, 6, 148, 286; science, 293
Carra, Jean-Louis, 142
Cassini, Dominique, 81, 185
Cavallo, Tiberius, 39, 65, 196
Cavendish, Henry, 44, 57, 65, 137, 167, 212–14, 218, 238, 241, 246
censorship, 4, 10, 44, 54, 56, 180, 231, 262, 278, 291
century of: balloon, 139; frivolity, 53; gold and pleasures, 118; light, 66; philosophy, 103
Chaillot steam engine, 39, 47, 56
chain of: gravity, 101, 124, 149, 192, 283; slavery, 283
Champ de Mars, 1, 5–6, 14, 55, 59–64, 88, 136–8, 213, 285–86, 293
characters, xxiv, 24, 77, 112; heroic, xxiii; literary, xxiv, 112, 115; mythological, xix, xxiv; pamphlet, 9, 112, 115, 119; theatrical, xxiv, 145
charlatan, 67, 229, 242, 257
Charles, Jacques-Alexandre-César, 74, 80, 190, 235–36; Academy and, 98–99; apotheosis of, 12, 96–101; ascents from Champ de Mars, 56–61, 77, 213, from Tuileries, 11, 84, 88–96, 110, 118, 124, 139, 218, 222, 244; disputes with Faujas, 63–69; at Louvre, 285; narrative of, 218
Charlière, 55, 59–62, 66, 89, 126, 140, 263–65
Charlists, 63–67, 93, 110, 117, 186, 214
chemistry, 33–38, 54–57, 63–66, 82, 118, 134, 239–40; courses in, 38, 82, 177; pneumatic, 38, 65, 134, 137, 167, 237, 240; theoretical, 49
Chénier, André Marie, 30
Cherbourg, 189, 235
chimera, 50, 128; of equality, 6; of flying (machine), xix, xxii, xxv, 41–44, 113, 190; of lunar voyage, 118
chimerical, 25, 63, 66, 127–28, 177
citizen, 7, 11, 24, 30, 47, 49, 105; of Calais, 243, 261; enlightened, 24, 124; happy, 13; mixed, 1, 5; provincial, 13; rational, xxiii; respectable, 276; royalist, 55; simple, 284; undesirable,

119; of the universe, 109, 122–23; universal, xxiii, 195; virtuous, 78, 124, 173, 193
Clovis au Champ de Mars, 285
Colisée, 41, 175–76
collective, xxv; alternative, 11; historical, xxiv; mass, 6, 101; transitional, 4
Collège: d'Annnonay, 33; d'Autun, 33–34; de Louis-le-Grand, 34; de Mazarin, 35; de Navarre, 35; de Sainte-Barbe, 34; de St. Nicaise, 41; royal de St. Louis, 82; de Tournon, 33
colonization, xxiv, 112, 123
Columbus, 13, 77–78, 101, 183; Christopher, xix, 66, 98, 220
Comédie: Française, 2–3, 132, 147; Italienne, 3, 191
commerce, 34, 41, 73, 79, 88, 120, 123, 170, 175, 196; of botanics, 57; national, 40; revolution in, 77, 118; slave, 78
Compagnie: des Eaux, 39; des Indes, 36, 68
compatriot, 69, 122, 142, 158, 236; earthly, 101, 121–22
comte d'Artois, 1, 58, 114, 142; patronage of balloon, 41, 72, 100, 117, 132, 153, 160, 243–44, of Mariage de Figaro, 147
comte: d'Adhémar, 241; d'Angiviller, 41, 44, 99; d'Antraigues, 36, 69; de Boissy d'Anglas, 69; de Dillon, 82; d'Eschermy, 277; de Laurencin, 161–64, 186; de Maillebois, 263–66; de Mirabeau, 279; de Paar, 277; de Périgord, 159; de Potocki, 158; de Provence, 1, 46, 243; de Romanzow, 271; de Vaudreuil, 58, 147
Condorcet, marquis de, 48–49, 53, 284
consensus, 9, 54, 106, 123, 173, 188, 192, 288
consumption, 32, 133; balloon, 136; emulative, 25, 133–34, 209; frivolous, 182; popular, 4; public, 142
contrarian, 113–15
Cordon: noir, 73; de Saint-Michel, 99, 141
cosmic: empire, 123; imagination, xxi; giant, 85; nations, 122; spectator, xviii, xxi–xxiii, 95, 111, 121; tales, xxi–xxii; view, xxii; visitor, xxiii; voyager, xxi, xxiii, 118, 121–23
Coulomb, Charles-Augustin de, 43, 47
court: displaced, 284; hierarchy, 2; masque, 11; rituals, 24; society, 2
Court de Gébelin, Antoine, 54, 143–44
Cradock, Ann Francesca, 93, 185
credit, 58, 63, 182, 213
Crosbie, Richard, 222, 244–46
crowd, 89, 186; for Blanchard's flying carriage, 41; composite, 1, 6, 10, 121, 173, 217; control, 265; curious, 111; disciplined as mass public, 4, 11, 109; of diverse composition, 6; of the French Revolution, 7, 284, 290; for guillotine, 288; mobilization of, 54–55, 59, 173, 198, 246, 290; as political actor, 4; at Pont Royal, 217; prerevolutionary, 4; race, 223; revolutionary, 3, 8, 11, 290; theater, 133, 147
culture, 164, 174, 221, 290; of calumny; 54, 65, 67, 192, 291; of consumption, 25, 215; of invention, 23, 31; material, 38; political, 111; public, 13, 23, 35, 54; of retribution, 149;

INDEX | 417

culture (*cont.*): of science, 35, 53–54; of spectacle, 23
curiosity, xix, 2, 25, 38, 49, 97, 111, 165, 174, 190, 211, 241, 270, 284; philosophical, 127; public, 213, 217–18; royal, 216; universal, 98, 152
Cyrano de Bergerac, xxi, 95, 122

Daedalus, xvii–xix, 43, 85, 245; French, 100; inheritors of, xxiv; provincial, 43
d'Alembert, Jean Baptiste le Rond, 23, 29, 144
Dauphiné, 37, 285
Deeker, James, 242–43, 257
De Luc, Jean André, 216, 221
Deparcieux, Antoine, 39, 213
Desmarest, Nicolas, 32, 40, 45, 49
despotism, 6, 126, 144, 191; academic, 142; aristocratic, 164; ministerial, 24, 114;
Diderot, Denis, 6, 24, 30–31, 34, 37, 78, 99, 117–18; *Encyclopédie*, 55, 104
Dijon, 13, 65, 137, 140, 152, 160, 164–71, 177, 181, 191, 220
Dinwiddie, James, 220–22
distinction, 2, 24, 73, 86, 93, 149, 154, 175, 284–85; balloon, 220; symbolic, 51, 196
domination, xxiv, 5, 109, 293
duc: de Castries, 183, 325n51; de Chaulnes, 47, 63, 89, 185, 214, 217; de Choiseul, 78; de Crillon, 74, 214; de Duras, 71, 86; de Fitz-James, 96; de La Rochefoucauld-Liancourt, 82; de Mouchy, 175–76; de Penthièvre, 41; de Richelieu, 1, 176
duc de Chartres, 1; Blanchard, 41, 243–44; Charlists, 117; competition with the queen's faction, 64, 66, 114, 126, 188; as duc d'Orléans, 284–86, 289; as Freemason, 153; as philosophical majesty, 106, 126; patronage of balloon, 82, 89, 93, 96–97; St. Cloud ascent, 174, 186–89; steam engine, 47
duc d'Orléans: Louis Philippe I (1752–1785), 39, 47, 56, 89, 126, 185, 244; Louis Philippe II (1785–1793, formerly Chartres), 114, 284–89; Philippe II (1701–1723), 53
duchesse de Polignac, 58, 84, 326n64; de Villeroy, 191
Duchess of: Buccleugh, 335n91; Devonshire, 201; Mecklenburg, 275; Richmond, 226
Duke of Cumberland, 93, 96, 217; Dorset, 244; Leinster, 222, 246, 334n62; Manchester, 217; Portland, 210, 223; Richmond, 226
Duret: Anne Catherine Manon, 34; Jean Jacques, 33; Mathieu-Louis-Pierre, 38
dystopia(n), xxi, 113–15, 283

Écrevisse, Jean Guillaume, 40, 68
egalitarian: drive, 181; fraternity, 6; justice, 113; monarchy, 113, 121; polity, 7, 43, 103, 106; power, 149; principle, 124; public space, 56; sociability, 67; society, 89, 105; vision, 6
emancipation, 6, 51, 283
emblem: constitutional, 286; malleable, 197; national, 109; royal, 2, 14, 81, 106, 149, 286
empire, xxv, 119, 122, 145, 163, 198; of Aeolus, 256; aerial, xviii, xxiv, 29–30, 77–78, 100, 112, 115–17, 132, 165; American, 103; benevolent, 113; boundless, 131; British, 8, 209, 223–24, 228–29, 236, 244, 249; Catholic, xviii, 8, 121; Christian, xxi; Clovis's, 285; cosmic, 123; cultural, 9, 14, 112, 152, 183, 195–98, 258, 261, 280; Eastern, xxi; European, xxii; of fashion, 143; French, 36, 183, 244, 280; global, 33, 78; glorious, 54, 112; Holy Roman, 195; human, xx, xxiii, 268; of human knowledge, 30; imagined, xix, 8; maritime, 29, 165, 257; material, 13, 198, 237, 293; moral, 8, 106, 293; Napoleonic, 11, 198, 262, 291; nationalistic, 14; peaceful, 7, 101; philosophical, 106, 110; republican, xxiv, 13, 26, 42–43, 104, 293; Roman, 104, 146; Russian, 103; scientific, xxiii, 8, 103, 283, 293; spiritual, xviii; of taste, 25; technological, 9, 112; world, 26, 237
emulation, 2, 151, 211, 249
emulator, xvii, 30, 99
encounter, 5, 25, 61, 111
Enlightenment, xviii, xxiii, xxv, 31, 67, 104–5, 128; balloon, 14; commerce, 57; contested, 280; French Revolution and, 3–4, 8, 198, 291; geography, 8, 14, 198; ideals, xxv, 8, 112, 291; industrial, 26; liberal, 4, 7; mass, 10, 106, 198, 281; military, xxii; Morellet's, 114; Neapolitan, 224; patriotic, 262, 278; patrons, 184; Parisian, 33, 110; popular, 197, 262, 278, 290; Prussian, 262, 278; public, xviii, 3–5, 8, 110; sciences, 280, 283; Scottish, 223; sensibility, 95; sites of, 53; utilitarian, xxii
enthusiasm: aristocratic, 239, 281; balloon, xviii, 89, 94–95, 118, 125, 136, 156, 170, 196; citizen's, 6; blind, 229; Linguet's, 88; lack of, 189; Lyonnais, 160; media, 215; multitude's, 227; for novelty, 119; Parisian, 214; patriotic, 285; persuasive, 154; of philosophical majesty, 124; popular, 13, 73, 175, 195, 217, 239, 257, 280; provincial, 220; public, 68, 98, 165, 168, 240; religious, 75; scientific, 35; vulgar, 49, 165, 211
entrée, 96, 111, 136, 169–70, 269
envelope, 31, 44, 70, 72, 84–86, 140, 153–54, 157, 166, 188, 249; impermeable, 39, 57–58, 161, 167, 307n29, 308n78; permeable, 186; silk, 221
equality, xxiv, 6–7, 78, 105–6, 113, 122–23, 177, 291
exclusion, 135, 284
experiment, 124; aerostatic, 237; balloon, 68, 125, 131, 137, 164, 167, 214–20, 227, 240, 267, 278–79; dangerous, 270; electrical, 45, 82, 117; epoch-making, 85; German, 262; haphazard, 38; Lavoisier's, 140; Lyon, 152, 154, 158, 161; magnetic, 190; meteorological, 219; papermaking, 45; parachute, 253; Périers, 39; Pilâtre's, 46, 183; pneumatic, 5, 38, 65, 137; political, 54, 174, 192; public, 47, 50, 168; scientific, 55, 166, 171, 184, 238, 241; sublime, 63; tethered, 213; useful, 60, 225; Versailles, 188; Windsor, 216

expert, 46, 55–56, 65, 95, 134–35, 181, 237
expertise, 4, 44, 48, 63–64, 66, 137, 151, 166, 261, 265

Fair Saint-Laurent, 73, 174
fashion, 115, 133, 143, 214, 220, 231, 237, 239, 249; balloon, 115, 134, 186, 195, 217, 224; of curiosity, 174; elite, 142; empire of, 143; frivolous, 114; geography of, 133; goddess and temple of, 115–16; intellectual, 114; madness, 220; material, 2, 134; philosophical, 214; queen, 115–18; Rousseauean, 122; scientific, 38–39, 51, 261, 271
fashionable: assembly, 217; attire, 1; botanist, 212; commodity, 53, 68, 135; displays, 2, 116; elite, 84, 209, 223, 241, 257; event, 74; inventor, 42; machine, 69; pastime, 247; pleasures, 195; public, 3, 54; socialite, 7; science, 4, 38; topic, 69; women, 187
fashionable society, 4, 41, 243, 288, 343; at contact zone, 149; criticism of, 112; Gustav III and, 181; London's, 219, 229, 251; *Mariage* and, 133; members of, 59, 82, 141; mesmerism and, 141, 146; Réveillon trials and, 69, 78, 80, 82, 284
faubourg St. Antoine, 45, 69, 79, 180–82
Faujas de Saint-Fond, Barthélemy, 31, 47, 73–74, 85–86, 160, 167, 212, 216; Champ de Mars ascent and, 55–59; *Description*, 60, 196, 215, 221, 265; dispute with Charlists, 63–67; nation and, 56, 72, 82–84, 89
fermentation, 13, 32, 64, 73, 133, 147–49, 178
festival, xix, 2–3, 7, 35, 94, 105; balloon, 5, 10, 88, 96, 151, 158, 169, 176, 245, 256, 268, 270–71, 286; of the nation, 286; revolutionary, 10, 290
Festival: of Constitution, 286; of Federation (Fête de la Fédération), 14, 285
fiction: dystopian, 114–15; material, 129; utopian, xxi, 104, 113
firework, 73–74, 99, 137, 167, 236, 254, 269, 274–75
fishwives, 6, 96
Flesselles, Jacques de, 83, 136, 152–58, 161, 285; *Le Flesselles*, 154–61
flying: carriage, 12, 41–42, 55, 66, 68, 190; as a chimera, xxv; dream, xix; boat (Lana's), xxii; ferry, 123; gladiators, xxiii; island, xxii; Medusa, 214; men, 115; nation, 165; observatory, 103; practical vs. theoretical, 218; thieves, 125; vessels, 118, 138
flying machine, 8; as a chimera, xxii, 45, 113–16; balloon as, 7; imaginary, xxi, 14; impractical, 40; metamorphosis of, xviii; philosophical, xviii; spectacular, 51; as weapon, 125
folly, 89, 136, 213, 229, 247
Fontana, Felice, 46, 65, 167
Fontenelle, Bernard Le Bovier de, 104
Fordyce, George, 224–26
Fox, Charles James, 209–10, 223–29, 246
Francophile, 182, 211, 262, 275
Franklin, Benjamin: American War, 78; balloons, 1, 59, 68, 70, 74, 80, 84, 94, 131; Blanchard, 244; Irish balloon, 212; letters to Banks, 212–13; mesmerism, 146; mini-balloons, 68, 213; Montgolfiers, 68; satire, 212
Frederick the Great, xxii, 29, 195, 263, 277–79
Freemason, 14, 54, 89, 136, 152–53, 217, 278; Gustav III, 162; Lyon, 153; Prince of Caramanico, 214, 224; Prince Gagarin, 277; prince de Ligne, 153; Potocki, 158; Watkin Lewes, 226; Willermoz, 143
friendship, 36–37, 125, 184
French Revolution, 3–14, 280–91; balloon and, xviii, xxv, 7; birthplace of, 56; causes of, 3, 7–8, 107, 287–88; Chartres and, 126; crowd of, 7; genealogy of, 10, 290–93; libels and, 116; Marat and, 257; publicity and, 192; violence and, 186, 192, 284
frivolity, 53, 68, 74, 115–16, 197, 214, 270, 275
fuseaux, 45, 167

Galien, Pierre-Joseph, 44
Galileo, xix, xxi, 220
gallantry, 78, 94, 115–16, 119
gas: deadly, 46; hydrogen, 58, 88, 91, 255–56, 280; inflammable, 46, 50, 65, 68, 166–68, 188, 225, 242, 246; lightest, 57; of Montgolfier, 48, 50, 72; patriotic, 223
Gazette littéraire de Berlin, 278
Gébelin, Antoine Court de, 54, 143–44
genealogy: of balloon spectator, 110–12; Foucault's, 10; of French Revolution, 10, 290–93; of mass public, 8; of material public sphere, 12; of Montgolfière, 75; scientific machines, 14; scientific-national artifacts, 291; scientific spectacle, 10; spectator, 110
geography: balloon, 13–14, 195–96; of cultural empire, 261, 280; of Enlightenment, 198, 262; family, 33; of fashion, 133; liminal, 14, 198; material, 2, 8, 135–36, 152; of pleasure, 25, 211; of science and spectacle, 13; universal, xxv
George III, 93, 187, 210–16, 223
genius, 100, 105, 112–13, 120, 137, 255–56; Beaumarchais's, 148; café, 8; Charles's, 98; Crosbie's, 245–46; Da Vinci's, xix; Diderot on, 30–31; mechanical, 245; Mesmer's, 191; Montgolfiers,' 12, 30–31, 158; Moret's, 220; Nature's, 77; Ovid's, xvii; philosophical, 144; Sadler's, 240
globe, xx, 78, 118, 120, 134, 178, 183, 212, 284; civilized, 237; copper, 50; glass, 102; taffeta, 59
glory, xxv, 44–49, 63–65, 92, 100, 122–28, 143–44, 154–58, 181, 243, 275; aeronaut's, 266; Blanchard's, 235–36, 275; Charles's, 92, 100; Flesselles's, 158; of France, 239; immortal, 127, 270; inventor's, xix, 30–31, 44–45, 47–48, 63–64; Jeffries's, 243; for men of science, 170; Mesmer's, 143; Mongtlfiers,' 63–65, 100; patriotic, 229, 245; royal, 49; sovereign's, xxv; as true Columbus, 183
goldbeater's skin, 68, 74, 136, 213, 216, 221, 308n78
golden age, xvii, xix, 7; visual satire of, 90
gomme, 46, 58, 307n36
Gordon, George 209, 229

Grand Tour, 14
Grenoble, 41, 55, 143, 153, 191
Grimm's Correspondance, 68, 84, 110, 139, 144–47, 161, 180–83, 189, 191
Gudin de la Brénellerie, 77
guild, 69, 267, 269
Guillotin, Joseph-Ignace, 146, 288
guillotine, 11, 14, 107, 158, 174, 193, 289–93
Guyton de Morveau, 140, 160, 164–70, 220; military ballooning and, 286

Hales, Stephen, 65, 137
Handel, George Frederick, 210, 232, 238
happiness, 34, 74, 78, 95, 101–5, 123, 151, 164, 244; chimerical, 128; citizens,' 11, 88, 104–7, 111, 151, 173; national, 113; people's, 105, 123, 142; subjects,' 114, 144, 163
Hardy, Siméon-Prosper, 68, 146
Harris, James, 261, 263
hegemonic: agency, 54; apparatus, 12; conduct, 75; consensus, 192; discourse, 113; façade, 149; function, 109, 133; intention, 25; operation, 10; process, 106; representation, 288; spectacle, 88; status, 11; system, 7, 10; theatricality, 23; transcript, 55
hegemony, xxiv–xxv; ancient regime, 35; cultural, 1, 25; scientific, 8, 63, 106, 109–10, 124–25, 132–33, 171, 173, 196, 231, 258, 283, 289–91
Helman, Isidore-Stanislas, 269, 282
hero, xviii, 2, 86, 100, 169, 196, 240, 243–44, 252; British, 261; Christian, 270; classical, 168; folk, 12, 97–99; Greek, 85; mythic, 290; national, 249, 283; subhuman, 126;
heroic: animal voyager, 126; ascent, 99; aeronaut, 86, 127; characters, xxiii; effort, 58; king, 146; narrative, 126; navigator, 85; poem, xxiii; prince, 182; tale, 85; voyage, 100
Hertzberg, Ewald Friedrich von, 278
Histoire comique, xxi, 122
Histoire de la dernière révolution de Suède, 146, 162
Histoire intéressante, 122
history: from below, 9; of the present, 8
honnête homme, honnêteté, 64, 120, 276
hope, 51, 137, 217, 226–27, 229, 247, 285; chimerical, 177; elite, 161; fading, 286; fleeting, 100; fragile, 37, 65; mass, 101, 174, 193; people's, 5, 99, 255, 290; public, 290; republican, 13; shared, 156; silent, 149; spectacle of, 8; transgressive, 78; utopian, 286
Houdon, Jean-Antoine, 81, 181
humanity, xviii, xx, xxiii, 36, 78, 110, 120, 131, 142–43, 145–46, 151, 192, 220, 289; condition of, xviii, xix; the good of, xxiii, 12, 39, 55, 78, 103, 112, 124, 142–43, 171, 257; sacred, 106, 128–29;
hybrid: identity, 175; perspective, 112; republicanism, 104, 123; zone, 196–97
hydraulic machine or pump, 39, 41–44, 51

Icaromenippus, Menippus, xvii, xx, 17
Icarus, xvii–xix, xxii, 41, 85, 254–55, 296n26; Parisian, 43

idleness, xviii, 53, 56, 67, 74, 100, 111, 155–56, 218, 247; curiosity, 165; spectator, 154, 161
ignorance, 84, 89, 163, 190, 266, 275; of people, 24, 61, 112, 123–24, 222; of science, 6, 36, 48, 60, 75, 154, 239
illusion of, xxv, 1, 101, 257; harmonious patrie, 170–71; material empire, 198; power, 132, 236; riches, 123; technoscientific utopia, 101; unity, 197
imaginary: historical, 5; political, 10, 88, 149; republican, 105; utopian, 5
imagination, xix, 31, 44, 64, 115, 145, 153, 190; audience's, 148; collective, 132; cosmic, xxi; cultural, 78; elite, 83, 283; European, xxiii, 78; Mercier's, 231
imperial: adventure, 36; aristocracy, 223, 271, 278; aspirations, 7; capital, 13, 35, 53, 133, 197, 211, 231, 257; center, 134; city, 236, 271; competition, 221, 235; cultures, 8; decline, 210; dreams, xxiii; economy, 33; elite, 224; excess, 78; garment, 285; government, 231; humiliation, 221; intentions, xxv; interest, 211; legacy, xxii, 293; look of power, 78; metropolis, 196, 215; nation, 293; order, 272; polities, 8; power, 229; province, xvii; rivalry, 218, 239; state, 221; use, 152, 198; voyage, 118
Inchbald, Elizabeth, 211, 224
industry, 63, 118, 210; paper, 31–32, 40, 45; silk, 136; textile, 267
inflammable air, 57, 65, 68, 80, 167, 239; in envelope, 57, 66, 96, 221, 225; levity, 39, 44, 57, 60, 65; research on, 134–35, 166–67; used in miniature balloon, 44, 50, 63, 68, 73, 216, 235, 256
Ingenhousz, Jan, 65, 316n3
integration, 2, 112–13, 152, 160, 173, 198, 210, 221
intrepid, 12, 83, 95, 118, 126, 183, 256; aeronaut, 12, 126, 266; philosopher, 255; physicist, 95; voyager, xxiii, 118
intrepidity, 85, 151, 188, 240
intrigue, 64, 67, 88, 92, 114, 119, 147, 174, 221
invention, xviii, 31, 65, 98–99, 166, 219, 257, 274, 283; French, 49, 75, 209–10, 216, 224, 231; patriotic, 30, 68, 135, 211; popular, 134; priority of, 45, 66; scientific, 103, 213; theoretical, 196; useful, 31, 241
inventor, 48, 64, 80, 98, 124, 126, 134, 156; ancient regime, 12, 25, 31, 35, 43–45, 47, 49, 67; Argand as, 69, 217; Bauer, xxiii; Blanchard as, 43, 137; Daedalus, xvii–xix; Da Vinci, xix, 19; fashionable, 42; as genius, 112–13; modern, 30; modest, 86, 158, 177; Montgolfiers as, 69, 100, 215; original, 48, 65–66, 68, 80, 91, 99, 140, 152, 235; patriotic, 51; Periers, 50; provincial, 42–43, 65; scientific, 31, 38, 63, 75, 80, 118, 141; royal, 66, 68, 83, 99, 171, 183–84; theoretical, 65, 97; virtuous, 65, 68, 75

Janinet, Jean-François, 140, 184–86
Jardin: des Tuileries, 99; du Luxembourg, 59, 184–85; du roi, 35, 55, 59; public (Bordeaux), 175, 178–80

Javel, 57, 91, 100, 131
Jefferson, Thomas, 131
Jeffries, John, 197, 240–44, 247, 281
Johannot, 32, 40, 49
Johnson, Samuel, xxiii, 210, 217
Jouffroy d'Abbans, steamboat, 47, 49, 56, 66
Journeyman, 33–34, 40, 133
justice, xxiv, 104–5, 121–22, 170, 179, 191, 284, 290–91; administration of, 114; democratic, 288, 290; egalitarian, 113; God's, 128; impartial, 164; mechanical, 290; scientific, 289

Keegan, Allen, 220, 228
Kempelen, Wolfgang von, 25
king-machine, 2, 6–7, 11, 23, 25, 86, 104, 173
Kirwan, Richard, 167, 213–14, 245

La Bruyère, Jean de, 3, 111
Laclos, Pierre Choderlos de, 114–15, 119
La Condamine, Charles Marie de, 57, 113
Lafayette, marquis de, 78, 82, 97, 143–44, 246, 286
La Folie, Louis-Guillaume de, xxiii, 102
La Harpe, Jean François, 105, 146–47
Lana de Terzi, Francesco, xxii–xxiii, 44, 50
Languedoc, 40, 68, 158
Lavoisier, Antoine-Laurent, 141; balloon commission and, 49, 66, 134–35, 140; education of, 35–36; Gunpowder and, 53–54; Mesmer commission and, 146; Meusnier and, 80; pneumatic research and, 38, 57, 65, 135, 140; Ruggieri and, 74; Tax Farm and, 289
Le Char volant, 114–18,
Le Dru, Nicolas-Philippe, 143
Lenoir, Jean-Charles-Pierre, 59, 69, 81, 88, 91–94, 132, 140–42, 267
Le philosophe sans prétention, xxiii, 102
Le Roy, Jean-Baptiste, 49, 81, 86, 91, 166, 184
lèse-majesté, 67
Le Triomphe, 124–26
Lettre à Mr. M. de Saint-Just, 117–18, 122–23
libel, 4, 63, 67, 88, 114, 116–17, 126, 133, 188, 262, 267; Hamburg, 273–77
liminal: court, 68–69; figure, 12; geography, 14, 198; moment, 14, 103, 173, 193; period, 283, 290; polity, 113; public sphere, 25; realm, 8, 101, 128; region, 40; space, 23; zone, 11, 198
liminality of, 23; inventor, 31; scientific spectacle, xxv
Linguet, Simon-Nicolas-Henri, 36, 135; balloon degeneration and, 174, 189–90, balloon's utility, 86–88, Bastille and, 78–79, Charles and, 97–98; Chartres and, 188
Lodge (Masonic), 53; Choix des Hommes Libres, 38; Grand Orient, 89, 187; La Parfaite Réunion, 161; La Vraie Vertu, 36, 38, 45, 303n71; Loge des Neuf Soeurs, 54, 81, 144, 177, 181, 288; Premier Grand Lodge of England, 94
London Hotel, 14, 243, 273, 277, 299n87
Lord: Bacon, 237; Bute, 213, 216, 223; Charlemont, 222, 245–46; Gordon, 209, 229; North, 225–26; Orford, 257

Louis XIV, 2, 126, 267,
Louis XVI, 54, 72, 114, 126, 132; balloonists and, 72, 84–85, 99; balloon ordinance and, 141; Bastille and, 79; Blanchard and, 243, 262; *Caisse*, 80; Cherbourg and, 235; entrée to Paris and, 169; French Revolution and, 286, 289; *Mariage de Figaro* and, 147; peace celebration and, 176; Navy and, 181
Louvre, 53, 285
Lucian, xvii–xxii, 95, 111, 121, 228
lunar voyage, xxi, 5–6, 117–18, 122, 125, 239
Lunardi, Vincenzo: criticism of, 247, 250; London ascent, 224–28, 237–40; other ascents, 224, 249–51
Lyon, 136; Academy, 165, 177; architect, 44; balloon, 13, 84, 100, 131–40, 140, 151–64, 170–71; elite, 161, 164; Federation, 285; intendant, 157; lawyer, 142; Masonic lodge, 38, 143, 153; mesmerism, 191; militia, 156; Montgolfiers and, 29; native, 163; paper and, 32; press, 163; prize, 140; silk merchant, 34; steam boat, 47, 49
Lyonnais, 155, 157, 160, 164

machine: aerostatic (*see* aerostatic globe); ascending, 39, 45, 48, 50, 56; astonishing, 183; diostatic, 48; chimerical, 33, 115; fashionable, 69; flying (*see* flying machine); frail, 256; humane, 290; hydraulic, 44; infernal, 125; majestic, 67, 163; monstrous, 186; patchy, 156; scientific, 11–12, 14, 48, 107, 276, 288, 290; sensational, 66
machine polity, 10, 23, 75, 112, 151, 173, 196
majestic: artifact, 55, 288; balloon, 5, 271; column, 267; machine, 67, 165; sight, 94; spectacle, 2, 26, 95, 97, 125, 173; thought, 24
malcontent, 43, 54, 75, 79, 109–10, 116, 122, 133, 147, 192, 288
Marat, Jean-Paul, 35, 135, 142, 255–57, 283
marquis: d'Argenson, xxiii; d'Arlandes, 83–86, 98–100, 165, 181, 185–86, 218; de Bacqueville, xxiii, 41; de Brantes, 160; de Breuilpont, 262, 265; de Castries, 68, 183, 325n51; de Champcenetz, 89, 93; de Ducrest, 39; de Lafayette, 78, 82, 97, 143–44, 246, 286; de La Maisonfort, 255–56; de Launay, 285; de Laurencin, 161–64, 186; de Marcien, 14; de Montesquiou-Fezensac, 182; d'Ormesson, 48, 80; de Vérac, 262–65; de Wignacourt, 269
mass, 24; absorption, 7, 157; action, 6, 8; adoration, 156; affect, 54; agency, 188; audience, 8; audience, 5, 8, 170; collective, 6, 101; control, 5, 129, 290; Enlightenment, 10, 106, 198, 281; gathering, 170, 175; hope, 101, 174, 193; mobilization, 291, 293; politics, 4, 9, 148, 173, 191; public, xxv, 2, 4, 8, 11–12, 14, 106, 109, 148, 192; theater, 12; silence, 8–9, 54, 129; spectacle, 6, 8, 11, 14, 117, 278, 289–90; subject, 3–4, 268; veneration, xxv, 2, 10, 14; violence, 209
material: agency, 8; appropriation, 109; constraint, 57, 153; culture, 7, 38, 237; domain, 12; empire, 13, 195, 198, 237, 293; fashion, 134;

INDEX | 421

material (*cont.*): fiction, 129; geography, 2, 8, 135–36, 151–52; knowledge, 161, 195–96, 261; nation, 133; performance, 9, 25; politics, 13, 110, 144, 149; power, 3–4, 192; public sphere, 12, 132; resources, 5, 66, 153, 196, 231, 262, 270; spectacle, 25, 262; theater, 148; translation, 261; utility, 106, 174
Maupeou, 24, 41, 89, 114
Mauritius, 34, 36
mécanicien, 38, 185, 235, 280
memory, 100, 107, 146, 159; collective, 5, 284, 290; cultural, xviii–xix, xxv, 121; European, xvii; of exclusion, 284; glorious, 9; historical, 5, 14; place of, 5
Menus-Plaisirs du roi, 71–72.
Mercier, Louis Sébastien, 7, 24, 31, 35, 39, 56, 78, 94, 103–6, 110, 115, 128–29, 145–47, 174, 231, 255,
merit, 7, 44, 48, 50, 64, 67, 114, 116, 170, 184, 188, 224, 239, 276
Mesmer, Franz Anton, 55, 132, 141–44, 149, 191, 288
mesmerism, 4, 13, 117, 132, 141–42, 144, 146–49, 190–92, 288
metamorphosis, xxv, 14, 25, 36, 78; animal, 126; in *Divine Comedy*, xix; Montgolfiers', 31; in Ovid, xvii–xviii, xxiii; revolutionary, 14, 283; of spectator, 109; temporal, 77, 129
Meusnier, Jean-Baptiste, 80, 97–98
Modern Atlantis, 211, 228–31
modernity, xviii, 8, 11–13, 195, 198
modesty, 73, 83, 216, 254
Mogul Tale, 211, 224
monarchy: absolutist, 56, 283; constitutional, 210, 285; egalitarian, 113; republican, xxv, 113, 123, 128, 144, 284–86, 289, 293; universal, xxii, 121
Montesquieu, 7, 100, 104, 111, 122, 126, 144–45, 231, 285
Montgolfier, Adélaïde, 37–38, 153; Antoine, 31; Antoine-François, 34, 40; Augustin, 301n35; Étienne, 301n35; Jacques, 33, 50; Jean-Pierre, 37, 73, 154–55; Maurice-Augustin, 34, 37, 158; Pierre, 29, 31–35, 37, 73, 99; Raymond, 33–34, 37–38
Montgolfier, Alexandre-Charles (the abbé), 33–34, 37–38, 70, 73, 83, 154–55, 158
Montgolfier, Étienne-Jacques, 34–41, 44–45, 47–50, 53, 64–65, 98–100, 140–41, 177, 235; collaboration with Joseph, 38; correspondence, 263, 265; disputes with Charles, 65–69, 94; Muette balloon, 78–87, 213; Versailles balloon, 68–74, 182–85
Montgolfier, Joseph-Michel, 29, 33–34, 37–40, 43–45, 47, 50, 69, 73, 84, 131, 154–61, 185, 285
Montgolfière, 35, 44, 48, 69–70, 83–84, 113, 131, 174, 184, 186; Aero-, 252; ban on, 171, 272; Bordeaux, 176–78, 180; Brantes, 160; vs. hydrogen balloon, 80, 126, 134–37, 166, 215, 239, 254–57, 263–65; Keegan's, 228; Lille, 267; Lyon, 153; Marseilles, 159; Moret's, 225; paper, 242; royal, 55, 66, 69, 71, 89, 183, 189, 218; Sadler's, 240; Strasbourg, 280; Tytler's, 223
Montgolfists, 63–67, 93, 110, 117, 186, 214
Montpellier, 29, 40, 158, 191
Monsieur, 29, 46, 72, 183
Morand, Jean-Antoine, 44, 153, 162–63
Moreau-Desproux, Pierre-Louis, 34–37
Moret, chavelier de, 219–20, 225, 264
multitude, 53, 65, 85, 151, 211, 225, 227, 253
Musée, 6, 46, 54, 152; Bordeaux, 177, 180–81; de Monsieur, 46, 54, 82, 160, 182–83, 243; de Paris, 54, 143
myth, xix, xxi, 5, 9, 30–31, 77–78, 112, 290

Napoleon, 132, 138, 197–98, 262, 281, 285, 290–93, 296n2
nation, 6–8, 11–14, 24, 29–31, 54–55, 64, 67, 69, 78, 88–89, 95, 97, 99–100, 154, 165, 184, 210, 226, 232, 245–46, 286; above kings, 5; administrative, 62; asymmetric, 171; British, 225; of citizens, 8, 54, 283–84; composite, 74; cultural, 2, 63, 134, 170, 283; as cultural empire, 258; distinct from royal court, 49; Dutch, 265; English, 145, 244; Faujas's, 56, 72, 82–84; free, 285; French, 5, 124–28, 137, 152, 173, 189, 243, 267; geography of, 135–36; vs. government, 49; great, 285; happy, 113; harmonious, 55, 112; ideal, 182; imagined, 5, 13, 83, 101, 103, 144, 278, 284; imperial, 293; integrated, 132; just, 284; material, 133; modern, 293; philosophical, 13, 63, 103–6, 171–74, 193, 284; progressive, 55; prosperous, 115; reformed; 283; republican, xxv, 2, 5, 13, 146, 284, 290; triumphant, 77, 131; unified, 149, 293; Versailles-centered, 175
nation-empire, xx, xxv, 8, 104, 106, 110, 113, 133, 212, 291; -state, 8, 198, 324n1
national: artifact, 2, 8, 11–12, 75, 103, 106, 139, 152, 173–74, 193, 288–93; assembly, 192, 262, 284, 286, 289; attention, 224; ballooning, 135; celebration, 129, 136; character, 114, 213–14, 231, 258; commerce, 40; consciousness, 24; culture, 136, 197, 279; emblem, 109; enterprise, 254; festival, 286; flag, 228; happiness, 113; heroes, 249, 283; identity, 258, 280; independence, 263; integration, 210; invention, 26; mores, 31; pastime, 136; pride, 136, 222; regeneration, 174; reputation, 235; sentiment, 74; spectacle, 24; spirit, 115; subscription, 52, 56, 63; superiority, 266; utility, 39
nationalism, 152, 237; enlightened, 197, 262
nations, x, 7, 13, 100, 106, 116, 122–23, 128–29, 142, 151–52, 220, 273; European, xxv, 14, 197–98; slaving, 128
natural philosophy, 30, 37–38, 62, 66, 103, 117, 124, 164, 184; and discovery, xxiii; Aristotelian, xviii, 8; merchant of, 64; modern, xviii; progress of, 215, 276
Nature, xvii, xxv, 31, 86, 104, 109, 125, 149, 190, 197; authority of, 149; benevolent, 121; conquest of, 88; control of, 63, 77, 256, 265, 283;

laws of, 143; marvels of, 122; secrets of, 127; truth of, 30–31, 78, 143
nobility, xxv, 31, 99, 103, 253, 275, 285, 288; English, 219, 242–43, 246, 253; French, 132–33, 156, 242; idle, 155; radical, 126; scientific, 31
novelty, 42, 69, 74, 95, 119–20, 124, 212, 214, 241, 253, 257
novelties, 53, 80, 117, 121, 128, 134, 209; fanciful, 123; merchants of, 59, 68; scientific, 55; useless, 278

Opéra, 2, 3, 35, 119, 244
ordinance: against gambling, 120; balloon, 132, 140–41, 161, 178, 220, 267; for theater, 175
Ovid, xvii–xix, xxiii, 85

Palais Royal, 35, 53, 56, 58, 63, 97, 117, 182, 185
pamphlet(s), 9–10, 67, 110–12, 117, 124–25, 168, 183, 190–91, 196, 212, 238, 242–43, 270; balloon, 213, 215, 270; Bertholon's, 221; Blanchard's, 237; characters, 9; Gébelin's, 143; literature, 9, 12, 110, 112, 117, 129, 285; mesmeric, 191; production, 110; warfare, 111
pamphleteer, 75, 103, 137, 155, 270
papermaking, 34, 45
papier vélin, 45, 69
Parisian: aeronautics, 139; amusements, 37; approbation or endorsement, 160, 164, 236; assembly, 284; attire, 37; balloon ascent, 38, 55, 59, 75, 83, 132, 171, 188, 246; balloon enthusiasm, 137, 181, 214, 228; balloon news, 214; cafés, 33–34; celebration, 285; chess players, 25; correspondent, 221; culture, 4, 42; curricula in architecture, 35; chronicler, 68, 146; debauchery, xxiii; elite, 140; Enlightenment, 33; entertainment, 51, 55; entrée, 96; fashion, 217; fashionable society, 243; folly, 136; geography of pleasure, 25; gossip, 94, 161, 187, 214; guild, 69; heads, 73, 131, 136; high society, 119; human ascents, 12, 154, 174, 218, 222, 267; Icarus, 41; institutions, 3, 35, 62; lectures, 38; life, 36, 95, 113; Loge, 177; luminaries, 50; manners, 231; mesmerism, 142; minds, 35; modernity, 12; newspapers, 165–66, 168, 170, 189, 213; notables, 83, 235; population, 93; press, 160, 170, 236; Prometheus, 88; public, 67, 174, 243; pleasure, 53; resources, 47; royal theaters, 2, 24–25; salons, 78; science, 4, 38, 53, 65, 161, 170; sentiments, 36; socialites, 50, 100; soirées, 61; supply of balloons, 135; world of power, 119
Pâris de l'Épinard, Joseph, 267–70
parlement, 5, 89, 111, 152; Bordeaux, 165, 170, 175–79; Grenoble, 55; Normandy, 236; Paris, 41, 78, 140, 190–91, 267
patrie, 12, 98, 101, 103, 120, 123, 162, 170–71, 184, 257, 285–86, 289; Lyonnais, 155, 158
patriotic: art, 133; character, 222; courage, 113; criticism, 263; desire, 198; duty, 278; Enlightenment, 262; enthusiasm, 285; festival, 245, 285; gas, 223; glory, 229, 245; heart, 257;
invention, 29–30, 68, 135, 211; inventor, 51; king, 164; Lyonnais, 157; maneuver, 222; men, 100; opposition, 24, 114, 274; paper, 221; passion, 30; philosopher, 245; philosopher-king, 163; resistance, 14, 195, 262, 280; revolution, 13, 261; science, 164; senator, 229; sentiment, 8, 276; virtue, 31; writer, 24; zeal, 37
patriots, 267; Dutch, 262–66; Irish, 220–22, 244–46; German, 197, 278; Hamburg, 262, 273, 275
patriotism: British, 210; Burgundy, 166; false, 229, 231; German, 197, 275; Irish, 222, 246; regional, 13, 155, 170–71; Swedish, 163; veritable, 24
passion, 37, 65, 95, 107, 119–23, 270; enlightened, xviii; patriotic, 30
patron, 66, 82, 89, 186, 196–97, 236, 240, 245, 286
patronage, 4, 41, 86, 195, 211, 231, 266; aristocratic, 262; court, 235; d'Orléans's, 39; elite, 192, 236; institutional, 177; networks, 12, 53; official, 225, 268; princely, 43, 47; public, 55, 240; royal, 236, 240; Viennay's, 139; Viviers's, 273
people, 6, 14, 63, 96, 105, 109, 117, 143, 158, 187, 189, 229, 231, 241, 251, 290; as balloon audience, 1, 5, 60, 62, 85, 93–95, 217, 238, 240–42, 250, 253, 266, 279, 285; as citizens, 24, 62, 263; common, 4, 12, 25, 30, 124–25, 149; of fashion, 249; ignorant, 24, 112, 123–24; illiterate, 3, 13, 25, 110, 129; innocent, 127; lowly (*menu peuple*), 24, 112; as mass collective, 101; as a nation, 145–46; of Paris, 7; powerless, 127–28; public-spirited, 241; riotous, 176, 186; as subjects, 113–14, 126, 136, 142, 163–64, 170, 218, 293; silence of, 55, 110; uneducated, 24, 54, 148; uninformed, 266; unsightly, 9; useless, 154
people-machine, 6, 26, 104, 106, 151
peoples (as nations), 66, 112, 123, 198, 231, 273
performance, xvii, xxv, 41, 290; benefit, 148; boundary, 280; cultural, 6, 280; literary, 9; material, 9, 25; marvelous, 25; miraculous, 67, 251; mixed, 25; of power, 148; private, 147; public, 24, 147; scientific, 2; theatrical, 25, 74, 105, 158, 175, 182, 191, 224, 277, 283; Versailles, 147
Périer, 39, 47, 50, 56, 126, 137, 283
Philippe Égalité, 1, 126, 285, 289
philosophe, 118, 120, 145
philosopher(s), xvii–xviii, 6, 105, 117, 121–23, 218; aerial, 124–25, 221; arrogant, 120; cautious, 239; English, 215, 224, 241; gentleman, 214; impractical, 83; intrepid, 255; modern, 220; moral, 120; patriotic, 245; pneumatic, 215; prophetic, 142, 237; rational, xxiii; subaltern, 111; true, 120
philosopher-king, 7, 13, 104–5, 123–25, 146, 161–64, 181
philosopher-voyager, xvii, 12, 78, 96–99, 116, 119, 125
philosophical majesty, 13, 75, 101, 106, 124, 126, 173, 193

INDEX | 423

physicien, 6, 31, 38, 65–68, 72–73, 80, 139, 184, 235, 256
Pilâtre de Rozier, 46, 54, 58, 73, 98–100, 181, 190, 221; Channel crossing and, 235, 241–46, 253–57, 268, 271, 278; Musée de Monsieur and, 46, 54, 243; Lyon ascent and, 154–60; Muette ascent and, 84–86, 165; Réveillon trials and, 81–83; Versailles ascent and, 182–83
Pitt, William, 210, 226
Place: de Louis XV, 59; de la Révolution, 289; des Victoires, 58, 96
pleasure, xviii, 2, 25, 36, 53, 73, 95, 105, 107, 111, 129, 188, 231; of aerial journey, 223; aristocratic, 14; artificial, 53; century of, 118; commodification of, 53, 221; economy of, 53, 251; fashionable, 195; garden, 223, 251; gentle, 122–23; immoral, 229; pursuit of, 114, 119, 121, 127; royal, 182, 277; ruinous, 229; small, 148; sublime, 228; vulgar, 251
pneumatic: chemistry, 38, 65; literacy, 137, 180, 196, 220; network, 137, 171, 196, 220
police, 54, 62, 69, 74–75, 86, 117, 245; Bordeaux, 178; instruction, 269; lack of, 245; London, 226; Lyon, 153, 162; Paris, 1, 79, 82, 138, 267; royal, 26, 59, 106, 128, 192, 231, 237
polity: absolutist, 2–3, 6–7, 13, 31, 55, 67, 73, 75, 104, 133, 151, 174, 290; alternative, 25, 54; baroque, 11; egalitarian, 7; electoral, 284; otherworldly, xxi; parliamentary, 11; republican, 101, 104–5, 113, 123, 290; science and, 3–4; theatrical (*see* theatrical)
populace, 99, 129, 149, 231, 281, 290; absence of, 9–10; as balloon audience, 1, 4, 7, 54, 72, 170, 195, 197, 211, 220, 227, 251, 274, 276; as citizens, 8, 24, 104, 110, 119; curious, 78; discipline of, 25, 60; Enlightenment, 8, xxv, 197, 278; erasure of, 9–10, 55–56; exclusion of, 3, 23, 104, 284–86; ignorant, 124; illiterate, 3–4, 63; inclusion of, 132, 192; in modern politics, 105; mass politics and, 4; mobilization of, 13, 54, 128, 173; public and, 5, 12;
popular: affection, 174; artifact, 283; consumption, 4; culture, 6, 9, 25, 107, 290; despair, 255; Enlightenment, 262, 278; entertainment, 1, 174; enthusiasm, 13, 73, 175, 195, 217, 239, 257; fascination, 211; fermentation, 147; hope, 255; opinion, 6; philosopher-voyager, 99; practice, 283; resentment, 155; revolution, 285; science, 4, 8, 132, 148, 190, 192; sovereignty, 104; unrest, 132, 149; violence, 149, 284
power, xix, 2, 63, 69, 83, 88, 94, 113, 120, 147–48, 152, 176, 278–80, 289; absolute, 3; active, 4; Anglican, 221; arbitrary, 174, 283; aristocratic, 5, 132, 262; British, 210; broker, 212; complex, 9–10; corrupting, 291; of crowd, 94; despotic, 78,104; elite, 9, 54, 106, 170–71, 177, 197, 268; emotive, 14; game, 119, 192; global, 132; human, 135; illusion of, 236; imperial, 229; machine, 11–12, 154, 221, 247, 253, 256; material, 3, 192; military, 104, 162; natural, 101, 148; of moral imagination, 104; of philosophy, 103; people, 12, 101, 162, 190; political, 290;

princely, 2; regime, 10, 290; relation, 5, 10, 117, 129, 152; royal, 11, 55, 67, 69, 115–16, 182, 211; sociopolitical, 10; sovereign, 30, 69; state, 8, 13, 123; system of, 7; topography of, 170; ultimate, 149
press, 54; British, 14, 225, 240–41, 247, 254, 257, 266; coverage, 41, 161, 253; French, 227; German, 275, 278; Irish, 222; London, 214–15, 220, 238, 258; Lyon 163; Parisian, 160, 170, 236; public, 12, 68, 74, 80, 250; radical, 54; underground, 56
Price, Richard, 68, 216
Priestley, Joseph, 38, 44, 65, 137, 167, 214–16, 221
prince, 1, 3, 54–56, 69, 74, 85, 93, 131–32, 173–74, 271–72; aerial, 261, 265–66; of blood, 1, 117, 153; enlightened, 106, 120; German, 279; Masonic, 106; philosophical, 101, 126; sensible, 114; territorial, 261, 265; virtuous, 146
prince de: Condé, 182–83; Conti, 96; Ligne, 155, 158, 277; Limbourg, 82; Louis-le-Grand, 125; Paar, 255, 277; Robecq, 268
Prince: of Brunswick, 275; of Caramanico, 214, 224; Charles, 156–58; Frederick of Hesse-Darmstadt, 271; Gagarin, 277; of Nassau-Weilbourg, 272; of Orange, William V, 261–67; of Wales, 217, 225–26, 238, 241, 251
Princess: Gagarin, 277; of Holstein, 274–75; of Mecklenburg, 274; of Nassau-Weilbourg, 272; of Orange, 265
privilege, 6, 31, 44–45, 68, 101, 120, 129, 151, 155, 177
progress, xxiii, 188–89, 197, 278; infinite, 54, 99; rational, 7; scientific, 125, 134, 174, 184, 192, 237, 257; slow, 7, 190, 215
public (n): fashion-driven, 237; homogenous, 237; layered, 236; literary, xxv; reading, xviii; vs. people, 3, 25, 54; plebian, 4; prerevolutionary, 7; science-minded, 6; scientific, 4;
public (adj): domain, 12, 56; vs. hidden, 54; lecture, 3, 35, 44, 165, 236, 240; space, 53, 56, 67, 133, 155, 243, 285; spirit, 284
public good, 7, 31–32, 47–48, 54, 63, 69, 75, 344; commitment to, 32, 48, 63, 110, 120, 124–25, 151, 190; as moral mandate, 7, 54, 69, 75, 95, 106–7, 112, 152, 182, 188; rhetoric of, 9, 31, 47, 111, 132, 171, 237, 245
publicity, 6, 44, 47, 145, 148, 174, 184, 192, 251, 276
public opinion, 65, 89, 109, 139, 160–62, 188, 220, 247; absolute, 110–11; as tribunal, 67, 147, 173, 190–91, 286, 288; consensual, 75, 106, 112, 115, 117, 139; engineering, 54, 85, 110, 118, 124, 164, 168; representative of, 86, 114
public sphere, 104–7, 111, 117, 186, 192, 196; Enlightenment, 110; harmonious, 63; material, 12, 132; liminal, 25; literary, 142–47, 192; Parisian, 67; representative, 55
public transcript, 9–10, 65–68, 75, 117, 214, 258; ballooning and, 12, 54, 75, 83, 109, 112, 119, 129, 189, 288; French, 101; hegemonic, 74, 146, 211, 227, 288; journalistic, 9, 97, 109, 170; localized, 152, 155, 163 (Lyon), 166 (Dijon), 265 (Dutch)
pump, 51, 83, 166–67, 180

Quinquet, Antoine, 69, 184

reason, xxi, 3, 24, 75, 104–5, 109, 124, 283–84, 291
Redoute: Chinois, 73, 79; de Wazemmes, 136; in Hamburg, 277
Réflexions amusantes, 113–15
Regnaud de Pleinchesne, 73–74
regime, 288, 290; absolutist, 5, 25, 114, 291; despotic, 4, 104; of etiquettes, 114; mixed, 43, 51; oppressive, 149; of publicity, 174, 192; of republican monarchy, 286; of science, 283; of scientific hegemony, 11, 106, 124; of scientific justice, 289; power-knowledge, 10; republican, 293
repertoire of: discourse, 288; flattery, 82; flying devices, xxiii; lectures, 39, 44; literature, 147; maladies, 135; performances, 9, 135, 251, 253; resources, 106, 173; technologies, 10, 107
republic: of America, 5, 281; imagined, 43, 106, 284; perfect, 129; of philosophers, 105; Rousseau's, 105; of Terror, 292; of universal citizens, 195
republic-empire, 26
Republic: Dutch, 196, 261–63; of Letters, 25, 49, 124, 195, 211; of Science, 165
République des philosophes, 104–5
reputation, 23, 92–95, 119, 126, 211, 229, 235; Academy's, 134; Banks's, 212; Blanchard's, 138, 239, 273, 275; Charles's, 66–67, 88; Dijon Academicians,' 168; Faujas's, 58; Houttuyn's, 265; Lunardi's, 251; Montgolfiers,' 32, 66, 80, 88, 155, 220; Pilâtre's, 154, 160; queen's, 142; scientific, 212, 247
restaurant, 53, 56, 74, 82
Rétif (or Restif) de la Bretonne, xxiii, 13, 34, 42–43, 106, 113, 145
Réveillon, Jean-Baptiste, 45, 68, 79; balloon and, 50, 80–85, 156, 182–83, 185, 213; Courtalin-en-Brie factory, 37, 40; faubourg St. Antoine factory, 69–71, 78, 80, 82; Franklin and, 68; riot, 284
revolution, 40, 77, 105–7, 114–18, 143–52, 164–69, 198, 285–88; by philosopher-king, 105–7, 118, 146, 151; consumer, 5; in English history, 145; great, 117, 182, 190; happy, 107, 118, 146, 151; machines and, 26; music and, 145; patriotic, 13, 261–63
Revolution: Chemical, 57, 291; Dutch, 262–65; French (*see* French Revolution); Glorious, 210; Maupeou, 24, 89, 114; Swedish, 146–48, 162–64
revolutionary, 35; actions, 290–91; agency, 290; cause, 11; consciousness, 143; crisis, 173, 193; crowd, 3; divide, 14, 290–91, 293; dreams, 149; Europe, 8, 197; events, 196; ferment, 198; festival, 10, 290; France, 4; government, 107; ideas, 53; ideology, 3; intention, 142; marches, 7; metamorphoses, 14, 283; play, 133, 147; politics, 4, 67; prince, 296n2; process, 285; rhetoric, 284, 291; role, 146; status, 290; thought, 4; turmoil, 266; violence, 7, 79; wars, 286

riot, 6, 13, 73, 79, 93–94, 151, 174, 178, 180, 184, 186, 225, 284
Rivarol, Antoine de, 66–67, 89, 101, 117, 315n38
Robert, Anne-Jean, 56; Marie-Noël, 10, 56, 77, 91, 94, 96, 108, 166; brothers, 52, 58, 63, 99, 138, 140, 150, 174, 186–87, 189
Robespierre, 284, 289
Rouen, 41, 236–39, 281; acid factory, 57
Rousseau, Jean-Jacques, xxiii, 2–3, 36–37, 106, 123, 141; *Discourses,* 156, 164; republicanism, 104, 285; on theater and spectacles, 7, 23–24, 105
Rousseauean: agitators, 143; fashion, 122; generation, 36, 43, 104, 177; idealism, 142; writers, 24
Royal Society of London, 11, 39, 59, 89, 211–15; Fellows of, 224, 231, 237–38
rubber, 46, 57–58, 63, 66, 88; solution (*gomme élastique*), 58
Ruggieri, Petroni Sauveur Balthasar, 73–74
rupture, 51, 81, 101, 283

Sadler, James, 239–42, 244, 257,
Sage, Baltazar-Georges, 82, 91, 93, 98
salon, 53, 83, 89
satire, xviii, 117, 126, 172, 174, 208, 225; dystopian, xxi, 113; of lost uncle, 80; of Charles's ascent, 91; visual, 186, 209, 223
satirist, 6, 83, 125, 276
savant, 1, 25, 65–66, 68, 72, 124–26, 132–37, 158, 184, 256; European, 58, 135; French, 60; true, 124
savant-chimiste, 31; -inventeur, 31, 118; -voyager, 125
scientific artifact, 13, 26, 231; alternative, 288, 291; authorship, 50; identity as, 48; imperial, xviii, 101, 195; useful, 12, 75, 174; venerated, 10, 75, 104, 109; virtuous, 211
sensibility: Enlightenment, 95; gentlemanly, 236; heightened, 85; republican, 104; utilitarian, xxii;
sentiment, 13, 66, 96, 139, 141, 148, 157, 170, 173, 210, 221, 227; aesthetic, 4; anti-aristocratic, 148; anti-ministerial, 148; civic, 25; elite, 247; humanitarian, 127; gentle, 120; genuine, 24, 106; mixed, 84; monarchial, 173; moral, 4, 95; national, 74; noble, 121; patriotic, 8, 276; political, 4; public, 242; religious, 86, 173; sincere, 36, 111; warm, 255
Seven Years' War, 30, 78
siege of: Calais, 30; of Gibraltar, 29–30, 215
silence, 69, 83; archeology of, 9, 54, 129; at ascent, 227; citizens,' 105; contemplative, 30, 186; of crowd, 94–95, 109, 169, 227; imposing, 96; of mass, 8–9, 54, 129, 157; of newspapers, 75; of people, 4, 10, 55, 110; respectful, 285; by ridicule, 133, 276; toward balloon, 190, 224, 239, 241
silk, 26, 34, 44, 66, 88, 89, 138, 153, 219, 222, 243, 253–54; balloon, 138, 160, 275; cords, 188; ensigns, 242; envelope, 221; *gros de Florence,* 166;

INDEX | 425

silk (*cont.*): India, 249; industry, 136; oiled, 215, 220, 225, 245, 249, 307n36; parachute, 160; patterned, 57; rubber-coated, 46
slave(ry), 2–3, 24, 30, 34, 75, 103–5, 127, 283, 285
sociability, 23, 56, 67, 119, 283
social: drama, 23; space, 5, 24, 56
Société: de Bonne Vente, 38; d'émulation de Reims, 82; de l'Harmonie universelle, 55, 143–44, 191; royal de médecine, 46; royale des Sciences, 158
Southern, John, 196, 221
sovereign, xxv, 30, 49, 69, 101, 104–5, 113–14, 133, 137, 182, 229, 231; magistrate, 104–5
sovereignty, 2, 11, 56, 67, 104–5, 111, 115
spectacle, xxiv–xxv, 1–26, 89, 99, 104–5, 111, 162, 169, 174, 187, 209, 211, 231, 268–70; aerial, 88; aerostatic, 266; amusing, 1; astonishing, 125, 157; balloon, 7, 13, 59, 262, 278; of cohesion, 251; confusing, 60; crowd-forming, 197; cultural hegemony and, xxiv; curious, 1; dazzling, 53; English, 145; ephemeral, 56–57; exquisite, 256; French, 195; frivolous, 85; grand, 88; hegemonic, 88; imposing, 183; magnificent, 183; majestic, 2, 26, 125, 173, 271; marvelous, 25, 176, 220; mass, 6, 8, 11, 14, 117, 278, 289–90; material, 25, 262, 300n11; miraculous, xxv, 124; national, 24; philosophical, 180; public, 12, 125, 181, 254, 258, 280; public-forming, 55; rare, 84; sensational, 51; scientific, xxv, 4–5, 10–11 149, 153, 261, 289; stately, 236, 261, 280; sublime, 227; substitute, 11; totalizing, 290–91; unifying, 129; useless, 188, 270; vain, 65
spectator, 86, 110; balloon, 109–12, 112, 121, 129, 268; cosmic, xviii, xxi–xxiii, 95, 111, 121; metamorphosis of, 109; spectrum of, 25; urban, 111
standpoint, 9, 112, 115, 129
state-body, xxv, 5, 128
state-machine, 2, 3, 11–12, 132
state, 4–5; absolutist, xxi, 151–52; civil, 123; control, 32; loss of, 117; medicine and, 143; sciences, 283; technologies, 291, 293
steam engine, 11, 26, 38–39, 47, 66, 231, 237
sublime, 14, 74, 85–86, 120, 123, 127, 144, 220, 228, 244–45, 268
subscription, 12, 49, 75, 81, 86; Blanchard's, 262 (Hague), 266 (Rotterdam), 268 (Lille), 272 (Frankfurt), 274 (Hamburg), 276–77 (Aix), 277 (Vienna), 279 (Nancy), 280 (Strasbourg), 281; Bordeaux, 177, 180–81; Calais, 248; Dijon, 165–66, 170; Dublin, 222, 246; Edinburgh, 223; Lyon, 153, 158, 161; Lunardi, 225, 228; Miolan, 184–85; national, 52, 56, 63; provincial, 135–37; public, 26, 55, 83, 136–37, 140, 153, 188; Zambeccari, 219
subversive, xvii, 6, 13, 53, 126, 132, 213
sulfuric acid, 57–58, 91, 137, 140, 271, 274, 281, 307n29
suffrage, 75; universal, 105, 137
symbolic: action, 6; body, 2; capital, 2, 25, 51, 67, 86, 161, 165, 168, 211; distinction, 51, 196;

economy of pleasure, 2; function, 3; hierarchy, 133; meaning, 188, 293; order, 82; possession, xix; reign, 141; resources, 67; value, 270

taffeta, 44, 59, 89, 166, 187, 307n36, 318n53, 321n17; *taffeta chiné,* 57; *taffeta gommé,* 46
technocracy, 4, 54, 112
technologies: administrative, 7–8, 11, 291; ancien régime, 283, 293; commercial, 8; of communication, 195, 291; of control, 10; of crowd mobilization, 290; literary, 10–11, 26, 47, 290; material, 13, 26, 47, 236, 290; of mass control, 5, 129, 290; of mass mobilization, 293; nation-forming, 54; of political machination, 11; public, 8; of publicity, 290; scientific, 11; social, 11, 26, 47, 107, 290; of social control, 11–12; stable, 10; of state maintenance, 290; state, 291, 293; of transport, 195; visual, 10
temple, 56, 73, 78, 181, 229
Temple, 117; Chinese, 219; of Fashion, 115–16; of Memory, 272
terror, 14, 24, 62–63, 94–95, 136, 145, 214, 225, 227, 228
Terror, 7, 67, 107, 289–90
theater, 24, 35, 53, 109, 121, 133, 147–49, 152, 217, 256; Aix-la- Chapelle, 277; alternative, 25; balloon, 13, 71–72; boom, 135; court, 2; crowd, 133; experimental, 26; fair, 24; Frankfurt, 272; Hamburg, 273–74; hidden, 148; as a laboratory of the nation, 24 ; Lille, 269; living, 2; Lyon, 158–59; mass, 12; material, 148; mechanical, 3; mobile, 95, 100; of natural fluids, 133; Parisian, 25; private, 147; public, 132, 147; respectable, 145; Rouen, 236; royal, 2, 7, 13, 24–25, 75, 105, 133, 148, 185, 243; small, 134; street, 61; veritable, 85
Theater: Covent Garden, 220, 239; Haymarket, 224
theater-nation, 2, 24; -state, 2; -temple, 175–76
theatrical: characters, xxiv; consciousness, 23, 231; occasion, 253; outlay, 224; performance, 74, 105, 144–45, 283; previews, 53; production, xix; relation, 4, 24, 109
theatricality, 7, 10, 23–24, 114, 124
theatrical polity, 2–3, 12–13, 23, 46, 69, 116–29, 231; absolutist, 3, 13, 31, 75, 151, 290; ancien régime, 148; camaraderie in, 86; distinction and, 284; entertainers in, 196, 242; French, 279; inventors in, 137; public sphere and, 192; Rousseau on, 36
theory, 35, 48–49, 64, 66, 80, 170, 177, 191, 257, 291; chimerical, 66; climate, 122; Mesmer's, 141; scientific, 66
toile de Rouen, 70, 140
tour de France, 33
transcript: balloon, 9–10, 54, 85, 113; hegemonic, 55; hidden 10, 13, 58, 63–67, 110, 117, 123, 139, 147, 192; journalistic, 55; public (*see* public transcript); self-serving, 59
transculturation, 13, 112
translation, xxv, 12, 14, 261; balloon, 195; cultural, xviii, 13, 152, 195–97, 258

tribunal, 49, 67, 111, 182
Tuileries, 53, 83, 89–93, 121–24, 139, 188, 244
Turgot, Anne-Robert-Jacques, 32, 39, 61

universal, 156; acclaim, 235; admiration, 12, 103; appeal, 129; applause, 94, 244, 266; approbation, 138; aspiration, 13; authority, 115; balloon enthusiasm, 170; beheading, 289; cheer, 94; citizens, xxiii, 101, 195; cure, 149; curiosity, 98, 152; desire for reputation and distinction, 23–24; fluid, 141, 148, 191; fraternity, 14; geography, xxv; harmony, 143; kingdom, 122; knowledge, 99, 121; laughter, 148; measure, 48; medicine, 146; monarchy, xxii, 121; morality, xxii; polity, 122; science, 54, 280; suffrage, 137; utility, 142; value, 109; veneration, 6; will, 3; wonder, 195
universe, xix–xx, 30, 74, 99, 101, 121, 127, 141, 190, 256; citizen (or man) of, 24, 109, 122–23
utility, 31, 44, 48, 105, 112, 115, 118, 124, 154, 158, 184; civic, 51; commercial, 247; distant, 192; general, 268; immediate, 173, 192; lack of, 277; material, 106, 174; medical, 192; national, 39; personal, 42; political, 174; potential, 118, 166, 192, 224; promise of, 54, 88; public, 44, 106, 212; pursuit of, 123; real, 83; rhetoric of, 117, 173; scientific, 247; uncertain, 44; universal, 142
utopia: scientific, 288; technoscientific, 101
utopian: desire, 221; dreams, xx–xxi, 12, 113, 283; fiction, xxi, 104, 113; hope, 286; ideal, 6; imaginary, 5; longings, 231, 290; Paris, 39, 106, 128; polity, 113; realm, 6; republic-empire, 26; vision, 8; world, xxv;

vainglory, 66, 89, 127–29, 143, 152–56, 166, 182, 190
Vallet, Jean-Baptiste, 100, 131
van Liender, Jan Daniël Huichelbos, 263, 265
van Noorden, Johannes, 263–66
veneration, xxv, 2, 6, 10, 14, 95–96, 109 (people's), 123, 163, 177
Versailles, 2, 33, 48, 78, 93–95, 105, 135, 147, 151–52, 174–75; administrators, 261; authorities, 243; balloon, 55, 68, 70–74, 78, 110, 126–27, 136, 139, 160, 174, 182, 185, 188, 213, 231; Blanchard's visit, 243–44; court, 171, 277; Montgolfière, 89

Vergennes, 61, 162, 176, 183, 236, 243, 263, 273, 279
Vidalon, 29, 31–40, 50, 136, 141, 154–55, 183
violence, 1, 7, 25, 120–21, 156, 198, 227; popular or mass, 149, 209, 284; revolutionary, 79, 192
virtue, 69, 83, 104–5, 119, 125, 144, 146, 151, 174, 275; genius's, 31; inventors,' 67, 75; magnetic, 141; moral, 121, patriotic, 31; philosophical, 8, 121, 293; public, 24; republican, 145; scientific, 280; sublime, 123; taste for, 121; unrewarded, 100
Vivarais, 47–48, 285
Volta, Alessandro, 38, 44, 65
Voltaire, xxiii, 31, 85, 95, 100, 141, 145, 231
voyage, 86, 158, 212, 241, 269; Blanchard's, 140, 242–44, 269, 272–73; Charles's, 98, 100, 140; Columbus's, 98; cosmic, 123; heroic, 100; imaginary, xxi; Indian, 36; lunar, xxi, 117, 125, 239; memorable, 158; Pacific, 212; Potain's, 246
voyager, 153–61, 246, 251, 275; animal, 112; cosmic, xxi, xxiii, 118, 121–23; intrepid, xxiii, 85, 118; philosophical, 101, 104

Watt, James, 39, 66, 216, 221
Watteau, Louis-Joseph, cover, 269, 279
Wilkes, John, 209, 229
Willermoz, Jean-Baptiste, 143, 153
wings, 77, 115–16, 177, 184, 215, 245; artificial, xvii, 41–42, 113; Blanchard's, 139, 238, 243; of Daedalus, 85, 217; fictional, 113; of Icarus, xxii; imaginary, xxii; iron, 217; Lunardi's, 225, 252; Sadler's, 240
witness, 7, 26, 48, 59, 67, 80–88, 94–95, 119, 153, 178–79, 183–85, 236, 241, 247
wonder, 95, 266; imposing, 127; scientific, 103; technological, 61; universal, 195

Zambeccari, Francesco, 215, 219, 224, 247–49, 257
zone: border, 14, 198, 266–67, 271, 279; contact, 5, 60–61, 112, 149, 258; hybrid, 196–97; liminal, 11, 198